DEVIN RAPPLEYE
801-422-1611

D1601802

Chemical Reactions and Chemical Reactors

Chemical Reactions and Chemical Reactors

George W. Roberts
North Carolina State University
Department of Chemical and Biomolecular Engineering

WILEY

John Wiley & Sons, Inc.

VICE PRESIDENT AND EXECUTIVE PUBLISHER	Don Fowley
ASSOCIATE PUBLISHER	Dan Sayre
ACQUISITIONS EDITOR	Jenny Welter
VICE PRESIDENT AND DIRECTOR OF MARKETING	Susan Elbe
EXECUTIVE MARKETING MANAGER	Chris Ruel
SENIOR PRODUCTION EDITOR	Trish McFadden
DESIGNER	Michael St. Martine
PRODUCTION MANAGEMENT SERVICES	Thomson Digital Limited
EDITORIAL ASSISTANT	Mark Owens
MARKETING ASSISTANT	Chelsee Pengal
MEDIA EDITOR	Lauren Sapira
COVER PHOTO	© Taylor Kennedy/NG Image Collection

Cover Description:
The firefly on the cover is demonstrating the phenomenon of "bioluminescence", the production of light within an organism (the *reactor*) by means of a chemical *reaction.* In addition to fireflies, certain marine animals also exhibit bioluminescence.

In the firefly, a reactant or substrate known as "firefly luciferin" reacts with O_2 and adenosine triphosphate (ATP) in the presence of an enzyme catalyst, luciferase, to produce a reactive intermediate (a four-member cyclic perester).

$$\text{Firefly luciferin} + \text{ATP} + O_2 \xrightarrow{\text{luciferase}} \text{Intermediate}$$

The intermediate then loses CO_2 spontaneously to form a heterocyclic intermediate known as "oxyluciferin". As formed, the oxyluciferin is in an excited state, i.e., there is an electron in an anti-bonding orbital.

$$\text{Intermediate} \rightarrow \text{Oxyluciferin}^* + CO_2$$

Finally, oxyluciferin decays to its ground state with the emission of light when the excited electron drops into a bonding orbital.

$$\text{Oxyluciferin}^* \rightarrow \text{Oxyluciferin} + h\nu\,(\text{light})$$

This series of reactions is of practical significance to both fireflies and humans. It appears that firefly larvae use bioluminescense to discourage potential predators. Some adult fireflies use the phenomenon to attract members of the opposite sex.

In the human world, the reaction is used to assay for ATP, a very important biological molecule. Concentrations of ATP as low as 10^{-11} M can be detected by measuring the quantity of light emitted. Moreover, medical researchers have implanted the firefly's light-producing gene into cells inside other animals and used the resulting bioluminescense to track those cells in the animal's body. This technique can be extended to cancer cells, where the intensity of the bioluminescense can signal the effectiveness of a treatment. Finally, the energy released by the bioluminescense-producing reactions is almost quantitatively converted into light. In contrast, only about 10% of the energy that goes into a conventional incandescent light bulb is converted into light.

This book was set in Times New Roman by Thomson Digital Limited and printed and bound by Quad/Graphics. The cover was printed by Quad/Graphics.

This book is printed on acid free paper. 

Copyright © 2009 John Wiley & Sons, Inc. All rights reserved. No part of this publication may be reproduced, stored in a retrieval system or transmitted in any form or by any means, electronic, mechanical, photocopying, recording, scanning or otherwise, except as permitted under Sections 107 or 108 of the 1976 United States Copyright Act, without either the prior written permission of the Publisher, or authorization through payment of the appropriate per-copy fee to the Copyright Clearance Center, Inc. 222 Rosewood Drive, Danvers, MA 01923, website www.copyright.com. Requests to the Publisher for permission should be addressed to the Permissions Department, John Wiley & Sons, Inc., 111 River Street, Hoboken, NJ 07030-5774, (201)748-6011, fax (201)748-6008, website http://www.wiley.com/go/permissions.

To order books or for customer service please, call 1-800-CALL WILEY (225-5945).

ISBN-13 978-0471-742203

Printed in the United States of America

V007448_071218

Contents

Preface

Intended Audience

This text covers the topics that are treated in a typical, one-semester undergraduate course in chemical reaction engineering. Such a course is taught in almost every chemical engineering curriculum, internationally. The last three chapters of the book extend into topics that may also be suitable for graduate-level courses.

Goals

Every engineering text that is intended for use by undergraduates must address two needs. First, it must prepare students to function effectively in industry with only the B.S. degree. Second, it must prepare those students that go to graduate school for advanced coursework in reaction kinetics and reactor analysis. Most of the available textbooks fall short of meeting one or both of these requirements. "Chemical Reactions and Chemical Reactors" addresses both objectives. In particular:

Focus on Fundamentals: The text contains much more on the fundamentals of chemical kinetics than current books with a similar target audience. The present material on kinetics provides an important foundation for advanced courses in chemical kinetics. Other books combine fundamentals and advanced kinetics in one book, making it difficult for students to know what's important in their first course.

Emphasis on Numerical Methods: The book emphasizes the use of numerical methods to solve reaction engineering problems. This emphasis prepares the student for graduate coursework in reactor design and analysis, coursework that is more mathematical in nature.

Analysis of Kinetic Data: Material on the analysis of kinetic data prepares students for the research that is a major component of graduate study. Simultaneously, it prepares students who will work in plants and pilot plants for a very important aspect of their job. These features are discussed in more detail below.

"Chemical Reactions and Chemical Reactors" is intended as a text from which to teach. Its objective is to help the student master the material that is presented. The following characteristics aid in this goal:

Conversational Tone: The tone of the book is conversational, rather than scholarly.

Emphasis on Solving Problems: The emphasis is on the solution of problems, and the text contains many example problems, questions for discussion, and appendices. Very few derivations and proofs are required of the student. The approach to problem-solving is to start each new problem from first principles. No attempt is made to train the student to use pre-prepared charts and graphs.

Use of Real Chemistry: Real chemistry is used in many of the examples and problems. Generally, there is a brief discussion of the practical significance of each reaction that is introduced. Thus, the book tries to teach a little industrial chemistry along with chemical kinetics and chemical reactor analysis. Unfortunately, it is difficult to find real-life examples to illustrate all of the important concepts. This is particularly true in a discussion of reactors in which only one reaction takes place. There are several important principles that must be illustrated in such a discussion, including how to handle reactions with different stoichiometries and how to handle changes in the mass density as the reaction takes place. It was not efficient to deal with all of these variations through real examples, in part because rate equations are not openly available. Therefore, in some cases, it has been necessary to revert to generalized reactions.

Motivation and Differentiating Features

Why is a new text necessary, or even desirable? After all, the type of course described in the first paragraph has been taught for decades, and a dozen or so textbooks are available to support such courses. "Chemical Reactions and Chemical Reactors" differs substantially in many important respects from the books that are presently available. On a conceptual level, this text might be regarded as a fusion of two of the most influential (at least for this author) books of the past fifty years: Octave Levenspiel's "Chemical Reaction Engineering" and Michel Boudart's "Kinetics of Chemical Processes." As suggested by these two titles, one of the objectives of this text is to integrate a fundamental understanding of reaction kinetics with the application of the principles of kinetics to the design and analysis of chemical reactors. However, this text goes well beyond either of these earlier books, both of which first appeared more than forty years ago, at the dawn of the computer era.

This text is differentiated from the reaction engineering books that currently are available in one or more of the following respects:

1. The field of chemical kinetics is treated in some depth, *in an integrated fashion* that emphasizes the fundamental tools of kinetic analysis, and challenges the student to apply these common tools to problems in many different areas of chemistry and biochemistry.
2. Heterogeneous catalysis is introduced early in the book. The student can then solve reaction engineering problems involving heterogeneous catalysts, in parallel with problems involving homogeneous reactions.
3. The subject of transport effects in heterogeneous catalysis is treated in significantly greater depth.
4. The analysis of experimental data to develop rate equations receives substantial attention; a whole chapter is devoted to this topic.
5. The text contains many problems and examples that require the use of numerical techniques.

The integration of these five elements into the text is outlined below.

Topical Organization

Chapter 1 begins with a review of the stoichiometry of chemical reactions, which leads into a discussion of various definitions of the reaction rate. Both homogeneous and heterogeneous systems are treated. The material in this chapter recurs throughout the book, and is particularly useful in Chapter 7, which deals with multiple reactions.

Chapter 2 is an "overview" of rate equations. At this point in the text, the subject of reaction kinetics is approached primarily from an empirical standpoint, with emphasis on power-law rate equations, the Arrhenius relationship, and reversible reactions (thermodynamic consistency). However, there is some discussion of collision theory and transition-state theory, to put the empiricism into a more fundamental context. The intent of this chapter is to provide enough information about rate equations to allow the student to understand the derivations of the "design equations" for ideal reactors, and to solve some problems in reactor design and analysis. A more fundamental treatment of reaction kinetics is deferred until Chapter 5. The discussion of thermodynamic consistency includes a "disguised" review of the parts of chemical thermodynamics that will be required later in the book to analyze the behavior of reversible reactions.

The definitions of the three ideal reactors, and the fundamentals of ideal reactor sizing and analysis are covered in Chapters 3 and 4. Graphical interpretation of the "design equations" (the "Levenspiel plot") is used to compare the behavior of the two ideal continuous reactors, the plug flow and continuous stirred-tank reactors. This follows the pattern of earlier texts. However, in this book, graphical interpretation is also used extensively in the discussion of ideal reactors in series and parallel, and its use leads to new insights into the behavior of *systems* of reactors.

In most undergraduate reaction engineering texts, the derivation of the "design equations" for the three ideal reactors, and the subsequent discussion of ideal reactor analysis and sizing, is based exclusively on *homogeneous* reactions. This is very unfortunate, since about 90 percent of the reactions carried out industrially involve *heterogeneous catalysis*. In many texts, the discussion of heterogeneous catalysis, and heterogeneous catalytic reactors, is deferred until late in the book because of the complexities associated with transport effects. An instructor who uses such a text can wind up either not covering heterogeneous catalysis, or covering it very superficially in the last few meetings of the course.

"Chemical Reactions and Chemical Reactors" takes a different approach. The design equations are derived in Chapter 3 for *both* catalytic and non-catalytic reactions. In Chapter 4, which deals with the use of the design equations to size and analyze ideal reactors, transport effects are discussed qualitatively and conceptually. The student is then able to size and analyze ideal, heterogeneous catalytic reactors, *for situations where transport effects are not important*. This builds an important conceptual base for the detailed treatment of transport effects in Chapter 9.

As noted previously, one major differentiating feature of "Chemical Reactions and Chemical Reactors" is its emphasis on the fundamentals of reaction kinetics. As more and more undergraduate students find employment in "non-traditional" areas, such as electronic materials and biochemical engineering, a strong grasp of the fundamentals of reaction kinetics becomes increasingly important. Chapter 5 contains a unified development of the basic concepts of kinetic analysis: elementary reactions, the steady-state approximation, the rate-limiting step approximation, and catalyst/site balances. These four "tools" then are applied to problems from a number of areas of science and engineering: biochemistry, heterogeneous catalysis, electronic materials, etc. In existing texts, these fundamental tools of reaction kinetics either are not covered, or are covered superficially, or are covered in a fragmented, topical fashion. The emphasis in "Chemical Reactions and Chemical Reactors" is on helping the student to understand and apply the *fundamental* concepts of kinetic analysis, so that he/she can use them to solve problems from a wide range of technical areas.

Chapter 6 deals with the analysis of kinetic data, another subject that receives scant attention in most existing texts. First, various techniques to test the suitability of a given rate equation are developed. This is followed by a discussion of how to estimate values of the unknown parameters in the rate equation. Initially, graphical techniques are used in order to provide a visual basis for the process of data analysis, and to demystify the subject for "visual learners". Then, the results of the graphical process are used as a starting point for statistical analysis. The use of non-linear regression to fit kinetic data and to obtain the "best" values of the unknown kinetic parameters is illustrated. The text explains how non-linear regression can be carried out with a spreadsheet.

Multiple reactions are covered in Chapter 7. This chapter begins with a qualitative, conceptual discussion of systems of multiple reactions, and progresses into the quantitative solution of problems involving the sizing and analysis of isothermal reactors in which more than one reaction takes place. The numerical solution of ordinary differential equations, and systems of ordinary differential equations, is discussed and illustrated. The solution of non-linear systems of algebraic equations also is illustrated.

Chapter 8 is devoted to the use of the energy balance in reactor sizing and analysis. Adiabatic batch and plug-flow reactors are discussed first. Once again, numerical techniques for solving differential equations are used to obtain solutions to problems involving these two reactors. Then, the CSTR is treated, and the concepts of stability and multiple steady states are introduced. The chapter closes with a treatment of feed/product heat exchangers, leading to a further discussion of multiplicity and stability.

The topic of transport effects in catalysis is revisited in Chapter 9. The structure of porous catalysts is discussed, and the internal and external resistances to heat and mass transfer are quantified. Special attention is devoted to helping the student understand the influence of transport effects on overall reaction behavior, including reaction selectivity. Experimental and computational methods for predicting the presence or absence of transport effects are discussed in some detail. The chapter contains examples of reactor sizing and analysis in the presence of transport effects.

The final chapter, Chapter 10, is a basic discussion of non-ideal reactors, including tracer techniques, residence-time distributions, and models for non-ideal reactors. In most cases, the instructor will be challenged to cover this material, even superficially, in a one-semester course. Nevertheless, this chapter should help to make the text a valuable starting point for students that encounter non-ideal reactors after they have completed their formal course of study.

Numerical Methods

"Chemical Reactions and Chemical Reactors" contains problems and examples that require the solution of algebraic and differential equations by numerical methods. By the time students take the course for which this text is intended, a majority of them will have developed some ability to use one or more of the common mathematical packages, e.g., Mathcad, Matlab, etc. This text does not rely on a specific mathematical package, nor does it attempt to teach the student to use a specific package. The problems and examples in the book can be solved with any suitable package(s) that the student may have learned in previous coursework. This approach is intended to free the instructor from having to master and teach a new mathematical package, and to reinforce the students' ability to use the applications they have already learned. Many of the numerical solutions that are presented in the text were developed and solved on a personal computer using a spreadsheet. Appendices are included to illustrate how the necessary mathematics can be carried out with a spreadsheet. This approach gives students a "tool" that they eventually might need in an environment where a specific mathematical package was not available. The spreadsheet approach also familiarizes the student with some of the mathematics that underlies the popular computer packages for solving differential equations.

In the Classroom

"Chemical Reactions and Chemical Reactors" is written to provide the instructor with flexibility to choose the order in which topics are covered. Some options include:

Applications Up Front: Lately, I have been covering the chapters in order, from Chapter 1 through Chapter 9. This approach might be labeled the "mixed up" approach because it switches back and forth between kinetics and reactor sizing/analysis. Chapter 2 provides just enough information about chemical kinetics to allow the student to understand ideal reactors, to size ideal reactors, and to analyze the behavior of ideal reactors, in Chapters 3 and 4. Chapters 5 and 6 then return to kinetics, and treat it in more detail, and from a more fundamental point of view. I use this approach because some students do not have the patience to work through Chapters 2 and 5 unless they can see the eventual application of the material.

Kinetics Up Front: Chapter 5 has been written so that it can be taught immediately after Chapter 2, before starting Chapter 3. The order of coverage then would be Chapters 1, 2, 5, 3, 4, 6, 7, 8, and 9. This might be referred to as the "kinetics up front" approach.

Reactors Up Front: A third alternative is the "reactors up front" approach, in which the order of the chapters would be either: 1, 2, 3, 4, 7, 8, 9, 5, 6 or 1, 2, 3, 4, 7, 8, 5, 6, 9. The various chapters have been written to enable any of these approaches. The final choice is strictly a matter of instructor preference.

Some important topics are not covered in the first version of this text. Two unfortunate examples are transition-state theory and reactors involving two fluid phases. An instructor that wished to introduce some additional material on transition-state theory could easily do so as an extension of either Chapter 2 or Chapter 5. Supplementary material on multiphase reactors fits well into Chapter 9.

Based on my personal experience in teaching from various versions of this text, I found it difficult to cover even the first nine chapters, in a way that was understandable to the majority of students. I seldom, if ever, got to Chapter 10. A student that masters the material in the first nine chapters should be very well prepared to learn advanced material "on the job," or to function effectively in graduate courses in chemical kinetics or chemical reaction engineering.

Instructor Resources

The following resources are available on the book website at www.wiley.com/college/roberts. These resources are available only to adopting instructors. Please visit the Instructor section of the website to register for a password:

Solutions Manual: Complete solutions to all homework exercises in the text.

Image Gallery: Figures from the text in electronic format, suitable for use in lecture slides.

Instructor's Manual: Contains the answers to all of the "Exercises" in the book.

Acknowledgements

This book is the culmination of a long journey through a subject that always held an enormous fascination for me. The trip has been tortuous, but never lonely. I have been accompanied by a number of fellow travelers, each of who helped me to understand the complexities of the subject, and to appreciate its beauty and importance. Some were teachers, who shared their accumulated wisdom and stimulated my interest in the subject. Many were collaborators, both industrial and academic, who worked with me to solve a variety of interesting and challenging problems. Most recently, my fellow travelers have been students, both undergraduate and graduate. They have challenged me to communicate my own knowledge in a clear and understandable manner, and have forced me to expand my comprehension of the subject. I hope that I can express the debt that I owe to all of these many individuals.

A summer internship started my journey through catalysis, reaction kinetics, and reactor design and analysis, before the term "chemical reaction engineering" came into popular use. For three months, with what was then the California Research Corporation, I tackled a very exciting set of problems in catalytic reaction kinetics. Two exceptional industrial practitioners, Drs. John Scott and Harry Mason, took an interest in my work, made the importance of catalysis in industrial practice clear to me, and had a great influence on the direction of my career.

I returned to Cornell University that fall to take my first course in "kinetics" under Professor Peter Harriott. That course nourished my developing interest in reaction kinetics and reactor design/analysis, and provided a solid foundation for my subsequent pursuits in the area.

In graduate school at the Massachusetts Institute of Technology, I had the privilege of studying catalysis with Professor Charles Satterfield, who became my thesis advisor. Professor Satterfield had a profound influence on my interest in, and understanding of, catalysts and catalytic reactors. My years with Professor Satterfield at MIT were one of the high points of my journey.

I began my professional career with the Rohm and Haas Company, working in the area of polymerization. In that environment, I had the opportunity to interact with a number of world-class chemists, including Dr. Newman Bortnick. I also had the opportunity to work with a contemporary, Dr. James White, in the mathematical modeling of polymerization reactors. My recent work in polymerization at North Carolina State University is an extension of what I learned at Rohm and Haas.

Next, at Washington University (Saint Louis), I had the opportunity to work and teach with Drs. Jim Fair and Ken Robinson. Jim Fair encouraged my study of gas/liquid/solid reactors, and Ken Robinson brought some valuable perspectives on catalysis to my teaching and research efforts.

The next stop in my travels was at what was then Engelhard Minerals and Chemicals Corporation, where I worked in a very dynamic environment that was focused on heterogeneous catalysts and catalytic processes. Four of my co-workers, Drs. John Bonacci, Larry Campbell, Bob Farrauto, and Ron Heck, deserve special mention for their contributions to my appreciation and understanding of catalysis. The five of us, in various combinations, spent many exciting (and occasionally frustrating) hours discussing various projects in which we were involved. I have continued to draw upon the knowledge and experience of this exceptional group throughout the almost four decades that have passed since our relationships began. I must also mention Drs. Gunther Cohn and Carl Keith, both extremely creative and insightful scientists, who helped me immeasurably and had the patience to tolerate some of my streaks of naivety.

I then spent more than a decade with Air Products and Chemicals, Inc. Although the primary focus of my efforts lay outside the area of chemical reaction engineering, there were some notable exceptions. These exceptions gave me the opportunity to work with another set of talented individuals, including Drs. Denis Brown and Ed Givens.

The last and longest stop in my travels has been my present position in the Department of Chemical Engineering (now Chemical and Biomolecular Engineering) at North Carolina State University. This phase of the journey led to four important collaborations that extended and deepened my experience in chemical reaction engineering. I have benefited greatly from stimulating interactions with Professors Eduardo Sáez, now at the University of Arizona, James (Jerry) Spivey, now at Louisiana State University, Ruben Carbonell, and Joseph DeSimone.

This book would not have been possible without the contributions of the Teaching Assistants that have helped me over the years, in both undergraduate and graduate courses in chemical reaction engineering. These include: Collins Appaw, Lisa Barrow, Diane (Bauer) Beaudoin, Chinmay Bhatt, Matt Burke, Kathy Burns, Joan (Biales) Frankel, Nathaniel Cain, "Rusty" Cantrell, Naresh Chennamsetty, Sushil Dhoot, Laura Beth Dong, Kevin Epting, Amit Goyal, Shalini Gupta, Surendra Jain, Concepcion Jimenez, April (Morris) Kloxin, Steve Kozup, Shawn McCutchen, Jared Morris, Jodee Moss, Hung Nguyen, Joan Patterson, Nirupama Ramamurthy, Manish Saraf, George Serad, Fei Shen, Anuraag Singh, Eric Shreiber, Ken Walsh, Dawei Xu, and Jian Zhou. Three graduate students: Tonya Klein, Jorge Pikunic, and Angelica Sanchez, worked with me as part of university-sponsored mentoring programs. Two undergraduates who contributed to portions of the book, Ms. Amanda (Burris) Ashcraft and Mr. David Erel, also deserve my special thanks.

I am indebted to Professors David Ollis and Richard Felder, who offered both advice and encouragement during the darker days of writing this book. I am also grateful to Professors David Bruce of Clemson University, Tracy Gardner and Anthony Dean of Colorado School of Mines, Christopher Williams of the University of South Carolina, and Henry Lamb and Baliji Rao of North Carolina State University for insightful comments and/or for "piloting" various drafts of the book in their classes. Professor Robert Kelly, also of North Carolina State University, contributed significantly to the "shape" of this book.

I would like to thank the following instructors who reviewed drafts of the manuscript, as well as those reviewers who wished to remain anonymous:

Pradeep K. Agrawal, Georgia Institute of Technology
Dragomir B. Bukur, Texas A&M University
Lloyd R. Hile, California State University, Long Beach
Thuan K. Nguyen, California State University, Pomona
Jose M. Pinto, Polytechnic University
David A. Rockstraw, New Mexico State University
Walter P. Walawender, Kansas State University

I fear that I may have omitted one or more important companions on my journey through reaction kinetics, reactor design and analysis, and heterogeneous catalysis. I offer my sincere apologies to those who deserve mention, but are the victims of the long span of my career and the randomness of my memory.

Dedication:
I am intensely grateful for the support of my family. I now realize that my wife, Mary, and my children, Claire and Bill, were the innocent victims of the time and effort that went into the preparation for, and the writing of, this book. Thank you, Mary, Claire, and Bill. This book is dedicated to the three of you, collectively and individually.

Chapter 1

Reactions and Reaction Rates

LEARNING OBJECTIVES

After completing this chapter, you should be able to

1. use stoichiometric notation to express chemical reactions and thermodynamic quantities;
2. use the extent of reaction concept to check the consistency of experimental data, and to calculate unknown quantities;
3. formulate a definition of reaction rate based on *where* the reaction occurs.

1.1 INTRODUCTION

1.1.1 The Role of Chemical Reactions

Chemical reactions[1] are an essential technological element in a huge range of industries, for example, fuels, chemicals, metals, pharmaceuticals, foods, textiles, electronics, trucks and automobiles, and electric power generation. Chemical reactions can be used to convert less-valuable raw materials into higher value products, e.g., the manufacture of sulfuric acid from sulfur, air, and water. Chemical reactions can be used to convert one form of energy to another, e.g., the oxidation of hydrogen in a fuel cell to produce electric power. A complex series of reactions is responsible for the clotting of blood, and the "setting" of concrete is a hydration reaction between water and some of the other inorganic constituents of concrete mix. Chemical reactions are also important in many pollution control processes, ranging from treatment of wastewater to reduce its oxygen demand to removal of nitrogen oxides from the flue gas of power plants.

Our civilization currently faces many serious technical challenges. The concentration of carbon dioxide in the earth's atmosphere is increasing rather rapidly. Reserves of crude oil and natural gas appear to be stagnant at best, whereas consumption of these fossil fuels is increasing globally. Previously unknown or unrecognized diseases are appearing regularly. Nonbiodegradable waste, such as plastic soda bottles, is accumulating in landfills. Obviously, this list of challenges is not comprehensive, and the items on it will vary from person to person and from country to country. Nevertheless, it is difficult to imagine that challenges such as these can be addressed without harnessing some known chemical reactions, plus some reactions that have yet to be developed.

[1] For the sake of brevity, the phrase "chemical reaction" is used in the broadest possible sense throughout this book. The phrase is intended to include biological and biochemical reactions, as well as organic and inorganic reactions.

The successful, practical implementation of a chemical reaction is not a trivial exercise. The creative application of material from a number of technical areas is almost always required. Operating conditions must be chosen so that the reaction proceeds at an acceptable rate and to an acceptable extent. The maximum extent to which a reaction can proceed is determined by *stoichiometry* and by the branch of thermodynamics known as *chemical equilibrium*. This book begins with a short discussion of the principles of stoichiometry that are most applicable to chemical reactions. A working knowledge of chemical equilibrium is presumed, based on prior chemistry and/or chemical engineering coursework. However, the book contains problems and examples that will help to reinforce this material.

1.1.2 Chemical Kinetics

The rate at which a reaction proceeds is governed by the principles of *chemical kinetics*, which is one of the major topics of this book. Chemical kinetics allows us to understand how reaction rates depend on variables such as concentration, temperature, and pressure. Kinetics provides a basis for manipulating these variables to increase the rate of a desired reaction, and minimize the rates of undesired reactions. We will study kinetics first from a rather empirical standpoint, and later from a more fundamental point of view, one that creates a link with the details of the reaction chemistry. *Catalysis* is an extremely important tool within the domain of chemical kinetics. For example, catalysts are required to ensure that blood clots form fast enough to fight serious blood loss. Approximately 90% of the chemical processes that are carried out industrially involve the use of some kind of catalyst in order to increase the rate(s) of the desired reaction(s). Unfortunately, the behavior of heterogeneous catalysts can be significantly and negatively influenced by the rates of *heat and mass transfer* to and from the "sites" in the catalyst where the reaction occurs. We will approach the interactions between catalytic kinetics and heat and mass transport conceptually and qualitatively at first, and then take them head-on later in the book.

1.1.3 Chemical Reactors

Chemical reactions are carried out in *chemical reactors*. Some reactors are easily recognizable, for example, a vessel in the middle of a chemical plant or the furnace that burns natural gas or heating oil to heat our house. Others are less recognizable—a river, the ozone layer, or a heap of compost. The development of a reactor (or a system of reactors) to carry out a particular reaction (or system of reactions) can require imagination and creativity. Today, catalysts are used in every modern refinery to "crack" heavy petroleum fractions into lighter liquids that are suitable for the production of high-octane gasoline. The innovation that brought "catalytic cracking" into such widespread use was the development of very large fluidized-bed reactors that allowed the cracking catalyst to be withdrawn continuously for regeneration. It is very likely that new reactor concepts will have to be developed for the optimal implementation of new reactions, especially reactions arising from the emerging realm of biotechnology.

The design and analysis of chemical reactors is built upon a sound understanding of chemical kinetics, but it also requires the use of information from other areas. For example, the behavior of a reactor depends on the nature of *mixing* and *fluid flow*. Moreover, since reactions are either endothermic or exothermic, thermodynamics comes into play once again, as *energy balances* are a critical determinant of reactor behavior. As part of the energy balance, *heat transfer* can be an important element of reactor design and analysis.

This book will help to tie all of these topics together, and bring them to bear on the study of *Chemical Reactions and Chemical Reactors*. Let's begin by taking a fresh look at stoichiometry, from the standpoint of how we can use it to describe the behavior of a chemical reaction, and systems of chemical reactions.

1.2 STOICHIOMETRIC NOTATION

Let's consider the chemical reaction

$$Cl_2 + C_3H_6 + 2NaOH \rightarrow C_3H_6O + 2NaCl + H_2O \qquad (1\text{-A})$$

The molecule C_3H_6O is propylene oxide, an important raw material in the manufacture of unsaturated polyesters, such as those used for boat bodies, and in the manufacture of polyurethanes, such as the foam in automobile seats. Reaction (1-A) describes the stoichiometry of the "chlorohydrin" process for propylene oxide manufacture. This process is used for about one-half of the worldwide production of propylene oxide.

The balanced stoichiometric equation for any chemical reaction can be written using a generalized form of stoichiometric notation

$$\sum_i \nu_i A_i = 0 \qquad (1\text{-}1)$$

In this equation, A_i represents a chemical species. For instance, in Reaction (1-A), we might choose

$$A_1 = Cl_2;\ A_2 = C_3H_6;\ A_3 = NaOH;\ A_4 = C_3H_6O;\ A_5 = NaCl;\ A_6 = H_2O$$

The stoichiometric coefficient for chemical species "i" is denoted ν_i. Equation (1-1) involves a convention for writing the stoichiometric coefficients. *The coefficients of the products of a reaction are positive, and the coefficients of the reactants are negative*. Thus, for Reaction (1-A):

$$\nu_1 = \nu_{Cl_2} = -1;\ \nu_2 = \nu_{C_3H_6} = -1;\ \nu_3 = \nu_{NaOH} = -2;$$
$$\nu_4 = \nu_{C_3H_6O} = +1;\ \nu_5 = \nu_{NaCl} = +2;\ \nu_6 = \nu_{H_2O} = +1$$

The sum of the stoichiometric coefficients, $\Delta\nu = \sum \nu_i$, shows whether the total number of moles increases, decreases, or remains constant as the reaction proceeds. If $\Delta\nu > 0$, the number of moles increases; if $\Delta\nu < 0$, the number of moles decreases; if $\Delta\nu = 0$, there is no change in the total number of moles. For Reaction (1-A), $\Delta\nu = 0$. As we shall see in Chapter 4, a change in the number of moles on reaction can have an important influence on the design and analysis of reactions that take place in the gas phase.

You may have used this stoichiometric notation in earlier courses, such as thermodynamics. For example, the standard Gibbs free energy change of a reaction (ΔG^0_R) and the standard enthalpy change of a reaction (ΔH^0_R) can be written as

$$\Delta G^0_R = \sum_i \nu_i \Delta G^0_{f,i} \qquad (1\text{-}2)$$

and

$$\Delta H^0_R = \sum_i \nu_i \Delta H^0_{f,i} \qquad (1\text{-}3)$$

In these equations, $\Delta G^0_{f,i}$ and $\Delta H^0_{f,i}$ are the standard Gibbs free energy of formation and standard enthalpy of formation of species i, respectively. For many reactions, values of ΔG^0_R and ΔH^0_R can be calculated from tabulated values of $\Delta G^0_{f,i}$ and $\Delta H^0_{f,i}$ for the reactants and products.

1.3 EXTENT OF REACTION AND THE LAW OF DEFINITE PROPORTIONS

Consider a closed system in which one chemical reaction takes place. Let

N_i = number of moles of species i present at time t
N_{i0} = number of moles of species i present at $t = 0$
$\Delta N_i = N_i - N_{i0}$

Alternately, consider an open system *at steady state*, in which one reaction takes place. For this case, let

N_i = number of moles of species i that leave the system in the time interval Δt
N_{i0} = number of moles of species i that enter the system in the same time interval Δt
$\Delta N_i = N_i - N_{i0}$

In both of these cases, the reaction is the only thing that causes N_i to differ from N_{i0}, i.e., the reaction is the only thing that causes ΔN_i to be nonzero.

The "extent of reaction," ξ is defined as

Extent of reaction for a single reaction in a closed system

$$\boxed{\xi = \Delta N_i/\nu_i} \tag{1-4}$$

The "extent of reaction" is a measure of how far the reaction has progressed. Since reactants disappear as the reaction proceeds, ΔN_i for every reactant is less than 0. Conversely, products are formed, so that ΔN_i for every product is greater than 0. Therefore, the sign convention for stoichiometric coefficients ensures that the value of ξ is always positive, as long as we have identified the reactants and products correctly.

When the extent of reaction is defined by Eqn. (1-4), ξ has units of moles.

The maximum value of ξ for any reaction results when the limiting reactant has been consumed completely, i.e.,

$$\xi_{max} = -N_{10}/\nu_1$$

where the subscript "1" denotes the limiting reactant. In fact, the extent of reaction provides a way to make sure that the limiting reactant has been identified correctly. For each reactant, calculate $\xi_{i0} = N_{i0}/\nu_i$. This is the value of ξ_{max} that would result if reactant "i" was consumed completely. The species with the *lowest* value of ξ_{i0} is the limiting reactant. This is the reactant that will disappear first if the reaction goes to completion.

If the reaction is reversible, equilibrium will be reached before the limiting reactant is consumed completely. In this case, the highest *achievable* value of ξ will be less than ξ_{max}.

The balanced stoichiometric equation for a reaction tells us that the various chemical species are formed or consumed in fixed proportions. This idea is expressed mathematically by the *Law of Definite Proportions*. For a single reaction,

Law of Definite Proportions for a single reaction in a closed system

$$\boxed{\begin{aligned}\Delta N_1/\nu_1 &= \Delta N_2/\nu_2 = \Delta N_3/\nu_3 = \cdots \\ &= \Delta N_i/\nu_i = \cdots = \xi\end{aligned}} \tag{1-5}$$

According to Eqn. (1-5), the value of ξ does not depend on the species used for the calculation. A reaction that obeys the Law of Definite Proportions is referred to as a "stoichiometrically simple" reaction. If the syntheses of propylene oxide (Reaction (1-A)) were stoichiometrically simple, we could write

$$\Delta N_{H_2O}/1 = \Delta N_{NaCl}/2 = \Delta N_{C_3H_6O}/1 = \Delta N_{C_3H_6}/-1 = \Delta N_{NaOH}/-2 = \Delta N_{Cl_2}/-1$$

The extent of reaction concept can be applied to open systems *at steady state* in a second way, by considering the *rates* at which various species are fed to and withdrawn from the system, instead of considering the *number of moles* fed and withdrawn in a specified interval of time. Let

F_i = *molar rate* at which species i flows out of the system (moles i/time)
F_{i0} = *molar rate* at which species i flows into the system (moles i/time)
$\Delta F_i = F_i - F_{i0}$

The extent of reaction now can be defined as

Extent of reaction for a single reaction in a flow system at steady state

$$\boxed{\xi = \Delta F_i/v_i} \tag{1-6}$$

When the extent of reaction is based on molar *flow rates* F_i, rather than on moles N_i, ξ has units of moles/time rather than moles. For this case, the Law of Definite Proportions is written as

Law of Definite Proportions for a single reaction in a flow system at steady state

$$\boxed{\begin{aligned}\Delta F_1/v_1 = \Delta F_2/v_2 = \Delta F_3/v_3 = \cdots \\ = \Delta F_i/v_i = \cdots = \xi\end{aligned}} \tag{1-7}$$

At first glance, the Law of Definite Proportions and the definition of a stoichiometrically simple reaction might seem trivial. However, Eqns. (1-5) and (1-7) can provide a "reality check" when dealing with an actual system. Consider Example 1-1.

EXAMPLE 1-1
Hydrogenolysis of Thiophene (C_4H_4S)

The thiophene hydrogenolysis reaction

$$C_4H_4S + 4H_2 \rightarrow C_4H_{10} + H_2S \tag{1-B}$$

takes place at about 1 atm total pressure and about 250 °C over a solid catalyst containing cobalt and molybdenum. This reaction sometimes is used as a model for the reactions that occur when sulfur is removed from various petroleum fractions (e.g., naphtha, kerosene, and diesel fuel) by reaction with hydrogen over a catalyst.

Suppose that the following data had been obtained in a continuous flow reactor, operating at steady state. The reactor is part of a pilot plant for testing new catalysts. Use these data to determine whether the system is behaving as though one, stoichiometrically simple reaction, i.e., Reaction (1-B), was taking place.

Pilot-plant data for test of thiophene hydrogenolysis catalyst

Species	Gram moles fed during third shift, 8 h	Gram moles in effluent during third shift, 8 h
C_4H_4S	75.3	5.3
H_2	410.9	145.9
C_4H_{10}	20.1	75.1
H_2S	25.7	95.7

APPROACH There are enough data in the preceding table to calculate ξ for each species. If the pilot-plant data are consistent with the hypothesis that one stoichiometrically simple reaction (Reaction (1-B)) took place, then by the Law of Definite Proportions (Eqn. (1-5)), the value of ξ should be the same for all four species.

SOLUTION The data for thiophene in the preceding table give the following value of the extent of reaction: $\xi = (5.3 - 75.3)/-1 = 70$. The complete calculations are shown in the following table.

Test for stoichiometrically simple reaction

Species	ΔN_i	ν_i	ξ
C_4H_4S	-70.0	-1	70.0
H_2	$145.9 - 410.9 = -265.0$	-4	66.25
C_4H_{10}	$75.1 - 20.1 = 55.0$	$+1$	55.0
H_2S	$95.7 - 25.7 = 70.0$	$+1$	70.0

The calculated extents of reaction show that the actual system did *not* behave as though only one stoichiometrically simple reaction took place. Clearly, our preconceived notion concerning Reaction (1-B) is not consistent with the facts.

What's going on in Example 1-1? The data provide some clues. The calculations show that the amount of hydrogen sulfide (H_2S) formed and the amount of thiophene consumed are in the exact proportion predicted by the stoichiometry of Reaction (1-B). However, less hydrogen is consumed than predicted by the balanced stoichiometric equation, given the consumption of thiophene. Moreover, less butane (C_4H_{10}) is produced.

As an aside, if we checked the elemental balances for C, H, and S, they would show that all sulfur atoms were accounted for (in = out), but that more hydrogen and carbon atoms entered than left.

It seems likely that the analytical system in the pilot plant failed to detect at least one hydrocarbon species. Moreover, the undetected species must have a lower H/C ratio than butane, since $\xi_{C_4H_{10}} < \xi_{H_2}$. If the behavior of the actual system cannot be described by one stoichiometrically simple reaction, perhaps more than one reaction is taking place. Can we postulate a *system* of reactions that is consistent with the data, which might help to identify the missing compound(s)?

1.3.1 Stoichiometric Notation—Multiple Reactions

If more than one reaction is taking place, then a given chemical species, say A_i, may participate in more than one reaction. This species will, in general, have a different stoichiometric coefficient in each reaction. It may be a product of one reaction and a reactant in another.

If the index "k" is used to denote one specific reaction in a system of "R" reactions, the generalized stoichiometric notation for a reaction becomes

$$\sum_i \nu_{ki} A_i = 0, \quad k = 1, 2, \ldots, R \tag{1-8}$$

Here, R is the total number of independent reactions that take place and ν_{ki} is the stoichiometric coefficient of species i in reaction k.

Each of the R reactions may contribute to ΔN_i, which is the change in the number of moles of species i. If the extent of reaction "k" is denoted by ξ_k, then the *total* change in the number of moles of species i is

Total change in moles—multiple reactions in a closed system

$$\Delta N_i = N_i - N_{i0} = \sum_{k=1}^{R} \nu_{ki}\xi_k \tag{1-9}$$

The term $\nu_{ki}\xi_k$ is the change in the number of moles of "i" that is caused by reaction "k." The *total* change in moles of species i, ΔN_i, is obtained by summing such terms over all of reactions that take place.

When the extent of reaction is defined in terms of molar flow rates, the equivalent of Eqn. (1-9) is

Total change in molar flow rate—multiple reactions in a flow system at steady state

$$\Delta F_i = F_i - F_{i0} = \sum_{k=1}^{R} \nu_{ki}\xi_k \tag{1-10}$$

EXAMPLE 1-2
Thiophene Hydrogenolysis—Multiple Reactions?

Suppose that the two reactions shown below were taking place in the thiophene hydrogenolysis pilot plant of Example 1-1.

$$C_4H_4S + 3H_2 \rightarrow C_4H_8 + H_2S \tag{1-C}$$

$$C_4H_8 + H_2 \rightarrow C_4H_{10} \tag{1-D}$$

Are the pilot-plant data stoichiometrically consistent with these reactions?

APPROACH

If Reactions (1-C) and (1-D) are sufficient to account for the behavior of the actual system, then *all* of the equations for ΔN_i, one equation for each species, must be satisfied by a *single* value of ξ_C, the extent of Reaction (1-C), plus a *single* value of ξ_D, the extent of Reaction (1-D). There are five chemical species in Reactions (1-C) and (1-D). However, the table in Example 1-1 lacks data for butene (C_4H_8), so only four equations for ΔN_i can be formulated with values for ΔN_i. Two of these equations will be used to calculate values of ξ_C and ξ_D. The two remaining equations will be used to check the values of ξ_C and ξ_D that we calculated.

SOLUTION

Let ν_{Ci} be the stoichiometric coefficient of species "i" in Reaction (1-C) and let ν_{Di} be the stoichiometric coefficient of species "i" in Reaction (1-D). For thiophene (T), from Eqn. (1-9),

$$\Delta N_T = \nu_{CT}\xi_C + \nu_{DT}\xi_D = -70$$

Since $\nu_{CT} = -1$ and $\nu_{DT} = 0$, $\xi_C = 70$.

For H_2 (H),

$$\Delta N_H = \nu_{CH}\xi_C + \nu_{DH}\xi_D = -265$$

Since $\nu_{CH} = -3$ and $\nu_{DH} = -1$,

$$(-3)(70) + (-1)\xi_D = -265; \quad \xi_D = 55$$

These values of ξ_C and ξ_D now must satisfy the remaining two equations for ΔN. For H_2S (S),

$$\Delta N_S = \nu_{CS}\xi_C + \nu_{DS}\xi_D = 70$$

$$(+1)(70) + (0)(55) = 70 \quad \text{Check!}$$

For butane (B),

$$\Delta N_B = \nu_{CB}\xi_C + \nu_{DB}\xi_D = 55$$

$$(0)(70) + (1)(55) = 55 \quad \text{Check!}$$

These calculations show that the data, as they exist, *are* consistent with the hypothesis that Reactions (1-C) and (1-D) are the only ones that take place.

This analysis does not *prove* that these two reactions are taking place. There are other explanations that might account for the experimental data. First, the data may be inaccurate. Perhaps only one reaction takes place, but the number of moles of both H_2 and C_4H_{10} was measured incorrectly. Perhaps more than two reactions take place. Clearly, additional data are required. The analysis of pilot-plant operations must be improved so that all the three species balances (carbon, hydrogen, and sulfur) can be closed within reasonable tolerances.

EXERCISE 1-1

What specific actions would you recommend to the team that is operating the pilot plant?

EXAMPLE 1-3
Thiophene Hydrogenolysis—Calculation of Butene

The extent of reaction concept also can be used to calculate the expected amounts of species that are not directly measured. Consider the previous example. Suppose that Reactions (1-C) and (1-D) are, in fact, the only independent reactions that occur in the pilot plant. What quantity of butene (C_4H_8) should have been found in the effluent from the pilot plant during the third shift?

APPROACH

Let the subscript "E" be used to denote butene. Equation (1-9) can be written for butene, as follows:

$$\Delta N_E = \nu_{CE}\xi_C + \nu_{DE}\xi_D$$

Since all of the quantities on the right-hand side of this equation are known, the value of ΔN_E can be calculated directly. The quantity of butene formed during the third shift is ΔN_E. If there was no butene in the feed to the reactor, ΔN_E also is the total amount of butene that would be collected from the effluent during the third shift.

SOLUTION

From Example 1-2, $\xi_C = 70$ and $\xi_D = 55$. From Reactions (1-C) and (1-D), $\nu_{CE} = +1$ and $\nu_{DE} = -1$. Therefore,

$$\Delta N_E = (+1)(70) + (-1)(55) = 15$$

If there was no butene in the feed to the reactor, we would expect to find 15 moles in the effluent that was collected during the third shift.

1.4 DEFINITIONS OF REACTION RATE

1.4.1 Species-Dependent Definition

In order to be useful in reactor design and analysis, the reaction rate must be an *intensive* variable, i.e., one that does not depend on the size of the system. Also, it is very convenient to define the reaction rate so that it refers explicitly to one of the chemical species that participates in the reaction. The reference species usually is shown as part of the symbol for the reaction rate, and the reference species should be specified in the units of the reaction rate.

Consider a system in which one stoichiometrically simple reaction is taking place. Let's define a reaction rate r_i as

$$r_i \equiv \frac{\text{rate of formation of product ``}i\text{'' (moles ``}i\text{'' formed/time)}}{\text{unit (something)}} \quad (1\text{-}11)$$

The subscript "i" refers to the species whose rate of formation is r_i. The denominator of the right-hand side of Eqn. (1-11) is what makes r_i an intensive variable. We will return to this denominator momentarily.

Several things are obvious about this definition of r_i. First, if "i" actually is being formed, r_i will be positive. However, we may want "i" to be a reactant, which is being consumed (disappearing). In this case, the value of r_i would be negative. An alternative, mathematically equivalent, definition can be used when "i" is a reactant:

$$-r_i \equiv \frac{\text{rate of disappearance of reactant ``}i\text{'' (moles ``}i\text{'' consumed/time)}}{\text{unit (something)}} \quad (1\text{-}12)$$

If "i" actually is being consumed, then $-r_i$ will be positive, i.e., the rate of *disappearance* will be positive.

In order to properly and usefully define "unit (something)," we need to know *where* the reaction actually takes place. Let's consider a few of the most important cases.

1.4.1.1 Single Fluid Phase

A chemical reaction may take place *homogeneously* throughout a single fluid phase. The reaction might result, for example, from collisions between molecules of the fluid or it might result from the spontaneous decomposition of a molecule of the fluid. In such cases, the *overall* rate at which "i" is generated or consumed, i.e., the number of molecules of "i" converted per unit time in the whole system, will be proportional to the volume of the fluid. Fluid volume is the appropriate variable for expressing the rate of a *homogeneous* reaction as an intensive variable. Thus,

Reaction rate—homogeneous reaction

$$-r_i \equiv \frac{\text{rate of disappearance of reactant ``}i\text{'' (moles ``}i\text{'' consumed/time)}}{\text{unit volume of fluid}} \quad (1\text{-}13)$$

In this case, r_i and $-r_i$ have the dimensions of moles i/time-volume.

1.4.1.2 Multiple Phases

Multiphase reactors are *much* more prevalent in industrial practice than single-phase reactors. The behavior of multiphase systems can be very complex. It is not always straightforward to determine whether the reaction takes place in one phase, more than one phase, or at the interface between phases. However, there is one very important case where the locus of reaction is well understood.

Heterogeneous Catalysis Approximately 90% of the reactions that are practiced commercially in fields such as petroleum refining, chemicals and pharmaceuticals manufacture, and pollution abatement involve solid, heterogeneous catalysts. The reaction takes place on the surface of the catalyst, not in the surrounding fluid phase(s). The *overall* reaction rate depends on the amount of catalyst present, and so the *amount of catalyst* must be used to make r_i and $-r_i$ intensive.

The amount of catalyst may be expressed in several valid ways, e.g., weight, volume, and surface area. The choice between these measures of catalyst quantity is one of convenience. However, weight is frequently used in engineering applications. For this choice,

Reaction rate—heterogeneous catalytic reaction

$$r_i \equiv \frac{\text{rate of formation of product ``}i\text{'' (moles ``}i\text{'' formed/time)}}{\text{unit weight of catalyst}} \tag{1-14}$$

In fundamental catalyst research, an attempt usually is made to relate the reaction rate to the number of atoms of the catalytic component that are in contact with the fluid. For example, if the decomposition of hydrogen peroxide (H_2O_2) is catalyzed by palladium metal, the rate of disappearance of H_2O_2 might be defined as,

$$-r_{H_2O_2} \equiv \frac{\text{rate of disappearance of } H_2O_2 \text{ (molecules reacted/time)}}{\text{atoms of Pd in contact with fluid containing } H_2O_2} \tag{1-15}$$

Expressed in this manner, $-r_{H_2O_2}$ has units of inverse time and is called a "turnover frequency." Physically, it is the number of molecular reaction events (i.e., H_2O_2 decompositions) that occur on a single atom of the catalytic component per unit of time.

Unfortunately, except in special cases, the symbol that is used to denote reaction rate is not constructed to tell the user what basis was used to make the rate intensive. This task usually is left to the units of the reaction rate.

Other Cases In some cases, a reaction takes place in one of the phases in a multiphase reactor but not in the others. Obviously, it is critical to know the phase in which the reaction occurs. If the definition of the reaction rate is based on the *total* reactor volume, serious trouble will result when the ratio of the phases changes. The ratio of the phases generally will depend on variables such as the reactor dimensions, the intensity of mechanical agitation, and the feed rates and compositions of the various fluids. Therefore, difficulty is inevitable, especially on scaleup, if the reaction rate is misdefined.

In a few industrial processes, the reaction occurs at the interface between two phases. The *interfacial area* then is the appropriate parameter to use in making the reaction rate intensive. The synthesis of poly(bisphenol A carbonate) (polycarbonate) from bisphenol A and phosgene is an example of a reaction that occurs at the interface between two fluid phases.

On occasion, a reaction takes place in more than one phase of a multiphase reactor. An example is the so-called "catalytic combustion." If the temperature is high enough, a hydrocarbon fuel such as propane can be oxidized catalytically, on the surface of a heterogeneous catalyst, at the same time that a homogeneous oxidation reaction takes place in the gas phase. This situation calls for two separate definitions of the reaction rate, one for the gas phase and the other for the heterogeneous catalyst.

1.4.1.3 Relationship Between Reaction Rates of Various Species (Single Reaction)

For a stoichiometrically simple reaction, that obeys the Law of Definite Proportions, the reaction rates of the various reactants and products are related through stoichiometry, i.e.,

$$r_1/\nu_1 = r_2/\nu_2 = r_3/\nu_3 = \cdots = r_i/\nu_i = \cdots \tag{1-16}$$

For example, in the ammonia synthesis reaction

$$N_2 + 3H_2 \rightleftarrows 2NH_3 \tag{1-E}$$

$$r_{N_2}/(-1) = r_{H_2}/(-3) = r_{NH_3}/(2)$$

In words, the molar rate of ammonia formation is twice the molar rate of nitrogen disappearance, and two-thirds the molar rate of hydrogen disappearance.

1.4.1.4 Multiple Reactions

If more than one reaction takes place, the rate of each reaction must be known in order to calculate the *total* rate of formation or consumption of a species. Thus,

Total rate of formation of "*i*" when multiple reactions take place

$$\boxed{r_i = \sum_{k=1}^{R} r_{ki}} \tag{1-17}$$

where R is the number of independent reactions that take place, and "k" again denotes a specific reaction. In words, the total rate of formation of species i is the sum of the rates at which "i" is formed in each of the reactions taking place.

1.4.2 Species-Independent Definition

The species-dependent definition of the reaction rate is used in a majority of published articles in the chemical engineering literature. The major disadvantage of this definition is that the reaction rates of the various species in one chemical reaction are different if their stoichiometric coefficients are different. The relationship of one rate to another is given by Eqn. (1-16). This disadvantage has led to the occasional use of an alternative, species-independent definition of reaction rate.

In the species-independent definition, the reaction rate is referenced to the *reaction* itself, rather than to a *species*. Consider the stoichiometrically simple reaction

$$\sum_i \nu_i A_i = 0$$

Equation (1-16) provides relationships between the various r_i for this reaction. However, we can define the rate of this *reaction* as $r \equiv r_i/\nu_i$, so that Eqn. (1-16) becomes

$$r = r_1/\nu_1 = r_2/\nu_2 = r_3/\nu_3 = \cdots = r_i/\nu_i = \cdots \tag{1-18}$$

With this definition, a species does not have to be specified in order to define the reaction rate. However, we *do* have to specify the *exact* way in which the balanced stoichiometric equation is written. For example, the value of r is not the same for

$$N_2 + 3H_2 \rightleftarrows 2NH_3 \tag{1-E}$$

as it is for

$$1/2N_2 + 3/2H_2 \rightleftarrows NH_3 \tag{1-F}$$

because the stoichiometric coefficients are not the same in these two stoichiometric equations.

EXERCISE 1-2

If $r = 0.45$ for Reaction (1-E) at a given set of conditions, what is the value of r for Reaction (1-F)?

Obviously, when using the species-independent definition of reaction rate, great care must be taken to write the balanced stoichiometric equation(s) at the beginning of the analysis, and to use the *same* stoichiometric equation(s) throughout the analysis.

The species-*dependent* definition of reaction rate will be used throughout the remainder of this text.

SUMMARY OF IMPORTANT CONCEPTS

- Sign convention for stoichiometric coefficients
  - products positive; reactants negative
- Extent of reaction
  - Single reaction
    - Closed system $\xi = \Delta N_i/\nu_i$
    - Open system at steady state $\xi = \Delta F_i/\nu_i$
  - Multiple reactions
    - Closed system $\Delta N_i = \sum_{k=1}^{R} \nu_{ki}\xi_k$
    - Open system at steady state $\Delta F_i = \sum_{k=1}^{R} \nu_{ki}\xi_k$
  - Applications
    - Checking consistency of data
      Single reaction?
      Multiple reactions? Which ones?
    - Calculating unknown quantities
- Defining the reaction rate
  - *Where* does the reaction occur?

PROBLEMS

Problem 1-1 (Level 1) A group of researchers is studying the kinetics of the reaction of hydrogen with thiophene (C_4H_4S). They have postulated that only one stoichiometrically simple reaction takes place, as shown below.

$$C_4H_4S + 4H_2 \rightarrow H_2S + C_4H_{10}$$

In one experiment, the feed to a continuous reactor operating at steady state was

C_4H_4S—0.65 g·mol/min
H_2—13.53 g·mol/min
H_2S—0.59 g·mol/min
C_4H_{10}—0.20 g·mol/min

The effluent rates were

C_4H_4S—0.29 g·mol/min
H_2—12.27 g·mol/min
H_2S—0.56 g·mol/min
C_4H_{10}—0.38 g·mol/min

Are the experimental data consistent with the assumption that only one stoichiometrically simple reaction (i.e., the above reaction) takes place?

Problem 1-2 (Level 2) A continuous reactor operating at steady state is being used to study the formation of methanol (CH_3OH) from mixtures of H_2 and CO according to the reaction

$$CO + 2H_2 \rightleftarrows CH_3OH$$

Some data from one particular run are shown below.

Species	Rate in (g·mol/min)	Rate out (g·mol/min)
CO	100	83
H_2	72	38
CO_2	9	9
CH_4	19	19
CH_3OH	2	13

Does the system behave as though only one stoichiometrically simple reaction, i.e., the above reaction, takes place? If not, develop a hypothesis that quantitatively accounts for any discrepancy in the data. How could your hypothesis be tested?

Problem 1-3 (Level 1) The following e-mail is in your in-box at 8 AM on Monday:

To: U. R. Loehmann
From: I. M. DeBosse
Subject: Quinoline Hydrogenation

U.R.,

I hope that you can help with the following:

When quinoline (C_9H_7N) is hydrogenated at about 350 °C over various heterogeneous catalysts, the three reactions shown below take place to varying extents. In one experiment in a batch reactor (closed system), the initial charge to the reactor was 100 mol of quinoline and 500 mol of hydrogen (H_2). After 10 h, the reactor contents were analyzed, with the following results:

Quinoline (C_9H_7N)—40 mol
Hydrogen (H_2)—290 mol
Decahydroquinoline ($C_9H_{17}N$)—20 mol

If the reactions shown below are the only ones that take place, how many moles of tetrahydroquinoline ($C_9H_{11}N$) and butylbenzylamine ($C_9H_{13}N$) should have been present after 10 h?

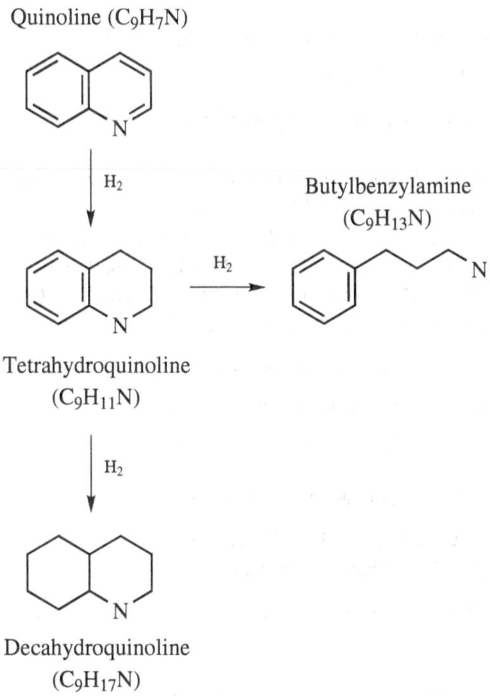

Quinoline (C_9H_7N)

H_2

Butylbenzylamine ($C_9H_{13}N$)

H_2

Tetrahydroquinoline ($C_9H_{11}N$)

H_2

Decahydroquinoline ($C_9H_{17}N$)

Please write me a short memo (not more than one page) containing the results of your calculations and explaining what you did. Attach your calculations to the memo in case someone wants to review the details.

Thanks, I. M.

Problem 1-4 (Level 2) The following memo is in your in-box at 8 AM on Monday:

To: U. R. Loehmann
From: I. M. DeBosse
Subject: Methanation

U.R.,

Hope you can help with the following:

The methanation of carbon monoxide

$$CO + 3H_2 \rightleftarrows CH_4 + H_2O$$

is an important step in the manufacture of ammonia, and in the manufacture of synthetic natural gas (SNG) from coal or heavy hydrocarbons. The reaction is very exothermic. Especially in the manufacture of SNG, a large quantity of heat must be removed from the methanation reactor in order to avoid catalyst deactivation and to maintain a favorable equilibrium.

A small research company, F. A. Stone, Inc., has offered to license us a novel methanation process. The reaction takes place in a slurry bubble-column reactor. Small particles of the catalyst are suspended in a hydrocarbon liquid (a mixture of heavy paraffins with an average formula of $C_{18}H_{38}$. A gas containing carbon monoxide (CO) and hydrogen (H_2) is sparged (bubbled) continuously through the slurry. The gas leaving the top of the reactor contains unreacted CO and H_2, as well as the products, CH_4 and H_2O. The heat of reaction is removed by water flowing through tubes in the reactor.

Please review the following pilot-plant data provided by F. A. Stone to be sure that the process is performing "as advertised." These data are from one particular continuous, steady-state run. F. A. Stone will not release additional data until we have made a downpayment on the license fee.

Species	Inlet flow rate (lb·mol/day)	Outlet flow rate (lb·mol/day)
CO	308	41
H_2	954	91
CH_4	11	327
H_2O	2	269
CO_2	92	92
N_2	11	11

You may assume that these flow rates are accurate, at least for now.

Specific questions

1. Does the system behave as though one stoichiometrically-simple reaction (the methanation of carbon monoxide) is taking place? Explain your answer.
2. If your answer is "no," what explanation(s) would you propose to account for the observed behavior?
3. Based on your hypotheses, what additional experiments or measurements should we require from F. A. Stone before we make a down payment?

Please write me a short memo (not more than one page) containing the answers to these questions, and explaining how you arrived at your conclusions. Attach your calculations to the memo in case someone wants to review the details.

Thanks, I. M.

Problem 1-5 (Level 2) Carbon monoxide (CO) and hydrogen (H_2) are fed to a continuous catalytic reactor operating at steady state. There are no other components in the feed. The outlet stream contains unconverted CO and H_2, along with the products methanol (CH_3OH), ethanol (C_2H_5OH), isopropanol (C_3H_7OH), and carbon dioxide (CO_2). These are the only species in the product stream.

The reactions occurring are

$$CO + 2H_2 \rightarrow CH_3OH$$
$$3CO + 3H_2 \rightarrow C_2H_5OH + CO_2$$
$$5CO + 4H_2 \rightarrow C_3H_7OH + 2CO_2$$

The feed rates of CO and H_2 to the reactor are 100 mol/h (each). The rates in the stream that leaves the reactor (in mols/h.)

are H_2-30; CO-30; C_2H_5OH-5. What is the *mole fraction* of each species in the product stream?

Problem 1-6 (Level 1) The hydrogenation of aniline at about 50 °C, over a Ru/carbon catalyst, is believed to involve the reactions[2]:

$$\text{Aniline (A)} + 3H_2 \longrightarrow \text{Cyclohexylamine (CHA)}$$

$$\text{Cyclohexylamine} + H_2 \longrightarrow \text{Cyclohexane (CH)} + NH_3$$

$$2\ \text{Cyclohexylamine} \longrightarrow \text{Dicyclohexylamine (DCHA)} + NH_3$$

In one particular experiment, aniline was hydrogenated in a closed vessel at 50 °C and 50 bar of H_2 pressure for 3 h. The following data were obtained:

Species	Moles after 3 h/mol of aniline charged
A	0.476
CHA	0.346
CH	0.080
DCHA	0.049

The amounts of ammonia formed and H_2 consumed were not measured.

1. Is the experimental data consistent with the assumption that these three reactions are the only ones that occur?
2. Estimate the amount of NH_3 formed (mols/mol A charged).
3. Estimate the amount of H_2 consumed (mol/mol A charged).

Problem 1-7 (Level 2) The gas-phase reactions

$$\text{Isobutanol (B)} \rightarrow \text{Isobutene (IB)} + H_2O$$
$$(CH_3)_2CHCH_2OH \rightarrow (CH_3)_2C{=}CH_2 + H_2O$$

$$2\ \text{Methanol} \rightarrow \text{Dimethyl ether (DME)} + H_2O$$
$$2CH_3OH \rightarrow CH_3OCH_3 + H_2O$$

$$\text{Isobutanol} + \text{Methanol} \rightarrow \text{Methylisobutyl ether (MIBE)} + H_2O$$
$$(CH_3)_2CHCH_2OH + CH_3OH \rightarrow (CH_3)_2\,CHCH_2OCH_3 + H_2O$$

$$2\ \text{Isobutanol} \rightarrow \text{Diisobutyl ether (DIBE)} + H_2O$$
$$2(CH_3)_2CHCH_2OH \rightarrow (CH_3)_2CHCH_2OCH_2CH(CH_3)_2 + H_2O$$

take place in a continuous reactor operating at steady state. The feed to the reactor consists of N_2—10,000 mol/h, isobutanol (B)—8333 mol/h, and methanol (M)—16,667 mol/h. The effluent flow rates are isobutene (IB)—2923 mol/h, dimethyl ether (DME)—3436 mol/h, methyl isobutyl ether (MIBE) = 5038 mol/h, and diisobutyl ether (DIBE)—22 mol/h.

1. What is the fractional conversion of isobutanol?
2. What is the fractional conversion of methanol?
3. What is the mole fraction of water leaving the reactor?

Problem 1-8 (K2-1) (Level 1) Consider the reaction

$$\underset{\text{(trichloroethane)}}{C_2H_3Cl_3} + 3H_2 \rightarrow \underset{\text{(ethane)}}{C_2H_6} + 3HCl$$

If the rate of formation of HCl (r_{HCl}) is 25×10^{-6} g·mol/g·cat-min

1. What is the rate of disappearance of trichloroethane?
2. What is the rate of formation of ethane?

Problem 1-9 (Level 1) Look carefully at Reaction (1-A). Refer to the literature as necessary. Prepare brief written answers to the following questions:

1. Is a process based on this reaction a good example of "green chemistry?"
2. What can be done with the NaCl that is produced?
3. Since Cl atoms do not appear in the final product (C_3H_6O), what role does chlorine play in this reaction?

Problem 1-10 (Level 1) Calculate the standard enthalpy change on reaction, ΔH_R^0, for Reaction (1-E) at 25 °C. Calculate the standard Gibbs free energy change on reaction, ΔG_R^0, for Reaction (1–E) at 25 °C. What are the units of ΔH_R^0 and ΔG_R^0?

Problem 1-11 (Level 2) Styrene, the monomeric building block for the polymer polystyrene, is made by the catalytic dehydrogenation of ethyl benzene. Ethyl benzene, in turn, is made by the alkylation of benzene with ethylene, as shown by Reaction (A) below. A common side reaction is that addition of another alkyl group to ethyl benzene to form diethyl benzene. This reaction is shown as Reaction (B). The second alkyl group may be in the *ortho*, *meta*, or *para* position.

[2] Cho, H.B. and Park, Y.H., *Korea J. Chem. Eng.*, 20(2), 262–267 (2003).

$$C_6H_6 + C_2H_4 \longrightarrow C_6H_5C_2H_5 \quad \text{(A)}$$

Benzene Ethylene Ethyl benzene

$$C_6H_5C_2H_5 + C_2H_4 \longrightarrow C_6H_4(C_2H_5)_2 \quad \text{(B)}$$

Diethyl benzene

Initially, 100 mol of benzene and 100 mol of ethylene are charged to a reactor. No material flows into or out of the reactor after this initial charge. After a very long time, the contents of the reactor are analyzed, with the following results:

Species	Benzene	Ethylene	Ethyl benzene	Diethyl benzene
Number of moles	35	2	39	19

1. Show that the behavior of the reactor is *not* consistent with the hypothesis that Reactions (A) and (B) are the only ones that take place.
2. Develop an alternative hypothesis that is consistent with *all* of the data, and demonstrate this consistency.

Problem 1-12 (Level 2) The overall reaction for the catalytic hydrodechlorination of 1,1,1-trichloroethane (111-TCA) is

$$C_2H_3Cl_3 + 3H_2 \rightarrow C_2H_6 + 3HCl$$

On certain catalysts, this overall reaction appears to take place via the following sequence of simpler reactions:

$$C_2H_3Cl_3 + H_2 \rightarrow C_2H_4Cl_2 + HCl \quad (1)$$

$$C_2H_4Cl_2 + H_2 \rightarrow C_2H_5Cl + HCl \quad (2)$$

$$C_2H_5Cl + H_2 \rightarrow C_2H_6 + HCl \quad (3)$$

A mixture of 111-TCA, H_2, and N_2 was fed to a continuous catalytic reactor operating at 523 K and 1 atm. total pressure at a rate of 1200 L(STP)/h. The feed contained 10 mol % H_2 and 1 mol % 111-TCA, and the reactor operated at steady-state.

It was not possible to accurately measure the outlet concentrations of H_2 and HCl. The flow rates of $C_2H_3Cl_3$, $C_2H_4Cl_2$, C_2H_5Cl, and C_2H_6 out of the reactor were 0.074 mol/h, 0.111 mol/h, 0.050 mol/h, and 0.301 mol/h, respectively.

1. Are these data consistent with the hypothesis that the overall reaction takes place via Reactions (1), (2) and (3) (and *only* Reactions (1), (2) and (3))? Justify your answer.
2. What is the molar flow rate of H_2 leaving the reactor?
3. What is the molar flow rate of HCl leaving the reactor?

Chapter 2

Reaction Rates—Some Generalizations

LEARNING OBJECTIVES

After completing this chapter, you should be able to

1. use the Arrhenius relationship to calculate how reaction rate depends on temperature;
2. use the concept of reaction order to express the dependence of reaction rate on the individual species concentrations;
3. calculate the frequency of bimolecular and trimolecular collisions;
4. determine whether the rate equations for the forward and reverse rates of a reversible reaction are thermodynamically consistent;
5. calculate heats of reaction and equilibrium constants at various temperatures (review of thermodynamics).

In order to design a new reactor, or analyze the behavior of an existing one, we need to know the rates of *all* the reactions that take place. In particular, we must know how the rates vary with temperature, and how they depend on the concentrations of the various species in the reactor. This is the field of chemical kinetics.

This chapter presents an overview of chemical kinetics and introduces some of the molecular phenomena that provide a foundation for the field. The relationship between kinetics and chemical thermodynamics is also treated. The information in this chapter is sufficient to allow us to solve some problems in reactor design and analysis, which is the subject of Chapters 3 and 4. In Chapter 5, we will return to the subject of chemical kinetics and treat it more fundamentally and in greater depth.

2.1 RATE EQUATIONS

A "rate equation" is used to describe the rate of a reaction quantitatively, and to express the functional dependence of the rate on temperature and on the species concentrations. In symbolic form,

$$r_A = r_A(T, \text{all } C_i)$$

where T is the temperature. The term "all C_i" is present to remind us that the reaction rate can be affected by the concentrations of the reactant(s), the product(s), and any other compounds that are present, even if they do not participate in the reaction.

The rate equation must be developed from experimental data. Unfortunately, we cannot make accurate *a priori* predictions of either the form of the rate equation or the constants that appear in the rate equation, at least for the present.[1] In Chapter 6, we will learn how to test rate equations against experimental data and to determine the unknown constants in a rate equation.

2.2 FIVE GENERALIZATIONS[2]

Based on more than a century of experimental and theoretical study of the kinetics of many different chemical reactions, some rules of thumb concerning the form of the rate equation have evolved. There are important exceptions to each of these rules-of-thumb. Nevertheless, the following five generalizations permit chemical engineers to attack many practical problems in reactor design and analysis.

Generalization I

For single reactions *that are essentially irreversible*, the rate of disappearance of reactant A can be expressed as

$$-r_A = k(T)F(\text{all } C_i) \tag{2-1}$$

This equation tells us that the effects of temperature and concentration frequently can be separated. The term $k(T)$ does not depend on any of the concentrations and is called a "rate constant." The term $F(\text{all } C_i)$ depends on the concentrations of the various species but not on temperature.

There are two major theories of chemical kinetics, collision theory (CT) and transition-state theory (TST). Both theories lead to rate equations that obey Generalization I, i.e., the effects of temperature and concentration are separable. Unfortunately, both CT and TST apply to a very limited category of reactions known as "elementary" reactions. An "elementary" reaction is one that occurs in a single step *on the molecular level exactly* as written in the balanced stoichiometric equation. The reactions that chemists and chemical engineers deal with on a practical level almost never are elementary. However, elementary reactions provide the link between molecular-level chemistry and reaction kinetics on a macroscopic level. Elementary reactions will be discussed in some depth in Chapter 5. For now, we must look at Eqn. (2-1) as an empirical attempt to extrapolate a key result of CT and TST to complex reactions that are outside the scope of the two theories.

Despite its lack of a strong theoretical justification, Eqn. (2-1) is very useful in a practical sense. It frequently provides a reasonable starting point for the analysis of experimental kinetic data as well, as for reactor design and analysis.

Equation (2-1) should not be applied directly to a *reversible* reaction, i.e., a reaction that stops well short of complete consumption of the limiting reactant. Rate equations for reversible reactions are the focus of Generalization V.

Generalization II

The rate constant can be written as

Arrhenius relationship

$$k(T) = A\exp(-E/RT) \tag{2-2}$$

[1] The key word in this sentence is "accurate." It is possible to predict rates reasonably well for very simple gas-phase reactions via quantum-mechanically based molecular simulations and to make order-of-magnitude predictions for more complex reactions. However, at this point in time, rate equations that are accurate enough for reactor design and analysis must be developed from experimental data.

[2] Adapted from Boudart, M., *Kinetics of Chemical Processes*, Prentice-Hall (1968).

where R is the gas constant and T is the *absolute* temperature. This relationship is called the "Arrhenius relationship" or "Arrhenius expression." The term "A" is known as the *preexponential factor* or alternatively as the "frequency factor." It does not depend on either temperature or concentration.

The symbol "E" represents the *activation energy* of the reaction. The value of E almost always is positive. Therefore, the rate constant increases with temperature. For chemical reactions, E usually is in the range of 40–400 kJ/mol (10–100 kcal/mol). This means that the rate of reaction is *very* sensitive to temperature. As a very rough approximation, the rate of a reaction doubles with every 10 K increase in temperature. Obviously, the exact change will depend on the values of E and T.

Equation (2-2) provides an accurate description of the effect of temperature on the rate constants of a very large number of chemical reactions. For a given reaction, the value of E usually is found to be constant over a reasonably wide range of temperature. In fact, a change of E with temperature can signal a change in the mechanism of the reaction, or a change in the relative rates of the various steps that make up the overall reaction.

EXAMPLE 2-1
Calculation of Ratio of Rate Constants at Two Different Temperatures

The activation energy of a particular reaction is 50 kJ/mol. What is the ratio of the rate constant at 100 °C to the rate constant at 50 °C?

APPROACH

The dependence of the rate constant on temperature is given by the Arrhenius expression, Eqn. (2-2). The preexponential factor, A, cancels out of the ratio of rate constants at two different temperatures. If the activation energy is known, the ratio depends only on the values of the two temperatures and can be calculated.

SOLUTION

From Eqn. (2-2)

$$\frac{k(T_2)}{k(T_1)} = \frac{A\exp(-E/RT_2)}{A\exp(-E/RT_1)} = \exp\left(\frac{-E}{R}\left[\frac{1}{T_2} - \frac{1}{T_1}\right]\right)$$

For $T_2 = 373$ K (100 °C) and $T_1 = 323$ K (50 °C),

$$\frac{k(373\text{ K})}{k(323\text{ K})} = \exp\left(\frac{-50{,}000\text{ (J/mol)}}{8.314\text{ (J/mol, K)}}\left[\frac{1}{373} - \frac{1}{323}\right](1/\text{K})\right) = 12.1$$

The Arrhenius relationship was developed in the late 1890s based on thermodynamic reasoning. However, there is a simple kinetic analysis that helps to explain the basis of this equation. This analysis is based on some elementary concepts from TST.

In order for a reaction to occur, the reactants must have enough energy to cross over an energy barrier that separates reactants from products, as illustrated in Figure 2-1. The height of the energy barrier is ΔE_k when the reaction proceeds in the forward direction, i.e., from reactants to products. The energy difference between the reactants and the products is ΔE_P. When the reaction goes in the reverse direction, an energy barrier of $\Delta E_k + \Delta E_P$ must be overcome. The units of these ΔEs are energy/mol, e.g., J/mol.

The individual molecules in a fluid at a temperature, T, will have different energies. Some will have enough energy to cross over the energy barrier and some will not. Let the energy of a single molecule be denoted "e". For simple molecular structures, the distribution of energies in a large population of molecules is given by Boltzmann's equation

$$f(e) = \frac{2\sqrt{e}}{\sqrt{\pi}(k_B T)^{3/2}}\exp(-e/k_B T)$$

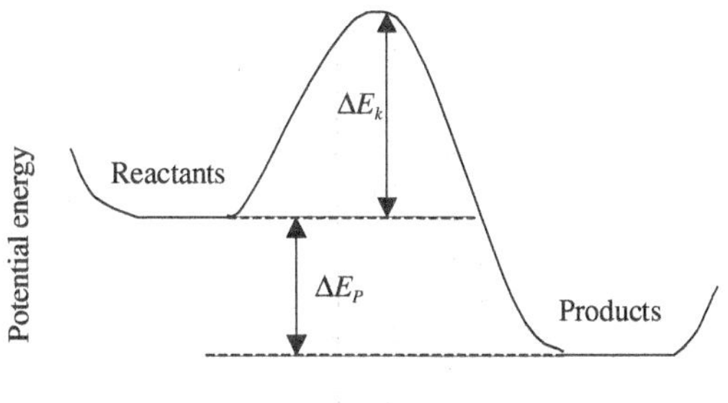

Figure 2-1 Illustration of the energy barrier that must be overcome in order for a reaction to take place. The "reaction coordinate" is a measure of the change in molecular geometry as the reaction progresses. "Molecular geometry" may include the distance between atoms, bond angles, bond lengths, etc., depending on the specific reaction.

In this equation, $f(e)$ is the *distribution function* for molecular energies. In words, $f(e)*de$ is the fraction of molecules with energies between e and $(e + de)$. This distribution function is normalized, so that $\int_0^\infty f(e)de = 1$. The other symbols in the Boltzmann equation are T, the *absolute* temperature (K), and k_B, the Boltzmann constant ($k_B = 1.38 \times 10^{-16}$ erg/molecule-K $= 1.38 \times 10^{-23}$ J/molecule-K $= 1.38 \times 10^{-16}$ g·cm²/s²-molecule-K).

Distribution functions are an important statistical tool, and they are used throughout the field of kinetics and reactor analysis. Distribution functions reappear in Chapter 9, as a means of characterizing porous catalysts, and in Chapter 10, as a means of describing fluid flow in nonideal reactors.

Suppose that a molecule must have a *minimum* energy in order to react, i.e., cross over the energy barrier. This minimum energy will be denoted e^*. For a gas that obeys the Boltzmann equation, the fraction of molecules that have *at least* this threshold value is denoted as $F(e > e^*)$ and is given by

$$F(e > e^*) = \int_{e^*}^{\infty} f(e)de = \frac{2}{\sqrt{\pi}(k_B T)^{(3/2)}} \int_{e^*}^{\infty} \sqrt{e}\, \exp(-e/k_B T)de$$

When $e^* > 3k_B T$, the above equation is closely approximated by

$$F(e > e^*) \cong \frac{2}{\sqrt{\pi}} \left(\frac{e^*}{k_B T}\right)^{(1/2)} \exp(-e^*/k_B T), \quad e^* > 3k_B T$$

According to the energy barrier concept shown in Figure 2-1, we would expect the reaction rate to be proportional to $F(e > e^*)$, i.e., to the fraction of molecules that have at least the minimum energy, e^*, required to cross the energy barrier.

In order to compare the above equation for $F(e > e^*)$ with the Arrhenius relationship, we must transform e (energy/molecule) to E (energy/mol). This is done by multiplying both e and k_B by N_{av}, Avogadro's number, and recognizing that $k_B N_{av} = R$. The above equation then becomes

$$F(E > E^*) \cong \frac{2}{\sqrt{\pi}} \left(\frac{E^*}{RT}\right)^{(1/2)} \exp(-E^*/RT), \quad E^* \geq 3RT, \tag{2-3}$$

For chemical reactions, the restriction that $E^* > 3RT$ is not important. At 500 K, the value of $3RT$ is about 12 kJ/mol. Typical activation energies for chemical reactions are at least three times this value.

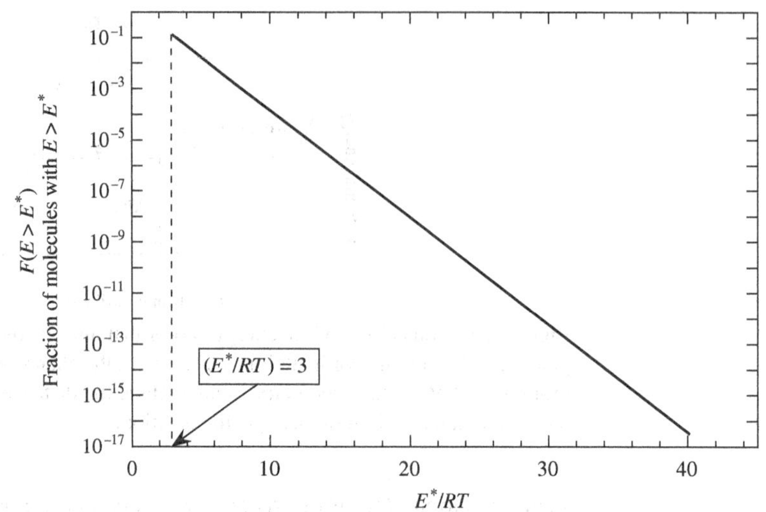

Figure 2-2 The fraction of molecules with an energy, E, greater than a threshold energy, E^*, for a gas that obeys the Boltzmann equation.

Figure 2-2 is a plot of $F(E > E^*)$ versus (E^*/RT). At a fixed value of E^*, E^*/RT gets smaller as T increases. Therefore, according to Figure 2-2, $F(E > E^*)$ increases as T increases. Physically, the fraction of molecules with enough energy to cross over the energy barrier increases as the temperature increases. For an energy of 80 kJ/mol and a temperature of 500 K, $E^*/RT \cong 19$. Figure 2-2 shows that the fraction of molecules that have at least this much energy is only about 2×10^{-8}.

The temperature dependence of $F(E > E^*)$, as shown in Eqn. (2-3), is very similar to the temperature dependence of k in the Arrhenius relationship. When the value of E^* is high, the $T^{(1/2)}$ term in the denominator of Eqn. (2-3) has very little effect on the overall temperature dependence. In this case, both Eqns. (2-2) and (2-3) predict that the reaction rate will increase exponentially with $(1/T)$.

Comparison of Eqns. (2-2) and (2-3) suggests that the activation energy, E, in the Arrhenius relationship can be interpreted as the *minimum* energy that the reactants must possess in order to cross over the energy barrier and react. Thus, $E = E^* = \Delta E_k$. This picture is supported by the results of the above analysis based on the Boltzmann equation, and by more sophisticated analyses based on TST.

Generalization III

The term $F(\text{all } C_i)$ decreases as the concentrations of the reactants decrease. This is equivalent to saying that the rate of reaction decreases as the concentrations of the reactants decrease, if the temperature is constant.

This generalization is violated occasionally. The oxidation of carbon monoxide to carbon dioxide, catalyzed by platinum metal,

$$CO + 1/2O_2 \rightarrow CO_2$$

is a well-known example. Carbon monoxide oxidation is used to reduce CO emissions to the atmosphere from a variety of combustion systems, ranging from automobile engines to the stationary gas turbine engines that are used for electric power generation. For this reaction, $-r_{CO}$ increases as the concentration of CO increases when C_{CO} is very low. In this "low concentration" region, Generalization III is obeyed. However, as the CO concentration

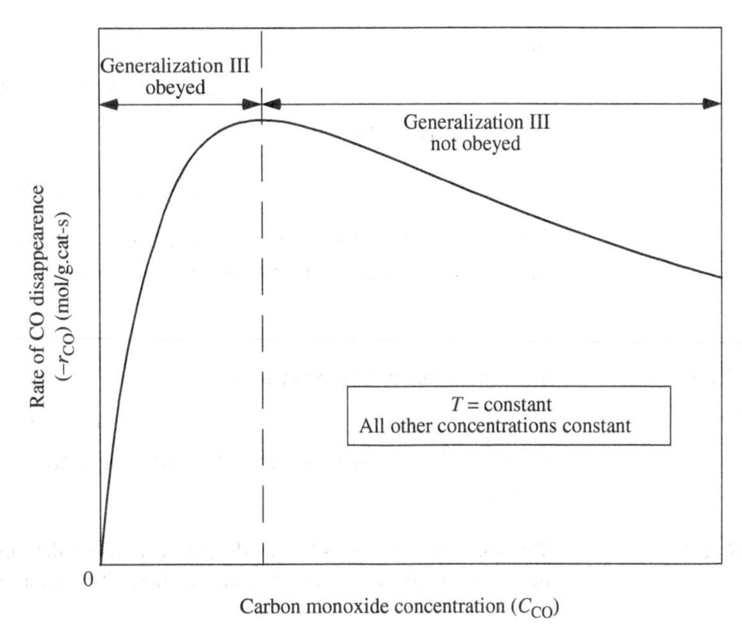

Figure 2-3 A schematic illustration of the effect of CO concentration (C_{CO}) on the rate of CO oxidation ($-r_{CO}$) over a heterogeneous Pt catalyst.

continues to increase, the rate goes through a maximum and then declines as C_{CO} increases further. In this "high concentration" region, Generalization III is not obeyed. This behavior is illustrated in Figure 2-3.

Platinum is a component of some of the catalysts that are used to remove pollutants such as CO from automobile exhaust. Therefore, the unusual variation of the rate with the CO concentration is of practical concern. In Chapter 5, we will explore the source of this behavior. Before that, in Chapter 4, we will see that rate equations that go through a maximum as a reactant concentration is increased can give rise to unusual reactor behavior.

Generalization IV

The term $F(\text{all } C_i)$ can be written as

Reaction orders

$$F(\text{all } C_i) = \prod_i C_i^{\alpha_i} \tag{2-4}$$

The symbol $\prod_i$ denotes the product over all values of *i*. *The* term α_i is called the *order* of the reaction with respect to species "*i*", or the *individual* reaction order with respect to *i*. Values of α_i generally are in the range $-2 \leq \alpha_i \leq 2$, and fractional values are permissible. The summation $\sum_i \alpha_i$ is referred to as the *overall* order of the reaction.

The order of the reaction with respect to *i*, α_i, reflects the sensitivity of the reaction rate to a change in the concentration of *i*. If $\alpha_i = 0$, the reaction rate does not depend on C_i. If $\alpha_i = 2$, the reaction rate quadruples when C_i is doubled.

Rate equations in which the concentration-dependent terms obey Eqn. (2-4) are called *power-law* rate equations. Power-law rate equations are very useful in chemical engineering and often are used as a starting point in the analysis of kinetic data.

Consider the reaction

$$\nu_A \text{A} + \nu_B \text{B} + \nu_C \text{C} + \nu_D \text{D} = 0$$

or

$$(-\nu_A)A + (-\nu_B)B \rightarrow \nu_C C + \nu_D D$$

If Eqn. (2-4) is obeyed

$$F(\text{all } C_i) = C_A^{\alpha_A} C_B^{\alpha_B} C_C^{\alpha_C} C_D^{\alpha_D}$$

Other species that do not participate in the reaction might have to be included in $F(\text{all } C_i)$, if their concentrations influence the reaction rate.

EXAMPLE 2-2
Reaction Orders

The rate equation for the reaction

$$\nu_A A + \nu_B B + \nu_C C + \nu_D D = 0$$

is $-r_A = kC_A C_B^{1/2}$. What are the orders with respect to A, B, C, and D? What is the overall order of the reaction?

APPROACH

The individual reaction orders are the powers to which the various concentrations are raised. These can be determined by examining the rate equation. The overall order is the sum of the individual orders.

SOLUTION

Order with respect to A: 1
Order with respect to B: 1/2
Order with respect to C: 0
Order with respect to D: 0
Overall order $= \sum_i \alpha_i = 1 + 1/2 = 3/2$

In general, the reaction orders, α_i, cannot be determined from the stoichiometry of the reaction. We certainly *cannot* presume that $\alpha_i = -\nu_i$. The order of the reaction with respect to component i is not necessarily equal to the "molecularity" of "i" in the balanced stoichiometric equation. There are an infinite number of ways to write a balanced stoichiometric equation. Therefore, there are an infinite number of permissible values of ν_i. However, the reaction order α_i reflects the *actual* behavior of the reaction and has to be determined from experimental data. For example, we might find that the reaction rate increases with the square of the concentration of reactant A ($\alpha_A = 2$) or perhaps with the square root of the concentration of reactant A ($\alpha_A = 0.5$). This dependency of $-r_A$ on C_A is not going to change simply because we choose to write the stoichiometric equation in one way instead of another.

There is one exception to the rule that "order $\neq$ molecularity." For "elementary" reactions, which were briefly mentioned in the discussion of Generalization I, the reaction order with respect to reactant i is equal to the negative of the stoichiometric coefficient of "i", and the reaction order is 0 for each product. Unfortunately, an overwhelming majority of the reactions that chemical engineers encounter are not elementary. Therefore, until elementary reactions have been discussed in detail in Chapter 5, no relationship between stoichiometry and reaction order should be presumed.

Both collision theory and transition-state theory provide some support for the use of power-law rate equations. Even though both theories apply only to elementary reactions, useful molecular insight can be obtained by examining some results from CT.

The central postulate of CT is that the rate of a reaction is proportional to the frequency of collisions between the reactants. Collision frequencies can be calculated for ideal-gas mixtures, provided that the molecules are spherical and their collisions are perfectly elastic.[3] For binary collisions between molecule A and molecule B, the result is

Frequency of bimolecular collisions

$$Z_{AB} = (\underline{r}_A + \underline{r}_B)^2 \left[8\pi k_B T \left(\frac{1}{m_A} + \frac{1}{m_B} \right) \right]^{(1/2)} (C_A N_{av})(C_B N_{av}) \tag{2-5}$$

In Eqn. (2-5), $\underline{r}_A$ is the radius of molecule "i," k_B is Boltzmann's constant, T is the absolute temperature (K), m_i is the mass of molecule "i," C_i is the concentration of molecule "i" (moles/volume), N_{av} is Avogadro's number, and Z_{AB} is the number of collisions between A and B per unit time per unit volume. Note that $m_i = (M_i/N_{av})$, where M_i is the molecular weight of "i". Also, $(C_i N_{av})$ is the *molecular* concentration (molecules/volume) of species "i."

For binary collisions of like molecules, say A with A, simply replace "B" with "A" in Eqn. (2-5) and divide the right-hand side by 2 to eliminate double counting of collisions.

Collision theory predicts simple, power-law rate equations for reactions that result directly from binary collisions between molecules. In fact, CT predicts that the reaction rate will be first order in each of the colliding species. For a reaction between A and B,

$$-\underline{r}_A \propto Z_{AB} \propto C_A C_B$$

For a reaction between two As,

$$-\underline{r}_A \propto Z_{AA} \propto C_A C_A \, (\propto C_A^2)$$

EXAMPLE 2-3
Frequency of Collisions between Oxygen Atoms

Calculate a numerical value of Z_{OO} at 300 K and 1 atm total pressure for the collision of two O atoms. Take the mole fraction of O atoms to be 10^{-5} and take the radius of the O atom to be 1.1 Å.

APPROACH

The frequency of oxygen-atom collisions can be calculated by substituting known values into Eqn. (2-5), after adapting it for collision between identical species.

SOLUTION

Because of the units of the Boltzmann constant, it usually is most convenient to use centimeters as the unit of length. From Eqn. (2-5), after dividing by 2 and simplifying,

$$Z_{OO} = 8(\underline{r}_O)^2 \left[\frac{\pi k_B T N_{av}}{M_O} \right]^{1/2} N_{av}^2 C_O^2$$

$(\underline{r}_O)^2 = 1.21 \times 10^{-16}\ \text{cm}^2$.

$C_O = y_O P/RT = 10^{-5} \times 1(\text{atm})/300(\text{K}) \times 0.0821(\text{l-atm/mol-K}) \times 1000(\text{cm}^3/\text{l}) = 4.06 \times 10^{-10}\ (\text{mol/cm}^3)$.

$$\left[\frac{\pi k_B T N_{av}}{M_O} \right]^{1/2} = (3.14 \times 1.38 \times 10^{-16}(\text{g-cm}^2/\text{s}^2\text{-molecule-K}) \times 300(\text{K}) \times 6.02 \times 10^{23}(\text{molecules/mol})/16(\text{g/mol}))^{1/2} = 6.99 \times 10^4\ \text{cm/s}.$$

$$Z_{OO} = 8 \times 1.21 \times 10^{-16}(\text{cm}^2) \times 6.99 \times 10^4(\text{cm/s}) \times (6.02 \times 10^{23})^2(\text{molecules/mol}) \times (4.06 \times 10^{-10})^2(\text{mol/cm}^3)^2 = 4.0 \times 10^{18}\ \text{collisions/s-cm}^3.$$

[3] Moelwyn-Hughes, E. A., *Physical Chemistry, 2nd revised edition*, Pergamon Press (1961), p. 51.

The units for Z_{OO} that result from the calculation are (molecules^2/s-cm^3). "Collisions" can be substituted for "molecules^2" because the number of collisions is proportional to the number of *pairs* of molecules.

The frequency of trimolecular (ternary) collisions also may be estimated for an ideal gas, subject to the restrictions mentioned earlier.[4] For the simultaneous collision of A, B, and C, the result is

Frequency of trimolecular collisions

$$Z_{ABC} = 4\pi(\underline{r}_A + \underline{r}_B + \underline{r}_C)^2 \Omega(T, A, B, C)(C_A N_{av})(C_B N_{av})(C_C N_{av}) \tag{2-6}$$

The function $\Omega(T, A, B, C)$ depends on the temperature and on the radii and masses of the three molecules. When the three molecules are reasonably similar in size and mass,

$$\Omega \cong 24\left(\frac{3k_B T}{2\pi \overline{m}}\right)^{1/2} (\bar{\underline{r}})^3 \tag{2-7}$$

Here, $\overline{m} = (m_A + m_B + m_C)/3$ and $\bar{\underline{r}} = (\underline{r}_A + \underline{r}_B + \underline{r}_C)/3$. Since many significant assumptions are already embedded in Eqn. (2-6), Eqn. (2-7) usually is a reasonable approximation for engineering calculations.

For ternary collisions involving two like molecules (A) and one unlike molecule (B), replace C with A in Eqn. (2-6) and divide the right-hand side by 2 to eliminate double counting collisions.

EXERCISE 2-1

Calculate a numerical value for Z_{OON_2} at 300 K and 1 atm total pressure for the reaction of two atoms of O with one molecule of N_2. Take the mole fraction of O atoms to be 10^{-5}, with N_2 comprising the balance of the gas mixture. Take $\underline{r}_O = 1.1$ Å and $\underline{r}_{N_2} = 1.6$ Å. (Answer: 1.6×10^{16} collisions/s-cm^3.)

For this example, the ternary collision rate is about two orders of magnitude less that the binary collision rate. The difference would have been much larger if the concentration of the third molecule (N_2 in this case) had been comparable to the concentrations of the first two species (O atoms, in this case).

Again, CT predicts simple, power-law rate equations for reactions that result directly from ternary collisions. The rate is first order in each of the colliding species. For a reaction between A, B, and C

$$-\underline{r}_A \propto Z_{ABC} \propto C_A C_B C_C$$

while for a reaction between two As and one B,

$$-\underline{r}_A \propto Z_{AAB} \propto C_A C_A C_B \; (= C_A^2 C_B)$$

Generalization V

If a reaction is *reversible*, the *net* rate is the *difference* between the rates of the forward and reverse reactions, i.e.,

Net rate of reversible reaction

$$-r_A(\text{net}) = (-r_{A,f}) - (r_{A,r}) \tag{2-8}$$

[4] Adapted from Moelwyn-Hughes, E. A., *Physical Chemistry, 2nd revised edition*, Pergamon Press (1961), p. 1149.

Here, $(-r_{A,f})$ represents the rate of the forward reaction, i.e., the rate at which reactant A is consumed in the forward reaction, $(r_{A,r})$ represents the rate at which A is formed by the reverse reaction, and $-r_A(\text{net})$ is the *net* rate of disappearance of A as a result of the two reactions. The previous generalizations apply to both $(-r_{A,f})$ and $(r_{A,r})$.

Suppose that

$$(-r_{A,f}) = k_f \prod_i C_i^{\alpha_{f,i}} \tag{2-9}$$

and

$$(r_{A,r}) = k_r \prod_i C_i^{\alpha_{r,i}} \tag{2-10}$$

Here, $\alpha_{r,i}$ is the order of the forward reaction with respect to species i, $\alpha_{r,i}$ is the order of the reverse reaction with respect to species i, k_f is the rate constant for the forward reaction, and k_r is the rate constant for the reverse reaction. From Eqn. (2-8), the rate equation for the *net* rate of disappearance of A is

$$-r_A(\text{net}) = (-r_{A,f}) - (r_{A,r}) = k_f \prod_i C_i^{\alpha_{f,i}} - k_r \prod_i C_i^{\alpha_{r,i}}$$

When the reaction reaches *chemical equilibrium*, the net rate of reaction, $-r_A(\text{net})$, must be zero. From the above equation, at equilibrium,

$$\frac{k_f}{k_r} = \prod_i C_i^{(\alpha_{r,i} - \alpha_{f,i})} \tag{2-11}$$

We reached Eqn. (2-11) strictly by using the principles of kinetics. Now, let's temporarily turn away from kinetics and consider thermodynamics. The rate of a reaction cannot be predicted from thermodynamics and neither can the form of the rate equation. However, thermodynamics *does* tell us how far a reaction can go before it stops, i.e., before it comes to equilibrium. Thermodynamics also tells us how the position of equilibrium depends on temperature and on initial composition.

For the reaction

$$\sum_i \nu_i A_i = 0 \tag{1-1}$$

the *equilibrium expression* is

Equilibrium expression (general form)

$$K_{eq} = \prod_i a_i^{\nu_i} \tag{2-12}$$

Here, K_{eq} is the equilibrium constant for the reaction, based on *activity*, and a_i is the activity of species i. Equation (2-12) *must* be satisfied when a reaction has reached equilibrium.

The value of K_{eq} will depend on the values of the stoichiometric coefficients, i.e., on how the stoichiometric equation is written. This is evident from Eqn. (2-12). It can also be seen from the equations

$$\Delta G_R^0(T) = \sum_i \nu_i \Delta G_{f,i}^0(T) \tag{1-2}$$

$$\ln K_{eq}(T) = -\Delta G_R^0(T)/RT \tag{2-13}$$

The value of $\Delta G_R^0(T)$ depends on the values of the ν_i. Therefore, the value of K_{eq} also must depend on the ν_i.

Rate equations almost always are written in terms of concentrations or partial pressures. It is rare to find a rate equation that is based on activities. Therefore, Eqn. (2-12) is not particularly useful in relating kinetics to thermodynamics. We need an equilibrium expression that is written in terms of concentrations, i.e.,

Equilibrium expression (based on concentration)

$$K_{eq}^C = \prod_i C_i^{\nu_i} \tag{2-14}$$

Here, K_{eq}^C is the equilibrium constant based on *concentration*. In general, K_{eq}^C will not have the same value as K_{eq}. Moreover, K_{eq}^C may depend on concentration to some extent, whereas K_{eq} does not.

Now we can consider the relationship between kinetics and thermodynamics. The form of the rate equations for a reversible reaction must be *thermodynamically consistent*. In other words, *the expression for the net rate of a reversible reaction must reduce to the equilibrium expression when the reaction has reached equilibrium, i.e., when the net rate is zero.*

At this point, we might be tempted to compare Eqns. (2-11) and (2-14) and conclude that $K_{eq}^C = k_f/k_r$ and $(\alpha_{r,i} - \alpha_{f,i}) = \nu_i$. This is a possibility, but not the only one.

We have enormous flexibility in writing the balanced stoichiometric equation for a chemical reaction. This flexibility may create ambiguity in writing the equilibrium expression. Suppose that the stoichiometric coefficients in Reaction (1-1) all are multiplied by some number, say N. Let's carry out this operation:

$$\sum_i N\nu_i A_i = 0 \tag{2-15}$$

The stoichiometric equation remains balanced and valid.

Let K_{eq}^C be the value of the equilibrium constant (based on concentration) when $N = 1$, i.e., for the original set of stoichiometric coefficients. Then the equilibrium expression for Reaction (2-15) is

$$\prod_i C_i^{N\nu_i} = \left(\prod_i C_i^{\nu_i}\right)^N = (K_{eq}^C)^N \tag{2-16}$$

Here, $(K_{eq}^C)^N$ is the equilibrium constant for Reaction (2-15), i.e., for the reaction whose stoichiometric coefficients are N times those of Reaction (1-1). The equilibrium constant for Reaction (2-15) is just the equilibrium constant for Reaction (1-1), raised to the Nth power. Now, when we compare Eqns. (2-11) and (2-16), we get the general results

$$\frac{k_f}{k_r} = (K_{eq}^C)^N \tag{2-17}$$

$$(\alpha_{r,i} - \alpha_{f,i}) = N\nu_i \tag{2-18}$$

Equation (2-18) provides a basis for analyzing the thermodynamic consistency of the rate equations for the forward and reverse reactions. The use of this equation is illustrated in the following example.

EXAMPLE 2-4
Formulation of Reverse Rate Equation—Phosgene Synthesis

Consider the reaction of carbon monoxide with chlorine to form phosgene.

$$CO + Cl_2 \rightleftarrows COCl_2 \tag{2-A}$$

Phosgene is a very important chemical intermediate. It is used to make the isocyanate monomers that go into products such as polyurethane foams and coatings. It also is used to make polycarbonate polymers. However, it is extremely toxic, so much so that it was used as a chemical warfare agent during World War I.[5]

Suppose we know from experiments that the rate equation for the forward reaction is

$$r_{COCl_2} = k_f[Cl_2]^{3/2}[CO]$$

What are the orders for the species in the rate equation for the reverse reaction?

APPROACH

A power-law rate equation will be used to describe the rate of the reverse reaction. Since the values of the $\alpha_{f,i}$ in Eqn. (2-18) are known, the values of the $\alpha_{r,i}$ can be calculated from that equation.

SOLUTION

Let's assume that the form of the rate equation for the reverse reaction is

$$-r_{COCl_2} = k_r[Cl_2]^{\beta_1}[CO]^{\beta_2}[COCl_2]^{\beta_3}$$

Let $N = 1$ correspond to the reaction as written in (2-A), so that $-\nu_{CO} = -\nu_{Cl_2} = \nu_{COCl_2} = 1$. From Eqn. (2-18)

$$\beta_1 - (3/2) = -N \tag{2-19a}$$

$$\beta_2 - 1 = -N \tag{2-19b}$$

$$\beta_3 = N \tag{2-19c}$$

Now we can generate thermodynamically-consistent sets of β_is by selecting values of N and computing the corresponding values of β_1, β_2, and β_3 from Eqns. (2-19). What we are doing by selecting different values of N is writing the balanced stoichiometric equation in different ways, as shown by Eqn. (2-15). The following table illustrates some of the results.

N	β_1	β_2	β_3
−1	5/2	2	−1
−1/2	2	1/2	−1/2
1/2	1	1/2	1/2
1	1/2	0	1
2	−1/2	−1	2
5	−7/2	−4	5

This exercise could be continued to include higher and lower values of N, and values in between those shown. However, the range of N between about −1 and +2 probably contains most of the sets of β that are of practical interest. When $N < -1$, the values of β_1 and β_2, i.e., the orders with respect to Cl_2 and CO, respectively, both exceed 2, and the order with respect to the reactant, $COCl_2$, is < -1, an unlikely situation. For values of $N > 2$, the order with respect to $COCl_2$ exceeds 2, and the orders with respect to the reactants are both <1, again unlikely.

This example shows that the form of the reverse rate equation does not automatically follow from the form of the forward rate equation, and *vice versa*. Even if the form of the forward rate equation is known, there are still an infinite number of forms of the reverse rate equation that are thermodynamically consistent with the forward rate equation, corresponding to all of the possible values of N.

[5] For those with a strong stomach, an interesting discussion of the use and potential use of chemical warfare agents during the first World War can be found in Vilensky, J. A. and Sinish, P. R., *Blisters as Weapons of War: The Vesicants of World War I*, Chemical Heritage 24:2 (Summer 2006), pp. 12–17.

EXERCISE 2-2

(a) Why doesn't the above table have an entry for $N = 0$?

(b) How are we going to determine which value of N is "correct?"

In earlier courses, you may have learned that

$$\frac{k_f}{k_r} = K_{eq}^C \tag{2-20}$$

We know that the value of K_{eq}^C depends on how the balanced stoichiometric equation is written, i.e., on the values of the stoichiometric coefficients. However, the values of the rate constants, k_f and k_r, must be determined from experiments. These values have nothing to do with how the balanced stoichiometric equation is written.

There is only one way of writing the balanced stoichiometric equation so that Eqn. (2-20) is satisfied. Suppose the rate equations for *both* the forward and reverse reactions are known, i.e., $\alpha_{f,i}$ and $\alpha_{r,i}$ have been experimentally determined for every species. Suppose further that these rate equations are thermodynamically consistent. Then Eqn. (2-20) will be valid only when K_{eq}^C is calculated using stoichiometric coefficients that are given by $\nu_i = \alpha_{r,i} - \alpha_{f,i}$. This can be seen by comparing Eqns. (2-11) and (2-14), as illustrated in the following example.

EXAMPLE 2-5
Thermodynamic Analysis of Phosgene Synthesis

Suppose that the rate of the forward reaction for phosgene synthesis

$$CO + Cl_2 \rightleftarrows COCl_2 \tag{2-A}$$

is given by

$$r_{COCl_2} = k_f[Cl_2]^{3/2}[CO]$$

and the rate of the reverse reaction is given by

$$-r_{COCl_2} = k_r[Cl_2][CO]^{1/2}[COCl_2]^{1/2}$$

A. Are the rate equations for the forward and reverse reactions thermodynamically consistent?

B. How must the balanced stoichiometric equation be written so that Eqn. (2-20) is satisfied?

C. What is the value of k_f/k_r at 298 K?

D. What is the value of k_f/k_r at 500 K?

E. Phosgene is manufactured by passing an equimolar mixture of CO and Cl_2 gases over a carbon catalyst at about 1 atm total pressure and a temperature of several hundred °C. The phosgene that is formed is a gas. What is the fractional conversion of CO if the reaction reaches equilibrium at 1 atm and 500 K?

In answering this question, you may assume that the ideal gas laws are applicable.

Parts:

Part A: Are the rate equations for the forward and reverse reactions thermodynamically consistent?

APPROACH There are several ways to check for thermodynamic consistency. If both the forward and reverse rate equations are power-law expressions, a value of N can be calculated from Eqn. (2-18) for *each* of the reactants and products. The values of ν_i for this calculation can come from any balanced stoichiometric equation for the reaction in question, e.g., Eqn. (2-A) for this example. If *all* of the calculated values of N are the same, the rate equations are thermodynamically consistent.

The most general way to check for thermodynamic consistency is to set the forward rate expression, $-r_{A,f}$, equal to the reverse rate expression, $-r_{A,r}$, to reflect the fact that the *net* rate must be zero at equilibrium. The resulting equation is then rearranged so that the ratio (k_f/k_r) is on one side of the equation and all of the remaining terms are on the other side. This manipulation should produce an equation that has the form of Eqn. (2-16). In particular, the terms on the side opposite to (k_f/k_r) must consist *only* of reactant concentrations and product concentrations, each raised to some power. These powers must be in the same ratios as the stoichiometric coefficients. In other words, the value of N in Eqn. (2-16) must be the same for each of the reactants and products. If this test is met, the rate equations are thermodynamically consistent.

The advantage of the procedure outlined in the preceding paragraph is that it can be used for rate equations that are not power-law expressions. We will use this approach to analyze the thermodynamic consistency of the proposed rate equations for phosgene synthesis.

SOLUTION For the present problem,

$$-r_{A,f} = k_f[Cl_2]^{3/2}[CO] = r_{A,r} = k_r[Cl_2][CO]^{1/2}[COCl_2]^{1/2}$$

Rearranging,

$$\frac{k_f}{k_r} = \frac{[COCl_2]^{1/2}}{[CO]^{1/2}[Cl_2]^{1/2}} \tag{2-21}$$

The left-hand side of Eqn. (2-21) depends on temperature but not on concentration. The opposite is true for the right-hand side. The exponents on the species on the right-hand side of the above equation are in the ratio 1:1:1. This is the ratio required by stoichiometry (see Reaction (2-A)). Therefore, the two rate equations *are* thermodynamically consistent.

Another way of looking at the same question is to recognize that the right-hand side of the above equation is exactly what we would have obtained by writing the equilibrium expression for the reaction

$$(1/2)CO + (1/2)Cl_2 \rightleftarrows (1/2)COCl_2 \tag{2-B}$$

$$K_{eq}^C = [COCl_2]^{1/2}/([CO]^{1/2}[Cl_2]^{1/2}) \tag{2-22}$$

This result supports the conclusion that the rate equations are thermodynamically consistent.

Part B: How must the balanced stoichiometric equation be written so that Eqn. (2-20) is satisfied?

APPROACH Equation (2-17) shows that the ratio (k_f/k_r) is equal to K_{eq}^C raised to a power, N. When $N = 1$, $(k_f/k_r) = K_{eq}^C$. According to Eqn. (2-18), N is equal to 1 when $\nu_i = (\alpha_{r,i} - \alpha_{f,i})$. In words, the equilibrium constant will be equal to the ratio of the rate constants when the stoichiometric equation is written so that the stoichiometric coefficient of species "i" is equal to the difference between the order of the reverse reaction with respect to "i" and the order of the forward reaction with respect to "i."

SOLUTION Applying Eqn. (2-18) with N set equal to 1 to Cl_2, CO, and $COCl_2$:

$$\begin{aligned} CO:&\quad \nu_{CO} = (1/2) - 1 = -(1/2) \\ Cl_2:&\quad \nu_{Cl_2} = 1 - (3/2) = -(1/2) \\ COCl_2:&\quad \nu_{COCl_2} = (1/2) - 0 = (1/2) \end{aligned}$$

Therefore, the stoichiometric equation that will lead to the value of K_{eq}^C that is equal to k_f/k_r is

$$1/2CO + 1/2Cl_2 \rightleftarrows 1/2COCl_2 \tag{2-B}$$

Part C: What is the value of k_f/k_r at 298 K?

APPROACH The equilibrium constant based on concentration for Reaction (2-B) at 298 K must be calculated, since we have shown that K_{eq}^C is equal to k_f/k_r when the reaction is written with that stoichiometry. To calculate K_{eq}^C at 298 K, K_{eq} must be calculated first. Equation (2-13) can be used for this calculation, provided that the standard Gibbs free energy change of the reaction at 298 K, $\Delta G_R^0(298\,K)$, is known. The standard Gibbs free energy change of the reaction at 298 K can be determined by using Eqn. (1-2) if thermochemical data (i.e., $\Delta G_{f,i}^0(298\,K)$) can be obtained for CO, Cl_2, and $COCl_2$. This kind of information is available from many sources. The following table contains the thermochemical data that are required for this calculation, and for subsequent parts of the problem.

Thermochemical data for phosgene synthesis[6]

Species	$\Delta G_f(298\,K)$ (kcal/mol)	$\Delta H_f(298\,K)$ (kcal/mol)	c_p (cal/mol, K)
CO	−32.8	−26.4	7.0
Cl_2	0	0	8.1
$COCl_2$	−48.9	−52.3	13.8

Once $\Delta G_R^0(298\,K)$ has been calculated from Eqn. (1-2), $K_{eq}(298\,K)$ can be determined from Eqn. (2-13). Finally, the value of K_{eq}^C can be calculated from K_{eq} by applying the ideal gas law.

SOLUTION

$$\Delta G_R^0(298\,K) = \sum_i \nu_i \Delta G_{f,i}(298\,K) \quad (1\text{-}2)$$

$$\Delta G_R^0(298\,K) = (1/2) \times (-48.9) + (-1/2) \times (0.0) + (-1/2) \times (-32.8) = -8.1 \text{ kcal/mol}$$

From Eqn. (2-13),

Relationship between equilibrium constant and free energy change on reaction

$$\ln K_{eq}(T) = -\Delta G_R^0(T)/RT \quad (2\text{-}13)$$

$$\ln K_{eq}(298\,K) = 8100(\text{cal/mol})/[1.99(\text{cal/mol, K}) \times 298(K)] = 14$$

$$K_{eq}(298\,K) = 1.2 \times 10^6$$

This equilibrium constant is based on *activity*, not concentration.[7] It now must be converted to a concentration-based equilibrium constant. For a gas,

$$a_i = (f_i/f_i^0)$$

where f_i is the fugacity of species "*i*." For the data in the preceding table, the standard-state values of $f_i(f_i^0)$ for all three compounds are 1 atm at 298 K. Thus, Eqn. (2-12) becomes

$$K_{eq} = \frac{a_{COCl_2}^{1/2}}{a_{CO}^{1/2} a_{Cl_2}^{1/2}} = \frac{(f_{COCl_2}/f_{COCl_2}^0)^{1/2}}{(f_{CO}/f_{CO}^0)^{1/2} \times (f_{Cl_2}/f_{Cl_2}^0)^{1/2}}$$

At 298 K,

$$\frac{(f_{COCl_2})^{1/2}}{(f_{CO})^{1/2} \times (f_{Cl_2})^{1/2}} = K_{eq}(298\text{ K}) \times \left(\frac{f_{COCl_2}^0}{f_{CO}^0 \times f_{Cl_2}^0}\right)^{1/2} = 1.2 \times 10^6 \text{ atm}^{-1/2}$$

[6] Weast, R.C. (ed.), *Handbook of Chemistry and Physics,* 64th edition, CRC Press, Boca Raton, FL (1983).

[7] An equilibrium constant that is calculated from thermochemical data, as illustrated above, always is an activity-based equilibrium constant, as shown in Eqn. (2-12).

We now will assume that the ideal gas laws are obeyed.[8] For an ideal gas, $f_i = p_i$. Therefore,

$$K_{eq}^{P} = 1.2 \times 10^{6}\ \text{atm}^{-1/2} = \frac{(p_{COCl_2})^{1/2}}{(p_{CO})^{1/2} \times (p_{Cl_2})^{1/2}}$$

Here, K_{eq}^{P} is the equilibrium constant based on pressure. Finally, for an ideal gas, $p_i = C_i(RT)$. Substituting this into the above equation

$$\frac{C_{COCl_2}^{1/2}}{C_{CO}^{1/2} C_{Cl_2}^{1/2}} = K_{eq}^{P}(RT)^{1/2} = K_{eq}^{C} = \frac{k_f}{k_r}$$

$$\frac{k_f}{k_r} = 1.2 \times 10^{6}(\text{atm})^{-1/2}\left(0.0821\left(\frac{\text{atm-l}}{\text{mol-K}}\right)298(\text{K})\right)^{1/2} = 5.9 \times 10^{6}(\text{l/mol})^{1/2}$$

Part D: What is the value of k_f/k_r at 500 K?

APPROACH

To determine k_f/k_r at 500 K, we must calculate the value of K_{eq}^{C} at this temperature. The variation of K_{eq} with temperature is given by

Variation of equilibrium constant with temperature

$$\left(\frac{\partial \ln K_{eq}}{\partial T}\right)_P = \frac{\Delta H_R^0(T)}{RT^2} \qquad (2\text{-}23)$$

The value of the heat of reaction at 298 K ($\Delta H_R^0(298\ K)$) can be calculated from

$$\Delta H_R^0(298\ \text{K}) = \sum_i \nu_i \Delta H_{f,i}^0(298\ \text{K}) \qquad (1\text{-}3)$$

The variation of ΔH_R^0 with temperature is given by

Variation of heat of reaction with temperature

$$\left(\frac{\partial \Delta H_R^0}{\partial T}\right)_P = \sum_i \nu_i c_{p,i} \qquad (2\text{-}24)$$

Equation (2-24) can be integrated from 298 K to an arbitrary temperature, T, to obtain an expression for ΔH_R^0 as a function of T. This expression can be substituted into Eqn. (2-23), which can then be integrated from 298 K to 500 K to obtain K_{eq} (500 K). Finally, K_{eq}^{C} (500 K) can be calculated from K_{eq} (500 K) by following the procedure in Part C of this example.

SOLUTION

Substituting the appropriate data from the preceding table into Eqn. (1-3) gives

$$\Delta H_R^0(298\ \text{K}) = -13.0\ \text{kcal/mol}$$

Clearly, the synthesis of phosgene is quite exothermic at 298 K. Again, substituting the data from the preceding table into Eqn. (2-24) gives

$$\left(\frac{\partial \Delta H_R^0}{\partial T}\right)_P = -0.70\ \text{cal/mol-K}$$

The heat of reaction is not a strong function of temperature. This variation could be neglected for practical purposes, especially since the range of temperature in this problem is not large. However, we will carry this term in order to illustrate the general calculation procedure.

Integrating the preceding equation from 298 K to T gives

$$\Delta H_R^0(T) = -13,000 - 0.70T\ \text{cal/mol}$$

Substituting this result into Eqn. (2-23) and integrating from 298 to 500 K

$$\ln K_{eq}(500\ \text{K}) = 5.5$$

$$K_{eq}(500\ \text{K}) = 240$$

[8] This assumption is reasonable since the reduced pressures of all three species are very low (<0.03) at the specified conditions.

The equilibrium constant decreases substantially as the temperature is increased because the reaction is strongly exothermic.

Since the ideal gas laws are obeyed,

$$K_{eq}(500\text{ K}) \times \left(\frac{f^0_{COCl_2}}{f^0_{CO} \times f^0_{Cl_2}}\right)^{1/2} = K^P_{eq}(500\text{ K}) = 240\text{ atm}^{-1/2}$$

From Part C

$$\frac{k_f}{k_r} = K^C_{eq} = K^P_{eq}(RT)^{1/2} = 240 \times (1.99 \times 500)^{1/2} = 7600\,(1/\text{mol})^{1/2}$$

Part E: What is the fractional conversion of CO if the reaction reaches equilibrium at 1 atm and 500 K?

APPROACH

The equilibrium fractional conversion of CO can be calculated from the equilibrium expression. Perhaps the most convenient starting point is

$$K_p = 240\text{ atm}^{-1/2} = \frac{(p_{COCl_2})^{1/2}}{(p_{CO})^{1/2} \times (p_{Cl_2})^{1/2}}$$

For an ideal gas, the partial pressure of species "i" is given by $p_i = y_i P$, where y_i is the mole fraction and P is the total pressure. Finally, the mole fractions of CO, Cl_2, and $COCl_2$ can be written as functions of the quantity of CO that has reacted. This permits the amount of CO reacted, and therefore the fractional conversion of CO, to be calculated from the equilibrium expression.

SOLUTION

Since the total pressure is 1 atm for this example, the preceding equation can be written as

$$\frac{(y_{COCl_2})^{1/2}}{(y_{CO})^{1/2} \times (y_{Cl_2})^{1/2}} = 240$$

To relate the mole fractions to the quantity of CO reacted, let's choose a basis of 1 mol of CO entering the reactor. Therefore, 1 mol of Cl_2 also enters, but there is no $COCl_2$ (or anything else) in the feed. Let ξ_e be the equilibrium extent of reaction, i.e., the number of moles of CO that are consumed when the reaction has reached equilibrium. By stoichiometry, the moles of each species in the equilibrium mixture are

$$\begin{aligned} CO &: \quad 1 - \xi_e \\ Cl_2 &: \quad 1 - \xi_e \\ COCl_2 &: \quad \xi_e \\ \text{Total moles} &: \quad 2 - \xi_e \end{aligned}$$

The mole fractions of the three species are

$$\begin{aligned} CO &: \quad (1 - \xi_e)/(2 - \xi_e) \\ Cl_2 &: \quad (1 - \xi_e)/(2 - \xi_e) \\ COCl_2 &: \quad \xi_e/(2 - \xi_e) \end{aligned}$$

With these relationships, the equilibrium expression becomes

$$\frac{\xi_e^{1/2}(2 - \xi_e)^{1/2}}{(1 - \xi_e)} = 240$$

The value of ξ_e that satisfies this equation is 0.996

Since 1 mol of CO initially was chosen as a basis, ξ_e is then the equilibrium fractional conversion of CO.

EXERCISE 2-3

Discuss the safety features that should be incorporated into a plant that manufactures phosgene.

2.3 AN IMPORTANT EXCEPTION

Rate equations of the form

$$-r_A = k(T)C_A/[1 + K_A(T) \times C_A] \quad (2\text{-}25)$$

describe the rates of many types of catalytic reactions. In the field of heterogeneous catalysis, this form of kinetic equation is known as a "Langmuir–Hinshelwood" rate equation. In biochemistry, a slight variation

$$-r_A = v_{max}(T) \times C_A/[K_m(T) + C_A] \quad (2\text{-}25a)$$

is referred to as a "Michaelis–Menten" rate equation.

The parameters k and K_A (or v_{max} and K_m) are functions of temperature. Therefore, the effects of concentration and temperature are not separated, in violation of Generalization I, (Eqn. (2-1)). Furthermore, Generalization IV does not apply either. At constant temperature, the effect of the concentration of A on the reaction rate is not well represented by C_A raised to a power. In fact, an examination of Eqns. (2-25) and (2-25a) shows that the apparent reaction order with respect to A varies from 1 at low concentrations of A to 0 at high concentrations.

EXERCISE 2-4

(a) Show that $v_{max}(T)$ in Eqn. (2-25a) is the maximum possible value of $-r_A$ at a given temperature.

(b) What is the physical interpretation of $K_m(T)$ in Eqn. (2-25a)?

The origin of Langmuir–Hinshelwood/Michaelis–Menten rate equations will be explored in Chapter 5. In the meanwhile, we will use this form of rate equation in Chapter 4, when we tackle some problems in sizing and analysis of ideal reactors. The next chapter is devoted to defining an *ideal* reactor and to providing the tools that are required for their sizing and analysis.

SUMMARY OF IMPORTANT CONCEPTS

- Reaction rates depend on temperature and on the various species concentrations.
- Arrhenius relationship (effect of temperature on the rate constant):

$$k(T) = A\exp(-E/RT)$$

- For a power-law rate equation, the order of the reaction with respect to species "i" (i.e., the *individual* order with respect of species "i") expresses the dependence of the rate on the concentration of species "i." The higher the absolute value of the individual order the stronger the dependence of the reaction rate on the concentration of this species.
- The net rate of a reversible reaction is the difference between the rates of the forward and the reverse reactions:

$$-r_A(\text{net}) = (-r_{A,f}) - (r_{A,r})$$

- The rate equations for the forward and reverse reactions must be consistent with the equilibrium expression, as formulated from thermodynamics.

PROBLEMS

Problem 2-1 (Level 1) Ozone decomposes to oxygen according to the stoichiometric equation

$$2O_3 \rightleftarrows 3O_2$$

The rate equation for the forward reaction is known to be

$$-r_{O_3} = \frac{k_1 p_{O_3}^2}{p_{O_2} + k_2 p_{O_3}}$$

1. "*The form of the rate equation for the reverse reaction must be thermodynamically consistent with the form of the rate equation for the forward reaction.*" Explain clearly and concisely what this statement means.
2. Give two forms of the rate equation for the *reverse* reaction that are thermodynamically consistent with this forward rate equation. Prove that they are thermodynamically consistent.

You may assume that O_2 and O_3 are ideal gases at the conditions for which the above rate equation applies.

Problem 2-2 (Level 1) The reaction A + B → Products is first order in A, one-half order in B, and has an activation energy of 90 kJ/mol.

What is the ratio of the rate of disappearance of A at Condition 2 below to the rate of disappearance of A at Condition 1?

Condition 1	Condition 2
$T = 300$ °C	$T = 350$ °C
$C_A = 1.5$ mol/l	$C_A = 1.0$ mol/l
$C_B = 2.0$ mol/l	$C_B = 2.5$ mol/l

Problem 2-3 (Level 1) Cyclohexane (C_6H_{12}) is made industrially by the hydrogenation of benzene (C_6H_6),

$$C_6H_6 + 3H_2 \rightarrow C_6H_{12}$$

1. What is the order of this reaction with respect to C_6H_6, H_2, and C_6H_{12}?
2. What is the overall order of the reaction?
3. Suppose that the order of the reaction with respect to C_6H_6 is known to be 1.0. What is the order with respect to H_2?

You may NOT assume that the reaction is elementary.

Problem 2-4 (Level 1) If the rate constant of a homogeneous reaction is 1 s^{-1} at 100 °C and 10,000 s^{-1} at 200 °C:

1. What is the activation energy of the reaction?
2. What is the overall order of the reaction?

Problem 2-5 (Level 2) Hinshelwood and Green[9] studied the kinetics of the reaction between NO and H_2.

$$2NO + 2H_2 \rightarrow N_2 + 2H_2O$$

They found that the reaction was second order in NO and first order in H_2. The rate constant was measured at various temperatures, with the results given in the following table.

Temperature (°C)	Rate constant (l^2/molecule^2-s)
826	476
788	275
751	130
711	59
683	25
631	5.3

1. Does the data follow the Arrhenius relationship? Justify your answer.
2. If so, what is the activation energy of the reaction?

[9] Hinshelwood, C. N. and Green, T. E., *Chem. Soc. J.*, 730 (1926).

Problem 2-6 (Level 2) The Temkin-Pyzhev rate equation for ammonia synthesis ($N_2 + 3H_2 \rightleftarrows 2NH_3$) on certain catalysts is

$$r_{NH_3} = k_1 p_{N_2}\left(\frac{p_{H_2}^3}{p_{NH_3}^2}\right)^{\alpha} - k_2\left(\frac{p_{NH_3}^2}{p_{H_2}^3}\right)^{\beta}$$

where p_i is the partial pressure of species "i".

Under what conditions is this rate equation thermodynamically consistent? In other words, does this equation reduce to the equilibrium expression when the reaction is at equilibrium? If so, under what circumstances?

Problem 2-7 (Level 1) Rate equations for heterogeneous catalytic reactions sometimes are written in terms of partial pressure rather than concentration. Suppose that the units of $-r_A$ are mol/g-s in the rate equation,

$$-r_A = \frac{k p_A p_B}{(1 + K_A p_A + K_B p_B)^2}$$

where p_i is the partial pressure of species "i".

What are the units of k, K_A, and K_B?

Problem 2-8 (Level 3) Air is composed of about 79 mol% N_2 and 21 mol% O_2. What is the frequency of binary collisions of all kinds at 300 K and 1 atm total pressure? The radius of N_2 is 1.6 Å and the radius of O_2 is 1.5 Å.

Think of an *approximate* and simpler way to calculate the frequency of binary collisions of all kinds. How does your approximate answer compare with the one you calculated initially?

Problem 2-9 (Level 1) The irreversible reaction A + B → C obeys the rate equation $-r_A = kC_A^2$. What is the order of the reaction with respect to B?

Problem 2-10 (Level 2) The irreversible reaction A + B → C obeys the rate equation

$$-r_A = \frac{kC_A C_B}{(1 + K_A C_A)^2}$$

Sketch a graph showing how the rate of disappearance of A depends on the concentration of A, at constant C_B and temperature. Identify all of the important features of the graph as quantitatively as possible.

Problem 2-11 (Level 2) Consider the reversible reaction

$$A + B \rightleftarrows C$$

This reaction takes place in the presence of a catalyst, D. Components A, B, C, and D are completely soluble. The rate equation for the *forward* reaction is known to be

$$-r_A = k_f C_A C_B^2$$

Are the following rate equations for the *reverse* reaction thermodynamically consistent with the forward rate equation? Explain your answer.

1. $r_A = k_r C_A^{-1}$
2. $r_A = k_r C_C$

3. $r_A = k_r C_C^2 C_A^{-1}$
4. $r_A = k_r C_C^2 C_D C_A^{-1}$
5. $r_A = k_r C_C^2 C_A^{-1}/(1 + K_B C_B)^2$

Problem 2-12 (Level 1) Refer to Problem 1-11. Use the data given in that problem statement to answer the following question.

With some catalysts and at some operating conditions, ethyl benzene can be formed by the disproportionation of benzene and diethyl benzene, as shown by the reaction below. If this reaction has reached equilibrium at the end of the experiment given in Problem 1-11, what is the value of the equilibrium constant (based on concentration) for this reaction, at the conditions of the experiment?

C_2H_5–C_6H_4–C_2H_5 + C_6H_6 → 2 C_6H_5–C_2H_5

Problem 2-13 (Level 2) Beltrame et al.[10] studied the oxidation of glucose to gluconic acid in aqueous solution at pH7 using Hyderase (a commercial enzyme catalyst system that is soluble in the reaction medium). The overall reaction is

$$C_5H_{10}O_6 + O_2 + H_2O \rightarrow C_5H_{10}O_7 + H_2O_2$$

The first step in the reaction mechanism is believed to be the formation of a complex between glucose and the enzyme, i.e.,

$$E + G \xrightarrow{k_1} RL$$

Here, E is the free (uncomplexed) enzyme, G is D-glucose, and RL is a complex between the reduced enzyme (R) and glucono-lactone. The authors studied the rate of this step and found the rate constant k_1 depended on temperature as shown in the following table:

Temperature (K)	Rate constant (L/g-h)
273.2	3.425
283.2	7.908
293.2	18.79
303.2	28.10

Do the above rate constants obey the Arrhenius relationship? If so, what is the value of the activation energy?

[10] Beltrame, P., Comotti, M., Della Pina, C., Rossi, M., Aerobic oxidation of glucose I. Enzymatic catalysis, *J. Catal.* 228–282, (2004).

Chapter 3

Ideal Reactors

LEARNING OBJECTIVES

After completing this chapter, you should be able to

1. explain the differences between the three ideal reactors: batch, continuous stirred tank, and plug flow;
2. explain how the reactant and product concentrations vary spatially in ideal batch, ideal continuous stirred tank, and ideal plug-flow reactors;
3. derive "design equations" for the three ideal reactors, for both homogeneous and heterogeneous catalytic reactions, by performing component material balances;
4. calculate reaction rates using the "design equation" for an ideal continuous stirred-tank reactor;
5. simplify the most general forms of the "design equations" for the case of constant mass density.

The next few chapters will illustrate how the behavior of chemical reactors can be predicted, and how the size of reactor required for a given "job," can be determined. These calculations will make use of the principles of reaction stoichiometry and reaction kinetics that were developed in Chapters 1 and 2.

There are many different types of reactor. One of the most important features that differentiates one kind of reactor from another is the nature of mixing in the reactor. The influence of mixing is easiest to understand through the material balance(s) on the reactor. These material balances are the starting point for the discussion of reactor performance.

3.1 GENERALIZED MATERIAL BALANCE

The reaction rate r_i is an intensive variable. It describes the rate of formation of species "i" at any *point* in a chemical reactor. However, as we learned in Chapter 2, the rate of any reaction depends on variables such as temperature and the species concentrations. If these variables change from point to point in the reactor, r_i also will change from point to point.

For the time being, to emphasize that r_i depends on temperature and on the various species concentrations, let's use the nomenclature

$$r_i = r_i(T, \text{all } C_i)$$

The term "all C_i" reminds us that the reaction rate may be influenced by the concentration of each and every species in the system.

Consider an arbitrary volume (V) in which the temperature and the species concentrations vary from point to point, as shown below.

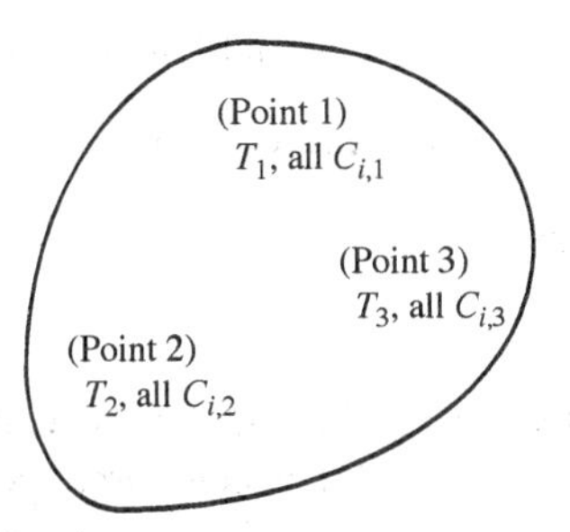

The rate at which "i" is formed in this control volume by a chemical reaction or reactions is designated G_i, the generation rate of "i". The units of G_i are moles/time. For a homogeneous reaction, G_i is related to r_i by

Generation rate—homogeneous reaction

$$G_i = \iiint_V r_i \, dV \tag{3-1}$$

For a heterogeneous catalytic reaction, where r_i has units of moles/time-weight of catalyst, G_i is given by

Generation rate—heterogenous catalytic reaction

$$G_i = \iiint_W r_i \, dW = \iiint_V r_i \rho_B \, dV \tag{3-2}$$

Here, ρ_B is the bulk density of the catalyst (weight/volume of reactor). In Eqns. (3-1) and (3-2), r_i is the net rate at which "i" is formed by all of the reactions taking place, as given by Eqn. (1-17).

Although they are formally correct, Eqns. (3-1) and (3-2) are not very useful in practice. This is because the reaction rate r_i is never known as an explicit function of position. Therefore, the indicated integrations cannot be performed directly. The means of resolving this apparent dilemma will become evident as we treat some specific cases.

Generalized component material balance
Consider the control volume shown below, with chemical reactions taking place that result in the formation of species "i" at a rate, G_i. Species "i" flows into the system at a molar flow rate of F_{i0} (moles i/time), and flows out of the system at a molar flow rate of F_i.

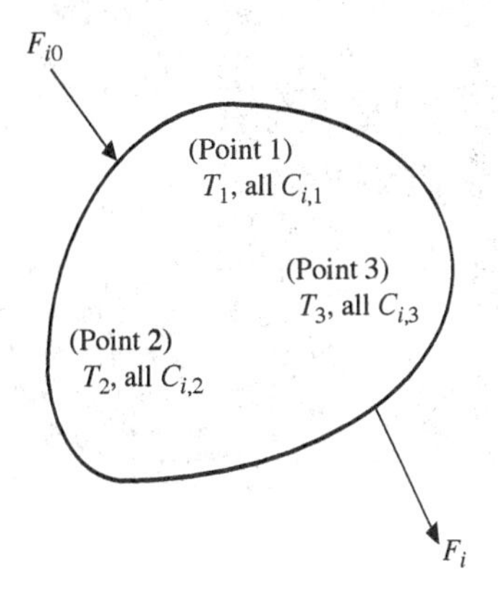

The molar material balance for species "i" for this control volume is

rate in − rate out + rate of generation by chemical reactions = rate of accumulation

Generalized material balance on component "i"

$$F_{i0} - F_i + G_i = \frac{dN_i}{dt} \tag{3-3}$$

Here "t" is the time and N_i is the number of moles of "i" in the system at any time.

Now let's consider three special cases that are of practical significance, and allow Eqn. (3-3) to be simplified to a point that it is useful.

3.2 IDEAL BATCH REACTOR

A batch reactor is defined as a reactor in which there is no flow of mass across the system boundaries, once the reactants have been charged. The reaction is assumed to begin at some precise point in time, usually taken as $t = 0$. This time may correspond, for example, to when a catalyst or initiator is added to the batch, or to when the last reactant is added.

As the reaction proceeds, the number of moles of each reactant decreases and the number of moles of each product increases. Therefore, the concentrations of the species in the reactor will change with time. The temperature of the reactor contents may also change with time. The reaction continues until it reaches chemical equilibrium, or until the limiting reactant is consumed completely, or until some action is taken to stop the reaction, e.g., cooling, removing the catalyst, adding a chemical inhibitor, etc.

Figure 3-1a Overall view of a nominal 7000 gallon batch reactor (in a plant of Syngenta Crop Protection, Inc.). This reactor is used to produce several different products. The reactor has a jacket around it to permit heat to be transferred into or out of the reactor contents via a heating or cooling fluid that circulates through the jacket. The hoists are used for lifting raw materials to higher levels of the structure. (Photo used with permission of Syngenta Crop Protection, Inc.)

Figure 3-1b The top of the reactor in Figure 3-1a. The view port in the left front of the picture permits the contents of the reactor to be observed, and can be opened to permit solids to be charged to the reactor. A motor that drives an agitator is located in the top center of the picture, and a charging line and valve actuator connected to a valve are on the left. (Photo used with permission of Syngenta Crop Protection, Inc.)

Batch reactors are used extensively throughout the chemical and pharmaceutical industries to manufacture products on a relatively small scale. Properly equipped, these reactors are very flexible. A single reactor may be used to produce many different products.

Batch reactors usually are mechanically agitated to ensure that the contents are well mixed. Agitation also increases the heat-transfer coefficient between the reactor contents and any heat-transfer surface in the reactor. In multiphase reactors, agitation may also keep a solid catalyst suspended, or may create surface area between two liquid phases or between a gas phase and a liquid phase.

Very few reactions are thermally neutral ($\Delta H_R = 0$), so it frequently is necessary to either supply heat or remove heat as the reaction proceeds. The most common means to transfer heat is to circulate a hot or cold fluid, either through a coil that is immersed in the reactor, or through a jacket that is attached to the wall of the reactor, or both.

For a batch reactor, $F_{i0} = F_i = 0$. Therefore, for a homogeneous reaction, Eqn. (3-3) becomes

$$G_i = \iiint_V r_i \, dV = \frac{dN_i}{dt} \tag{3-4}$$

Ideal batch reactor

Now, consider a limiting case of batch-reactor behavior. Suppose that agitation of the reactor contents is vigorous, i.e., mixing of fluid elements in the reactor is very intense. Then *the temperature and the species concentrations will be the same at every point in the reactor, at every point in time*. A batch reactor that satisfies this condition is called an *ideal* batch reactor. Many laboratory and commercial reactors can be treated as ideal batch reactors, at least as a first approximation.

For an *ideal* batch reactor, r_i is not a function of position. Therefore, $\iiint_V r_i \, dV = r_i V$ and Eqn. (3-4) becomes

$$r_i V = \frac{dN_i}{dt}$$

Rearranging

Design equation—
ideal batch reactor—
homogeneous reaction
(moles)

$$\boxed{\frac{1}{V}\frac{dN_i}{dt} = r_i} \tag{3-5}$$

Equation (3-5) is referred to as the *design equation* for an ideal batch reactor, in differential form. This equation is valid no matter how many reactions are taking place, provided that Eqn. (1-17) is used to express r_i, and provided that all of the reactions are homogeneous.

The subject of multiple reactions is treated in Chapter 7. Until then, we will be concerned with the behavior of one, stoichiometrically simple reaction. For that case, r_i in Eqn. (3-5) is just the rate equation for the formation of species "i" in the reaction of concern.

The variable that describes composition in Eqn. (3-5) is N_i, the total moles of species "i". It sometimes is more convenient to work problems in terms of either the extent of reaction ξ or the fractional conversion *of a reactant*, usually the limiting reactant. Extent of reaction is very convenient for problems where more than one reaction takes place. Fractional conversion is convenient for single-reaction problems, but can be a source of confusion in problems that involve multiple reactions. The use of all three compositional variables, moles (or molar flow rates), fractional conversion, and extent of reaction, will be illustrated in this chapter, and in Chapter 4.

If "i" is a reactant, say A, then the number of moles of A in the reactor at any time can be written in terms of the fractional conversion of A.

$$x_A = \frac{N_{A0} - N_A}{N_{A0}}, \quad N_A = N_{A0}(1 - x_A)$$

In terms of fractional conversion, Eqn. (3-5) is

Design equation—
ideal batch reactor—
homogeneous reaction
(fractional conversion)

$$\boxed{\frac{N_{A0}}{V}\frac{dx_A}{dt} = -r_A} \tag{3-6}$$

If "A" is the limiting reactant, the value of x_A that is stoichiometrically attainable will lie between 0 and 1. However, as discussed in Chapter 1, chemical equilibrium may limit the value of x_A that can actually be achieved to something less than 1.

Equation (3-6) should not be applied to a product. First, N_A will be greater than N_{A0} if "A" is a product. Moreover, if $N_{A0} = 0$, x_A is infinite. However, Eqn. (3-5) can be used for either a product or a reactant.

The design equation can also be written in terms of the extent of reaction. If only one stoichiometrically simple reaction is taking place

$$\xi = \frac{\Delta N_i}{\nu_i} = \frac{N_i - N_{i0}}{\nu_i} \tag{1-4}$$

Design equation—
ideal batch reactor—
homogeneous reaction
(extent of reaction)

$$\boxed{\frac{\nu_i}{V}\frac{d\xi}{dt} = r_i} \tag{3-7}$$

Equations (3-5)–(3-7) are alternative forms of the design equation for an ideal batch reactor with a homogeneous reaction taking place. Despite the somewhat pretentious name, design equations are nothing more than component material balances, i.e., molar balances on "i", "A", etc.

The volume V in Eqns. (3-5)–(3-7) is that portion of the overall reactor volume in which the reaction *actually takes place*. This is not necessarily the whole geometrical volume of the reactor. For example, consider a reaction that takes place in a liquid that partially fills a vessel. If no reaction takes place in the gas-filled "headspace" above the liquid, then V is the volume of liquid, not the geometrical volume of the vessel, which includes the "headspace."

Equations (3-5)–(3-7) apply to a homogeneous reaction. For a reaction that is catalyzed by a solid, the design equation that is equivalent to Eqn. (3-5) is

Design equation—
ideal batch reactor—
heterogeneous catalytic reaction
(moles)

$$\boxed{\frac{1}{W}\frac{dN_i}{dt} = r_i} \tag{3-5a}$$

EXERCISE 3-1

Derive this equation.

The equivalents to Eqns. (3-6) and (3-7) for heterogeneously catalyzed reactions are given in Appendix 3 at the end of this chapter, and are labeled Eqns. (3-6a) and (3-7a). Be sure that you can derive them.

Temperature variation with time

In developing Eqns. (3-5)–(3-7), we did *not* assume that the temperature of the reactor contents was constant, independent of time. Only one assumption was made concerning temperature, i.e., there are no *spatial* variations in the temperature at any time. An ideal batch reactor is said to be *isothermal* when the temperature does not vary with time. The design equations for an ideal batch reactor are valid for both isothermal and nonisothermal operation.

Constant volume

If V is constant, independent of time, Eqn. (3-5) can be written in terms of concentration as

$$\boxed{\frac{dC_i}{dt} = r_i} \tag{3-8}$$

where C_i is the concentration of species i. Similarly, if V is constant, Eqn. (3-6) can be written as

$$\boxed{C_{A0}\frac{dx_A}{dt} = -r_A} \tag{3-9}$$

where C_{A0} is the initial concentration of A. Equations (3-8) and (3-9) are alternative forms of the design equations for an ideal, *constant-volume* batch reactor, in differential form. The lighter boxes around these equations indicate that they are not as general as Eqns. (3-5) and (3-6) because they contain the assumption of constant volume.

If the volume V is constant, then the mass density of the system, ρ (mass/volume), must also be constant, since the *mass* of material in a batch reactor does not change with time. We could have specified that the mass density was constant instead of specifying that the reactor volume was constant. These two statements are equivalent. However, for a batch reactor, constant volume probably is easier to visualize than constant mass density. For constant-volume (constant mass density) systems, the design equations can be written directly in terms of concentrations, which can easily be measured. For systems where the mass density is not constant, we must work with the most general forms of the design equations, using moles, fractional conversion, or extent of reaction.

For a heterogeneous catalytic reaction, derivation of the constant-volume version of Eqn. (3-5a) requires a bit of manipulation.

$$\frac{1}{W}\frac{dN_i}{dt} = r_i \tag{3-5a}$$

Dividing by the reactor volume V and multiplying by W

$$\frac{1}{V}\frac{dN_i}{dt} = \left(\frac{W}{V}\right) r_i$$

If V is constant,

Design equation—
ideal batch reactor—
heterogeneous catalytic reaction
(constant volume)

$$\boxed{\frac{dC_i}{dt} = C_{cat} r_i} \tag{3-8a}$$

Equation (3-8a) is the design equation for an ideal, constant-volume, batch reactor for a reaction that is catalyzed by a solid catalyst. The symbol C_{cat} represents the *mass* concentration (mass/volume) of the catalyst. The catalyst concentration does not change with time if V is constant.

The equivalent of Eqn. (3-9) for a heterogeneous catalytic reaction is given in Appendix 3.I at the end of this chapter and is labeled Eqn. (3-9a).

The assumption of constant volume is valid for most industrial batch reactors. *The mass density is approximately constant for a large majority of liquids, even if the temperature changes moderately as the reaction proceeds. Therefore, the assumption of constant volume is reasonable for batch reactions that take place in the liquid phase.* Moreover, if a rigid vessel is filled with gas, the gas volume will be constant because the dimensions of the vessel are fixed and do not vary with time.

Variable volume

If V changes with time, Eqn. (3-5) must be written

$$\frac{1}{V}\frac{dN_i}{dt} = \frac{1}{V}\frac{d(C_i V)}{dt} = \frac{dC_i}{dt} + \frac{C_i}{V}\frac{dV}{dt} = r_i$$

Clearly, this equation is more complex, and harder to work with, than Eqn. (3-8).

EXERCISE 3-2

There are a few batch reactors where the assumption of constant volume is not appropriate. Can you think of one? *Hint*: You probably come within 10 ft of this reactor at least once a week, perhaps even every day.

Integrated forms of the design equation

The design equation must be integrated in order to solve problems in reactor design and analysis. In order to actually perform the integration, the temperature must be known as a function of either time or composition. This is because the rate equation r_i contains one or more constants that depend on temperature.

As we shall see in Chapter 8, the energy balance determines how the reactor temperature changes as the reaction proceeds. Broadly, there are three possibilities:

1. The energy balance is so complex that the design equation and the energy balance must be solved simultaneously. We shall leave this case for Chapter 8.
2. The reactor can be heated or cooled such that the temperature changes, but is known as a function of time. An example of this case is treated in Chapter 4.
3. The reactor is adiabatic, or is heated or cooled so that it is isothermal. If the reactor is isothermal, the parameters in the rate equation are constant, i.e., they do not depend on either time or composition. In the adiabatic case, the temperature can be expressed as a function of composition. Therefore, the parameters in the rate equation can also be written as functions of composition. This will be illustrated in Chapter 8.

For the third case, i.e., an isothermal or adiabatic reactor, r_i depends only on concentration. If V is constant, or can be expressed as a function of concentration, Eqn. (3-5) can be symbolically integrated from $t = 0,\ N_i = N_{i0}$ to $t = t,\ N_i = N_i$. The result is

$$\boxed{\int_{N_{i0}}^{N_i} \frac{1}{V}\frac{dN_i}{r_i} = \int_0^t dt = t} \tag{3-10}$$

When the reactor temperature varies with time *in a known manner*, then r_i depends on time as well as concentration. In such a case, Eqn. (3-5) must be used as a starting point instead of Eqn. (3-10). This will be illustrated in the next chapter.

The integrated forms of Eqns. (3-5) through (3-9) for Case 3 above are given in Appendix 3.I, and are labeled as Eqns. (3-10) through (3-14), respectively. Appendix 3.I also contains the integrated forms of Eqns. (3-5a) through (3-9a) for Case 3.

Once the integrations of the design equations have been performed, the time required to reach a concentration C_A, or a fractional conversion x_A, or an extent of reaction ξ can be calculated. Conversely, the value of C_A, x_A, or ξ that results for a specified reaction time can also be calculated. Chapter 4 illustrates the solution of some batch-reactor problems where the reactor is isothermal, or where the temperature is known as a function of time. The simultaneous solution of the design equation and the energy balance is considered in Chapter 8.

3.3 CONTINUOUS REACTORS

When the demand for a single chemical product reaches a high level, in the region of tens of million pounds per year, there generally will be an economic incentive to manufacture the product continuously, using a reactor that is dedicated to that product. The reactor may operate at steady state for a year or more, with planned shutdowns only for regular maintenance, catalyst changes, etc.

Almost all of the reactors in a petroleum refinery operate continuously because of the tremendous annual production rates of the various fuels, lubricants, and chemical intermediates that are manufactured in a refinery. Many well-known polymers such as polyethylene and polystyrene are also produced in continuous reactors, as are many large-volume chemicals such as styrene, ethylene, ammonia, and methanol.

Figure 3-3 is a simplified flowsheet showing some of the auxiliary equipment that may be associated with a continuous reactor. In this example, the feed stream is heated to the desired inlet temperature, first in a feed/product heat exchanger and then in a fired heater. The stream leaving the reactor contains the product(s), the unconverted reactants, and any inert components.

Figure 3-2 A continuous reactor, with associated equipment, for the catalytic isomerization of heavy normal paraffins, containing about 35 carbon atoms, to branched paraffins. The catalyst is comprised of platinum on an acidic zeolite that has relatively large pores. The reaction produces lubricants that have a high viscosity at high temperatures, but retain the characteristics of a liquid at low temperatures. Without the isomerization reaction, the lubricant would become "waxy" and would not flow at low temperatures. This unit is located at the ExxonMobil refinery in Fawley, UK. (Photo, ExxonMobil 2003 Summary Annual Report.)

This stream is cooled in the feed/product heat exchanger and then is cooled further to condense some of its components. The gas and liquid phases are separated. The liquid phase is sent to a separations section (fractionation unit), where the product is recovered. A purge is taken from the gas that leaves the separator, in part to prevent buildup of impurities in the recycle loop. The remainder of the gas is recycled.

Most of the unconverted reactants from a continuous reactor will be recycled back into the feed stream, unless the fractional conversion of the reactants is very high. Some of the product and/or inert components also may be recycled, to aid in control of the reactor temperature, for example.

Continuous reactors normally operate at steady state. The flowrate and composition of the feed stream do not vary with time, and the reactor operating conditions do not vary with time. We will assume steady state in developing the design equations for the two ideal

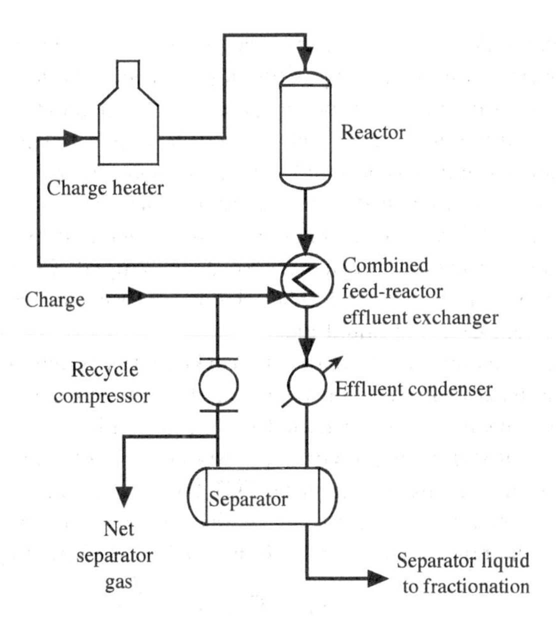

Figure 3-3 A typical flow scheme for the reactor section of a continuous plant.[1] Several items of heat exchange equipment, a recycle compressor, and a phase separator are required to support the steady-state operation of the reactor. (Figure Copyright 2004 UOP LLC. All rights reserved. Used with permission.)

continuous reactors, the ideal continuous stirred-tank reactor (CSTR), and the ideal plug-flow reactor (PFR).

3.3.1 Ideal Continuous Stirred-Tank Reactor (CSTR)

Like the ideal batch reactor, the ideal CSTR is characterized by intense mixing. The temperature and the various concentrations are the same at every point in the reactor. The feed stream entering the reactor is mixed *instantaneously* into the contents of the reactor, immediately destroying the identity of the feed. Since the composition and the temperature are the same everywhere in the CSTR, it follows that *the effluent stream must have exactly the same composition and temperature as the contents of the reactor.*

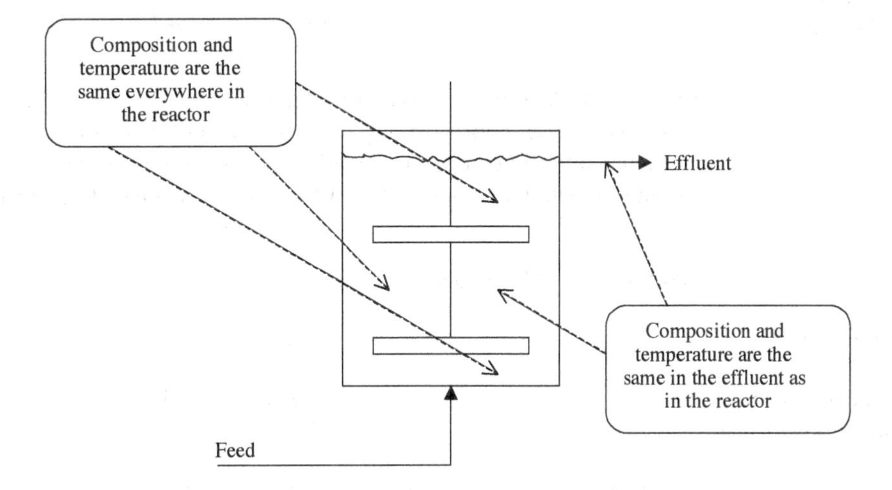

[1] Stine, M. A., *Petroleum Refining*, presented at the North Carolina State University AIChE Student Chapter Meeting, November 15, 2002.

On a small scale, e.g., a laboratory reactor, mechanical agitation is usually required to achieve the necessary high intensity of mixing. On a commercial scale, the required mixing sometimes can be obtained by introducing the feed stream into the reactor at a high velocity, such that the resulting turbulence produces intense mixing. A bed of catalyst powder that is fluidized by an incoming gas or liquid stream, i.e., a *fluidized-bed reactor*, might be treated as a CSTR, at least as a first approximation. Another reactor configuration that can approximate a CSTR is a *slurry bubble column reactor*, in which a gas feed stream is sparged through a suspension of catalyst powder in a liquid. Slurry bubble column reactors are used in some versions of the Fischer–Tropsch process for converting synthesis gas, a mixture of H_2 and CO, into liquid fuels.

The continuous stirred-tank reactor is also known as a continuous *backmix, backmixed*, or *mixed flow* reactor. In addition to the catalytic reactors mentioned in the preceding paragraph, the reactors that are used for certain continuous polymerizations, e.g., the polymerization of styrene monomer to polystyrene, closely approximate CSTRs.

Because of the intense mixing in a CSTR, temperature and concentration are the same at every point in the reactor. Therefore, as with the ideal batch reactor, r_i does not depend on position. For a homogeneous reaction, Eqns. (3-1) and (3-3) simplify to

$$F_{i0} - F_i + r_i V = \frac{dN_i}{dt} \tag{3-15}$$

Equation (3-15) describes the *unsteady-state* behavior of a CSTR. This is the equation that must be solved to explore strategies for starting up the reactor, or shutting it down, or switching from one set of operating conditions to another.

At steady state, the concentrations and the temperature of the CSTR do not vary with time. The exact temperature of operation is determined by the energy balance, as we shall see in Chapter 8. At steady state, the right-hand side of Eqn. (3-15) is zero.

$$F_{i0} - F_i + r_i V = 0$$

Design equation—
ideal CSTR—
homogeneous reaction
(molar flow rates)

$$\boxed{V = \frac{F_{i0} - F_i}{-r_i}} \tag{3-16}$$

Equation (3-16) is the *design equation* for an ideal CSTR. It can be applied to a reactor where more than one reaction is taking place, if Eqn. (1-17) is used to express r_i.

For a single reaction, it frequently is convenient to write Eqn. (3-16) in terms of either extent of reaction or fractional conversion of a reactant. If "i" is a reactant, say A,

$$F_A = F_{A0}(1 - x_A)$$

and Eqn. (3-16) becomes

Design equation—
ideal CSTR—
homogeneous reaction
(fractional conversion)

$$\boxed{\frac{V}{F_{A0}} = \frac{x_A}{-r_A}} \tag{3-17}$$

Alternatively, if only one, stoichiometrically simple reaction is taking place,

$$\xi = \frac{F_i - F_{i0}}{\nu_i}$$

and Eqn. (3-16) becomes

Design equation—
ideal CSTR—
homogeneous reaction
(extent of reaction)

$$\boxed{V = \frac{\nu_i \xi}{r_i}} \tag{3-18}$$

Equations (3-16)–(3-18) are equivalent forms of the *design equation* for an ideal CSTR. In these equations, $-r_A$ (or r_i) is always evaluated at the *exit* conditions of the reactor, i.e., at the temperature and concentrations that exist in the effluent stream, and therefore in the whole reactor volume. *Once again, the design equation is simply a molar component material balance.*

For a heterogeneous catalytic reaction, the equivalent form of Eqn. (3-16) is

Design equation—
ideal CSTR—
heterogeneous catalytic reaction
(molar flowrates)

$$\boxed{W = \frac{F_{i0} - F_i}{-r_i}} \tag{3-16a}$$

EXERCISE 3-3

Derive this equation.

Appendix 3.II gives the forms of the design equation for a heterogeneous catalytic reaction that are equivalent to Eqns. (3-17) and (3-18). These equations are labeled Eqns. (3-17a) and (3-18a).

Space time and space velocity

The molar feed rate, F_{A0}, is the product of the inlet concentration C_{A0} and the inlet volumetric flow rate υ_0, i.e.,

$$F_{A0} = \upsilon_0 C_{A0} \tag{3-19}$$

For a homogeneous reaction, the *space time at inlet conditions* τ_0 is defined as

$$\tau_0 \equiv V/\upsilon_0 \tag{3-20}$$

This definition of space time applies to any continuous reactor, whether it is a CSTR or not.

For a homogeneous reaction, space time has the dimension of time. It is *related to* the average time that the fluid spends in the reactor, although it is not necessarily *exactly equal* to the average time. However, the space time and the average residence time behave in a similar manner. If the reactor volume V increases and the volumetric flow rate υ_0 stays constant, both the space time and the average residence time increase. Conversely, if the volumetric flow rate υ_0 increases and the reactor volume stays constant, both the space time and the average residence time decrease.

Space time influences reaction behavior in a continuous reactor in the same way that real time influences reaction behavior in a batch reactor. In a batch reactor, if the time that the reactants spend in the reactor increases, the fractional conversion and the extent of reaction will increase, and the concentrations of the reactants will decrease. The same is true for space time and a continuous reactor. If a continuous reactor is at steady state, the conversion and the extent of reaction will increase, and the reactant concentrations will decrease, when the space time is increased.

Using Eqns. (3-19) and (3-20), Eqn. (3-17) can be written as

Design equation—ideal CSTR—homogeneous reaction (in terms of space time)

$$\tau_0 = \frac{C_{A0}x_A}{-r_A} \tag{3-21}$$

The concept of space time is also applicable to heterogeneously catalyzed reactions. In this case, τ_0 is defined by

$$\tau_0 \equiv W/\upsilon_0 \tag{3-22}$$

Here, the units of τ_0 are (wt. catalyst-time/volume of fluid). With this definition, Eqn. (3-21) applies to both homogeneous reactions and reactions catalyzed by solids.

The inverse of space time is known as *space velocity*. Space velocity is designated in various ways, e.g., SV, GHSV (gas hourly space velocity), and WHSV (weight hourly space velocity). "Space velocity" is commonly used in the field of heterogeneous catalysis, and there can be considerable ambiguity in the definitions that appear in the literature. For example, GHSV may be defined as the volumetric flowrate of gas entering the catalyst divided by the weight of catalyst. In this case, the units of space velocity are (volume of fluid/time-wt. catalyst). The volumetric flowrate may correspond to inlet conditions or to STP. However, it is not uncommon to find space velocity defined as the volumetric flowrate of gas divided by the *volume* of catalyst bed, or by the volume of catalyst particles. With either of these definitions, the units of space time are inverse time, even though the reaction is catalytic.

When the term "space velocity" is encountered in the literature, it is important to pay very careful attention to how this parameter is defined! Analysis of the units may help.

This book will emphasize the use of space time, since it is analogous to real time in a batch reactor. Space velocity can be a bit counterintuitive. Conversion increases as space time increases, but conversion decreases as space velocity increases.

Constant fluid density

If the mass density (mass/volume) of the fluid flowing through the reactor is constant, i.e., if it is the same in the feed, in the effluent, and at every point in the reactor, then the subscript "0" can be dropped from both τ and υ. In this case Eqn. (3-21) can be written as

$$\tau = \frac{C_{A0}x_A}{-r_A} \tag{3-23}$$

When the fluid density is constant, then $\tau(=V/\upsilon)$ is the *average* residence time that the fluid spends in the reactor. This is true for the CSTR, and for any other continuous reactor operating at steady state.

If (*and only if*) the fluid density is constant,

$$x_A \equiv \frac{F_{A0} - F_A}{F_{A0}} = \frac{\upsilon C_{A0} - \upsilon C_A}{\upsilon C_{A0}} = \frac{C_{A0} - C_A}{C_{A0}}$$

so that Eqn. (3-23) becomes

$$\boxed{\tau = \frac{C_{A0} - C_A}{-r_A}} \tag{3-24}$$

Equations (3-23) and (3-24) are design equations for an ideal CSTR with a constant-density fluid. The lighter box around these equations indicates that they are not as general as Eqn. (3-21), which is not restricted to a constant-density fluid. Equations (3-23) and (3-24) apply to both homogeneous and heterogeneously catalyzed reactions, provided that τ is calculated from the appropriate equation, either Eqn. (3-20) or Eqn. (3-22).

Calculating the reaction rate

The various forms of the design equation for an ideal CSTR (Eqns. (3-16) through (3-18), (3-21), (3-23), and (3-24)) can be used to calculate a numerical value of the rate of reaction, if all of the other parameters in the equation are known. The following example illustrates this use of the CSTR design equation.

EXAMPLE 3-1
Calculation of Rate of Disappearance of Thiophene

The catalytic hydrogenolysis of thiophene was carried out in a reactor that behaved as an ideal CSTR. The reactor contained 8.16 g of "cobalt molybdate" catalyst. In one experiment, the feed rate of thiophene to the reactor was 6.53×10^{-5} mol/min. The fractional conversion of thiophene in the reactor effluent was measured and found to be 0.71. Calculate the value of the rate of disappearance of thiophene for this experiment.

APPROACH

Equation (3-16a) is the most fundamental form of the design equation for a heterogeneous catalytic reaction in an ideal CSTR.

$$\boxed{W = \frac{F_{i0} - F_i}{-r_i}} \tag{3-16a}$$

Using the subscript "T" for thiophene and rearranging,

$$-r_T = \frac{F_{T0} - F_T}{W}$$

From the definition of fractional conversion, $F_{T0} - F_T = F_{T0}x_T$. Therefore, all of the parameters on the right-hand side of the above equation are known and $-r_T$ can be calculated.

SOLUTION

$$-r_T = \frac{F_{T0}x_T}{W} = \frac{6.53 \times 10^{-5}(\text{mol/min}) \times 0.71}{8.16(\text{g})} = 0.57 \times 10^{-5}(\text{mol/min-g})$$

3.3.2 Ideal Continuous Plug-Flow Reactor (PFR)

The plug-flow reactor is the third and last of the so-called "ideal" reactors. It is frequently represented as a tubular reactor, as shown below.

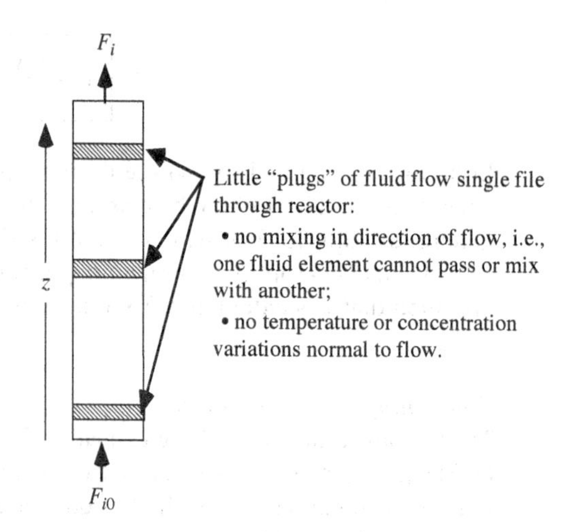

The ideal plug-flow reactor has two defining characteristics:

1. *There is no mixing in the direction of flow.* Therefore, the concentrations of the reactants decrease in the direction of flow, from the reactor inlet to the reactor outlet. In addition, the temperature may vary in the direction of flow, depending on the magnitude of the heat of reaction, and on what, if any, heat transfer takes place through the walls of the reactor. Because of the variation of concentration, and possibly temperature, the reaction rate, r_i, varies in the direction of flow;

2. *There is no variation of temperature or concentration normal to the direction of flow.* For a tubular reactor, this means that there is no radial or angular variation of temperature or of any species concentration at a given axial position z. As a consequence, the reaction rate r_i does not vary normal to the direction of flow, at any cross section in the direction of flow.

The plug-flow reactor may be thought of as a series of miniature batch reactors that flow through the reactor in single file. Each miniature batch reactor maintains its integrity as it flows from the reactor inlet to the outlet. There is no exchange of mass or energy between adjacent "plugs" of fluid.

In order for a real reactor to approximate this ideal condition, the fluid velocity cannot vary normal to the direction of flow. For a tubular reactor, this requires a flat velocity profile in the radial and angular directions, as illustrated below.

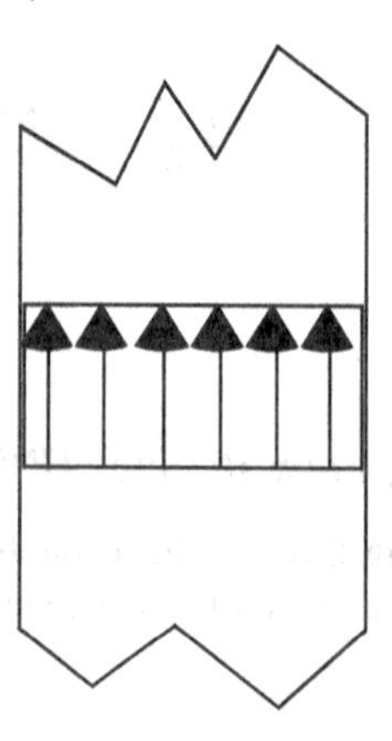

For flow through a tube, this flat velocity profile is approached when the flow is highly turbulent, i.e., at high Reynolds numbers.

Let's analyze the behavior of an ideal plug-flow reactor. We might be tempted to choose the whole reactor as a control volume as we did with the ideal batch reactor and the ideal CSTR and apply Eqn. (3-3),

$$F_{i0} - F_i + G_i = \frac{\mathrm{d}N_i}{\mathrm{d}t} \tag{3-3}$$

Assuming a homogeneous reaction, setting the right-hand side equal to 0 to reflect steady state, and substituting Eqn. (3-1),

$$F_{i0} - F_i + \iiint_V r_i \, \mathrm{d}V = 0 \tag{3-25}$$

For a PFR, the reaction rate varies with position in the direction of flow. Therefore, r_i is a function of V and the above integral cannot be evaluated directly.

We can solve this problem in two ways, the easy way and the hard way.

3.3.2.1 The Easy Way—Choose a Different Control Volume

Let's choose a different control volume over which to write the component material balance. More specifically, let's choose the control volume such that r_i does not depend on V.

From the above discussion, it should be clear that the new control volume must be differential in the direction of flow, since r_i varies in this direction. However, the control volume can span the whole cross section of the reactor, normal to flow, since there are no temperature or concentration gradients normal to flow. Therefore, r_i will be constant over any such cross section.

For a tubular reactor, the control volume is a slice through the reactor perpendicular to the axis (z direction), with a differential thickness dz, as shown below.

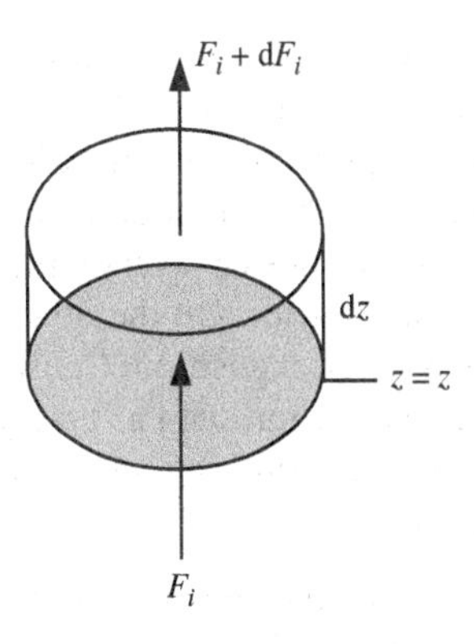

The inlet face of the control volume is located at an axial position, z. The molar flow rate of "i" into the element is F_i and the molar flow rate of "i" leaving the element is $F_i + \mathrm{d}F_i$. For this element, the steady-state material balance on "i" is

$$F_i - (F_i + \mathrm{d}F_i) + r_i \, \mathrm{d}V = 0$$

Design equation—ideal PFR—homogeneous reaction (molar flow rate)

$$\boxed{\mathrm{d}V = \frac{\mathrm{d}F_i}{r_i}} \tag{3-26}$$

Equation (3-26) is the *design equation* for an ideal PFR, in differential form. This equation applies to a PFR where more than one reaction is taking place, provided that r_i is expressed using Eqn. (1-17).

For a single reaction, it may be convenient to write Eqn. (3-26) in terms of either extent of reaction or fractional conversion. If "i" is a reactant, say A, the molar flow rates may be written in terms of the fractional conversion x_A, i.e., $F_A = F_{A0}(1 - x_A)$, and $dF_A = -F_{A0}\, dx_A$. With these transformations, Eqn. (3-26) becomes

Design equation—ideal PFR—homogeneous reaction (fractional conversion)

$$\boxed{\frac{dV}{F_{A0}} = \frac{dx_A}{-r_A}} \tag{3-27}$$

If only one, stoichiometrically simple reaction takes place,

$$\xi = \frac{F_i - F_{i0}}{\nu_i}; \quad dF_i = \nu_i d\xi$$

Design equation—ideal PFR—homogeneous reaction (extent of reaction)

$$\boxed{dV = \frac{\nu_i d\xi}{r_i}} \tag{3-28}$$

Equations (3-26)–(3-28) are various forms of the *design equation* for a homogeneous reaction in an ideal, plug-flow reactor, in differential form. The equivalents of Eqns. (3-26)–(3-28) for a heterogeneous catalytic reaction are given in Appendix 3.IIIA as Eqns. (3-26a), (3-27a), and (3-28a). Be sure that you can derive them.

Temperature variation with position

In developing Eqns. (3-26)–(3-28), we assumed that the temperature was constant in any cross section normal to the direction of flow. We did *not* assume that the temperature was constant in the direction of flow. For a PFR, the reactor is said to be *isothermal* if the temperature does not vary with position in the direction of flow, e.g., with axial position in a tubular reactor. On the other hand, for *nonisothermal* operation, the temperature will vary with axial position. Consequently, the rate constant and perhaps other parameters in the rate equation such as an equilibrium constant will also vary with axial position. The design equations for an ideal PFR are valid for both isothermal and nonisothermal operation.

Space time and space velocity

As noted in the discussion of the ideal CSTR, the space time at inlet conditions, τ_0, is defined by

$$\tau_0 \equiv \frac{V}{\upsilon_0} \tag{3-20}$$

Using Eqn. (3-20), Eqn. (3-27) can be written in terms of C_{A0} and τ_0 as

$$\boxed{d\tau_0 = C_{A0} \frac{dx_A}{(-r_A)}} \tag{3-29}$$

This equation is also valid for a heterogeneous catalytic reaction, if Eqn. (3-22) is used to define τ_0.

The concept of "space velocity," discussed in connection with the ideal CSTR, also applies to ideal PFRs.

Constant density

If the mass density of the flowing fluid is the same at every position in the reactor, the subscript "0" can be dropped from τ and υ. As noted in the discussion of the ideal CSTR, *for the case of constant density*, $\tau(=V/\upsilon)$ is the average residence time of the fluid in the reactor. This is true for any continuous reactor operating at steady state. However, for the ideal PFR, τ has an even more exact meaning. Not only is τ the average residence time of the fluid in the reactor, it is also the *exact* residence time that each and every fluid element spends in the reactor. For an ideal PFR, there is no mixing in the direction of flow, i.e., adjacent fluid elements cannot mix with or pass each other. Therefore, every element of fluid must spend exactly the same time in the reactor. That time is τ, when the mass density is constant.

For the case of constant density, Eqn. (3-29) becomes

$$\boxed{d\tau = C_{A0}\frac{dx_A}{(-r_A)}} \tag{3-30}$$

For constant density, $x_A = (C_{A0} - C_A)/C_{A0}$ and $dx_A = -dC_A/C_{A0}$, so that Eqn. (3-30) can be written as

$$\boxed{d\tau = \frac{-dC_A}{(-r_A)}} \tag{3-31}$$

Equations (3-30) and (3-31) are also valid for a PFR with a heterogeneous catalytic reaction taking place, provided that Eqn. (3-22) is used to define τ.

Integrated forms of the design equation

As with the batch reactor, the design equations in differential form for the PFR must be integrated to solve engineering problems. The same three possibilities that were discussed for the batch reactor also exist here, except that the variable of time for the batch reactor is replaced by position in the direction of flow for the ideal PFR. For Case 3, where the reactor is either isothermal or adiabatic, Eqns. (3-26) and (3-27) can be integrated symbolically to give

$$\boxed{V = \int_{F_{i0}}^{F_i} \frac{dF_i}{r_i}} \tag{3-32}$$

$$\boxed{\frac{V}{F_{A0}} = \int_0^{x_A} \frac{dx_A}{-r_A}} \tag{3-33}$$

The initial conditions for these integrations are $V = 0$, $F_i = F_{i0}$, $x_A = 0$.

Appendix 3.III contains the equivalents of these equations for different variables (e.g., ξ and τ), for heterogeneous catalytic reactions, and for the constant-density case. The numbering of the equations in Appendix 3.IIIA continues from Eqn. (3-33) above.

3.3.2.2 The Hard Way—Do the Triple Integration

Let's return to Eqn. (3-25) and again focus on a tubular reactor, with flow in the axial direction.

$$F_{i0} - F_i + \iiint_V r_i \, \mathrm{d}V = 0 \tag{3-25}$$

The triple integral may be written in terms of three coordinates z (axial position), θ (angular position), and R (radial position).

$$F_{i0} - F_i + \int_0^{2\pi} \int_0^{R_0} \int_0^{L} r_i \, \mathrm{d}\theta \, R \, \mathrm{d}R \, \mathrm{d}z = 0 \tag{3-38}$$

In this equation, R_0 is the inside radius of the tube and L is its length. Since there are no temperature or concentration gradients normal to the direction of flow, r_i does not depend on either θ or R. Since $\int_0^{2\pi} \int_0^{R_0} \mathrm{d}\theta \, R \, \mathrm{d}R = A$, where A is the cross-sectional area of the tube ($A = \pi R_0^2$), Equation (3-38) may be written as

$$F_{i0} - F_i + A \int_0^L r_i \, \mathrm{d}z = 0$$

Differentiating this equation with respect to z gives $\mathrm{d}F_i/\mathrm{d}z = Ar_i$, which can be rearranged to

$$A\mathrm{d}z = \mathrm{d}V = \frac{\mathrm{d}F_i}{r_i} \tag{3-26}$$

Equation (3-26) has been recaptured. Therefore, all of the equations derived from it can be obtained via the triple integration in Eqn. (3-38).

3.4 GRAPHICAL INTERPRETATION OF THE DESIGN EQUATIONS

Figure 3-4 is a plot of $(1/-r_A)$, the *inverse* of the rate of disappearance of Reactant A, versus the fractional conversion of reactant A (x_A). The shape of the curve in Figure 3-4 is based on the assumption that $-r_A$ decreases as x_A increases. In this case, $(1/-r_A)$ will increase as x_A increases. We will refer to this situation as "normal kinetics."

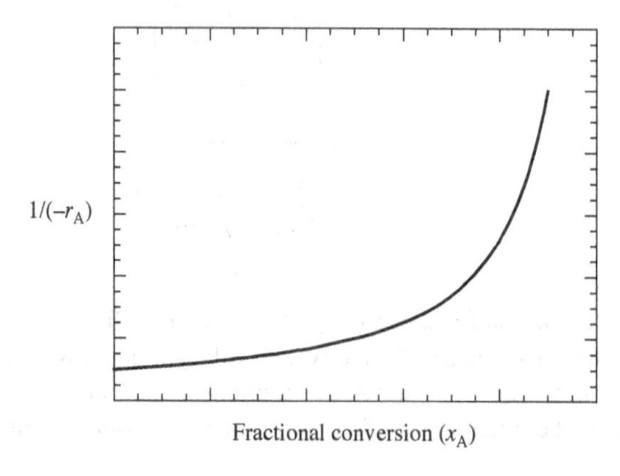

Figure 3-4 Inverse of reaction rate (rate of disappearance of reactant A) versus fractional conversion of A.

"Normal kinetics" will be observed in a number of situations, e.g., if the reactor is isothermal and the concentration-dependent term in the rate equation obeys Generalization III from Chapter 2. Recall that Generalization III stated that the concentration-dependent term F (all C_i) decreases as the concentrations of the reactants decrease, i.e., as the reactants are consumed.

In the discussion of graphs of $(1/-r_A)$ versus x_A, the term "isothermal" will be used to mean that the temperature does not change as x_A changes. This definition is consistent with the definitions, given previously, for isothermal ideal batch reactors and isothermal ideal plug-flow reactors. However, this definition of "isothermal" is more general and can apply to a CSTR or to a series of reactors.

"Normal kinetics" will also be observed if Generalization III applies and the reaction temperature decreases as x_A increases. The temperature will decrease as x_A increases, for example, when an endothermic reaction is carried out in an adiabatic reactor.

The shape of the $(-1/r_A)$ versus x_A curve is not always "normal." This curve can be very different if the reaction is exothermic and the reactor is adiabatic, or if the rate equation does not obey Generalization III.

Now, let's reexamine one form of the design equation for an ideal CSTR:

$$\frac{V}{F_{A0}} = \frac{x_A}{-r_A} \tag{3-17}$$

In order to discriminate between the variable x_A and the outlet conversion from the CSTR, let's call the latter $x_{A,e}$ ("e" for "effluent"), and write the design equation as

$$\frac{V}{F_{A0}} = \frac{x_{A,e}}{-r_A(x_{A,e})}$$

This equation tells us that (V/F_{A0}) for an ideal CSTR is the product of the fractional conversion of A in the reactor *outlet* stream ($x_{A,e}$) and the inverse of the reaction rate, *evaluated at the outlet conditions* $[1/-r_A(x_{A,e})]$. This product is shown graphically in Figure 3-5. The length of the shaded area is equal to $x_{A,e}$, and the height is equal to $1/-r_A(x_{A,e})$. The area is equal to V/F_{A0}, according to the above equation.

Now, let's examine the comparable design equation for an ideal PFR:

$$\frac{V}{F_{A0}} = \int_0^{x_{A,e}} \frac{dx_A}{-r_A} \tag{3-33}$$

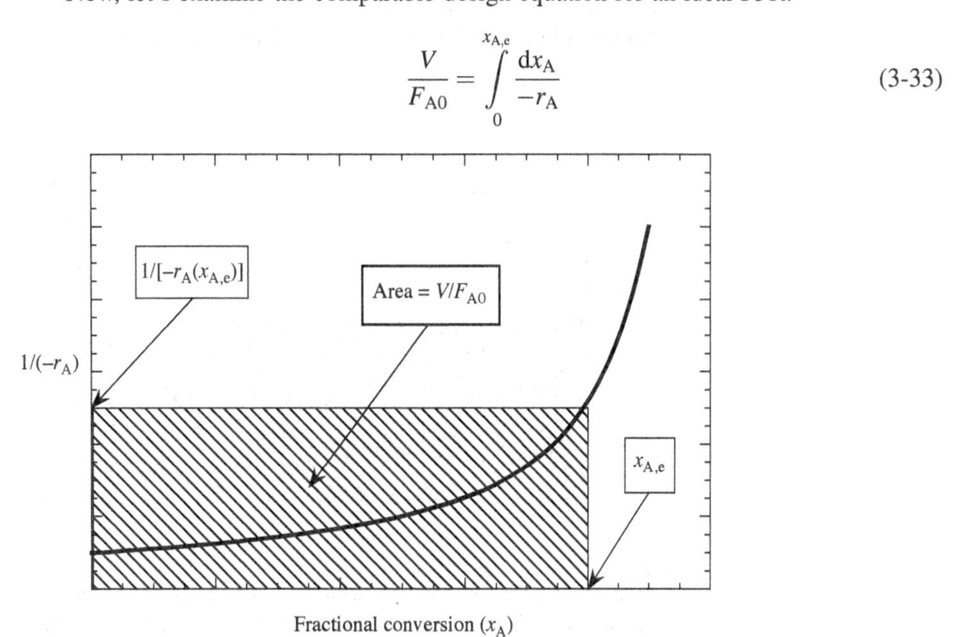

Figure 3-5 Graphical representation of the design equation for an ideal CSTR.

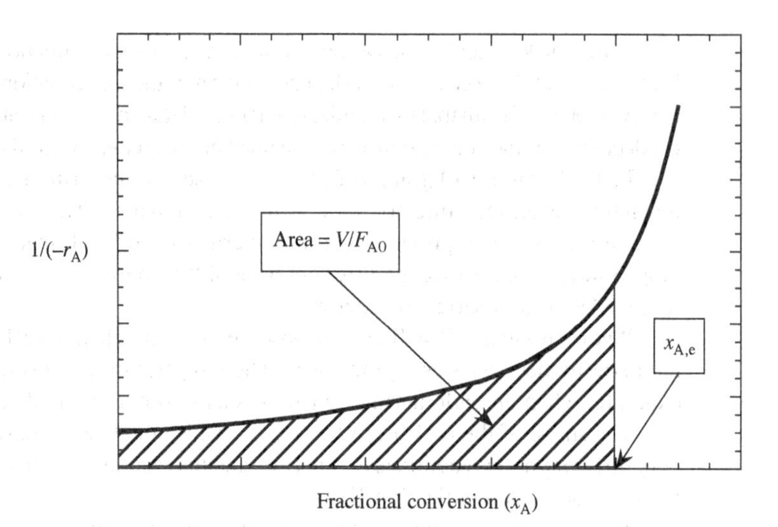

Figure 3-6 Graphical representation of the design equation for an ideal PFR.

This equation tells us that (V/F_{A0}) for an ideal PFR is the *area under the curve* of $(1/-r_A)$ versus x_A, between the inlet fractional conversion $(x_A = 0)$ and the outlet fractional conversion $(x_{A,e})$. This area is shown graphically in Figure 3-6.

Now we can compare the volumes (or weights of catalyst) required to achieve a specified conversion in each of the two ideal, continuous reactors. Suppose we have an ideal CSTR and an ideal PFR. The same reaction is being carried out in both reactors. The PFR is isothermal and operates at the same temperature as the CSTR. The molar feed rate of Reactant A to both reactors is F_{A0}. If the kinetics are "normal," which reactor will require the smaller volume to produce a specified conversion, $x_{A,e}$, in the effluent stream?

Figure 3-7 shows the graphical answer to this question. For a given F_{A0}, the required volume for an ideal CSTR is proportional to the *entire* shaded area (both types of shading). The required volume for an ideal PFR is proportional to the area under the curve. Clearly, the required volume for the PFR is substantially less than the required volume for the ideal CSTR.

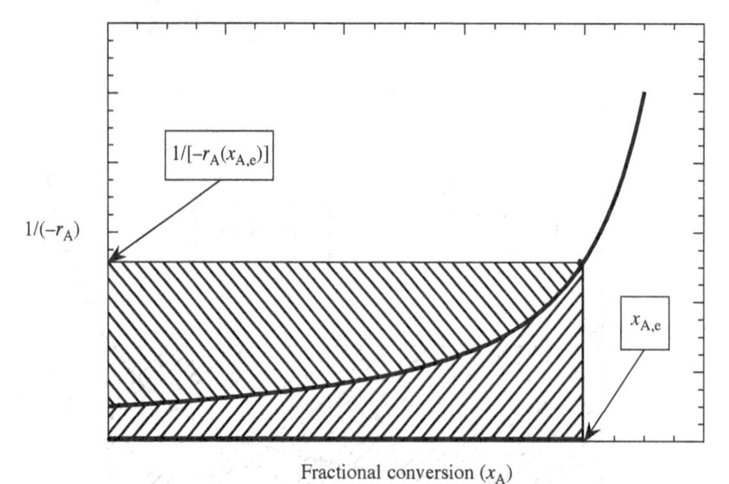

Figure 3-7 Comparison of the volumes required to achieve a given conversion in an ideal PFR and an ideal CSTR, for a given feed rate, F_{A0}. The required volume for a PFR is proportional to the area under the curve. The required volume for a CSTR is proportional to the area of the rectangle (the sum of the two cross-hatched areas).

EXERCISE 3-4

Explain this result qualitatively. What is there about the operation of an ideal CSTR with "normal kinetics" that causes it to need a larger volume than an ideal PFR to achieve a specified outlet conversion for a fixed F_{A0}?

Hint: Using the above figure, compare the average reaction rate in the PFR with the rate in the CSTR. Why are these rates different?

The graphical interpretation of the design equations for the two ideal continuous reactors has been illustrated using fractional conversion to measure the progress of the reaction. The analysis could have been carried out using the extent of reaction ξ with Eqns. (3-18) and (3-34). Moreover, for a constant-density system, the analysis could have been carried out using the concentration of Reactant A, C_A, with Eqns. (3-24) and (3-37).

Plots such as those in Figures 3-5–3-7 are often referred to as "Levenspiel" plots. Octave Levenspiel, a pioneering figure in the field of chemical reaction engineering, popularized the use of this type of plot as a pedagogical tool more that 40 years ago.[2] "Levenspiel plots" will recur in Chapter 4, as a means of analyzing the behavior of "systems" of ideal reactors.

SUMMARY OF IMPORTANT CONCEPTS

- Design equations are nothing more than component material balances.
- There are no spatial variations of temperature or concentration in an ideal batch reactor or in an ideal continuous stirred-tank reactor (CSTR).
- There are no spatial variations of temperature or concentration normal to flow in an ideal plug-flow reactor (PFR). However, the concentrations, and perhaps the temperature, do vary in the direction of flow;
- If (and only if) the mass density is constant, the design equations can be simplified and written in terms of concentration.

PROBLEMS

Problem 3-1 (Level 1) Radial reactors are sometimes used in catalytic processes where pressure drop through the reactor is an important economic parameter, for example, in ammonia synthesis and in naphtha reforming to produce high-octane gasoline.

Top and cross-sectional views of a simplified radial catalytic reactor are shown below.

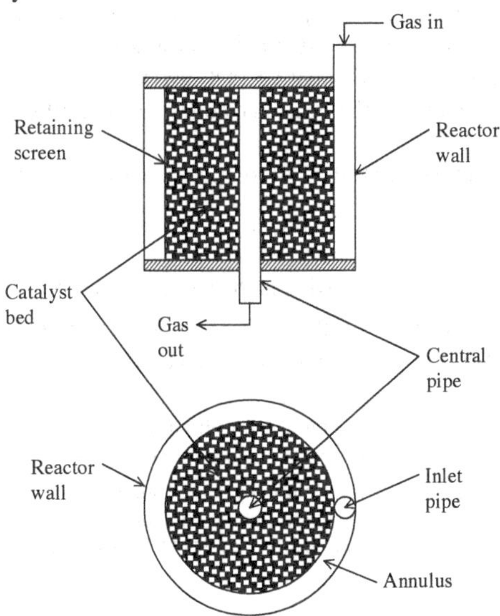

In this configuration, the feed to the reactor, a gas, is introduced through a pipe into an outer annulus. The gas distributes evenly throughout the annulus, i.e., the total pressure is essentially constant at every position in the annulus. The gas then flows radially inward through a uniformly packed catalyst bed, in the shape of a hollow cylinder with outer radius R_o and inner radius R_i. The total pressure is essentially constant along the length of the central pipe. There is no fluid mixing in the radial direction. There are no temperature or concentration gradients in the vertical or angular directions. The catalyst bed contains a total of W pounds of catalyst.

Reactant A is fed to the reactor at a molar feed rate F_{A0} (moles A/time), and the average final fractional conversion of A in the product stream is x_A.

Derive the "design equation," i.e., a relationship among F_{A0}, x_A, and W, for a radial reactor operating at steady state.

A more detailed design of a radial fixed-bed reactor is shown below.[1]

[1] Stine, M. A., *Petroleum Refining*, presented at the North Carolina State University AIChE Student Chapter Meeting, November 15, 2002.

[2] Levenspiel, O., *Chemical Reaction Engineering*, 1st edition, John Wiley & Sons, Inc., New York (1962).

Conventional radial flow reactor

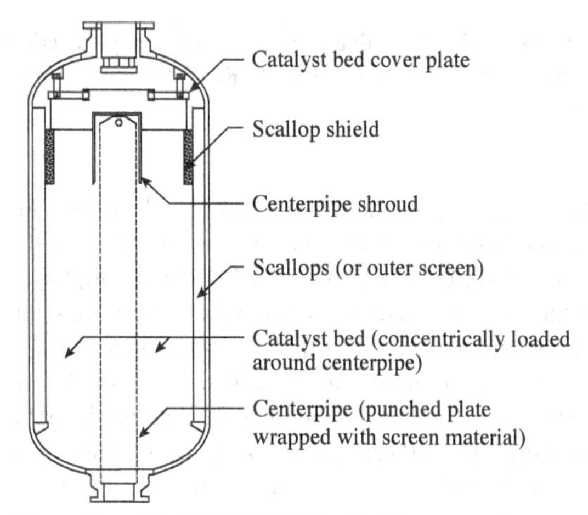

(Figure Copyright 2004 UOP LLC. All rights reserved. Used with permission.)

Problem 3-2 (Level 2) Rate equations of the form

$$-r_A = \frac{kC_AC_B}{(1+K_AC_A)^2}$$

are required to describe the rates of some heterogeneous catalytic reactions. Suppose that the reaction A + B → products occurs in the liquid phase. Reactant B is present in substantial excess, so that C_B does not change appreciably as reactant A is consumed.

1. The value of $-r_A$ goes through a maximum as C_A is increased. At what value of C_A does this maximum occur?
2. The concentration of A in the feed to a continuous reactor is $C_{A0} = 1.5/K_A$. The concentration of A in the effluent is $0.50/K_A$. Make a sketch of $(1/-r_A)$ versus C_A that covers this range of concentration. One ideal continuous reactor will be used to carry out this reaction. Should it be a CSTR or PFR? Explain your answer.
3. Would your answer to Part b be different if the inlet concentration was $C_{A0} = 1.5/K_A$ and the outlet concentration was $C_A = 1.0/K_A$? Explain your answer.

Problem 3-3 (Level 2) In an ideal, *semi-batch* reactor, some of the reactants are charged initially. The remainder of the reactants are fed, either continuously or in "slugs," over time. The contents of the reactor are mixed vigorously, so that there are no spatial gradients of temperature or concentration in the reactor at any time.

Consider the single liquid-phase reaction

$$A + B \rightarrow \text{products}$$

which takes place in an ideal, semi-batch reactor. The initial volume of liquid in the reactor is V_0 and the initial concentrations of A and B in this liquid are C_{A0} and C_{B0} respectively. Liquid is fed continuously to the reactor at a volumetric flow rate υ. The concentrations of A and B in this feed are C_{Af} and C_{Bf}, respectively.

1. Derive a design equation for this system by carrying out a material balance on "A". Work in terms of C_A, not x_A or ξ.
2. Ultimately, we would like to determine C_A as a function of time. Under what conditions is the design equation that you derived sufficient to do this? Assume that the rate equation for the disappearance of A is known.

Problem 3-4 (Level 1) Plot $(1/-r_A)$ versus x_A for an isothermal, zero-order reaction with a rate constant of k. If the desired outlet conversion is $x_A = 0.50$, which of the two ideal continuous reactors requires the smaller volume, for a fixed value of F_{A0}?

What is the outlet conversion from a PFR when $V/F_{A0} = 2/k$?

Problem 3-5 (Level 1) Develop a graphical interpretation of the design equation for an ideal batch reactor.

Problem 3-6 (Level 1) The kinetics of the catalytic reaction[3]

$$SO_2 + 2H_2S \rightarrow 3S + 2H_2O$$

are being studied in an ideal CSTR. Hydrogen sulfide (H_2S) is fed to the reactor at a rate of 1000 mol/h. The rate at which H_2S leaves the reactor is measured and found to be 115 mol/h. The feed to the reactor is a mixture of SO_2, H_2S, and N_2 in the molar ratio 1/2/7.5. The total pressure and temperature in the reactor are 1.1 atm and 250 °C, respectively. The reactor contains 3.5 kg of catalyst.

What is the rate of disappearance of H_2S? What are the corresponding concentrations of H_2S, SO_2, S, and H_2O?

Problem 3-7 (Level 2) The homogeneous decomposition of the free-radical polymerization initiator diethyl peroxydicarbonate (DEPDC) has been studied in supercritical carbon dioxide using an ideal, continuous, stirred-tank reactor (CSTR).[4] The concentration of DEPDC in the feed to the CSTR was 0.30 mmol. Because of this very low concentration, constant fluid density can be assumed. At 70 °C and a space time of 10 min, the fractional conversion of DEPDC was 0.21. At 70 °C and a space time of 30 min, the fractional conversion of DEPDC was 0.44.

The rate equation for DEPDC decomposition is believed to have the form

$$-r_{\text{DEPDC}} = k[\text{DEPDC}]^n$$

1. What is the value of n, and what is the value of k, at 70 °C?
2. The activation energy of the decomposition reaction is 132 kJ/mol. What is the value of the rate constant k at 85 °C?

---

[3] This reaction is the catalytic portion of the well-known Claus process for converting H_2S in waste gas streams into elemental sulfur.

[4] Adapted from Charpentier, P. A., DeSimone, J. M., and Roberts, G. W. Decomposition of polymerization initiators in supercritical CO_2: a novel approach to reaction kinetics in a CSTR, *Chem. Eng. Sci.*, 55, 5341–5349 (2000).

Problem 3-8 (Level 1) An early study of the dehydrogenation of ethylbenzene to styrene[5] contains the following comments concerning the behavior of a palladium-black catalyst:

"A fair yield (of styrene) was obtained at 400 °C. A 12-g sample of catalyst produced no greater yield than the 8-g sample. By passing air with the ethylbenzene vapor, the dehydrogenation occurred at even lower temperatures and water was produced."

The Pd-black catalyst was contained in a quartz tube and the flow rate of ethylbenzene through the tube was 5 cc (liquid) per hour for all experiments. The measured styrene yields (moles styrene formed/mole ethylbenzene fed) were about 0.2 with both 8 and 12 g of catalyst. For the purposes of this problem, assume that "yield" of styrene is the same as the conversion of ethylbenzene, and assume that the dehydrogenation of ethylbenzene is stoichiometrically simple. Also assume that the reaction took place at 1 atm total pressure. In answering the questions below, assume that the experimental reactor was an ideal PFR.

1. In view of the design equation for an ideal PFR in which a heterogeneously catalyzed reaction takes place, how would you expect the yield of styrene to have changed when the amount of catalyst was increased from 8 to 12 g?
2. How do you explain the fact that the yield of styrene did not change when the catalyst weight was increased from 8 to 12 g?
3. How do you explain the behavior of the reaction when air was added to the feed?

Problem 3-9 (Level 1) The hydrogenolysis of thiophene (C_4H_4S) has been studied at 235–265 °C over a cobalt–molybdate catalyst, using a CSTR containing 8.16 g of catalyst. The stoichiometry of the system can be represented by

$$C_4H_4S + 3H_2 \rightarrow C_4H_8 + H_2S \qquad \text{Reaction 1}$$

$$C_4H_8 + H_2 \rightarrow C_4H_{10} \qquad \text{Reaction 2}$$

All species are gaseous at reaction conditions.

The feed to the CSTR consisted of a mixture of thiophene, hydrogen, and hydrogen sulfide. The mole fractions of butene (C_4H_8), butane (C_4H_{10}), and hydrogen sulfide in the reactor effluent were measured. The mole fractions of hydrogen and thiophene were not measured.

The data from one particular experimental run are given below:

Total pressure in reactor = 832 mmHg

Feed rate

Thiophene = 0.653×10^{-4} g·mol/min
Hydrogen = 4.933×10^{-4} g·mol/min
Hydrogen sulfide = 0

Mole fractions in effluent

$H_2S = 0.0719$
Butenes (total) = 0.0178
Butane = 0.0541

[5] Taylor, H. S. and McKinney, P. V., Adsorption and activation of carbon monoxide at palladium surfaces, *J. Am. Chem. Soc.*, 53, 3604 (1931).

Calculate $-r_T$ (the rate of disappearance of thiophene) and the partial pressures of thiophene, hydrogen, hydrogen sulfide, butenes (total), and butane in the effluent. You may assume that the ideal gas laws are valid.

Problem 3-10 (Level 2) The hydrolysis of esters, i.e.,

$$R\text{-}C(=O)\text{-}OR' + H_2O \longrightarrow R\text{-}C(=O)\text{-}OH + R'OH$$

(E) (W) (A)

frequently is catalyzed by acids. As the above hydrolysis reaction proceeds, more acid is produced and the concentration of catalyst increases. This phenomenon, known as "autocatalysis," is captured by the rate equation

$$-r_E = (k_0 + k_1[A])[E][H_2O]$$

Here k_0 is the rate constant in the absence of the organic acid that is produced by the reaction.

To illustrate the behavior of autocatalytic reactions, let's arbitrarily assumed the following values:

E_0 (ester concentration in feed) = 1.0 mol/l
W_0 (water concentration in feed) = 1.0 mol/l
A_0 (acid concentration in feed) = 0

The concentration of the alcohol (R′OH) in the feed also is zero. The rate constants are

$k_0 = 0.01$ l/mol-h
$k_1 = 0.20$ l^2/mol^2-h

In answering the following questions, assume that the reaction takes place in the liquid phase.

1. Calculate values of $-r_E$ at fractional conversions of ester $(x_E) = 0,\ 0.10,\ 0.20,\ 0.30,\ 0.40,\ 0.50,\ 0.60,$ and 0.70.
2. Plot $1/-r_E$ versus x_E.
3. What kind of *continuous* reactor system would you use if a final ester conversion of 0.60 were desired, and if the reaction were to take place isothermally. Choose the reactor or combination of reactors that has the smallest volume. Justify your answer.

Problem 3-11 As pointed out in Chapter 2, kinetics are not always "normal" (e.g., see Figure 2-3). Consider a liquid-phase reaction that obeys the rate equation: $-r_A = kC_A^{-1}$, over some range of concentration. Suppose this reaction was to be carried out in a continuous, isothermal reactor, with feed and outlet concentrations within the range where the rate equation is valid.

1. Make a sketch of $(1/-r_A)$ versus C_A for the range of concentration where the rate equation is valid.
2. What kind of continuous reactor (or system of continuous reactors) would you use for this job, in order to minimize the volume required? Explain your answer.

APPENDIX 3 SUMMARY OF DESIGN EQUATIONS

Warning: Careless use of the equations in this appendix can be damaging to perfomance. In extreme cases, improper use of these materials can be academically fatal. Always carry out a careful analysis of the problem being solved before using this appendix.

When in doubt, first carefully choose the *type* of reactor for which calculations are to be performed. Then decide whether the reaction is homogeneous or heterogeneous. Finally, begin with the most general form of the appropriate design equation and make any simplifications that are warranted.

I. Ideal batch reactor

A. General—differential form	**Design equation**	
Variable	**Homogeneous reaction**	**Heterogeneous catalytic reaction**
Moles of species "i", N_i	$\frac{1}{V}\frac{dN_i}{dt} = r_i$ (3-5)	$\frac{1}{W}\frac{dN_i}{dt} = r_i$ (3-5a)
Fractional conversion of reactant A, x_A	$\frac{N_{A0}}{V}\frac{dx_A}{dt} = -r_A$ (3-6)	$\frac{N_{A0}}{W}\frac{dx_A}{dt} = -r_A$ (3-6a)
Extent of reaction, ξ	$\frac{\nu_i}{V}\frac{d\xi}{dt} = r_i$ (3-7)	$\frac{\nu_i}{W}\frac{d\xi}{dt} = r_i$ (3-7a)
B. Constant volume—differential form	**Homogeneous reaction**	**Heterogeneous catalytic reaction**
Concentration of species "i", C_i	$\frac{dC_i}{dt} = r_i$ (3-8)	$\frac{dC_i}{dt} = C_{cat}(r_i)$ (3-8a)
Fractional conversion of reactant A, x_A	$C_{A0}\frac{dx_A}{dt} = -r_A$ (3-9)	$C_{A0}\frac{dx_A}{dt} = C_{cat}(-r_A)$ (3-9a)
C. General—integrated form (see Note 1)	**Homogeneous reaction**	**Heterogeneous catalytic reaction**
Moles of species "i", N_i	$\int_{N_{i0}}^{N_i} \frac{1}{V}\frac{dN_i}{r_i} = t$ (3-10)	$\frac{1}{W}\int_{N_{i0}}^{N_i} \frac{dN_i}{r_i} = t$ (3-10a)
Fractional conversion of reactant A, x_A	$N_{A0}\int \frac{1}{V}\frac{dx_A}{(-r_A)} = t$ (3-11)	$\frac{N_{A0}}{W}\int_0^{x_A} \frac{dx_A}{(-r_A)} = t$ (3-11a)
Extent of reaction, ξ	$\nu_i\int_0^{\xi} \frac{1}{V}\frac{d\xi}{r_i} = t$ (3-12)	$\frac{\nu_i}{W}\int_0^{\xi} \frac{d\xi}{r_i} = t$ (3-12a)
D. Constant volume—integrated form (see Note 1)	**Homogeneous reaction**	**Heterogeneous catalytic reaction**
Concentration of species "i", C_i	$\int_{C_{i0}}^{C_i} \frac{dC_i}{r_i} = t$ (3-13)	$\int_{C_{i0}}^{C_i} \frac{dC_i}{r_i} = C_{cat}t$ (3-13a)
Fractional conversion of reactant A, x_A	$C_{A0}\int_0^{x_A} \frac{dx_A}{(-r_A)} = t$ (3-14)	$C_{A0}\int_0^{x_A} \frac{dx_A}{(-r_A)} = C_{cat}t$ (3-14a)

Note 1: For Case 3 (p. 43).

II. Ideal continuous stirred-tank reactor (CSTR)

A. General—in terms of *V* or *W*	Homogeneous reaction	Heterogeneous catalytic reaction
Molar flow rate of species "i", F_i	$V = \frac{F_{i0} - F_i}{-r_i}$ (3-16)	$W = \frac{F_{i0} - F_i}{-r_i}$ (3-16a)
Fractional conversion of reactant A, x_A	$\frac{V}{F_{A0}} = \frac{x_A}{(-r_A)}$ (3-17)	$\frac{W}{F_{A0}} = \frac{x_A}{(-r_A)}$ (3-17a)
Extent of reaction, ξ	$V = \frac{\nu_i \xi}{r_i}$ (3-18)	$W = \frac{\nu_i \xi}{r_i}$ (3-18a)
B. General—in terms of τ_0 (see Note 1)	**Homogeneous reaction**	**Heterogeneous catalytic reaction**
Fractional conversion of reactant A, x_A	$\tau_0 = \frac{C_{A0} x_A}{-r_A}$ (3-21)	$\tau_0 = \frac{C_{A0} x_A}{-r_A}$ (3-21)
C. Constant density—in terms of τ (see Note 2)	**Homogeneous reaction**	**Heterogeneous catalytic reaction**
Fractional conversion of reactant A, x_A	$\tau = \frac{C_{A0} x_A}{-r_A}$ (3-23)	$\tau = \frac{C_{A0} x_A}{-r_A}$ (3-23)
Concentration of reactant A, C_A	$\tau = \frac{C_{A0} - C_A}{-r_A}$ (3-24)	$\tau = \frac{C_{A0} - C_A}{-r_A}$ (3-24)

Note 1:

For a homogeneous reaction, $\tau_0 = V/\upsilon_0 = VC_{A0}/F_{A0}$.
For a heterogeneous reaction, $\tau_0 = W/\upsilon_0 = WC_{A0}/F_{A0}$.

Note 2:

For a homogeneous reaction, $\tau = V/\upsilon = VC_{A0}/F_{A0}$.
For a heterogeneous reaction, $\tau = W/\upsilon = WC_{A0}/F_{A0}$.

III. Ideal continuous plug-flow reactor (PFR)

A. General—differential form—in terms of *V* or *W*	Homogeneous reaction	Heterogeneous catalytic reaction
Molar flow rate of species "i", F_i	$dV = \frac{dF_i}{r_i}$ (3-26)	$dW = \frac{dF_i}{r_i}$ (3-26a)
Fractional conversion of reactant A, x_A	$\frac{dV}{F_{A0}} = \frac{dx_A}{-r_A}$ (3-27)	$\frac{dW}{F_{A0}} = \frac{dx_A}{-r_A}$ (3-27a)
Extent of reaction, ξ	$dV = \frac{\nu_i d\xi}{r_i}$ (3-28)	$dW = \frac{\nu_i d\xi}{r_i}$ (3-28a)
B. General—differential form–in terms of τ_0 (see Note 1)	**Homogeneous reaction**	**Heterogeneous catalytic reaction**
Fractional conversion of reactant A, x_A	$d\tau_0 = C_{A0} \frac{dx_A}{(-r_A)}$ (3-29)	$d\tau_0 = C_{A0} \frac{dx_A}{(-r_A)}$ (3-29)
C. Constant density—differential form in terms of τ (see Note 2)	**Homogeneous reaction**	**Heterogeneous catalytic reaction**
Fractional conversion of reactant A, x_A	$d\tau = C_{A0} \frac{dx_A}{(-r_A)}$ (3-30)	$d\tau = C_{A0} \frac{dx_A}{(-r_A)}$ (3-30)
Concentration of reactant A, C_A	$d\tau = \frac{-dC_A}{(-r_A)}$ (3-31)	$d\tau = \frac{-dC_A}{(-r_A)}$ (3-31)

D. General—integrated form (see Note 3)	**Homogeneous reaction**	**Heterogeneous catalytic reaction**
Molar flow rate of species "i", F_i	$V = \int_{F_{i0}}^{F_i} \frac{dF_i}{r_i}$ (3-32)	$W = \int_{F_{i0}}^{F_i} \frac{dF_i}{r_i}$ (3-32a)
Fractional conversion of reactant A, x_A	$\frac{V}{F_{A0}} = \int_0^{x_A} \frac{dx_A}{(-r_A)}$ (3-33)	$\frac{W}{F_{A0}} = \int_0^{x_A} \frac{dx_A}{(-r_A)}$ (3-33a)
Extent of reaction, ξ	$V = \int_0^{\xi} \frac{\nu_i d\xi}{r_i}$ (3-34)	$W = \int_0^{\xi} \frac{\nu_i d\xi}{r_i}$ (3-34a)
Fractional conversion of reactant A, x_A—in terms of τ_0 (see Note 1)	$\tau_0 = C_{A0}\int_0^{x_A} \frac{dx_A}{(-r_A)}$ (3-35)	$\tau_0 = C_{A0}\int_0^{x_A} \frac{dx_A}{(-r_A)}$ (3-35)
E. Integrated form—constant density—in terms of τ (see Note 2 and 3)	**Homogeneous reaction**	**Heterogeneous catalytic reaction**
Fractional conversion of reactant A, x_A	$\tau = C_{A0}\int_0^{x_A} \frac{dx_A}{(-r_A)}$ (3-36)	$\tau = C_{A0}\int_0^{x_A} \frac{dx_A}{(-r_A)}$ (3-36)
Concentration of reactant A, C_A	$\tau = -\int_{C_{A0}}^{C_A} \frac{dC_A}{(-r_A)}$ (3-37)	$\tau = -\int_{C_{A0}}^{C_A} \frac{dC_A}{(-r_A)}$ (3-37)

Note 1:

For a homogeneous reaction, $\tau_0 = V/\upsilon_0 = VC_{A0}/F_{A0}$.
For a heterogeneous reaction, $\tau_0 = W/\upsilon_0 = WC_{A0}/F_{A0}$.

Note 2:

For a homogeneous reaction, $\tau = V/\upsilon = VC_{A0}/F_{A0}$.
For a heterogeneous reaction, $\tau = W/\upsilon = WC_{A0}/F_{A0}$.

Note 3:

For the PFR equivalent of Case 3 (p. 43). See the discussion on p. 53.

Chapter 4

Sizing and Analysis of Ideal Reactors

LEARNING OBJECTIVES

After completing this chapter, you should be able to

1. use the design equations for the three ideal reactors: batch, continuous stirred-tank, and plug-flow, to size or analyze the behavior of a single reactor;
2. use the design equations for the CSTR and PFR to size or analyze the behavior of *systems* of continuous reactors;
3. explain the meaning of the phrases "the reaction is controlled by intrinsic kinetics" and "transport effects are negligible";
4. use "Levenspiel plots" to qualitatively evaluate the behavior of *systems* of reactors.

This chapter will illustrate the use of the design equations that were developed in Chapter 3. Broadly speaking, the design equations can be used for two purposes: to predict the performance of an existing reactor or to calculate the size of reactor that is needed to produce a product at a specified rate and concentration.

Sizing is an essential element of reactor design. However, it is only one of many elements. Reactor design includes, for example, the heat-transfer system, the agitation system, the vessel internals, e.g., baffles and fluid distributor(s), as well as the materials of construction. Reactor design also includes ensuring compliance with the many codes that govern the mechanical design of reactors. All of these elements of design require the use of tools that go beyond the design equations. For example, Figure 4-1 illustrates some of the design features that are required to achieve an even distribution of flow in a gas-phase catalytic reactor.

In this chapter, we will presume that the rate equation, including the values of the constants, is known. Chapter 6 will deal with the use of ideal reactors to obtain the data needed to create rate equations.

4.1 HOMOGENEOUS REACTIONS

4.1.1 Batch Reactors

4.1.1.1 Jumping Right In

EXAMPLE 4-1
Aspirin Manufacture

Aspirin (acetylsalicylic acid) has been used as a pain reliever since the start of the 20th century. Despite competition from newer products, worldwide consumption is about 70 million pounds per year.

The last step in the manufacture of acetylsalicylic acid is the reaction of salicylic acid with acetic anhydride:

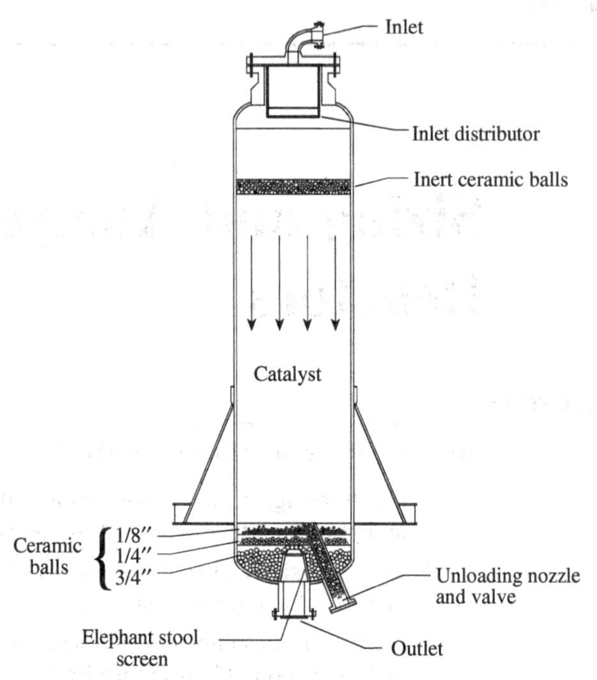

Figure 4-1 A fixed-bed catalytic reactor with vapor flowing down through a packed catalyst bed.[1] The design incorporates a number of devices that promote an even distribution of flow through the catalyst bed. (Copyright 2004 UOP LLC. All rights reserved. Used with permission.)

Salicylic acid + Acetic anhydride → Acetylsalicylic acid (aspirin) + Acetic acid

This reaction is carried out in batch reactors at about 90 °C. Salicylic acid (SA), acetic acid (HOAc), and acetic anhydride (AA) are charged to the reactor initially. After 2–3 h of reaction, the contents of the reactor are discharged to a crystallizer to recover the acetylsalicylic acid (ASA).

If we were designing or operating a plant to manufacture aspirin, it would be important to know how much acetylsalicylic acid was in the reactor at the time that it was discharged to the crystallizer, and how much unconverted salicylic acid and acetic anhydride were present in the discharge. These values would be required to design and operate the crystallizer, and to design and operate the system for recycling unconverted reactants. We also might want to know how much time was required to reduce the concentration of salicylic acid to some specified value. Finally, we would want to know how much acetylsalicylic acid could be produced in a reactor with a known volume, or conversely, what volume of reactor was required to produce a specified quantity of aspirin.

If the rate equation for the reaction is known, and if the reactor behaves as an *ideal* batch reactor, these quantities can be calculated by solving the design equation, as illustrated below.

Suppose that the rate equation for salicylic acid disappearance is

$$-r_{SA} = kC_{SA}C_{AA} \tag{4-1}$$

[1] Stine, M. A., *Petroleum Refining*, presented at the North Carolina State University AIChE Student Chapter Meeting, November 15, 2002.

Suppose that the reactor is isothermal, i.e., the temperature is constant at 90 °C over the whole course of the reaction, and that the value of the rate constant k is 0.30 l/mol-h at 90 °C. The initial concentrations of SA, HOAc, and AA are 2.8, 2.8, and 4.2 mol/l, respectively.

Let's see if we can answer the following questions:

1. What will the concentrations of SA, AA, HOAc, and ASA be after 2 h?
2. How much time will it take for the concentration of SA to reach 0.14 mol/l?
3. How many pounds of ASA can be produced in a 10,000 l reactor in 1 year, if 95% conversion of salicylic acid is required before the reactor is discharged to the crystallizer?

Part A: What will the concentrations of SA, AA, HOAc, and ASA be after 2 h?

APPROACH

For a homogeneous reaction in an ideal batch reactor, the most basic form of the design equation is

$$\frac{1}{V}\frac{dN_i}{dt} = r_i \tag{3-5}$$

If we write Eqn. (3-5) for salicylic acid and insert the rate equation, we get

$$-\frac{1}{V}\frac{dN_{SA}}{dt} = -r_{SA} = kC_{SA}C_{AA} \tag{4-2}$$

This equation must be integrated with respect to time. In order to perform the integration, N_{SA}, C_{AA}, and C_{SA} must be written as functions of a single variable. To begin, let's use C_{SA} as that variable. Later, we will use the extent of reaction ξ instead of C_{SA}.

The number of moles of SA in the reactor at any time, N_{SA}, is just VC_{SA}. The concentration of acetic anhydride C_{AA} can be written in terms of the concentration of salicylic acid C_{SA} using the principles of stoichiometry. Once N_{SA} and C_{AA} have been written as functions of C_{SA}, Eqn. (4-2) can be integrated and the questions posed above can be answered.

SOLUTION

Let's begin by relating N_{SA} and C_{AA} to C_{SA}. This requires a little "bookkeeping," for which we will use a *stoichiometric table*. To construct the stoichiometric table for a batch reactor, we list *all* of the species in the reactor in the far-left column, as shown in Table 4-1. The *number of moles* of each species at zero time, i.e., at the time the reaction starts, are listed in the next column. Finally, in the third column, the *number of moles* of each species at some arbitrary time t are listed. In filling in the third column, the *stoichiometry* of the reaction is used, i.e., for every mole of SA that is consumed, one mole of AA is consumed, one mole of acetic acid (HOAc) is formed, and one mole of acetylsalicylic acid (ASA) is formed.

The quantity of SA that is in the reactor initially is $C_{SA,0} \times V_0$, where V_0 is the volume of solution at $t = 0$ and $C_{SA,0}$ is the initial concentration of SA (at $t = 0$). The quantity of SA in the reactor at $t = t$ is $C_{SA} \times V$, where V is the volume of solution at $t = t$ and C_{SA} is the concentration of SA at $t = t$. Therefore, the number of moles of SA that react between $t = 0$ and $t = t$ is $[C_{SA,0} \times V_0 - C_{SA} \times V]$.

The stoichiometric table for this problem, with C_{SA} as the independent variable, is given in Table 4-1.

Table 4-1 Stoichiometric Table for Example 4-1 Using the Whole Reactor as a Control Volume and Using C_{SA} as the Composition Variable

Species	Moles at $t = 0$	Moles at $t = t$
Salicylic acid (SA)	$C_{SA,0} \times V_0$	$C_{SA} \times V$
Acetic anhydride (AA)	$C_{AA,0} \times V_0$	$C_{AA,0} \times V_0 - [C_{SA,0} \times V_0 - C_{SA} \times V]$
Acetic acid (HOAc)	$C_{HOAc,0} \times V_0$	$C_{HOAc,0} \times V_0 + [C_{SA,0} \times V_0 - C_{SA} \times V]$
Acetylsalicylic acid (ASA)	$C_{ASA,0} \times V_0$	$C_{ASA,0} \times V_0 + [C_{SA,0} \times V_0 - C_{SA} \times V]$

In this table, $C_{AA,0}$, $C_{HOAc,0}$, and $C_{ASA,0}$ are the initial concentrations of acetic anhydride, acetic acid, and acetylsalicylic acid, respectively.

Since this reaction occurs in the liquid phase, constant mass density, i.e., constant volume, may be assumed. Therefore, $V_0 = V$ and

$$\frac{1}{V}\frac{dN_{SA}}{dt} = \frac{dC_{SA}}{dt}$$

From the stoichiometric table, $C_{AA} = (N_{AA}/V) = C_{AA,0} - [C_{SA,0} - C_{SA}]$, since $V = V_0$. Substituting these two relationships into Eqn. (4-2) gives

$$-\frac{dC_{SA}}{dt} = kC_{SA}[C_{AA,0} - (C_{SA,0} - C_{SA})] = kC_{SA}[(C_{AA,0} - C_{SA,0}) + C_{SA}] \tag{4-3}$$

Equation (4-3) can be rearranged to

$$\frac{dC_{SA}}{C_{SA}[(C_{AA,0} - C_{SA,0}) + C_{SA}]} = -kdt$$

Integrating from $t = 0$ to $t = t$ and from $C_{SA} = C_{SA,0}$ to $C_{SA} = C_{SA}$ (do not forget the lower limit!),

$$\int_{C_{SA,0}}^{C_{SA}} \frac{dC_{SA}}{C_{SA}[(C_{AA,0} - C_{SA,0}) + C_{SA}]} = -\int_0^t kdt$$

Since the temperature is constant, k is constant. This equation then can be integrated to give

$$\ln\left\{\frac{C_{SA} \times C_{AA,0}}{C_{SA,0}[(C_{AA,0} - C_{SA,0}) + C_{SA}]}\right\} = -(C_{AA,0} - C_{SA,0})kt \tag{4-4}$$

Substituting the given values of t, k, $C_{SA,0}$, and $C_{AA,0}$ into Eqn. (4-4) and solving for C_{SA} gives $C_{SA} = 0.57$ mol/l.

The concentrations of AA, HOAc, and ASA now can be calculated from the stoichiometric table. The results are $C_{AA} = 1.96$ mol/l, $C_{HOAc} = 5.03$ mol/l, and $C_{ASA} = 2.23$ mol/l.

For this problem, the stoichiometric table could have been written in a simpler form, if the assumption of constant density had been made *before* the table was constructed. Table 4-1 is based on using the whole reactor as a control volume. For a constant-density (constant-volume) system, and *only* for such a system, the stoichiometric table can be written in terms of concentrations. If the control volume is taken to be 1 liter of solution, the stoichiometric table becomes as given in Table 4-2.

Table 4-2 Stoichiometric Table for Example 4-1 Using 1 liter of Solution as a Control Volume* (C_{SA} is the Composition Variable)

Species	Moles at $t = 0$ (initial concentration)	Moles at $t = t$ (concentration at $t = t$)
Salicylic acid (SA)	$C_{SA,0}$	C_{SA}
Acetic anhydride (AA)	$C_{AA,0}$	$C_{AA,0} - [C_{SA,0} - C_{SA}]$
Acetic acid (HOAc)	$C_{HOAc,0}$	$C_{HOAc,0} + [C_{SA,0} - C_{SA}]$
Acetylsalicylic acid (ASA)	$C_{ASA,0}$	$C_{ASA,0} + [C_{SA,0} - C_{SA}]$

**This basis can only be used for constant-volume problems.*

Part B: How much time will it take for the concentration of SA to reach 0.14 mol/l?

APPROACH An equation that relates C_{SA} to time (Eqn. (4-4)) was derived in Part A. The known values of k, $C_{SA,0}$, $C_{AA,0}$, and the known final value of C_{SA} (0.14 mol/l) can be substituted into this equation and the value of time t can be calculated.

SOLUTION The result is $t = 4.60$ h. Note that $C_{SA} = 0.14$ mol/l corresponds to 95% conversion.

Part C: How many pounds of ASA can be produced in a 10,000-liter reactor in 1 year, if 95% conversion of salicylic acid is required before the reactor is discharged to the crystallizer?

APPROACH The time required to reach 95% conversion of SA, about 4.6 h, was calculated in Part B. However, additional time is required to charge the reactants, to heat the reactor from ambient to 90 °C (during which time some reaction will occur), to cool the reactor to close to ambient after 4.6 h at 90 °C, to discharge the contents of the reactor to the crystallizer, and perhaps to clean the reactor prior to starting the next batch. As a guess, a complete cycle might require 16 h (two shifts). Thus, at maximum, about 545 batches could be made in one full year.

The number of moles of ASA in the reactor at the end of the batch can be calculated from stoichiometry, since C_{SA} is known at the end of the batch. The weight of ASA produced per year is this number of moles × the number of batches per year (545) × the molecular weight of ASA (180), provided that there is no loss of ASA in the crystallizer and/or in packaging.

SOLUTION From the stoichiometric table (Table 4-1), the number of moles of ASA in the reactor at the end of reaction $= C_{ASA,0} \times V_0 + [C_{SA,0} \times V_0 - C_{SA} \times V] = 10,000\,(l) \times (2.80 - 0.14)(\text{mol/l}) = 2.66 \times 10^4$ mol.

Annual production of ASA = 545 (batches/year) × 180 (g/mol) × 2.66×10^4 (mol) = 2.61×10^9 g/year = 5.75×10^6 pounds/year.

Alternative approach to Part A (using extent of reaction)

This problem could have been set up using a variable other than C_{SA} to keep track of the composition in the reactor. For example, we could have used the extent of reaction ξ instead of C_{SA}. For the reaction of salicylic acid with acetic anhydride

$$\xi = \frac{N_{SA} - N_{SA,0}}{\nu_{SA}} = \frac{N_{AA} - N_{AA,0}}{\nu_{AA}} = \frac{N_{HOAc} - N_{HOAc,0}}{\nu_{HOAc}} = \frac{N_{ASA} - N_{ASA,0}}{\nu_{ASA}}$$

Recognizing that $\nu_{ASA} = \nu_{HOAx} = -\nu_{SA} = -\nu_{AA} = 1$, the equivalent of Table 4-1, i.e, Table 4-3, can be constructed.

Table 4-3 Stoichiometric Table for Example 4-1 Using the Whole Reactor as a Control Volume and the Extent of Reaction ξ as the Composition Variable

Species	Moles at $t = 0$	Moles at $t = t$
Salicylic acid (SA)	$N_{SA,0}$	$N_{SA,0} - \xi$
Acetic anhydride (AA)	$N_{AA,0}$	$N_{AA,0} - \xi$
Acetic acid (HOAc)	$N_{HOAc,0}$	$N_{HOAc,0} + \xi$
Acetylsalicylic acid (ASA)	$N_{ASA,0} (= 0)$	ξ

Using these relationships, Eqn. (4-2) becomes

$$\frac{1}{V}\frac{d\xi}{dt} = \frac{k}{V}(N_{SA,0} - \xi)(N_{AA,0} - \xi)$$

Since k is constant,

$$\int_0^{\xi} \frac{d\xi}{(N_{SA,0} - \xi)(N_{AA,0} - \xi)} = k\int_0^t dt = kt$$

$$\ln\left\{\frac{N_{SA,0}(N_{AA,0} - \xi)}{N_{AA,0}(N_{SA,0} - \xi)}\right\} = (N_{AA,0} - N_{SA,0})kt$$

Finally, since the volume is constant

$$\ln\left\{\frac{C_{SA,0}(C_{AA,0} - (\xi/V))}{C_{AA,0}(C_{SA,0} - (\xi/V))}\right\} = (C_{AA,0} - C_{SA,0})kt$$

Substituting values of $C_{SA,0}$, $C_{AA,0}$, k, and t into this equation gives $(\xi/V) = 2.23$. However, since the volume is constant, $C_{SA} = (N_{SA}/V) = (N_{SA}^0 - \xi)/V = C_{SA}^0 - (\xi/V)$, so that $C_{SA} = 2.8 - 2.23 = 0.57$ mol/l. As expected, this is the same answer we obtained when the problem was solved using C_{SA} as the composition variable.

Example 4-1 is a rather straightforward illustration of the solution of the design equation for an ideal batch reactor. In particular, the volume was constant so that the problem could have been solved directly in terms of concentration, by *starting* with a constant-volume form of the design equation, e.g., Eqn. (3-8). Moreover, the stoichiometry and the rate equation were relatively simple.

Now we need to consider a number of complications. In the discussion that follows, we will analyze the use of the design equation for two "made up" reactions that incorporate many variations of the batch-reactor design equations. These examples will also permit a more detailed discussion of some of the procedures used in Example 4-1.

4.1.1.2 General Discussion: Constant-Volume Systems

As noted earlier, the assumption of constant volume is valid for a vast majority of liquid-phase reactions. This is because the mass density (mass/volume) of most of the liquids is not very sensitive to either composition or temperature.

The constant-volume assumption is also valid for certain gas-phase reactions. One obvious example is when the dimensions of the vessel are fixed. In this case, the pressure in the reactor may rise or fall as the reaction proceeds. The exact change in pressure will depend on the change in the number of moles on reaction, and on the change in the temperature of the system as the reaction proceeds.

Consider the irreversible homogeneous reaction

$$\mathrm{A} + 2\mathrm{B} \rightarrow 3\mathrm{C} + \mathrm{D} \tag{4-A}$$

that obeys the rate equation

$$-r_{\mathrm{A}} = kC_{\mathrm{A}}^{2}C_{\mathrm{B}}C_{\mathrm{D}}^{-1} \tag{4-5}$$

Writing the design equation, Eqn. (3-5), for reactant A and substituting Eqn. (4-5) gives

$$-\frac{1}{V}\frac{\mathrm{d}N_{\mathrm{A}}}{\mathrm{d}t} = kC_{\mathrm{A}}^{2}C_{\mathrm{B}}C_{\mathrm{D}}^{-1} \tag{4-6}$$

In order to integrate Eqn. (4-6), N_{A} and each of the three concentrations on the right-hand side must be expressed as functions of a single variable that measures the progress of the reaction, i.e., that defines the composition of the reactor contents at any time. Moreover, if the volume V is not constant (independent of time and composition), we will need to express V as a function of either time or composition. First, we will revisit the constant-volume case. Then we will extend our analysis to the case of a variable-volume reactor.

Describing the Progress of a Reaction There are three variables that are commonly used to describe the composition of a reacting system, when a single reaction takes place. These three variables are the concentration of a species (usually the limiting reactant), the fractional conversion of a species (usually the limiting reactant), and the extent of reaction. The concentration of salicylic acid and the extent of reaction were used in Example 4-1. The application of each of these variables to Reaction (4-A) is discussed below.

Concentration

Let's assume that A is the limiting reactant, and write the concentrations of B, C, and D as functions of C_{A}. Let the initial number of moles of each species in the reactor, i.e., the number of moles at $t = 0$, be N_{A0}, N_{B0}, N_{C0}, and N_{D0}. The corresponding initial concentrations are C_{A0}, C_{B0}, C_{C0}, and C_{D0}. If A is the limiting reactant, $N_{\mathrm{A0}} < 2N_{\mathrm{B0}}$ and $C_{\mathrm{A0}} < 2C_{\mathrm{B0}}$. The number of moles of A at any time t is N_{A} and the corresponding concentration is C_{A}.

For a stoichiometrically simple reaction, the Law of Definite Proportions is

$$\Delta N_1/\nu_1 = \Delta N_2/\nu_2 = \Delta N_3/\nu_3 = \cdots = \Delta N_i/\nu_i = \cdots = \xi \tag{1-4}$$

Equation (1-4) can be used to construct a stoichiometric table showing the number of moles of B, C, and D at any time t in terms of the number of moles of A at that time. For example,

$$\frac{N_A - N_{A0}}{\nu_A} = \frac{N_B - N_{B0}}{\nu_B} = \frac{N_A - N_{A0}}{-1} = \frac{N_B - N_{B0}}{-2}$$

$$N_B = N_{B0} - 2(N_{A0} - N_A)$$

In Table 4-4, the number of moles of B, C, and D have been written in terms of N_A, N_{B0}, N_{C0}, and N_{D0}.

Table 4-4 Stoichiometric Table for Reaction (4-A) Using the Whole Batch Reactor as a Control Volume and Using N_A as the Composition Variable

Species	Initial number of moles at $t = 0$	Number of moles at $t = t$
A	N_{A0}	N_A
B	N_{B0}	$N_{B0} - 2(N_{A0} - N_A)(=N_B)$
C	N_{C0}	$N_{C0} + 3(N_{A0} - N_A)(=N_C)$
D	N_{D0}	$N_{D0} + (N_{A0} - N_A)(=N_D)$

In general, it is always desirable to create a stoichiometric table in terms of moles, or molar flow rates for continuous reactors. You may (or may not) be able to easily convert moles to concentrations, as illustrated below.

If (and only if) the volume of the system is constant, then each term in the second and third column of Table 4-4 can be divided by V to obtain Table 4-5.

Table 4-5 Stoichiometric Table for Reaction (4-A) Using the Whole Batch Reactor as a Control Volume and Using C_A as the Composition Variable

Species	Initial concentration at $t = 0$	Concentration at $t = t$
A	C_{A0}	C_A
B	C_{B0}	$C_{B0} - 2(C_{A0} - C_A)(=C_B)$
C	C_{C0}	$C_{C0} + 3(C_{A0} - C_A)(=C_C)$
D	C_{D0}	$C_{D0} + (C_{A0} - C_A)(=C_D)$

EXERCISE 4-1

Explain why the above table is not valid unless the volume of the system is constant.

For a constant-volume system, $dN_A/dt = VdC_A/dt$, so that Eqn. (4-6) can be written as

$$-\frac{dC_A}{dt} = kC_A^2(C_{B0} - 2C_{A0} + 2C_A)(C_{D0} + C_{A0} - C_A)^{-1} \tag{4-7}$$

The concentration of A, C_A, is the only composition variable in this equation. The equation can be integrated (one way or another), provided that the rate constant k is known as a function of either t or C_A, or is constant.

The concentration of one particular species, usually the limiting reactant, is a very convenient variable to use for constant-volume (constant-density) systems. Note that Eqn. (4-7) could have been obtained by substituting the rate equation, Eqn. (4-5), into the design equation for a constant-volume batch reactor, Eqn. (3-8). However, the stoichiometric table still would have been required to relate the various concentrations in the rate equation.

Fractional conversion

The fractional conversion of reactant A is defined as

$$x_A \equiv (N_{A0} - N_A)/N_{A0}$$

Therefore, the number of moles of A at any time t is given by

$$N_A = N_{A0}(1 - x_A)$$

The number of moles of A that have reacted at time t is $N_{A0}x_A$.

The number of moles of the other species, B, C, and D, can be calculated as a function of x_A by using Eqn. (1-4). The results are presented in Table 4-6.

Table 4-6 Stoichiometric Table for Reaction (4-A) Using the Whole Batch Reactor as a Control Volume and Using Fractional Conversion of Reactant A, x_A, as the Composition Variable

Species	Initial number of moles at $t = 0$	Number of moles at $t = t$
A	N_{A0}	$N_{A0}(1 - x_A)$
B	N_{B0}	$N_{B0} - 2N_{A0}x_A$
C	N_{C0}	$N_{C0} + 3N_{A0}x_A$
D	N_{D0}	$N_{D0} + N_{A0}x_A$

If each term in the second and third columns of this table is divided by V, and if V is constant, we obtain the results given in Table 4-7.

Table 4-7 Stoichiometric Table for Reaction (4-A) Using the Whole Batch Reactor as a Control Volume and Using Fractional Conversion of Reactant A, x_A, as the Composition Variable, for a Constant-Volume System

Species	Initial concentration at $t = 0$	Concentration at $t = t$
A	C_{A0}	$C_{A0}(1 - x_A)(=C_A)$
B	C_{B0}	$C_{B0} - 2C_{A0}x_A(=C_B)$
C	C_{C0}	$C_{C0} + 3C_{A0}x_A\ (=C_C)$
D	C_{D0}	$C_{D0} + C_{A0}x_A\ (=C_D)$

Since V is constant and $C_A = C_{A0}(1 - x_A)$,

$$-\frac{1}{V}\frac{dN_A}{dt} = -\frac{dC_A}{dt} = C_{A0}\frac{dx_A}{dt}$$

Using this equation, plus the last column of Table 4-7, permits Eqn. (4-6) to be written as

$$\frac{dx_A}{dt} = kC_{A0}(1 - x_A)^2(\Theta_{BA} - 2x_A)(\Theta_{DA} + x_A)^{-1} \tag{4-8}$$

where

$$\Theta_{BA} = C_{B0}/C_{A0} \tag{4-8a}$$

$$\Theta_{DA} = C_{D0}/C_{A0} \tag{4-8b}$$

The symbol Θ will be used throughout this book to denote the ratio of two initial concentrations (or two concentrations in the feed to a continuous reactor). The first letter of the subscript refers to the species in the numerator, and the second letter refers to the species in the denominator.

Once again, the design equation has been written in terms of a single composition variable, this time x_A. This permits the design equation to be integrated.

Extent of reaction
The extent of reaction ξ may also be used to describe the composition of the system, as illustrated in Table 4-8.

Table 4-8 Stoichiometric Table for Reaction (4-A) Using the Whole Batch Reactor as a Control Volume and Using Extent of Reaction ξ as the Composition Variable

Species	Initial number of moles at $t = 0$	Number of moles at $t = t$
A	N_{A0}	$N_{A0} - \xi$
B	N_{B0}	$N_{B0} - 2\xi$
C	N_{C0}	$N_{C0} + 3\xi$
D	N_{D0}	$N_{D0} + \xi$

The extent of reaction is a convenient way to describe the composition of a reacting system, even when the volume is not constant. Moreover, it probably is the easiest way to describe the composition of a system when more than one reaction is taking place. We shall see this in Chapter 7, which deals with multiple reactions.

Using the results in the last column of this table, Eqn. (4-6) can be written as

$$V\frac{\mathrm{d}\xi}{\mathrm{d}t} = k(N_{A0} - \xi)^2(N_{B0} - 2\xi)(N_{D0} + \xi)^{-1} \tag{4-9}$$

Equation (4-9) is valid for both constant- and variable-volume systems. However, if the volume of the system varies, V must be known as a function of either t or ξ. If V is constant, Eqn. (4-9) can be integrated.

Solving the Design Equation Equations (4-7)–(4-9) are equivalent forms of the design equation for an ideal, constant-volume, batch reactor. The only difference between them is that three different variables, C_A, x_A, and ξ, have been used to describe the composition of the system at any time.

The time and composition variables are separable in Eqns. (4-7)–(4-9). If the rate constant k is constant, or if it can be written as a function of either time or composition, the equations can be integrated directly.

Integration of the design equation will be illustrated by working with Eqn. (4-7). However, you should convince yourself that the same operations can be carried out beginning with Eqns. (4-8) and (4-9), and that the same results are obtained.

Equation (4-7) can be rearranged and symbolically integrated to give

$$\int_{C_{A0}}^{C_A} \frac{(\delta - C_A)\mathrm{d}C_A}{C_A^2(\beta + 2C_A)} = -\int_0^t k\mathrm{d}t \tag{4-10}$$

where

$$\beta = (C_{B0} - 2C_{A0}) \tag{4-10a}$$

$$\delta = (C_{D0} + C_{A0}) \tag{4-10b}$$

The left-hand side of Eqn. (4-10) can be evaluated by using a standard table of integrals.

$$\left[\frac{-\delta(C_{A0} - C_A)}{\beta C_A C_{A0}} + \frac{(2\delta + \beta)}{\beta^2}\ln\frac{C_{A0}(\beta + 2C_A)}{C_A(\beta + 2C_{A0})}\right] = -\int_0^t k\mathrm{d}t \tag{4-11}$$

If the reactor is isothermal, k does not vary with time and Eqn. (4-11) becomes

$$\left[\frac{-\delta(C_{A0}-C_A)}{\beta C_A C_{A0}}+\frac{(2\delta+\beta)}{\beta^2}\ln\frac{C_{A0}(\beta+2C_A)}{C_A(\beta+2C_{A0})}\right]=-kt \tag{4-12}$$

The following problems illustrate the use of Eqns. (4-11) and (4-12).

EXAMPLE 4-2
Use of the Design Equation for an Ideal Batch Reactor; Reaction (4-A) in the Liquid Phase

Consider Reaction (4-A), taking place in the liquid phase in an ideal batch reactor. The rate equation is given by Eqn. (4-5). At 350 K, the value of the rate constant is $k = 1.05$ l/mol-h. The activation energy of the reaction is 100 kJ/mol. The initial concentrations are $C_{A0} = 1.0$ g-mol/l, $C_{B0} = 4.0$ g-mol/l, $C_{C0} = 0$ g-mol/l and $C_{D0} = 1.0$ g-mol/l. The values of β and δ are, from Eqns. (4-10a) and (4-10b),

$$\beta = (C_{B0} - 2C_{A0}) = 2 \text{ g-mol/l}; \quad \delta = (C_{D0} + C_{A0}) = 2 \text{ g-mol/l}$$

We will neglect any reaction that takes place while the initial charge is being added to the reactor, and while the reactor and contents are being heated to reaction temperature. In a real situation, these assumptions would have to be examined carefully.

Part A: How much time is required for the concentration of A to reach 0.10 mol/l if the reactor is run isothermally at 350 K? What is the concentration of *C* at this time?

APPROACH

These two questions are independent. The time t required to reach $C_A = 0.10$ mol/l is the only unknown in Eqn. (4-12), and can be calculated by substituting known values of C_{A0}, C_A, β, δ, and k into this equation. The concentration of C when $C_A = 0.10$ mol/l can be obtained from stoichiometry, without knowing the value of t. The value of C_C can be calculated from the expression for C_C in the stoichiometric table, Table 4-5.

SOLUTION

Substituting $C_A = 0.10$ mol/l into Eqn. (4-10) gives $kt = 6.5$ l/mol. For $k = 1.05$ l/mol-h, $t = 6.1$ h. From Table 4-5, $C_C = C_{C0} + 3(C_{A0} - C_A) = 0 + 3(1.0 - 0.10)$ (mol/l) $= 2.70$ mol/l.

Part B: The reactor will be run isothermally at 350 K. The concentration of A in the final product must be less than 0.20 mol/l, and the molecular weight of *C* is 125. An average of 16 h is required between batches in order to empty and clean the reactor, and prepare for the next batch. How large must the reactor be in order to produce 200,000 kg of *C* annually (with 8000 h per year of operation)?

APPROACH

The reaction time required to reach $C_A = 0.20$ mol/l can be calculated by substituting known values into Eqn. (4-12). The total batch time is 16 h plus the calculated reaction time. The number of batches per year then can be calculated by dividing the time required for one batch by the total available time in a year (8000 h in this case). The concentration of C at the end of the reaction can be calculated from stoichiometry. Finally, the required reactor volume can be calculated from this concentration, the number of batches per year, and the annual production rate.

SOLUTION

Substituting $C_A = 0.20$ mol/l into Eqn. (4-12) gives $kt = 5.65$ l/mol. For $k = 1.05$ l/mol-h, $t = 5.4$ h. The total batch time therefore is $t_{tot} = 16 + 5.4 = 21.4$ h. The number of batches per year is batches/year $= 8000$ (h/year)$/21.4$ (h/batch) $= 374$ (batches/year). Let the working volume of the reactor be V. From the stoichiometric table (Table 4-5), for $C_A = 0.20$ mol/l, $C_C = 2.40$ mol/l. The amount of C produced per batch $= V$ (l) $\times 2.40$ (mol C/l). The annual production of $C = 374$ (batches/year) $\times 2.40V$ (g-mol/batch) $\times 125$ (g/g-mol)$/1000$ (g/kg) $=$ $112\ V$ (kg/year) $= 200,000$ (kg/year). Therefore, $V = 1780$ l.

Part C: The annual production of *C* must be 200,000 kg, and the final concentration of A must be 0.20 mol/l or less. The only reactor available has a working volume of 1500 l. At what temperature does the reactor have to be operated, if it is operated isothermally? The activation energy of the reaction is 100,000 J/mol. Once again, an average of 16 h is required between batches to empty and clean the reactor, and to prepare for the next batch.

APPROACH The concentration of C when $C_A = 0.20$ mol/l can be calculated from stoichiometry. The total allowable batch time (t_{tot}) can be calculated from the required annual production. The allowable *reaction* time t_{react} is $t_{tot} - 16$. A value of kt_{react} can be calculated from Eqn. (4-12) since the reactor is isothermal and the final value of C_A (0.20 mol/l) is known. Since t_{react} is known, the necessary value of k can be calculated. The Arrhenius expression then can be used to calculate the temperature required to produce the calculated value of the rate constant k.

SOLUTION From the stoichiometric table (Table 4-5), $C_C = C_{C0} + 3(C_{A0} - C_A) = 0 + 3(1.0 - 0.20) =$ 2.40 mol/l. The annual production of C (mol/year) = 200, 000 (kg/year) × 1000 (g/kg)/ 125 (g/mol) = 1500 (l) × 2.40 (mol/l) × 8000 (h/year)/t_{tot} (h). Therefore, $t_{tot} = 18.0$ h. The allowable reaction time is $t_{reaction} = 18.0 - 16.0 = 2.0$ h. From Eqn. (4-12), for $C_A = 0.20$ mol/l, $kt = 5.65$ l/mol. Therefore, $k = 5.65$ (l/mol)/2.0 (h) = 2.83 l/mol-h. According to the Arrhenius relationship,

$$\frac{k(T)}{k(350\,\mathrm{K})} = \exp\left\{-\frac{E}{R}\left(\frac{1}{T} - \frac{1}{350}\right)\right\} = \frac{2.83}{1.05} = 2.69$$

Taking the natural log of both sides

$$-\frac{E}{R}\left(\frac{1}{T} - \frac{1}{350}\right) = 0.990$$

For $E = 100{,}000$ J/mol and $R = 8.314$ J/mol-K, $T = 360$ K.

EXERCISE 4-2

What concerns might there be about operating the reactor at a higher temperature?

Part D: What will the concentration of A be if the reactor is operated isothermally at 350 K for 12 h?

APPROACH Since the reactor is isothermal, Eqn. (4-12) still is valid. The value of k is known and the value of t is specified. Therefore, Eqn. (4-12) can be solved for C_A.

SOLUTION The value of kt is 12 (h) × 1.05 (l/mol-h) = 12.6 (l/mol). The solution of Eqn. (4-12) is a "trial-and-error" problem that can be solved in several ways. For example, the value of the left-hand side of Eqn. (4-12) could be calculated for various values of C_A over a range that bracketed the value of −12.6 (l/mol). The desired value of C_A then could be found by interpolation, either graphical or numerical. A simpler approach is to use the GOALSEEK function that is a "Tool" in the Microsoft Excel spreadsheet application. GOALSEEK is a root-finding technique, i.e., it finds the value of a variable (C_A in this case) that makes a specified function of that variable equal to zero. In this case, we want to make the function

$$f(C_A) = \left[\frac{-\delta(C_{A0} - C_A)}{\beta C_A C_{A0}} + \frac{(2\delta + \beta)}{\beta^2}\ln\frac{C_{A0}(\beta + 2C_A)}{C_A(\beta + 2C_{A0})}\right] + kt$$

equal to zero. Using GOALSEEK to find the value of C_A that makes this function equal to zero gives $C_A = 0.059$ (mol/l).

Part E: The initial reactor temperature is 350 K. Heat is added to the reactor so that the temperature increases linearly at a rate of 10 K/h. What is the concentration of A after 5.0 h of operation?

APPROACH Since the reactor is not isothermal, Eqn. (4-12) is not valid. This is because isothermality (k = constant) was assumed when the right-hand side of Eqn. (4-11) was integrated. In order to solve this problem, we will integrate the right-hand side of Eqn. (4-11) numerically, taking into account the variation of k with temperature, and therefore time. We then will use GOALSEEK to determine the value of C_A that makes the left-hand side of Eqn. (4-11) equal to the value of $\int_0^t k\,dt$ that resulted from the numerical integration.

In order to calculate values of the rate constant k at various times, we first must calculate the temperature at various times. This can be done with the relationship between temperature and time that is given in the problem statement, i.e., $T = 350 + 10t$, where T is in Kelvin and t is in hour. The Arrhenius relationship then can be used to calculate the rate constant at each time.

SOLUTION

The numerical integration of kdt from $t = 0$ ($T = 350$ K) to $t = 5$ h ($T = 400$ K) is shown in the spreadsheet labeled as Table 4-9.

Table 4-9 Example 4-2, Part E—Numerical Integration of Right-Hand Side of Eqn. (4-11)

Time (h)	Temperature (K)	Rate constant, k (l/mol h)	Factor	Factor × k
0.0	350	1.050	1	1.050
0.5	355	1.704	4	6.815
1.0	360	2.728	2	5.455
1.5	365	4.311	4	17.242
2.0	370	6.729	2	13.457
2.5	375	10.379	4	41.517
3.0	380	15.829	2	31.658
3.5	385	23.877	4	95.509
4.0	390	35.639	2	71.279
4.5	395	52.659	4	210.637
5.0	400	77.051	1	77.051
				Sum = 571.670

The value of $\int_0^t kdt$ is obtained by numerical integration, using Simpson's One-Third Rule. The "factors" in Table 4-9 are specific to this rule. The value of $\int_0^t kdt$ is given by the sum of the (factor × k) column multiplied by the interval of Δt, i.e., 0.50 h, divided by 3. Thus, $\int_0^t kdt = 572 \times 0.50/3 = 95.3$. Using GOALSEEK to find the value of C_A that satisfies Eqn. (4-11) for $\int_0^t kdt = 95.3$ gives $C_A = 9.8 \times 10^{-3}$ mol/l.

Important note: When doing a numerical integration, the interval (or step size) must be small enough so that the value of the integral does not depend on the value of the interval. For this example, the value of Δt must be small enough so that the calculated value of $\int_0^t kdt$ is independent of Δt.

EXERCISE 4-3

Let Δt be 0.25 h instead of 0.50 h. What is the value of $\int_0^t kdt$ for this smaller interval? Is the value of the integral in the table above independent of step size?

4.1.1.3 General Discussion: Variable-Volume Systems

It is unusual for a chemical engineer to encounter a variable-volume batch reactor. This would require that a gas-phase reaction be carried out in a vessel whose dimensions changed with time. One important example of such a system is the cylinder of an automobile engine. In a fuel-injected car, air is drawn into the cylinder as the piston moves down. The air is compressed as the piston moves up and fuel is injected as the piston approaches the top of its stroke. The spark plug fires when the piston is near top-dead-center. The combustion reaction then drives the piston down, producing work, which is transferred to the crankshaft. Finally, the products of combustion are discharged from the cylinder when the piston rises again to complete the four-stroke cycle. Depending on the compression ratio of the engine, the volume in which the combustion reactions take place changes by about a factor of 10 as the cylinder moves from the top of its stroke, where the spark fires, to the bottom of the stroke, where the combustion reactions are essentially complete.

An example of a variable-volume batch reactor is treated below. This example will introduce the methodology that will be required later, for variable-density flow reactors.

EXAMPLE 4-3

Consider the gas-phase decomposition reaction,

$$\text{A} \rightarrow \text{B} + \text{C} + 2\text{D} \tag{4-B}$$

This reaction takes place in a variable-volume batch reactor *at constant total pressure.*

If the pressure is constant, the volume of the reactor may change because either (1) the number of moles in the reactor changes as the reaction proceeds, and/or (2) the temperature changes as the reaction proceeds. Both of these phenomena occur in the automobile engine. There is an increase in the number of moles in the cylinder as the fuel is burned, and the temperature of the burning gases increases because heat is not removed through the cylinder walls as rapidly as it is "produced" by combustion.

In the present example, we will ignore any temperature change and assume that the reactor is isothermal.

The rate equation for Reaction (4-B) is

$$-r_{\text{A}} = kC_{\text{A}}/(1 + K_{\text{A}}C_{\text{A}}) \tag{4-13}$$

The initial volume of the reactor is V_0 and the reactor initially contains N_{A0} moles of A, N_{I0} moles of an inert gas, and N_{D0} moles of D. There is no B and no C in the reactor initially, so $N_{\text{B0}} = N_{\text{C0}} = 0$. The sum of N_{A0}, N_{I0}, and N_{D0} will be designated N_{T0}. We will assume that the mixture obeys the ideal gas laws over the whole course of the reaction.

Part A: Derive a relationship between the composition of the gas mixture and the time.

APPROACH

It is most convenient to solve variable-volume (variable mass density) problems in terms of either the fractional conversion x_{A} or the extent of reaction ξ. The solution to this problem is developed using the fractional conversion. Be sure that you can solve the problem using the extent of reaction.

First, a stoichiometric table will be constructed to help with the "molecular bookkeeping." Using the stoichiometric table, the variables C_{A}, N_{A}, and V can be written as functions of x_{A}. Finally, the design equation for an ideal batch reactor can be written and integrated to give the desired relationship between x_{A} and t. If x_{A} is known, the complete composition of the gas mixture can be calculated using the relationships in the stoichiometric table.

SOLUTION

Table 4-10 is the stoichiometric table for this problem.

Table 4-10 Stoichiometric Table for Reaction (4-B) Using the Whole Batch Reactor as a Control Volume and Using Fractional Conversion of Reactant A, x_{A}, as the Composition Variable

Species	Initial number of moles at $t = 0$	Number of moles at $t = t$
A	N_{A0}	$N_{\text{A0}}(1 - x_{\text{A}})\ (=N_{\text{A}})$
B	0	$N_{\text{A0}}x_{\text{A}}\ (=N_{\text{B}})$
C	0	$N_{\text{A0}}x_{\text{A}}\ (=N_{\text{C}})$
D	N_{D0}	$N_{\text{D0}} + 2N_{\text{A0}}x_{\text{A}}\ (=N_{\text{D}})$
Inert (I)	N_{I0}	N_{I0}
Total	$N_{\text{A0}} + N_{\text{D0}} + N_{\text{I0}}\ (=N_{\text{T0}})$	$N_{\text{T0}} + 3N_{\text{A0}}x_{\text{A}}\ (=N_{\text{T}})$

Notice that the inert gas is included in the stoichiometric table, and that a "Total" row has also been included. The "Total" row was not needed to solve constant-volume problems. However, this row is essential for variable-volume (variable-density) problems.

Using the above table, we can write the mole fraction of any species and the volume of the system at any time in terms of the fractional conversion. For example, the mole fraction of D, y_D, is given by the moles of D divided by the total number of moles, i.e.,

$$y_D = \frac{N_D}{N_T} = \frac{N_{D0} + 2N_{A0}x_A}{N_{T0} + 3N_{A0}x_A} = \frac{y_{D0} + 2y_{A0}x_A}{1 + 3y_{A0}x_A}$$

where y_{A0}, the initial mole fraction of A, $= N_{A0}/N_{T0}$, and y_{D0}, the initial mole fraction of D, $= N_{D0}/N_{T0}$. Similarly,

$$y_A = \frac{N_A}{N_T} = \frac{N_{A0}(1 - x_A)}{N_{T0} + 3N_{A0}x_A} = \frac{y_{A0}(1 - x_A)}{1 + 3y_{A0}x_A}$$

For an ideal gas,[2]

$$C_A = \frac{Py_A}{RT} = \left(\frac{Py_{A0}}{RT}\right)\frac{(1 - x_A)}{(1 + 3y_{A0}x_A)} = C_{A0}\left(\frac{1 - x_A}{1 + 3y_{A0}x_A}\right)$$

For an ideal gas at constant temperature and pressure, $V/V_0 = N_T/N_{T0}$. Therefore,

$$V = V_0\frac{(N_{T0} + 3N_{A0}x_A)}{N_{T0}} = V_0(1 + 3y_{A0}x_A) \tag{4-14}$$

The design equation for an ideal batch reactor is

$$\frac{1}{V}\frac{dN_i}{dt} = r_i \tag{3-5}$$

Writing this equation for reactant A and substituting Eqn. (4-13) gives

$$-\frac{1}{V}\frac{dN_A}{dt} = \frac{kC_A}{(1 + K_AC_A)}$$

The volume V is expressed in terms of x_A by Eqn. (4-14). The expression for N_A in the stoichiometric table can be differentiated to give $dN_A = -N_{A0}dx_A$. The concentration of A for an ideal gas is $C_A = Py_A/RT$, where P is the total pressure. These substitutions transform the above equation into

$$\frac{N_{T0}}{V_0}\frac{dx_A}{dt} = \frac{k\left(\frac{P}{RT}\right)(1 - x_A)}{1 + K_A\left(\frac{P}{RT}\right)\left(\frac{y_{A0}(1 - x_A)}{1 + 3y_{A0}x_A}\right)}$$

Rearranging and using the relationship $N_{T0} = PV_0/RT$ produces

$$\frac{dx_A}{(1 - x_A)} + \left(\frac{y_{A0}K_AP}{RT}\right)\frac{dx_A}{(1 + 3y_{A0}x_A)} = kdt$$

Integrating this expression from $x_A = 0$, $t = 0$ to $x_A = x_A$, $t = t$ for an isothermal reactor gives

$$-\ln(1 - x_A) + \left(\frac{K_AP}{3RT}\right)\ln(1 + 3y_{A0}x_A) = kt \tag{4-15}$$

[2] The ideal gas laws will be used frequently throughout this text, primarily for reasons of conceptual and algebraic simplicity. The assumption of ideal gas behavior permits more attention to be focused on reaction engineering concepts, at the expense of actual gas behavior. The ideal gas equation, $PV = nRT$ is just one of many equations of state. If the ideal gas equation is not valid, any other (valid) equation of state could be used to express a concentration C_A. For example, using the compressibility factor equation of state, $C_A = p_A/ZRT$. Here, Z is the compressibility factor. If the temperature, pressure, and composition of a mixture do not vary significantly as a reaction takes place, the value of Z will not change significantly. However, if Z does change, its variation must be taken into account, making the solution of a problem more complicated than if the ideal gas equation of state were valid.

This is the relationship between composition and time that was required by the problem statement.

Part B: Suppose that $y_{A0} = 1.0$, $(K_A P/RT) = 1.5$, and $k = 0.010\ \text{min}^{-1}$. How long will it take for the conversion of A to reach 50%?

APPROACH

The time required to reach $x_A = 0.50$ can be calculated by substituting known values of x_A, y_{A0}, $(K_A P/RT)$, and k into Eqn. (4-15).

SOLUTION

Substituting $x_A = 0.50$, $y_{A0} = 1.0$, and $(K_A P/RT) = 1.5$ and $k = 0.010\ \text{min}^{-1}$ into Eqn. (4-15) gives $t = 115$ min.

EXERCISE 4-4

Suppose that the reaction was $A \rightarrow B$ instead of Reaction (4-B). All of the other parameters of the problem remain the same. Will the conversion of A be larger, the same, or smaller at a given time for $A \rightarrow B$ than it is for Reaction (4-B)? First, use qualitative, physical reasoning to answer this question. Explain in words how you arrived at your answer. Then, derive the equivalent of Eqn. (4-15) for $A \rightarrow B$. Take the values for $(K_A P/RT)$, y_{A0}, and kt that are given above. Calculate x_A from your rederived equation and compare it with $x_A = 0.50$. If your qualitative analysis was not correct, identify the flaw in your reasoning and reformulate your answer.

4.1.2 Continuous Reactors

As noted in Chapter 3, continuous reactors are used for the production of most large-volume chemicals, fuels, and polymers.

For many reactions, the number of moles of products is different from the number of moles of reactants. Polymerizations are an extreme example of reactions with a large change in the number of moles. A single polymer molecule may contain as many as 10,000 monomer molecules. The stoichiometry of a chain-growth polymerization reaction can be represented as

$$n\text{M} \rightarrow -(\text{M})_{\overline{n}}-$$

where M is the monomer that is being polymerized, and "n" is the number of monomer units in the polymer molecule. For example, M might represent ethylene monomer (C_2H_4), with a molecular weight of 28. If the number-average molecular weight of the polyethylene being produced were 200,000, the average value of "n" would be approximately 7000. A commercial reactor for producing polyolefins such as polyethylene is shown in Figure 4-2.

Sizing and analyzing continuous reactors can be particularly challenging when there is a change in the number of moles on reaction. For gas-phase reactions, a change in the number of moles leads to a change in the mass density of the system. In other words, the volume occupied by a given mass of gas will change as the reaction proceeds if there is a net change in the number of moles as the reaction proceeds. The treatment of continuous reactors where the mass density varies is analogous to the treatment of variable-volume batch reactors.

We will begin the discussion of continuous reactors with the ideal CSTR, first with a constant-density example and then with a variable-density example. The ideal PFR then will be treated, for the constant-density and then the variable-density case. To point out the differences between constant- and variable-density systems, and the differences between how the different reactors are treated, all of our analysis will be based on Reaction (4-B) and the rate equation given by Eqn. (4-13).

Figure 4-2 A Unipol® reactor for the manufacture of polyolefins such as polyethylene and polypropylene. This fluidized-bed reactor operates continuously at steady state for long periods of time. The feed is an olefin or a mixture of olefins, an inert hydrocarbon, and hydrogen. A solid catalyst is used to increase the rate of the polymerization reaction. The catalyst is fed continuously or semi-continuously, and is removed continuously along with the polymer. Unreacted olefin, inert hydrocarbon, and hydrogen are separated from the polymer and recycled. (Photo, The Lamp, Spring 2002.)

4.1.2.1 Continuous Stirred-Tank Reactors (CSTRs)

Constant-Density Systems

EXAMPLE 4-4
Liquid-Phase Decomposition of A in a CSTR

Suppose that the irreversible decomposition reaction

$$A \rightarrow B + C + 2D \tag{4-B}$$

is taking place in the *liquid phase* in an ideal CSTR operating at steady state. Because the reaction occurs in the liquid phase, constant mass density can be assumed. At steady state, the mass flow rate of the feed to the reactor and the mass flow rate of the effluent from the reactor must be the same. Therefore, if the mass density is constant, the inlet volumetric flow rate υ_0 must be the same as the outlet volumetric flow rate υ. This is true for *any* flow reactor operating at steady state.

Reaction (4-B) obeys the rate equation

$$-r_A = kC_A/(1 + K_A C_A) \tag{4-13}$$

The value of k is 8.6 h^{-1} and the value of K_A is 0.50 l/mol A. The feed to the CSTR is a mixture of A and an inert solvent I. The concentration of A in the feed (C_{A0}) is 0.75 mol/l. There is no B, C, or D in the feed. The volumetric flow rate of the feed (υ_0) is 1000 l/h.

What reactor volume is required for the net production rate of D to be at least 1200 mol/h? (The net production rate is the rate at which the product leaves the reactor minus the rate at which it enters the reactor.)

APPROACH First, the fractional conversion of A in the effluent from the CSTR will be calculated from the specified production rate of D and the stoichiometry of the reaction. A stoichiometric table will be constructed to facilitate this calculation. Next, the design equation for the CSTR will be solved to determine the required volume, using the calculated fractional conversion.

SOLUTION First, set up a stoichiometric table (Table 4-11). Rather than working with moles *per se*, as in the case of batch reactors, stoichiometric tables for steady-state flow reactors should be constructed in terms of *molar flow rates*. For a CSTR, the second column contains the *inlet* molar flow rates and the third column contains the molar flow rates in the reactor *effluent*.

Table 4-11 Stoichiometric Table for Reaction (4-B) in an Ideal CSTR Using Fractional Conversion of Reactant A, x_A, as the Composition Variable

Species	Molar feed rate (mol/time)	Molar effluent rate (mol/time)
A	F_{A0}	$F_{A0}(1-x_A)\ (=F_A)$
B	0	$F_{A0}x_A\ (=F_B)$
C	0	$F_{A0}x_A\ (=F_C)$
D	0	$2F_{A0}x_A\ (=F_D)$
Inert (I)	F_{I0}	F_{I0}

The fractional conversion is given by

$$x_A = (F_{A0} - F_A)/F_{A0}$$

The value of F_{A0} is $\upsilon_0 C_{A0} = 1000\ (\text{l/h}) \times 0.75\ (\text{mol A/l}) = 750\ \text{mol A/h}$. The net production rate of D $(F_D - F_{D0})$ is 1200 mol/h. Since $F_{D0} = 0$, $F_D = 1200\ \text{mol/h}$. From the stoichiometric table, $F_D = 2F_{A0}x_A$. Therefore,

$$x_A = \frac{F_D}{2F_{A0}} = \frac{1200}{2 \times 750} = 0.80$$

From Chapter 3, the design equation for an ideal CSTR with a homogeneous reaction taking place is

$$\frac{V}{F_{A0}} = \frac{x_A}{-r_A} \tag{3-17}$$

Substituting the expression for $-r_A$ (Eqn. (4-13)) gives

$$\frac{V}{F_{A0}} = \frac{x_A(1 + K_A C_A)}{kC_A} \tag{4-16}$$

At this point, x_A must be expressed as a function of C_A, or C_A must be expressed as a function of x_A. Alternatively, both x_A and C_A might be written as functions of the extent of reaction ξ. All the three alternatives work equally well for this problem. Here, let's use x_A to describe the system composition. Since the mass density is constant $C_A = F_A/\upsilon$ and $C_{A0} = F_{A0}/\upsilon$. From the stoichiometric table, $F_A = F_{A0}(1 - x_A)$. Dividing both sides of this equation by υ,

$$\frac{F_A}{\upsilon} = C_A = \frac{F_{A0}}{\upsilon}(1 - x_A) = C_{A0}(1 - x_A)$$

Therefore, Eqn. (4-16) can be written as

$$\frac{V}{F_{A0}} = \frac{x_A[1 + K_A C_{A0}(1 - x_A)]}{kC_{A0}(1 - x_A)} \tag{4-17}$$

Substituting $F_{A0} = 750\ \text{mol/h}$, $x_A = 0.80$, $K_A = 0.50$, $C_{A0} = 0.75\ \text{mol/l}$, and $k = 8.6\ \text{h}^{-1}$ into the above equation gives $V = 500$ l.

When the mass density is constant, the volumetric flow rate out of the CSTR (υ) must be the same as the volumetric flow rate into the CSTR (υ_0), at steady state. Dividing each term in the second and third columns of Table 4-11 by υ (or υ_0) gives the results shown in Table 4-12.

Table 4-12 Stoichiometric Table for Reaction (4-B) in an Ideal CSTR Using Fractional Conversion of Reactant A, x_A, as the Composition Variable (Constant Density)

Species	Feed concentration (mol/volume)	Effluent concentration (mol/volume)
A	C_{A0}	$C_A = C_{A0}(1 - x_A)$
B	0	$C_B = C_{A0}x_A$
C	0	$C_C = C_{A0}x_A$
D	0	$C_D = 2C_{A0}x_A$
Inert (I)	C_{I0}	C_{I0}

As noted in the discussion of ideal batch reactors, since the density is constant, we could have begun this problem by constructing a stoichiometric table based on concentration.

Variable-Density (Variable-Volume) Systems

EXAMPLE 4-5 ***Gas-Phase Decomposition of A in a CSTR***

The same reaction

$$A \rightarrow B + C + 2D \tag{4-B}$$

is taking place *in the gas phase* in an ideal CSTR at constant total pressure. The rate equation is

$$-r_A = kC_A/(1 + K_AC_A) \tag{4-13}$$

The volume of the CSTR is V. The molar flow rates to the reactor are F_{A0}, F_{I0}, and F_{D0}. There is no B and no C in the reactor feed, i.e., $F_{B0} = F_{C0} = 0$. The sum of F_{A0}, F_{I0}, and F_{D0} will be designated F_{T0}. The gas mixture obeys the ideal gas laws. The feed is at an absolute temperature T_0 and total pressure, P_0, and the inlet concentrations of A, I, and D, C_{A0}, C_{I0}, and C_{D0}, are known at these feed conditions. The reactor operates at a known temperature T and a known pressure P.

Part A: Derive a relationship between the composition of the gas mixture leaving the reactor and the parameter (V/F_{A0}), for an ideal CSTR.

APPROACH The solution to this problem will be developed using the fractional conversion x_A as the composition variable. Be sure that you can solve the problem using the extent of reaction ξ. First, a stoichiometric table (Table 4-13) will be constructed to express C_A as a function of x_A. Then the design equation will be solved to obtain the required relationship between (V/F_{A0}) and the composition of the stream leaving the reactor.

SOLUTION Table 4-13 is the stoichiometric table for this problem.

Table 4-13 Stoichiometric Table for Reaction (4-B) in an Ideal CSTR Using Fractional Conversion of Reactant A, x_A, as the Composition Variable (Variable Density)

Species	Molar feed rate (mol/time)	Molar effluent rate (mol/time)
A	F_{A0}	$F_{A0}(1 - x_A)$
B	0	$F_{A0}x_A$
C	0	$F_{A0}x_A$
D	F_{D0}	$F_{D0} + 2F_{A0}x_A$
Inert (I)	F_{I0}	F_{I0}
Total	$F_{A0} + F_{D0} + F_{I0}\ (=F_{T0})$	$F_{T0} + 3F_{A0}x_A$

Note the "Total" row in the stoichiometric table. Using this table, the mole fraction of any species in the effluent can be written in terms of the fractional conversion. For example, the mole fraction of A in the effluent y_A is the molar flow rate of A in the effluent divided by the *total* molar effluent rate, i.e.,

$$y_A = \frac{F_{A0}(1-x_A)}{F_{T0}+3F_{A0}x_A} = \frac{y_{A0}(1-x_A)}{1+3y_{A0}x_A}$$

where y_{A0} is the mole fraction of A in the feed F_{A0}/F_{T0}.

As we learned in Chapter 3, the composition of the effluent from an ideal CSTR is the same as the composition in the reactor. Therefore, the above expression for y_A also gives the mole fraction of A *in the reactor* For an ideal gas, $C_A = Py_A/RT$ so that

$$C_A = \left(\frac{P}{RT}\right)\frac{y_{A0}(1-x_A)}{(1+3y_{A0}x_A)}$$

The quantity (Py_{A0}/RT) is the concentration of A in the feed to the CSTR, *if the feed is at the temperature and pressure of the reactor.* Therefore,

$$C_A = C_{A0}\frac{(1-x_A)}{(1+3y_{A0}x_A)} \tag{4-18}$$

where $C_{A0} = (Py_{A0}/RT)$ and P and T are the pressure and temperature in the CSTR.

Equation (4-18) shows that *two* phenomena contribute to the difference between C_{A0} and C_A: (1) consumption of A by the reaction, and (2) dilution caused by the increase in total moles as the reaction proceeds. For Reaction (4-B), *at constant pressure*, the mass density decreases as the reaction takes place, i.e., the volume occupied by a given mass increases as the reaction proceeds, because the number of moles increases as the reaction proceeds. This dilutes the unreacted A, lowering its concentration, and lowering the reaction rate.

The design equation for an ideal CSTR with a homogeneous reaction taking place is

$$\frac{V}{F_{A0}} = \frac{x_A}{-r_A} \tag{3-17}$$

We now substitute Eqn. (4-18) into the rate equation, Eqn. (4-13), and substitute the resulting expression into Eqn. (3-17) to obtain

$$\frac{V}{F_{A0}} = \frac{x_A}{k}\left[\frac{RT}{y_{A0}P}\left\{\frac{1+3y_{A0}x_A}{1-x_A}\right\} + K_A\right] \tag{4-19}$$

This is the relationship required by the problem statement. Compare it with Eqn. (4-17) for the constant-density case.

Now suppose that the feed concentration is specified at conditions different from those at which the reactor operates. For example, the feed concentration might be given at a temperature T_0 and a pressure P_0 that are not the same as T and P. To avoid confusion, let's designate the feed concentration at these conditions as $C_{A0}(T_0, P_0)$, whereas the feed concentration at the temperature and pressure of the CSTR is C_{A0}. If the feed mixture obeys the ideal gas laws at both T, P and T_0, P_0, then

$$C_{A0} = C_{A0}(T_0, P_0)\frac{P}{P_0}\frac{T_0}{T} \tag{4-20}$$

Substituting this into Eqn. (4-19) gives

$$\frac{V}{F_{A0}} = \frac{x_A}{k}\left[\left(\frac{P_0T}{PT_0}\right)\left\{\frac{1+3y_{A0}x_A}{C_{A0}(T_0,P_0)(1-x_A)}\right\} + K_A\right]$$

This equation can be used in place of Eqn. (4-19) if the concentrations in the feed are specified at conditions other than the temperature and pressure of the CSTR.

Part B: Reaction (4-B) is taking place in the gas phase in an ideal CSTR at steady state. The reactor operates at 400 K and 1 atm total pressure. The feed enters the reactor at 300 K and 1 atm total pressure. The volume of the reactor is 1000 l, and the molar feed rate of A (F_{A0}) is 500 mol A/h. The mole fraction of A in the feed stream is 0.50. The rate

of reaction is given by Eqn (4-13), with $k = 45\,h^{-1}$ and $K_A = 50$ l/mol A at 400 K. What is the fractional conversion of A in the stream leaving the CSTR?

APPROACH

The value of C_{A0} can be calculated from the specified values of y_{A0}, T, and P. Then Eqn. (4-19) can be solved for x_A.

SOLUTION

$C_{A0} = (y_{A0}P/RT) = 0.50 \times 1.0\,(\text{atm})/0.0821\,(\text{atm-l/mol-K}) \times 400\,(\text{K}) = 0.0152\,(\text{mol/l})$. Equation (4-19) can be rearranged to

$$x_A^2\left(\frac{3y_{A0}}{C_{A0}} - K_A\right) + x_A\left(K_A + \frac{1}{C_{A0}} + \frac{Vk}{F_{A0}}\right) - \frac{Vk}{F_{A0}} = 0$$

$$x_A^2(3\alpha y_{A0} - K_A) + x_A(\alpha + \beta + K_A) - \beta = 0$$

where

$$\alpha = RT/y_{A0}P = 65.7\,\text{l/mol}$$

$$\beta = kV/F_{A0} = 90\,\text{l/mol}$$

Using the quadratic formula

$$x_A = \frac{-(\alpha + \beta + K_A) \pm \sqrt{(\alpha + \beta + K_A)^2 + 4\beta(3\alpha y_{A0} - K_A)}}{2(3\alpha y_{A0} - K_A)}$$

Substituting the values of α, β, K_A, and y_{A0} gives $x_A = 0.40$. We choose the solution with the "+" sign in front of the square root since x_A must be positive.

4.1.2.2 Plug-Flow Reactors

Constant-Density (Constant-Volume) Systems

EXAMPLE 4-6
Liquid-Phase Decomposition of A in a PFR

Suppose that Reaction (4-B)

$$A \rightarrow B + C + 2D \tag{4-B}$$

is taking place in the *liquid phase* in an ideal, isothermal plug-flow reactor (PFR), operating at steady state. The reaction obeys the rate equation

$$-r_A = kC_A/(1 + K_AC_A) \tag{4-13}$$

The value of k is $8.6\,h^{-1}$ and the value of K_A is 0.50 l/mol A at the operating temperature of the reactor. The feed to the PFR is a mixture of A and an inert solvent I. The concentration of A in the feed (C_{A0}) is 0.75 mol/l. There is no B, C, or D in the feed. The volumetric flow rate of the feed (υ_0) is 1000 l/h.

What reactor volume is required for the net production rate of D to be at least 1200 mol/l ? (*Note*: This problem is exactly the same as Example 4-4, except that the reactor is a PFR instead of a CSTR.)

APPROACH

First, the fractional conversion of A in the effluent from the PFR will be calculated from stoichiometry. Then, this conversion will be used to solve the PFR design equation to obtain the required volume. Once again, a stoichiometric table will be used to relate C_A to x_A.

SOLUTION

Since the inlet flow rates and concentrations are the same as in Example 4-4, and the required production rate of D is also the same, the fractional conversion of A in the stream leaving the PFR must also be the same, i.e., $x_A = 0.80$.

From Chapter 3, the design equation for an ideal PFR with a homogeneous reaction taking place is, in integrated form,

$$\frac{V}{F_{A0}} = \int_0^{x_A} \frac{dx_A}{-r_A} \tag{3-33}$$

Substituting the rate equation gives

$$\frac{V}{F_{A0}} = \int_0^{x_{A,e}} \frac{[1 + K_A C_A] dx_A}{k C_A} \tag{4-21}$$

The symbol $x_{A,e}$ designates the conversion of A in the effluent from the reactor, i.e., $x_{A,e} = 0.80$.

As in Example 4-4, C_A must be written as a function of x_A, with the aid of a stoichiometric table (Table 4-14). For a PFR, the last column in the stoichiometric table will contain the molar flow rates *at an arbitrary position in the direction of flow*. The resulting entries will be valid at every point along the direction of flow, including the outlet.

Table 4-14 Stoichiometric Table for Reaction (4-B) in an Ideal PFR Using Fractional Conversion of Reactant A, x_A, as the Composition Variable

Species	Molar feed rate (mol/time)	Molar flow rate (any position in direction of flow) (mol/time)
A	F_{A0}	$F_{A0}(1 - x_A)\ (=F_A)$
B	0	$F_{A0}x_A\ (=F_B)$
C	0	$F_{A0}x_A\ (=F_C)$
D	0	$2F_{A0}x_A\ (=F_D)$
Inert (I)	F_{I0}	F_{I0}

Because the reaction occurs in the liquid phase, the mass density is constant. Therefore, the inlet volumetric flow rate υ_0 is the same as the volumetric flow rate, υ at every point along the direction of flow. Dividing each term in the second and third columns of the above table by υ gives the results in Table 4-15.

Table 4-15 Stoichiometric Table for Reaction (4-B) in an Ideal PFR Using Fractional Conversion of Reactant A, x_A, as the Composition Variable (Constant Density)

Species	Feed concentration (mol/volume)	Concentration (any position in direction of flow) (mol/volume)
A	C_{A0}	$C_A = C_{A0}(1 - x_A)$
B	0	$C_B = C_{A0}x_A$
C	0	$C_C = C_{A0}x_A$
D	0	$C_D = 2C_{A0}x_A$
Inert (I)	C_{I0}	C_{I0}

Since $C_A = C_{A0}(1 - x_A)$, Eqn. (4-21) can be written as

$$\frac{V}{F_{A0}} = \int_0^{x_{A,e}} \frac{[1 + K_A C_{A0}(1 - x_A)] dx_A}{k C_{A0}(1 - x_A)}$$

This equation can be rearranged to

$$\frac{k C_{A0} V}{F_{A0}} = k\tau_0 = \int_0^{x_{A,e}} \frac{dx_A}{1 - x_A} + K_A C_{A0} \int_0^{x_{A,e}} dx_A$$

The parameters k and K_A were taken outside the integral since the reactor is isothermal. This operation would not have been legitimate if the temperature were not the same at every point in the reactor.

Integrating,

$$k\tau = -\ln(1 - x_{A,e}) + K_A C_{A0} x_{A,e} \tag{4-22}$$

For $C_{A0} = 0.75$ mol/l, $k = 8.6\,h^{-1}$, and $K_A = 0.50$ l/mol, the solution to Eqn. (4-22) is $\tau = 0.22$ h. Since $\tau = V/\upsilon$, and $\upsilon = 1000$ l/h, $V = 220$ l. This value is significantly smaller than the 500 l that were required to reach the same conversion with a CSTR.

EXERCISE 4-5

Explain in physical terms why the conversion is significantly lower in a CSTR ($x_A = 0.80$) than it is in a PFR ($x_A = 0.98$), even though the same reaction with the same kinetics is taking place, and the two reactors have the same volume, the same inlet concentration, and the same flow rate.

Variable-Density (Variable-Volume) Systems

EXAMPLE 4-7
Gas-Phase Decomposition of A in a PFR

The same reaction

$$A \rightarrow B + C + 2D \tag{4-B}$$

is taking place *in the gas phase* in an ideal, isothermal PFR at constant total pressure. The reactor temperature and pressure are T and P, respectively. At these conditions, the gas mixture obeys the ideal gas laws. Once again, the rate equation is

$$-r_A = kC_A/(1 + K_A C_A) \tag{4-13}$$

The volume of the PFR is V. The molar flow rates into the reactor are F_{A0}, F_{I0}, and F_{D0}. There is no B and no C in the reactor feed, so that $F_{B0} = F_{C0} = 0$. The sum of F_{A0}, F_{I0}, and F_{D0} is designated as F_{T0}. The concentrations of A, D, and I in the feed are C_{A0}, C_{D0}, and C_{I0}, respectively.

Part A: Derive a relationship between the composition of the gas mixture leaving the reactor and (V/F_{A0}), for an ideal, isothermal PFR.

APPROACH

The solution to this problem will be developed using the fractional conversion x_A as the composition variable. Be sure that you can solve the problem using the extent of reaction ξ.

First, a stoichiometric table will be constructed in order to express C_A as a function of x_A. Then the design equation will be solved to obtain the required relationship between (V/F_{A0}) and the composition of the stream leaving the reactor.

SOLUTION

The stoichiometric table for this problem is Table 4-16.

Table 4-16 Stoichiometric Table for Reaction (4-B) in an Ideal PFR Using Fractional Conversion of Reactant A, x_A, as the Composition Variable (Variable Density)

Species	Molar feed rate (mol/time)	Molar flow rate (any position in direction of flow) (mol/time)
A	F_{A0}	$F_{A0}(1 - x_A)$
B	0	$F_{A0}x_A$
C	0	$F_{A0}x_A$
D	F_{D0}	$F_{D0} + 2F_{A0}x_A$
Inert (I)	F_{I0}	F_{I0}
Total	$F_{A0} + F_{D0} + F_{I0}\ (=F_{T0})$	$F_{T0} + 3F_{A0}x_A\ (=F_T)$

The mole fraction of A, y_A, at any position in the reactor is given by the moles of A divided by the total number of moles, i.e.,

$$y_A = \frac{F_{A0}(1-x_A)}{F_{T0}+3F_{A0}x_A} = \frac{y_{A0}(1-x_A)}{1+3y_{A0}x_A}$$

where y_{A0} is the mole fraction of A in the feed F_{A0}/F_{T0}.

For an ideal gas, $C_A = Py_A/RT$ and $C_{A0} = P_0 y_{A0}/RT_0$. When we analyzed the ideal CSTR, we introduced the variable $C_{A0}(T_0, P_0)$ to account for the possibility that the feed entering the CSTR might be at a different temperature and pressure than the contents of the reactor. However, the feed that enters an isothermal PFR must be at the temperature of the reactor, and almost always must be at the pressure of the reactor. In a PFR, there is no mixing in the axial direction that would instantaneously bring the feed to the operating temperature of the reactor. Therefore, for an isothermal PFR that operates at constant pressure,

$$C_A = C_{A0}\frac{y_A}{y_{A0}}$$

For the present problem,

$$C_A = C_{A0}\left\{\frac{1-x_A}{1+3y_{A0}x_A}\right\} \tag{4-23}$$

The design equation for an ideal PFR with a homogeneous reaction taking place, in integrated form, is

$$\frac{V}{F_{A0}} = \int_0^{x_{A,e}} \frac{dx_A}{-r_A} \tag{3-33}$$

Substituting Eqns. (4-13) and (4-23) into the design equation gives

$$\frac{V}{F_{A0}} = \int_0^{x_{A,e}} \frac{[(1+3y_{A0}x_A)+K_A C_{A0}(1-x_A)]dx_A}{kC_{A0}(1-x_A)}$$

Performing the indicated integration and substituting $C_{A0} = Py_{A0}/RT$ gives

$$k\tau_0 = -(1+3y_{A0})\ln(1-x_{A,e}) + y_{A0}\left(\frac{K_A P}{RT}-3\right)x_{A,e} \tag{4-24}$$

Note the difference between this equation and Eqn. (4-22) for the constant-density case.

Part B: The PFR operates at 400 K and 1 atm total pressure. The volume of the reactor is 1000 l, and the molar feed rate of A (F_{A0}) is 500 mol A/h. The mole fraction of A in the feed stream is 0.50. At 400 K, $k = 45\ \text{h}^{-1}$ and $K_A = 50$ l/mol A. What is the fractional conversion of A in the stream leaving the PFR?

APPROACH All of the parameters in Eqn. (4-24) are known, except for $x_{A,e}$. The value of $x_{A,e}$ can be obtained by solving this equation.

SOLUTION The value of τ_0 is

$$\tau_0 = \frac{VC_{A0}}{F_{A0}} = \frac{V}{F_{A0}}\left(\frac{y_{A0}P}{RT}\right) = \frac{1000\,(1)\times 0.50\times 1\,(\text{atm})}{500\,(\text{mol/h})\times 0.0821\,(\text{l-atm/mol-K})\times 400\,(\text{K})}$$

$$\tau_0 = 0.0305\ \text{h}$$

Substituting the values of τ_0, k, y_{A0}, K_A, P, R, and T into Eqn. (4-24) and solving with GOALSEEK gives $x_{A,e} = 0.50$.

EXERCISE 4-6

Review the results of Examples 4-5, Part B and 4-7, Part B. Explain in physical terms why the conversion is significantly lower in the CSTR ($x_A = 0.40$) than it is in the PFR ($x_A = 0.50$).

EXERCISE 4-7

The problem statement specified that the total pressure in the reactor is constant. In order for this to be true, the pressure drop in the direction of flow, e.g., along the length of a tubular PFR, must be very small. Go back over the solution to this problem and find where the assumption of constant pressure was used. Describe how you would solve this problem if the pressure drop through the reactor could *not* be neglected.

4.1.2.3 Graphical Solution of the CSTR Design Equation

EXAMPLE 4-8
Cell Growth in a CSTR

The Monod equation, Eqn. (4-25), frequently provides a reasonable description of the growth rate of cells, such as yeast cells or the activated sludge that is formed during wastewater treatment.

$$r_C(\text{mass of cells/volume-time}) = \frac{kC_A C_C}{C_A + K_S} \tag{4-25}$$

In this equation, C_A is the *mass*[3] concentration of the growth-limiting reactant (mass A/volume), C_C is the *mass* concentration of cells (mass C/volume), and k and K_S are constants. The stoichiometry of the reaction is such that $Y(C/A)$ is the mass of cells produced per mass of A consumed. Therefore,

$$(-r_A) = r_C/Y(C/A)$$

Since cells are produced by the reaction, the Monod equation predicts *autocatalytic* behavior, i.e., the higher the concentration of product C, the faster the reaction goes. The Monod equation also shows that the rate of cell production is zero when either $C_A = 0$ or $C_C = 0$.

Part A: An ideal CSTR with a volume of V is operating at steady state. The mass concentrations of A and C in the feed are C_{A0} and C_{C0}, respectively. The volumetric flow rate of the feed to the reactor is v (volume/time). What is the mass concentration of A in the reactor effluent, for the following values: $C_{A0} = 10$ g/l, $C_{C0} = 0$ g/l, $V = 1.0$ liter, $v = 0.5$ l/h, $Y(C/A) = 0.50$, $k = 1.0\,\text{h}^{-1}$, $K_S = 0.20$ g/l. What is the mass concentration of cells in the reactor effluent for this condition? The reaction takes place in the liquid phase.

---

[3] Up to this point in the text, all concentrations have been molar concentrations, with units of *moles*/volume. In some reactions, one or more of the reactants and/or products may be so complex structurally that they cannot be characterized via a simple molecular weight. In such cases, molar concentrations are impossible to calculate.

Coal is a good example. Reactions of coal are very important. Coal can be combusted to generate heat, as in electric power plants. Coal also can be "liquefied" by reaction with hydrogen to form liquid fuels, and it can be "gasified" by reaction with steam and oxygen to form various gases and light liquids. Although the ratios of the various elements (C, H, O, N, S, etc.) in coal can be measured, coal itself is a complex mixture of many different molecules. "Coal" cannot be characterized by a single molecular weight, and we cannot calculate the number of moles of coal that disappear when a given mass of coal reacts. Crude oil and heavy petroleum fractions such as "gas oil" are additional examples.

"Biomass," e.g., the cells that are formed in this example, is another example of a material whose structure frequently is so complex that the concepts of moles and molecular weight cannot be employed. The atomic ratios of the elements in a specific kind of cell often are constant. However, live cells are constantly growing and dividing, so the molecular weight is not constant from cell to cell.

Nevertheless, materials such as "coal" and "biomass" are important in a practical sense, and scientists and engineers must deal with reactions involving these materials. In these cases, *mass* concentrations are used instead of molar concentrations. The use of mass concentrations in rate equations is less fundamental, perhaps, than the use of molar concentrations. This is because theories such as collision theory and transition-state theory teach that reaction rates depend on *molar* concentrations. Nevertheless, the use of mass concentrations in problems involving complex materials has proven to be a practical approach is solving such problems.

APPROACH

The design equation for an ideal CSTR with a liquid-phase rection taking place will be written for the growth-limiting reactant A. The rate equation for the disappearance of A $(-r_A)$ contains the concentration of cells C_C. This concentration can be written in terms of C_A by using the definition of $Y(C/A)$. The CSTR design equation then can be solved for C_A, and C_C can be calculated from the definition of $Y(C/A)$.

SOLUTION

The design equation for the CSTR, with the rate equation (Eqn. (4-25)) substituted, is

$$\frac{V}{F_{A0}} = \frac{x_A}{-r_A} = \frac{V}{\upsilon C_{A0}} = \frac{(C_{A0} - C_A)/C_{A0}}{kC_A C_C/[(C_A + K_S)Y(C/A)]}$$

Since C_{A0} is the mass concentration of A, F_{A0} is the *mass* flow rate of A (mass/time).

From the definition of the yield, $Y(C/A)$,

$$Y(C/A) = \frac{C_C - C_{C0}}{C_{A0} - C_A}$$

$$C_C = C_{C0} + Y(C_{A0} - C_A) \tag{4-26}$$

For simplicity, the parenthesis (C/A) has been dropped from Y.

Substituting the expression for C_C into the design equation, substituting $\tau = V/\upsilon$, and simplifying,

$$k\tau C_A[C_{C0} + Y(C_{A0} - C_A)] = Y(C_{A0} - C_A)(C_A + K_S) \tag{4-27}$$

For $C_{C0} = 0$, the indicated terms can be canceled, i.e.,

$$k\tau C_A \cancel{Y(C_{A0} - C_A)} = \cancel{Y(C_{A0} - C_A)}(C_A + K_S) \tag{4-28}$$

$$C_A = K_S/(k\tau - 1)$$

Substituting the values given in the problem statement,

$$C_A = 0.20\,(\mathrm{g/l})/\{1.0\,(\mathrm{h}^{-1}) \times 2.0\,(\mathrm{h}) - 1\} = 0.20\,\mathrm{g/l}$$

The concentration of cells can be calculated from Eqn. (4-26)

$$C_C = C_{C0} + Y(C_{A0} - C_A) = 0 + 0.5(10 - 0.20) = 4.90\,\mathrm{g/l}$$

Do you believe that these answers are a complete solution to the problem? Look at the Monod equation (Eqn. (4-25)). Suppose that the feed, as defined in the problem statement, flows into a CSTR that has no cells in it initially. The reaction will never get started. The concentration of A leaving the reactor will be the same as the concentration entering, i.e., $C_A = C_{A0}$. Take a look at Eqn. (4-27). It is satisfied by $C_A = C_{A0}$, if $C_{C0} = 0$. However, we canceled out this solution in Eqn. (4-28).

In this problem, the reactor can have *two* steady states, i.e., there are *two* solutions to the design equation. The one that actually occurs will depend on the initial concentration of cells in the reactor. If the initial concentration is zero, the reaction will never get started, and the steady-state solution will be $C_A = C_{A0}$. However, if the initial concentration of cells is high enough, the steady-state solution will be $C_A = 0.20$ g/l.

Why didn't we see this "problem" in previous examples? How can we avoid getting "tricked" in the future? The problem arose here because the Monod rate equation has a very unusual characteristic: $-r_A$ goes through a maximum as C_A increases. For the feed specified in the problem statement, the product $C_A C_C$ is equal to $YC_A(C_{A0} - C_A)$. As C_A decreases, $(C_{A0} - C_A)$ increases and the product of the two goes through a maximum.

The design equation for a CSTR has a graphical interpretation that makes the existence of more than one steady state easier to understand. For a constant-density system, the design equation can be rearranged to

$$V(-r_A) = \upsilon(C_{A0} - C_A)$$

Rate of consumption of A in reactor = flow of A into reactor − flow of A out of reactor

The term on the left is the rate at which A is consumed in the whole CSTR. The term on the right is the difference between the rate at which A enters the reactor (υC_{A0}) and the rate at which it flows

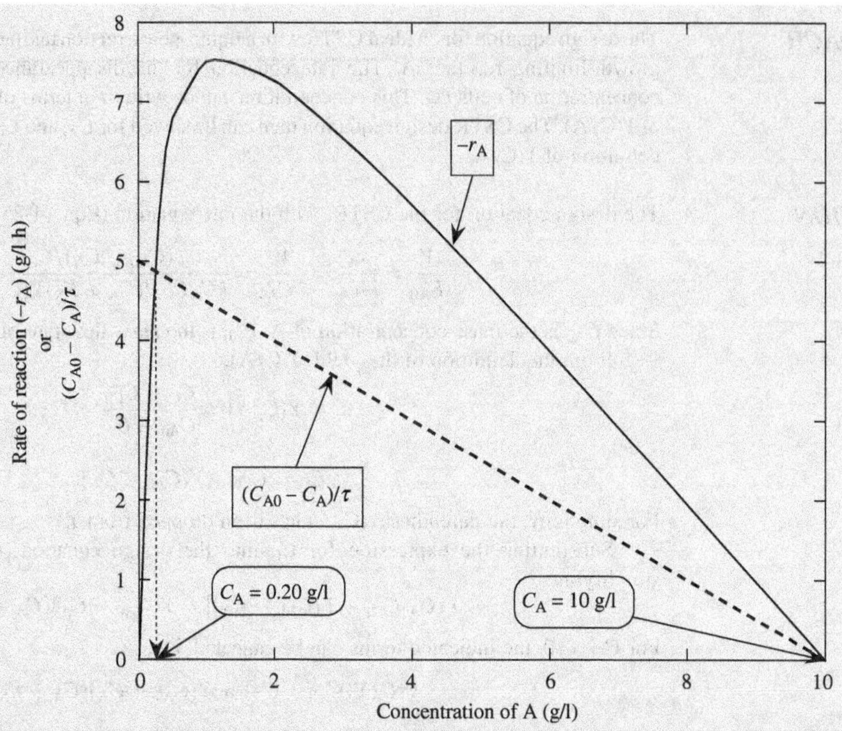

Figure 4-3 Graphical solution of the design equation for an ideal CSTR for the Monod rate equation and the parameter values given in Example 4-8, Part A. Two different steady-state outlet concentrations are possible, $C_A = 0.20$ and 10 g/l.

out of the reactor (υC_A). At steady state, the left-hand side of this equation must be equal to the right-hand side.

If we make a plot of the right-hand side of the above equation ($\upsilon(C_{A0} - C_A)$) versus C_A, it will be a straight line with an intercept of υC_{A0} and a slope of $-\upsilon$. We then can plot $V(-r_A)$ on the same graph. The steady-state solution(s) to the design equation will be the point(s) at which the two functions intersect. These points of intersection are the *only* points at which the design equation is satisfied.

In making this type of plot, it is more common to divide through by V and plot $-r_A$ versus $(C_{A0} - C_A)/\tau$. The straight line $(C_{A0} - C_A)/\tau$ has an intercept of C_{A0}/τ and a slope of $1/\tau$. This plot is shown as Figure 4-3 and was constructed using the values given in the problem statement. The curve for $-r_A$ intersects the straight line for $(C_{A0} - C_A)/\tau$ at two points: $C_A = 0.20$ and 10 g/l. These are the solutions identified above. Using Eqn. (4-26), $C_C = 4.90$ g/l for $C_A = 0.20$ g/l, and $C_C = 0$ when $C_A = 10$ g/l.

Part B: Suppose that the reactor described in the previous example is operating at the point, $C_A = 0.20$ g/l, $C_C = 4.9$ g/l. The volumetric feed rate υ is increased to 1.5 l/h. Use the graphical technique to find the steady-state operating point(s) for this new condition.

APPROACH For this case, the curve of $-r_A$ versus C_A remains the same. However, the value of τ decreases from 2 h to (2/3) h. This increases the intercept of the $(C_{A0} - C_A)/\tau$ versus C_A line to 15 g/l-h, and increases the slope of this line to 1.5 h^{-1}. The new $(C_{A0} - C_A) \times \tau$ line will be constructed. Its intersection with the $-r_A$ versus C_A curve will determine the outlet concentration of A. The outlet concentration of cells can then be calculated from the values of $Y(C/A)$ and C_A.

SOLUTION The plot for the new value of τ is shown as Figure 4-4.

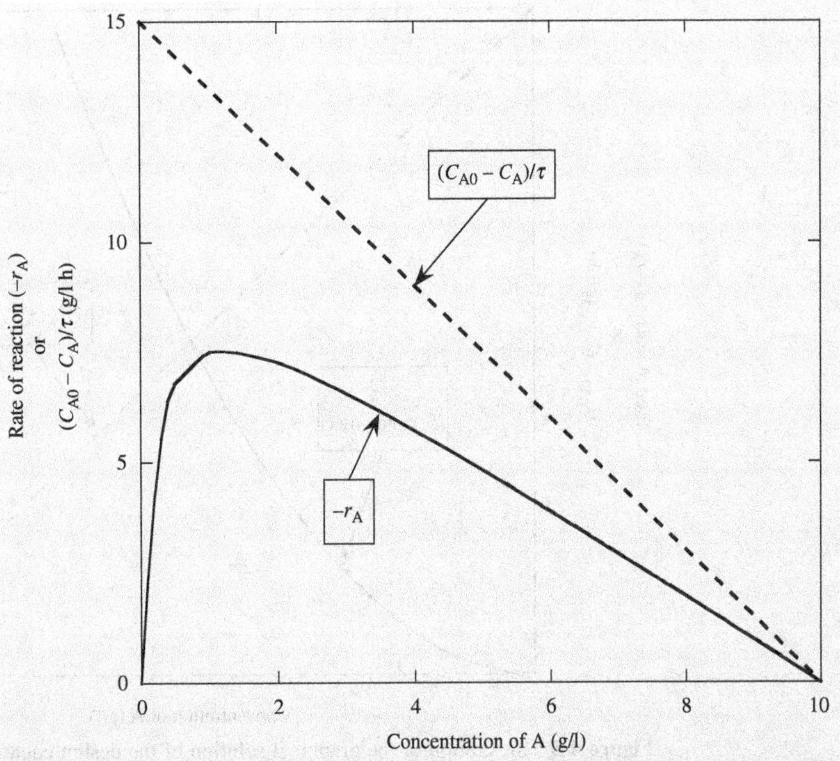

Figure 4-4 Graphical solution of the design equation for an ideal CSTR for the Monod rate equation and the parameter values given in Example 4-8, Part B. The only steady-state outlet concentration that is possible is $C_A = 10$ g/l. No reaction takes place in the CSTR.

The *only* intersection of the $-r_A$ curve with the $(C_{A0} - C_A)/\tau$ line is at $-r_A = 0$, $C_A = 10$ g/l. The corresponding concentration of cells is $C_C = 0$. No reaction takes place in the CSTR.

The phenomenon that is illustrated in Example 4-8, Part B is known as "washout." Even if we start with a high concentration of cells in the reactor, cells are carried out of the reactor by the flowing fluid faster than they are produced by the reaction. Ultimately, at steady state, no cells remain in the reactor, and no reaction takes place.

The only difference between Part A and Part B is the flow rate through the reactor. When the flow rate is low enough, the rate at which cells are carried out of the reactor can be balanced by the rate at which cells are produced in the reactor. However, if the flow rate is too high, the rate of production of cells at steady state cannot match the rate at which they flow out of the reactor.

Why didn't we see this bizarre behavior, e.g., "washout" and multiple steady states, previously? The reason is that our previous examples involved "normal" kinetics. In all of our past work, the rate increased monotonically as the reactant concentration increased. In such a case, the graphical solution to the CSTR design equation looks as shown in Figure 4-5.

Although Figure 4-5 shows the $-r_A$ curve for a second-order reaction, the result that it illustrates is general. *As long as the reaction rate increases monotonically with C_A, the two functions that we have been plotting always will intersect, and there will be only one* intersection. For "normal kinetics," it is perfectly satisfactory to solve the CSTR design equation algebraically. However, when the rate equation goes through a maximum as C_A

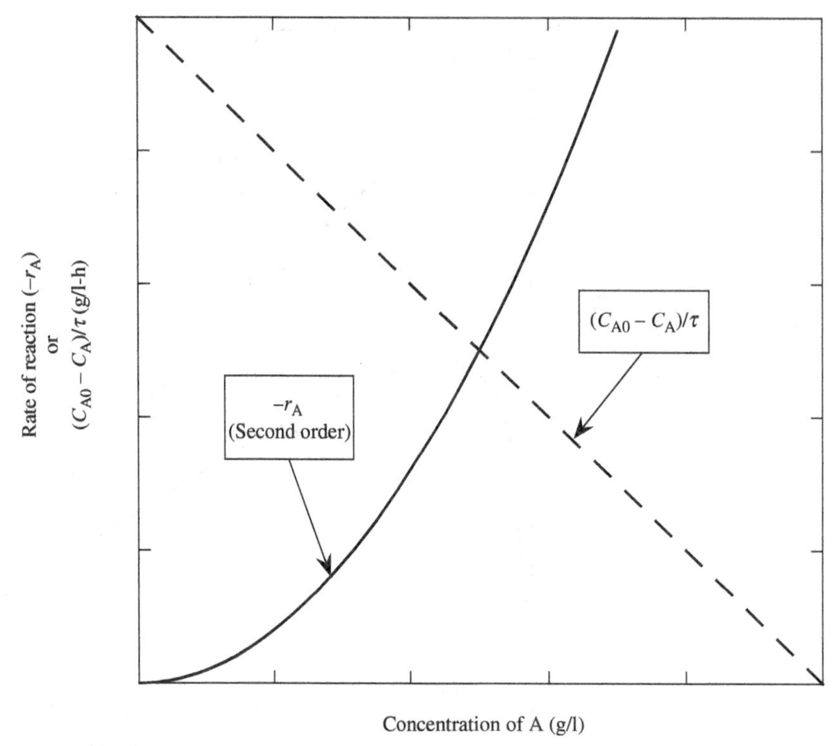

Figure 4-5 Illustration of the graphical solution of the design equation for an ideal CSTR for a second-order rate equation. There is only one possible intersection between the $-r_A$ curve and the $(C_{A0} - C_A)/\tau$ line.

increases, it can be valuable to examine the behavior of the reactor by means of the graphical technique.

4.1.2.4 Biochemical Engineering Nomenclature

As noted earlier, different nomenclature has grown up in different branches of science and engineering. For example, in biochemistry and biochemical engineering, a reactant is referred to as a "substrate." Continuous stirred-tank reactors frequently are used to study cell growth and to produce commercial quantities of cells. However, it is very likely that the reactor will be called a "chemostat, " not a CSTR.

Let's write the design equation for a reaction taking place in a "chemostat" (i.e., a CSTR).

$$V = \frac{F_{i0} - F_i}{-r_i} \tag{3-16}$$

For a liquid-phase reaction, such as cell growth in an aqueous medium, $F_{i0} = \upsilon C_{i0}$ and $F_i = \upsilon C_i$. Let "i" be the cells C. Then,

$$\frac{V}{\upsilon} = \frac{C_{C0} - C_C}{-r_C} = \frac{C_C - C_{C0}}{r_C}$$

Inverting and multiplying by $(C_C - C_{C0})$,

$$\frac{\upsilon}{V}(C_{C0} - C_C) = r_C$$

Earlier, in Chapter 3, we called (υ/V) the "space velocity." However, in biochemical engineering, (υ/V) is known as the "dilution rate" or "dilution" and often is designated "D." Thus,

$$D(C_C - C_{C0}) = r_C$$

If there are no cells in the feed to the "chemostat," i.e., if $C_{C0} = 0$, the feed is "sterile." For this case,

$$DC_C = r_C$$

Now, the Monod equation, and other alternative rate equations that are used to describe cell growth, can be written in the form

$$r_C = \mu C_C$$

The parameter μ is known as the "specific growth rate." For the Monod equation, $\mu = kC_A/(C_C + K_s)$.

For a sterile feed,

$$DC_C = \mu C_C = r_C$$

or

$$D = \mu$$

"dilution" = "specific growth rate"

This is a shorthand way of saying that for a sterile feed

$$\begin{Bmatrix} \text{rate at which cells flow} \\ \text{out of chemostat (CSTR)} \end{Bmatrix} = \begin{Bmatrix} \text{rate at which cells are produced} \\ \text{in the chemostat (CSTR)} \end{Bmatrix}$$

4.2 HETEROGENEOUS CATALYTIC REACTIONS (INTRODUCTION TO TRANSPORT EFFECTS)

At least two phases are involved when a heterogeneous catalytic reaction takes place. The reaction itself occurs on the surface of the solid catalyst. However, the reactants are contained in a fluid phase (or phases) that surrounds the solid catalyst particles. The reactants must be supplied to the catalyst surface from the fluid phase(s).

The participation of two or more phases in the overall reaction creates complications that are not present with homogeneous reactions. Perhaps the best way to point out the complexities of heterogeneous catalysis is through an illustration.

The reversible isomerization of normal pentane (n-C_5) to isopentane (i-C_5)

$$n\text{-C}_5 \rightleftarrows i\text{-C}_5$$

is often used as a model for the family of isomerization reactions that are involved in the production of high-octane motor gasoline. In general, branched paraffins have much higher octane numbers than their straight-chain counterparts. A common catalyst for this reaction is platinum deposited on a porous alumina particle.

Suppose that the reaction will be carried out in a fluidized-bed reactor. The temperature of the gas in the reactor is 750 °F. For the purpose of a preliminary analysis, the fluidized-bed reactor will be assumed to operate as an ideal stirred-tank reactor (CSTR). The feed to the reactor will be a $H_2/n\text{-}C_5$ mixture containing 4 mol H_2/mol n-C_5. An approximate rate equation at this temperature is

$$-r_n = k\left(C_n - \frac{C_i}{K_{eq}^C}\right)$$

where $-r_n$ is the rate of disappearance of normal pentane (lb·mol/lb·cat-h), C_n is the concentration of normal pentane (lb·mol/ft^3), C_i is the concentration of isopentane (lb·mol/ft^3), and K_{eq}^C is the equilibrium constant for the reaction, based on concentration. At 750 °F, $k = 6.09$ ft^3/lb·cat-h and $K_{eq}^C = 1.63$. The total pressure in the reactor is 500 psia, and the feed rate of n-C_5 is 280 lb·mol/h. The H_2/pentane mixture behaves as an ideal gas.

Suppose that we were asked to estimate the amount of catalyst required to reach a final n-C_5 conversion of 55%. From Chapter 3, the design equation for an ideal stirred-tank reactor with a heterogeneous catalytic reaction taking place is

$$\frac{W}{F_{A0}} = \frac{x_A}{-r_A} \tag{3-17a}$$

Writing the design equation for n-C_5 and substituting the rate equation gives

$$\frac{W}{F_{n0}} = \frac{x_n}{k(T_{cat})\left(C_n(\text{cat}) - \dfrac{C_i(\text{cat})}{K_{eq}^C}\right)}$$

The concentrations in the rate equation have been labeled (cat) in order to emphasize an important point. The reaction rate is determined by the concentrations in the *immediate vicinity* of the "site" on the catalyst where the reaction takes place. Similarly, the rate constant has been labeled $k(T_{cat})$ to emphasize that this constant must be evaluated at the temperature in the *immediate vicinity* of the catalytic "site."

The fractional conversion of n-pentane is

$$x_n = \frac{F_{n0} - F_n}{F_{n0}}$$

Since the reactor is isothermal and there is no change in the number of moles on reaction

$$x_n = \frac{C_{n0} - C_n}{C_{n0}}$$

where C_{n0} is the concentration of n-pentane in the feed, at the temperature and pressure of the reactor. The concentration C_n is the concentration of n-pentane in the *bulk gas* in the CSTR and leaving the CSTR. To emphasize this point, let's label C_n as C_n(bulk). The design equation then becomes

$$\frac{WC_{n0}}{F_{n0}} = \frac{C_{n0} - C_n(\text{bulk})}{k(T_{cat})\left(C_n(\text{cat}) - \dfrac{C_i(\text{cat})}{K_{eq}^C}\right)} \tag{4-29}$$

The isomerization reaction occurs on the surface of the solid catalyst. Typical catalysts are composed of a network of fine, interconnecting pores that run though the interior of the catalyst particle. Most of the surface on which the reaction takes place is located on the walls of these pores. Therefore, most of the reaction takes place inside the particle. Many commercial heterogeneous catalysts have surface areas in the range of 100–1000 m^2/g.

Only 5–50 g of such a catalyst would provide a surface area equivalent to that of a football field. If we had a spherical catalyst particle with a diameter of 1 mm and a density of 3 g/cc, the external (geometric) surface area of the particle would be only 10^{-3} m^2/g. If this catalyst had a total surface area in the vicinity of 100–1000 m^2/g, essentially all of the area would have to be in the *interior* of the particle, on the walls of many pores with a very small diameter.

For the reaction to occur, the reactant (*n*-pentane) must be transported by convective diffusion through a boundary layer of stagnant gas that surrounds the catalyst particle. The *n*-pentane then must diffuse through the fine pores, into the interior of the catalyst particle. Both the boundary layer and the porous interior offer resistances to mass transfer. *A driving force is required in order to have a flux of the reactant through these resistances. The driving force is a gradient (or difference) in the reactant concentration.*

In the following figure, species A is a reactant. The concentration of A declines from the bulk fluid stream through the boundary layer. The concentration difference $(C_{A,b} - C_{A,s})$ is the driving force that causes the flux of A through the boundary layer. Reactant A then diffuses into the interior of the catalyst particle, where reaction takes place. The concentration of A continues to decline as the reactant penetrates deeper and deeper into the particle.

The extent of the concentration decrease will depend on the rate of the reaction and on the transport coefficients, i.e., the diffusivity of A in the pores of the catalyst and the mass-transfer coefficient between the bulk fluid stream and the external surface of the catalyst particle. The concentration differences will be most pronounced for fast reactions and low transport coefficients.

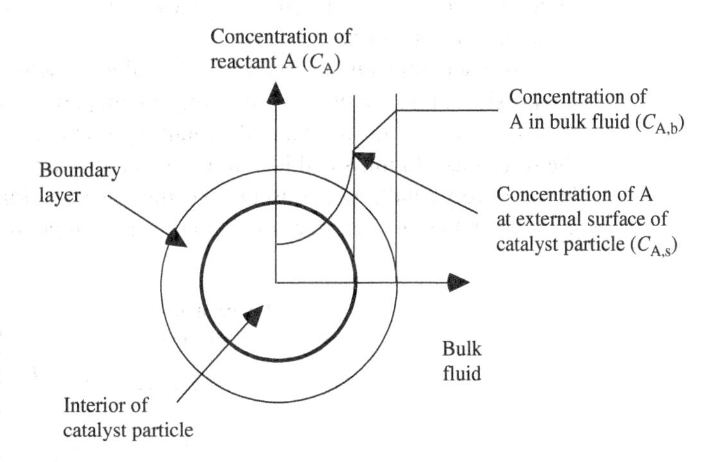

EXERCISE 4-8

Sketch the concentration profile of a product P starting at the centerline of the catalyst particle and ending in the bulk fluid stream.

In general, there will also be a temperature difference between the interior of the catalyst particle and the bulk fluid stream. Unless the reaction is thermally neutral, i.e., $\Delta H_R = 0$, heat will have to be transported into or out of the catalyst particle in order to keep the particle at steady state. The following figure shows the temperature profile for an exothermic reaction.

Since the reaction is exothermic, heat must be conducted through the catalyst particle to the external surface, and then transported through the boundary layer. Temperature gradients must be present in order for these fluxes to exist. The temperature declines from the interior of the particle through the boundary layer and to the bulk fluid stream.

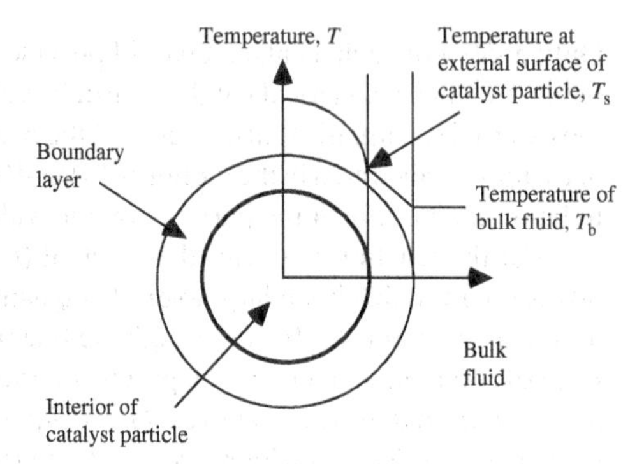

EXERCISE 4-9

Sketch the temperature profile between the interior of the catalyst particle and the bulk fluid stream for an endothermic reaction.

We know that the reaction rate depends on temperature and concentration. If the temperature and concentration differences between the interior of the catalyst particles and the bulk fluid are significant, then these differences must be taken into account in solving the design equation. In essence, this would require simultaneously solving the design equation and equations that describe heat transport, mass transport, and reaction kinetics in the interior of the catalyst particle, using the equations for transport through the boundary layer as boundary conditions.

To concentrate on the principles of catalytic reactor design, we will temporarily ignore the possible presence of concentration and temperature differences between the bulk fluid and the "sites" in the interior of the catalyst particle. For the time being, we will *assume* that the resistances to mass and heat transfer in the catalyst particle and through the boundary layer are very small. As a consequence, the concentration and temperature gradients will be very small. For this case, the concentration and temperature profiles will be as shown below.

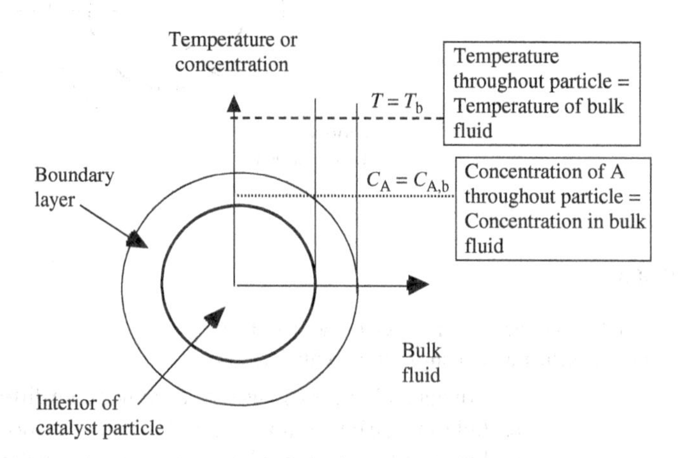

When the concentrations throughout the catalyst particle are the same as those in the bulk fluid, and when the temperature throughout the particle is the same as that of the bulk fluid, we say that transport effects are negligible, or that transport resistances can be neglected. For this situation, the reaction rate is *controlled* by the *intrinsic kinetics* of the reaction.

In Chapter 9, we will learn how to estimate whether transport effects are significant, and how to take them into account when they are. Until we reach Chapter 9, we will assume that transport effects are insignificant. The results of calculations based on this assumption *normally* are optimistic. That is, the weight of catalyst required to do a given "job" will be underestimated, or the "job" that can be done with a given weight of catalyst will be overestimated. However, there are some situations where the opposite will be true. In any event, the assumption of negligible transport resistance permits a very important limiting case of catalyst behavior to be calculated.

Now, let's return to the problem of pentane isomerization, specifically to Eqn. (4-29). If transport effects are neglected,

$$\frac{WC_{n0}}{F_{n0}} = \frac{C_{n0} - C_n(\text{bulk})}{k(T_{\text{bulk}})\left(C_n(\text{bulk}) - \dfrac{C_i(\text{bulk})}{K_{\text{eq}}^{\text{C}}}\right)}$$

In this equation, all concentrations are bulk concentrations, and the rate constant is evaluated at the bulk fluid temperature. Therefore, there is no longer any need to carry the label "bulk."

Table 4-17 Stoichiometric Table for Pentane Isomerization in an Ideal CSTR Using Fractional Conversion of n-Pentane, x_A, as the Composition Variable (Constant Density)

Species	Inlet concentration	Outlet concentration
H_2	$4C_{n0}$	$4C_{n0}$
n-C_5	C_{n0}	$C_{n0}(1 - x_n)$
i-C_5	0	$C_{n0}x_n$

We now use a stoichiometric table to relate C_n and C_i to x_n. Since this gas-phase reaction takes place at constant pressure and temperature, with no change in the number of moles, the mass density is constant. Therefore, the stoichiometric table (Table 4-17) can be constructed directly in terms of concentrations, instead of starting with molar flow rates.

The concentrations of n-pentane and i-pentane in the gas leaving the reactor are

$$C_n = C_{n0}(1 - x_n) \qquad \text{and} \qquad C_i = C_{n0}x_n$$

where $C_{n0} = y_{n0}P/RT = 7.70 \times 10^{-3}$ lb·mol/ft^3.

The design equation becomes

$$\frac{kWC_{n0}}{F_{n0}} = \frac{x_n}{\left((1 - x_n) - \dfrac{x_n}{K_{\text{eq}}^{\text{C}}}\right)}$$

All of the values in this equation are known with the exception of the catalyst weight W. For $x_n = 0.55$, $W = 30{,}000$ lb·cat.

Now that a means for estimating the performance of heterogeneous catalytic reactors has been developed, we are in a position to explore some additional applications of that methodology.

EXAMPLE 4-9 ***Isomerization of n-Pentane***

The catalytic isomerization of *n*-pentane

$$n\text{-}C_5H_{12} \rightleftarrows i\text{-}C_5H_{12}$$

is being carried out in a fluidized-bed reactor using a Pt/Al_2O_3 catalyst. The reactor can be approximated as an ideal CSTR. The feed to the reactor is a $H_2/n\text{-}C_5$ mixture; the feed rate of $n\text{-}C_5$ is 280 lb·mol/h and the H_2 feed rate is 1120 lb·mol/h. The reactor operates at 750 °F. At this temperature, the rate equation is

$$-r_n = k\left(C_n - \frac{C_i}{K_{eq}^C}\right)$$

At 750 °F, $k = 6.09\ \text{ft}^3/\text{lb}\cdot\text{cat-h}$ and $K_{eq}^C = 1.63$. The total pressure in the reactor is 500 psia, and the H_2/pentane mixture behaves as an ideal gas. Assume that the reaction is controlled by intrinsic kinetics.

Part A: How much catalyst is required in an ideal CSTR to reach a $n\text{-}C_5$ conversion of 70% in the reactor effluent?

APPROACH

This problem appears to be a slight variation of the preceding illustration, where only the desired outlet conversion is different. All of the values in the design equation

$$\frac{kWC_{n0}}{F_{n0}} = \frac{x_n}{\left((1-x_n) - \dfrac{x_n}{K_{eq}^C}\right)}$$

are known, except for the weight of catalyst, *W*. The design equation can be solved for *W*.

SOLUTION

For $x_n = 0.70$, the above equation gives $W = -32{,}000$ lb·cat. Obviously, this answer is nonsense, but why?

The *equilibrium* conversion of *n*-pentane at 750 °F, x_{eq}, can be calculated from the equilibrium expression:

$$\frac{C_i}{C_n} = K_{eq}^C = \frac{C_{n0}x_{eq}}{C_{n0}(1-x_{eq})} = \frac{x_{eq}}{(1-x_{eq})}$$

For $K_{eq}^C = 1.63$, $x_{eq} = 0.62$. Therefore, the required conversion of 0.70 exceeds the maximum conversion permitted by thermodynamics. Mathematically, a negative catalyst weight is obtained because the rate of disappearance of *n*-pentane, $-r_n$, is negative when x_n exceeds 0.62.

The message here is that the reaction equilibrium should *always* be understood *before* carrying out an analysis based on kinetics.

Part B: How much catalyst is required in an ideal plug-flow reactor (PFR) to reach a 55% conversion of $n\text{-}C_5$ in the reactor effluent?

APPROACH

From Part A, we know that a $n\text{-}C_5$ conversion of 55% is less than the equilibrium conversion. Therefore, the rate equation can be substituted into the design equation and the resulting expression can be solved for the required weight of catalyst.

SOLUTION

The design equation for a heterogeneous catalytic reaction in an ideal PFR, in integrated form, is

$$\frac{W}{F_{n0}} = \int_0^x \frac{dx_n}{-r_n} \tag{3-33a}$$

Substituting the rate equation gives

$$\frac{W}{F_{n0}} = \int_0^x \frac{dx_n}{k\left(C_n - \dfrac{C_i}{K_{eq}^C}\right)} = \int_0^x \frac{dx_n}{kC_{n0}\left((1-x) - \dfrac{x}{K_{eq}^C}\right)}$$

For an isothermal reactor,

$$\frac{kC_{n0}W}{F_{n0}} = \int_0^{0.55} \frac{dx_n}{\left[1 - \left(1 + \frac{1}{K_{eq}^C}\right)x_n\right]}$$

Integrating,

$$-\ln\left[1 - \left(1 + \frac{1}{K_{eq}^C}\right)x_n\right]_0^{0.55} = \frac{kC_{n0}W}{F_{n0}}\left(1 + \frac{1}{K_{eq}^C}\right)$$

Substituting values for K_{eq}^C, k, C_{n0}, F_{n0}, and x_n results in $W = 8100$ lb.

EXERCISE 4-10

The amount of catalyst required to produce the same final conversion is about a factor of 4 less in the PFR than in the CSTR. Does this difference seem reasonable? If so, explain why the ideal PFR requires so much less catalyst to do the same "job."

EXERCISE 4-11

Do you think that the ratio (catalyst required in the CSTR/catalyst required in the PFR), will depend on the final conversion? If so, how? (Will the ratio go up or down as the final conversion increases?) Explain your reasoning.

4.3 SYSTEMS OF CONTINUOUS REACTORS

A single reactor is not always the optimum design for a reaction that is run continuously. Suppose you had a single CSTR with a volume of V that was processing F_{A0} moles of reactant A per unit time with a feed concentration of C_{A0}, and achieving an outlet conversion of x_A. Using *only* the original CSTR, could you double the production rate by feeding $2F_{A0}$ moles per unit time, without changing the final conversion or the feed concentration?

You might think of raising the temperature at which the reactor operates. This usually will increase the reaction rate, and permit the feed rate to be increased. However, there might be reasons why the reactor temperature cannot be increased. Perhaps the rate of a side reaction would be too high at the higher temperature. Perhaps the catalyst, if a catalyst were involved, might deactivate too rapidly. Perhaps the reactor is not rated to operate at the higher temperature.

EXERCISE 4-12

Consider a single reaction. Under what circumstances will raising the temperature *not* increase the rate of reaction?

If the reactor temperature cannot be increased, the existing reactor will not be able to handle the new requirements. A larger reactor will be required. However, it would not make practical sense to scrap the old reactor and install a new one that had enough volume to handle the higher production rate. The cheapest solution usually is to keep the first reactor in use, and add a second reactor so that the *combination* of the two can handle the new production rate.

A whole new set of questions now arises. Should the new reactor be a CSTR or a PFR? Should it be in series or in parallel with the original reactor? If in series, should the new reactor precede or follow the original? The answers to these questions will depend on the kinetics of the reaction, and on whether the original reactor was a CSTR or a PFR. Mechanical considerations might also enter into the decision.

Systems of continuous reactors can result from the need to expand capacity, but they also might be the best alternative for a "grass roots" design. For example, it may be desirable for thermodynamic and/or kinetic reasons to change the reactor temperature as the conversion increases. When this is the case, the simplest design usually involves several reactors in series, with heat exchangers between reactors to make the necessary temperature changes. Other reasons to use more than one reactor can arise when the best catalyst at low conversions is not the same as the best catalyst at high conversions, or when a second feed or recycle stream must be added as the reaction progresses.

Figure 4-6 shows a commercial reactor that is used for the production of high-octane gasoline by the dehydrogenation, dehydrocyclization, isomerization, and cracking of petroleum naphthas. The reactions that occur are endothermic. The feed is reheated in external exchangers between reactors, and the feed temperature generally is increased from reactor to reactor.

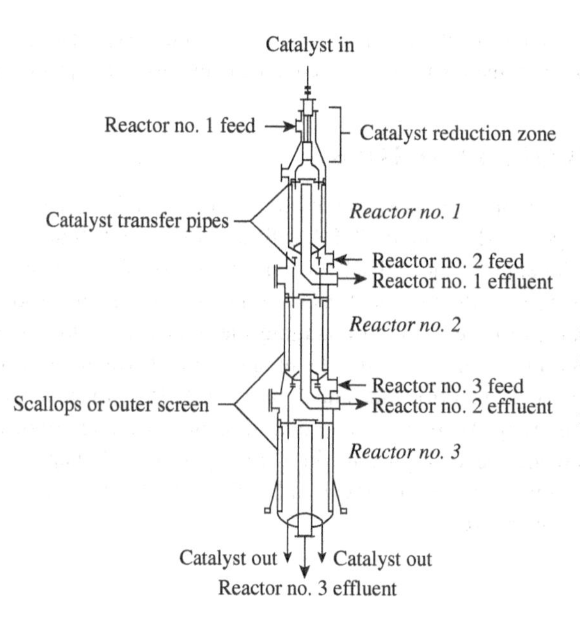

Figure 4-6 Three catalytic reactors in series.[4] These units are part of UOP's proprietary process for the continuous catalytic reforming of petroleum naphthas, a key element in the process of producing gasoline for automobiles. The catalyst is comprised of platinum and another metallic component on an acidic support. The catalyst slowly moves downward through the reactors and is regenerated after it leaves reactor 3 at the bottom of the picture. The reactors themselves are radial-flow reactors, as discussed in Chapter 3. (Copyright 2004 UOP LLC. All rights reserved. Used with permission.)

4.3.1 Reactors in Series

4.3.1.1 CSTRs in Series

Suppose three CSTRs each with the same volume V are arranged in series, as shown in the following figure. The molar flow rate of reactant A into the first reactor is F_{A0}. The fractional conversion of A in the effluent from the first reactor is $x_{A,1}$, the conversion of A in the effluent from the second reactor is $x_{A,2}$, and the conversion of A leaving the third reactor, and the system as a whole, is $x_{A,3}$.

[4] Stine, M. A., *Petroleum Refining*, presented at the North Carolina State University AIChE Student Chapter Meeting, November 15, 2002.

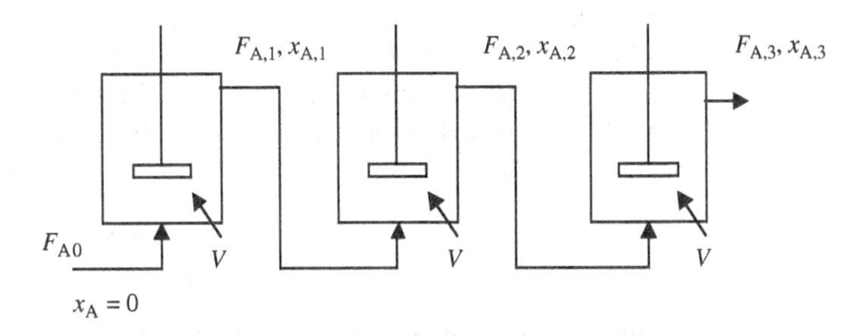

It is very important to be consistent in defining the fractional conversions for a series of reactors. The easiest definition, which will be followed in this book, is to base the conversion on the molar flow rate to the *first* reactor, i.e., F_{A0}. Let $F_{A,1}$ be the molar flow rate of A out of the first reactor, $F_{A,2}$ be the molar flow rate out of the second reactor, and $F_{A,3}$ be the molar flow rate out of the third reactor. Then

Definition of fractional conversions for reactors in series

$$x_{A,1} = (F_{A0} - F_{A,1})/F_{A0}$$
$$x_{A,2} = (F_{A0} - F_{A,2})/F_{A0} \qquad (4\text{-}30)$$
$$x_{A,3} = (F_{A0} - F_{A,3})/F_{A0}$$

The conversion $x_{A,1}$ is the fractional conversion of A in the stream leaving the first reactor. This is the same definition that is used for a single reactor. The conversion $x_{A,2}$ is the *overall* conversion of A in the stream leaving the second reactor. In other words, $x_{A,2}$ is the conversion for the *first and second reactors combined.* Finally, $x_{A,3}$ is the conversion of A in the stream leaving the third (last) reactor. It is the overall conversion for the *series* of three reactors.

With this basis, the fractional conversion of reactant A in the stream leaving reactor $N + 1$ is always greater that the fractional conversion of A in the stream leaving the reactor immediately upstream, i.e., reactor N. Moreover, the fractional conversion of the stream entering reactor $N + 1$ is the same as the conversion in the stream leaving reactor N.

Another way to analyze a series of reactors is to "reset the clock" after each reactor. In this approach, the fractional conversion is set back to zero in the stream that enters each reactor. At the same time, a new value of F_{A0} is calculated for the stream entering the reactor. In other words, the value of F_{A0} for the second reactor is the molar flow rate of A leaving the first reactor, i.e., $F_{A,1}$ in the above figure. This second approach frequently is harder to use for calculations, and it requires much more careful "bookkeeping." A common mistake when using this approach is to set x_A back to zero without resetting the value of F_{A0}, or to reset F_{A0} without resetting x_A. Because of its complexity, and the potential for error, this approach will not be mentioned again.

Let's carry out a material balance on the *second* reactor in the figure above. This balance will illustrate how to use the preferred approach to defining conversions and molar flow rates. At steady state, the material balance for A, using the whole second reactor as a control volume, is

$$\text{rate in} - \text{rate out} + \text{rate generation} = 0$$
$$\text{rate in} = F_{A,1} = F_{A0}(1 - x_{A,1})$$
$$\text{rate out} = F_{A,2} = F_{A0}(1 - x_{A,2})$$
$$\text{rate generation} = r_A(x_{A,2}) \times V_2$$

The relationships between $F_{A,1}$, $F_{A,2}$, $x_{A,1}$, $x_{A,2}$, and F_{A0} in the "rate in" and "rate out" terms follow directly from Eqns. (4-30). In the "generation" term, the symbol V_2 denotes the volume of the second reactor. We have written $r_A(x_{A,2})$ to emphasize that the reaction rate must be evaluated at the exit conditions, for a CSTR.

Using these relationships reduces the material balance to

$$\frac{V_2}{F_{A0}} = \frac{(x_{A,2} - x_{A,1})}{-r_A(x_{A,2})}$$

This equation can be generalized to apply to the Nth reactor in a series of CSTRs:

Design equation for *N*th CSTR in a series

$$\boxed{\frac{V_N}{F_{A0}} = \frac{(x_{A,N} - x_{A,N-1})}{-r_A(x_{A,N})}} \tag{4-31}$$

Equation (4-31) can be regarded as a generalized ***design equation*** for one CSTR in a series of reactors. It provides a basis for using the graphical technique that was developed in Chapter 3 to analyze a series of reactors. Equation (4-31) shows that (V/F_{A0}) for the *N*th reactor is the area of a rectangle that has a base of $(x_{A,N} - x_{A,N-1})$ and a height of $[1/-r_A(x_{A,N})]$. In words, (V/F_{A0}) for the *N*th reactor is the difference between the outlet and inlet conversions, $x_{AN} - x_{A,N-1}$, multiplied by the inverse of the reaction rate, evaluated at the exit conditions for the *N*th reactor $[1/-r_A(x_{A,N})]$. Thus, V_2/F_{A0} is equal to the area of the rectangle labeled "second reactor" in Figure 4-7.

Three equal-volume CSTRs in series are compared with a single PFR in Figure 4-7. The areas that represent each of the three CSTRs are equal, since each reactor has the same volume *V*, and the value of F_{A0} does not change from reactor to reactor.

For the three reactors in series, the total volume (or weight of catalyst) required for a specified value of $x_{A,3}$ is proportional to the sum of the three areas labeled "first reactor," "second reactor," and "third reactor." For "normal" kinetics, the series of three CSTRs requires less volume (or catalyst) to do a given "job" than a single CSTR. For a single CSTR, the volume required is proportional to the area $x_{A,3} \times (1/-r_A(x_{A,3}))$. However, the volume required by the series of three CSTRs is still greater than for an ideal PFR.

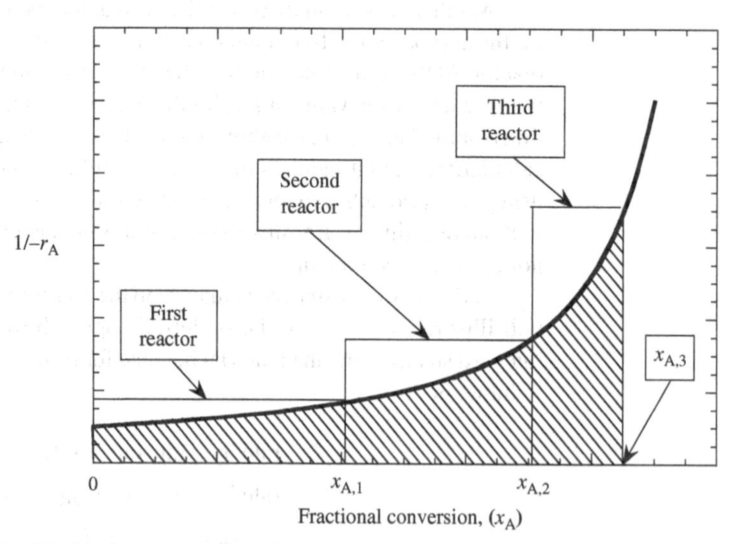

Figure 4-7 Graphical representation of the design equations for three ideal CSTRs in series, and comparison with a single PFR (shaded area).

As the number of ideal CSTRs in series approaches infinity, the total volume requirement approaches that of an ideal PFR. For first-order kinetics, a very close approach to plug-flow performance will be obtained when the number of CSTRs in series N is 10 or more.

In a practical situation, total capital cost is the important parameter rather than total reactor volume. Increasing the number of CSTRs in series will tend to reduce the total cost by reducing the total volume requirement. However, it will also tend to increase the cost, since more agitators, valves, piping, reactor heads, etc. will be required. The economic optimum usually occurs at a value of N that is substantially less than 10, perhaps as low as 2 or 3, depending on the operating pressure of the reactors and the final conversion.

The comparison in Figure 4-7 is based on the assumption that the curve of $1/-r_A$ versus x_A does not change from reactor to reactor. This will be true if the temperature of each reactor is the same, and if the composition of the stream leaving reactor N is identical to that of the stream entering reactor $N+1$. *These are important restrictions.* It is common to operate reactors in series at different temperatures. Moreover, it is not unusual to add a feed or a recycle stream between reactors, thus changing the composition of the stream entering the downstream reactor.

Calculations for CSTRs in series

Case 1: Suppose that you are asked to calculate the performance of the three CSTRs in series. The volume of each reactor, V_i, is given, as is the inlet molar flow rate to the first reactor, F_{A0}. The rate equation for the disappearance of A is also specified. The conversion of A leaving the third (last) reactor is to be calculated.

The final conversion can be calculated by "marching" from the first reactor to the second and then to the third. For each reactor, the design equation is used to calculate the exit conversion, which then becomes the inlet conversion to the next reactor. For reactor 1

$$\frac{V_1}{F_{A0}} = \frac{x_{A,1}}{-r_A(x_{A,1})}$$

Since the value of V_1/F_{A0} is known, and all of the constants in the rate equation are given, the value of $x_{A,1}$ can be calculated.

The design equation for the second reactor is

$$\frac{V_2}{F_{A0}} = \frac{x_{A,2} - x_{A,1}}{-r_A(x_{A,2})}$$

Since the value of $x_{A,1}$ was calculated previously, the only unknown in this equation, $x_{A,2}$, can be calculated. The value of $x_{A,3}$ then can be calculated from the design equation for the third reactor, and the problem is solved.

Case 2: Suppose that you are asked to calculate the volume required to achieve a specified conversion leaving the third (last) reactor. Each of the three reactors will have the same volume V. The rate equation for the disappearance of A is given. The inlet molar flow rate to the first reactor, F_{A0}, is specified.

As soon as we write the design equation for the first reactor, we see that this variation of the problem is more challenging. For example, for reactor 1

$$\frac{V}{F_{A0}} = \frac{x_{A,1}}{-r_A(x_{A,1})} \tag{4-32}$$

This equation contains two unknowns, V and $x_{A,1}$. The design equations for the second and third reactors

$$\frac{V}{F_{A0}} = \frac{x_{A,2} - x_{A,1}}{-r_A(x_{A,2})} \tag{4-33}$$

$$\frac{V}{F_{A0}} = \frac{x_{A,3} - x_{A,2}}{-r_A(x_{A,3})} \tag{4-34}$$

contain an additional unknown, $x_{A,2}$.

Equations (4-32)–(4-34) form a set of three algebraic equations containing three unknowns, V, $x_{A,1}$, and $x_{A,2}$. These equations have to be solved simultaneously, generally using a numerical technique.

EXAMPLE 4-10
Three Equal-Volume CSTRs in Series

The liquid-phase, irreversible reaction

$$A + B \rightarrow C + D$$

is to be carried out in a series of three, equal-volume CSTRs. The temperature will be the same for each reactor, and the effluent from one reactor will flow directly into the next. The volumetric flow rate to the first reactor is 10,000 l/h and the concentrations of A and B in the feed to the reactor are $C_{A0} = C_{B0} = 1.2$ mol/l. The reaction obeys the rate equation

$$-r_A = kC_AC_B$$

The value of k is 3.50 l/mol-h at the operating temperature of the reactors.

The final conversion must be at least 0.75. What reactor volume is required?

APPROACH

Design equations for each of the three reactors can be written. These three algebraic equations will contain three unknowns, V (the volume of one reactor), $x_{A,1}$, and $x_{A,2}$, the fractional conversions after the first and second reactors, respectively. The three design equations can be solved simultaneously to obtain the three unknowns. The total required reactor volume is $3V$.

SOLUTION

The design equation for the first CSTR, with the above rate equation inserted and written in terms of x_A, is

$$\frac{V}{F_{A0}} = \frac{x_{A,1}}{kC_{A0}^2(1 - x_1)^2}$$

This can be rearranged to

$$kC_{A0}\tau = \frac{x_{A,1}}{(1 - x_{A,1})^2} \tag{4-35}$$

where $\tau = VC_{A0}/F_{A0}$. The value of τ is unknown since V is unknown.

The design equations for the second and third CSTRs are, respectively,

$$kC_{A0}\tau = \frac{x_{A,2} - x_{A,1}}{(1 - x_{A,2})^2} \tag{4-36}$$

$$kC_{A0}\tau = \frac{x_{A,3} - x_{A,2}}{(1 - x_{A,3})^2} \tag{4-37}$$

Equations (4-35)–(4-37) may be solved for the three unknowns, x_1, x_2, and $kC_{A0}\tau$. The solution is

$$x_{A,1} = 0.460$$

$$x_{A,2} = 0.651$$

$$kC_{A0}\tau = 1.577$$

$$3V = 3kC_{A0}\tau \times (v/kC_{A0}) = 3 \times 1.577 \times (10,000(\text{l/h})/3.5(\text{l/mol-h}) \times 1.2(\text{mol/l})$$

$$3V = 11,300\text{l}$$

The procedure that was used to solve this system of equations is based on GOALSEEK and is explained in Appendix 4 at the end of this chapter.

4.3.1.2 PFRs in Series

Suppose we had two PFRs in series, as shown below.

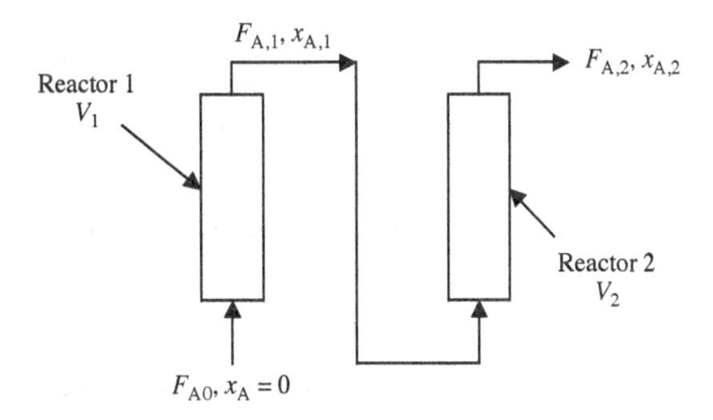

The conversions and flow rates for PFRs in series are defined as specified in the previous section. With these definitions, the design equations for the two PFRs are

$$\text{First reactor:} \quad \frac{V_1}{F_{A0}} = \int_0^{x_{A,1}} \frac{dx}{-r_A(x)}$$

$$\text{Second reactor:} \quad \frac{V_2}{F_{A0}} = \int_{x_{A,1}}^{x_{A,2}} \frac{dx}{-r_A(x)}$$

If heat is not added or removed between the two reactors, and if there are no sidestreams entering between reactors, then the two PFRs in series may be represented graphically as shown in Figure 4-8.

Figure 4-8 shows that the volume required to reach a final conversion of $x_{A,2}$ with two PFRs in series is the same as the volume required for a single PFR. However, the restriction of no temperature change and no sidestream introduction between the two reactors is important. One of the most common reasons for breaking a single PFR into two separate reactors is to add or remove heat between reactors. Moreover, it is not unusual for a recycle stream or a second feed stream to be introduced between two PFRs.

4.3.1.3 PFRs and CSTRs in Series

When a brand-new reactor system is being designed, it is uncommon (but not impossible) to encounter a situation that calls for using CSTRs and PFRs in series. However, if existing equipment is being used to satisfy an interim need, to establish production quickly, or to expand an existing plant, there may be good reason to consider such combinations.

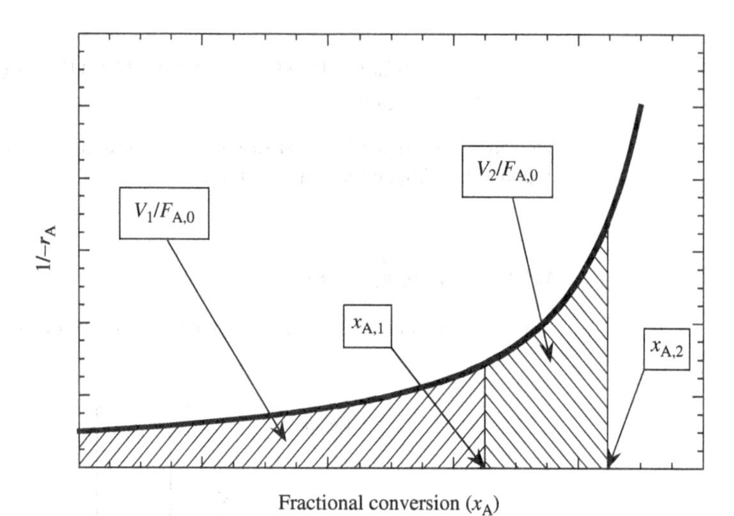

Figure 4-8 Graphical representation of the design equations for two PFRs in series.

If CSTRs and PFRs are to be used in series, one obvious question is which type of reactor should come first? For a given feed rate, inlet concentration, and inlet temperature, and for fixed reactor volumes, will the final conversion depend on how the reactors are ordered?

Since CSTRs and PFRs represent the extremes of mixing, the best order will depend on whether it is better to mix when the reactant concentrations are high or when they are low. Stated differently, is it better to mix early in the reaction (when the conversion is low) or to mix late in the reaction (when the conversion is high)? The answer depends on the rate equation.

The generalizations that apply are as follows:

1. If the effective order of the reaction is greater than 1 ($n > 1$), avoid mixing for as long as possible. Keep the reactant concentration as high as possible for as long as possible. For example, suppose that three reactors were to be used in series, a small CSTR, a large CSTR, and a PFR. For $n > 1$, mixing should be delayed for as long as possible. The optimum reactor arrangement is the PFR first, followed by the small CSTR, with the large CSTR last.
2. If the effective order of the reaction is less than 1 ($n < 1$), mix as soon as possible. Drop the reactant concentration as low as possible as soon as possible. In the above example, for $n < 1$, the large CSTR should be first, followed by the small CSTR, with the PFR last.
3. If the effective order is exactly 1 ($n = 1$), then the earliness or lateness of mixing does not matter. The order of the reactors will not affect the final conversion.

These generalizations apply to a situation where the "amount" of mixing is fixed, and the only question is whether to mix early or late in the reaction. The generalizations do *not* mean that mixing is beneficial. Obviously, based *only* on the total volume required, mixing is undesirable in all three of these cases. We would use a PFR in all three situations, if that were allowable.

The phrase "effective reaction order" requires some explanation. For a single reaction, all of the species concentrations can be written as a function of one variable, say the concentration of reactant A (C_A). If we do this for a given feed composition, the reaction rate

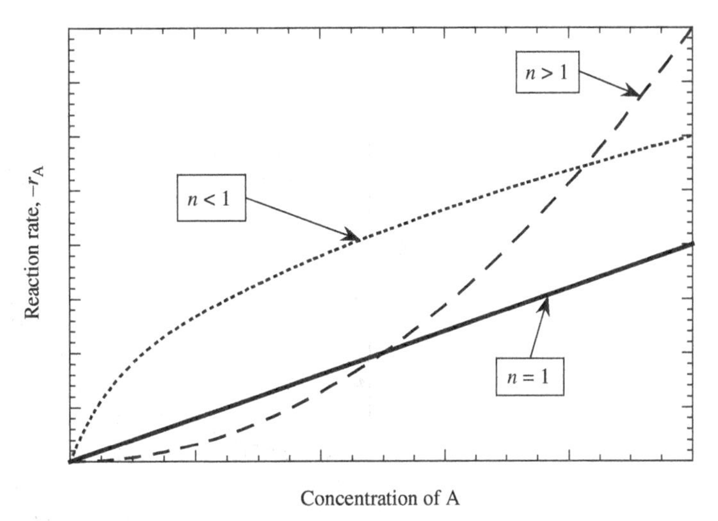

Figure 4-9 Illustrations of effective reaction order.

is a function only of C_A. Suppose we then made a plot of $-r_A$ versus C_A. Some of the possibilities are shown in Figure 4-9.[5]

The plot of reaction rate versus reactant concentration is concave upward when the effective reaction order is greater than 1 $(n > 1)$. When the effective reaction order is less than 1 $(n < 1)$, the curve is concave downward. The effective order is exactly equal to 1 $(n = 1)$ when the $-r_A$ versus C_A curve is a straight line, through the origin. We can make a plot such as the one above for many rate equations and determine which of the three classifications describes the reaction kinetics.

The curves in Figure 4-9 can help us to understand the generalizations stated above. Let's use the $n > 1$ curve to illustrate.

Suppose we have two elements of fluid, as pictured below. One element has a very large volume (V_l) and contains a concentration of reactant A, C_A. The second element has a very small volume (V_s) and contains a higher concentration of A, $C_A + \Delta C_A$. In the following exercise, we will assume that ΔC_A is small compared to C_A.

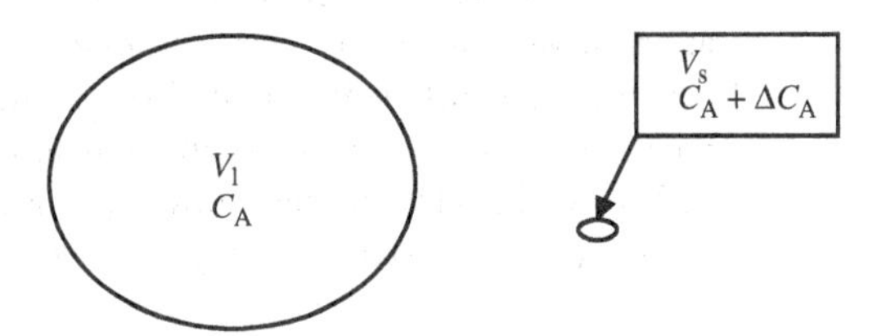

We now mix the two elements of fluid, at the same time allowing just enough reaction to take place so that the concentration of A in the combined elements is C_A. In other words, we will allow a total of $V_s \times \Delta C_A$ moles of A to react.

This mixing/reaction process could be carried out in two different ways: (1) mix first and then allow reaction to occur; (2) allow reaction to occur and then mix. Consider the second approach. Suppose that reaction takes place in the small element until the concentration of A has declined to C_A. The total number of moles of A reacted will be

[5] Of course, there are possibilities other than the three shown in Figure 4-9. For example, if the reaction exhibited autocatalytic behavior, as discussed in Example 4-8 of this chapter, the plot of $-r_A$ versus C_A would go through a maximum. This type of rate behavior would require a separate analysis, and the following generalizations would not apply.

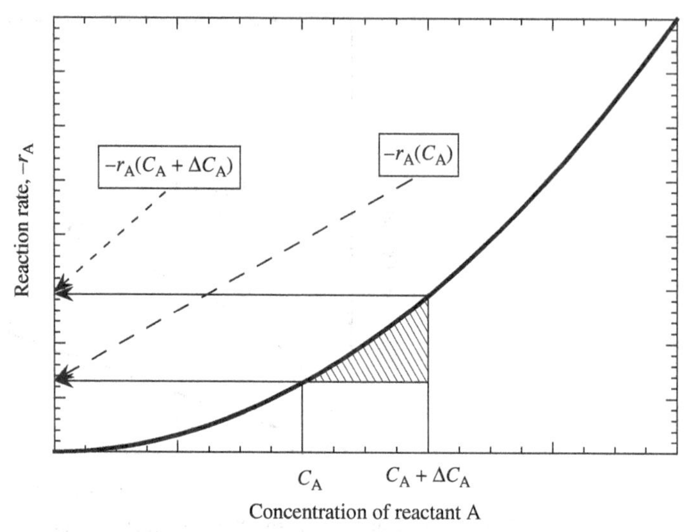

Figure 4-10 Effect of mixing for $n > 1$.

$V_s \times \Delta C_A$, as desired. As the reaction takes place, the rate will decrease along the $-r_A$ versus C_A curve, from $-r_A(C_A + \Delta C_A)$ to $-r_A(C_A)$, as shown in Figure 4-10.

At this point, the small element is mixed into the large element.

Now consider the first approach, where the large and small elements are mixed *before* any reaction takes place, and then $V_s \Delta C_A$ moles of A are allowed to react. Since $V_l \ggg V_s$, and ΔC_A is small compared to C_A, the concentration in the mixed system is very close to C_A, and all of the reaction takes place at a rate that corresponds to C_A, i.e., $-r_A(C_A)$.

The average rate for the second approach (react, then mix) is larger than the average rate for the first approach (mix, then react). In fact, the difference in the rates is proportional to the cross-hatched area in the above figure. This area represents the reaction-rate penalty that is associated with mixing.

How can this penalty be minimized? Should we mix at high C_A or low C_A? Figures 4-9 and 4-10 show that the slope of the $-r_A$ versus C_A curve increases as C_A increases, when $n > 1$. This leads to the comparison shown in the figures below. The rate versus concentration curve is shown as a straight line in these figures since this curve is approximately linear if ΔC_A is sufficiently small. Both triangles have the same base, ΔC_A. The height of the triangle is the difference between the reaction rate at $(C_A + \Delta C_A)$ and the reaction rate at C_A. The height is less for the "low C_A" case than for the "high C_A" case, because the $-r_A$ versus C_A curve is steeper at high C_A when $n > 1$.

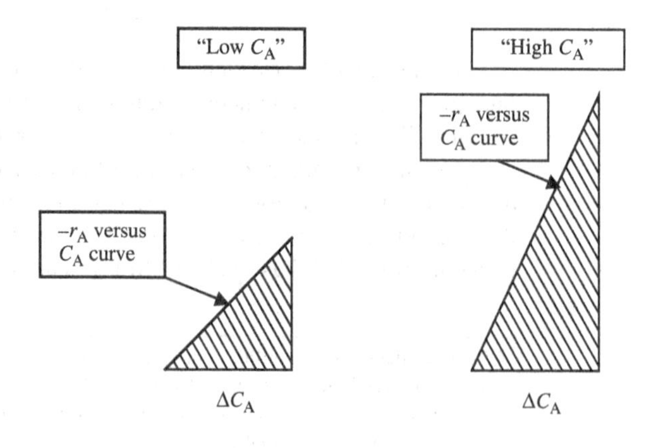

Clearly, for $n > 1$, the penalty for mixing is higher at high C_A than at low C_A. Therefore, for $n > 1$, we want to avoid mixing for as long as possible, i.e., mix late in the reaction, at the lowest possible C_A. This logic is consistent with the first of the three previous generalizations concerning PFRs and CSTRs in series.

The same approach can be used to analyze the $n < 1$ case. In that case, the penalty for mixing is largest when C_A is *low*. This is consistent with the second generalization, which tells us to mix as soon as possible if $n < 1$, i.e., mix when the concentration is as high as possible.

EXERCISE 4-13

Carry out the analysis for $n < 1$ and verify the above assertion.

The third case, $n = 1$, is the easiest to understand. The earliness or lateness of mixing has no effect when the relationship between $-r_A$ and C_A is linear. The slope of the $-r_A$ versus C_A curve does not depend on C_A. It does not matter whether mixing occurs at high C_A or at low C_A.

It is important to recognize that the graphical technique employed above is not a *proof* of the three generalizations. It is merely a convenient way to remember and rationalize these rules.

4.3.2 Reactors in Parallel

4.3.2.1 CSTRs in Parallel

Suppose that two CSTRs are operating in parallel, as shown below.

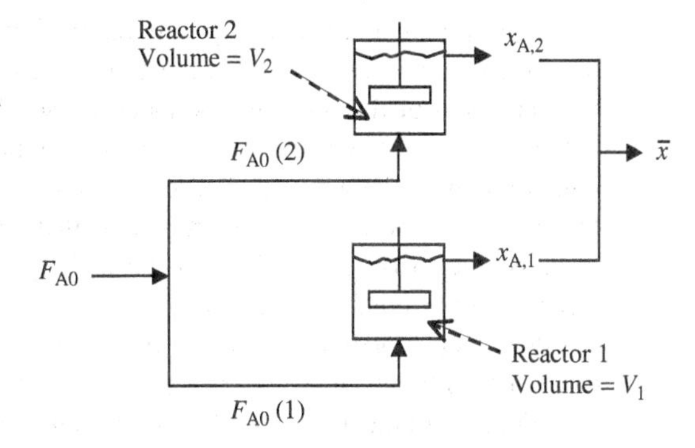

Reactor 1 has a volume of V_1; V_2 is the volume of reactor 2. The feed stream is split between the two reactors, such that the molar flow rate of A to reactor 1 is F_{A0} (1) and the molar flow rate to reactor 2 is F_{A0} (2). The fractional conversion of A in the effluent from reactor 1 is $x_{A,1}$, and the conversion of A in the effluent from reactor 2 is $x_{A,2}$. The average conversion of A in the combined effluent stream is $\bar{x}$.

You probably are suspicious that this configuration is not very practical. We have already learned that the performance of two CSTRs in series is better than the performance of a single CSTR with the same total volume. Is there any reason to believe that two CSTRs in parallel will perform better than a single CSTR with the same total volume?

Let's tackle this question by analyzing a simplified version of the above configuration. Suppose that the feed is split into two equal streams, so that $F_{A0}(1) = F_{A0}(2) = F_{A0}/2$. Consider the case where both reactors are operated at the same temperature and with the same feed concentrations, and where the kinetics are "normal." If the average conversion $\bar{x}$ is fixed, will the total reactor volume be lower if the two reactors each have the same volume, or will the total volume be lower if the two reactors have different volumes?

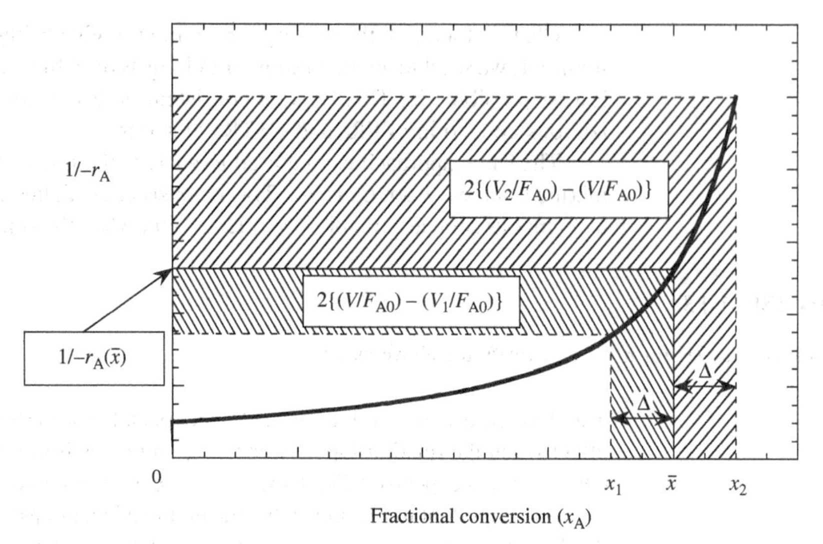

Figure 4-11 Graphical representation of the design equations for two CSTRs in parallel.

Let reactor 1 be smaller than reactor 2, i.e., $V_1 < V_2$. Since both reactors have the same molar inlet flow rate of A $(F_{A0}/2)$, the fractional conversion of A in the stream leaving reactor 1 will be lower than the conversion leaving reactor 2, i.e., $x_{A,1} < x_{A,2}$. The conversion of A in the combined effluent is $\bar{x}$, and the molar feed rate of A to each reactor is the same. Therefore, if $\Delta = \bar{x} - x_{A,1}$, then $x_{A,2} = \Delta + \bar{x}$.

A graphical analysis comparing the performance of the two unequal-size reactors with that of two equal-size reactors is shown in Figure 4-11.

The area of the rectangle bounded by solid lines, i.e., $(1/-r_A(\bar{x})) \times \bar{x}$, is the value of $V/(F_{A0}/2)$, where V is the volume required to produce a conversion of $\bar{x}$ when the molar flow rate of A to the reactor is $F_{A0}/2$. The area of the lower unfilled region is $V_1/(F_{A0}/2)$. Therefore, the area of the lower "L-shaped" region, with diagonals running from upper left to lower right, is the difference between $V/(F_{A0}/2)$ and $V_1/(F_{A0}/2)$, i.e., $2\{(V/F_{A0}) - (V_1/F_{A0})\}$. This area is directly proportional to the difference $(V - V_1)$. It is the volume "savings" associated with the smaller reactor, operating at a conversion $x_{A,1}$, compared to a reactor with a volume of V, operating with the same feed rate $(F_{A0}/2)$, but at a higher conversion $\bar{x}$.

The area of the upper "L-shaped" region, with diagonals running from lower left to upper right, is the difference between $V_2/(F_{A0}/2)$ and $V/(F_{A0}/2)$, i.e., $2\{(V/F_{A0}) - (V_1/F_{A0})\}$. This area is directly proportional to $(V_2 - V)$. This is the volume "penalty" associated with the larger reactor, operating at a conversion $x_{A,2}$, compared to a reactor with a volume of V, with the same feed rate $(F_{A0}/2)$, operating at a lower conversion $\bar{x}$.

Clearly, the upper "L-shaped" area is larger than the lower "L-shaped" area. Therefore,

$$2\left\{\frac{V_2}{F_{A0}} - \frac{V}{F_{A0}}\right\} > 2\left\{\frac{V}{F_{A0}} - \frac{V_1}{F_{A0}}\right\}$$

or

$$V_1 + V_2 > 2V$$

For two CSTRs in parallel, with the same feed to each reactor, the required total volume is *greater* if the CSTRs have different volumes and operate at different conversions than if the two CSTRs have the same volume and operate at the same conversion.

Suppose we adjusted $F_{A0}(1)$ and $F_{A0}(2)$ so that the conversion was $\bar{x}$ in the streams leaving both CSTRs. Then $V_1/F_{A0}(1) = V_2/F_{A0}(2) = \bar{x}/-r_A(\bar{x}) = V/(F_{A0}/2)$. If $F_{A0}(1) + F_{A0}(2) = (F_{A0})$, then $V_1 + V_2 = 2V$. This shows that the *best* that can be done with two CSTRs in parallel is to match the performance of a single CSTR with the same total volume as the two CSTRs in parallel.

This analysis confirms that operating two CSTRs in series would give a better performance than operating the same reactors in parallel. In fact, it is difficult to imagine a situation where one would deliberately choose to operate two CSTRs in parallel.

4.3.2.2 PFRs in Parallel

Now let's consider two PFRs in parallel. This case is not quite as obvious. We learned previously that operating two PFRs in series gives the same performance as a single PFR operating at the same space time. Here we will analyze the case where both of the parallel PFRs have the same feed rate, feed composition, and temperature. Again, "normal" kinetics will be assumed. The situation is shown in the following figure.

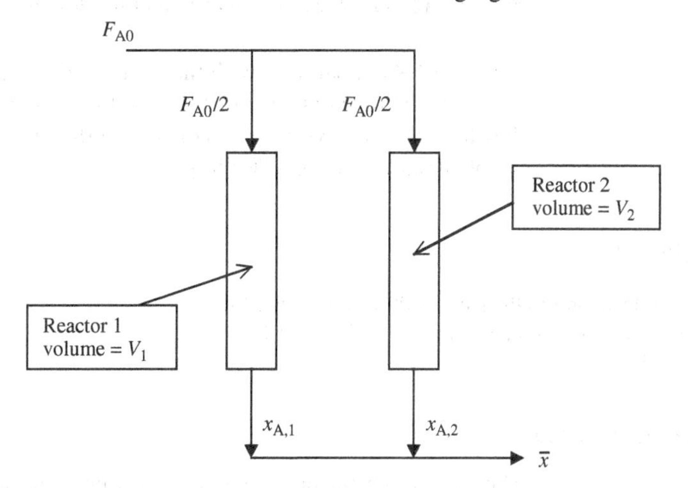

Let reactor 1 be the smaller of the two reactors, i.e., $V_1 < V_2$. Since both reactors have the same molar inlet flow rate of A, $F_{A0}/2$, and the same feed composition and the same temperature, the fractional conversion of A in the stream leaving reactor 1 will be lower than the conversion in the stream leaving reactor 2, i.e., $x_{A,1} < x_{A,2}$. However, the conversion of A in the combined effluent must be $\bar{x}$, and the molar feed rates of A to each reactor are the same. If $\Delta = \bar{x} - x_{A,1}$, then $x_{A,2} = \Delta + \bar{x}$.

This case can also be analyzed using the graphical technique, as shown in Figure 4-12.

The shaded area on the left, with diagonals running from lower left to upper right, is equal to $2\{(V/F_{A0}) - (V_1/F_{A0})\}$. This area is proportional to $(V - V_1)$, the volume "savings" associated with the smaller reactor, operating at a conversion $x_{A,1}$, relative to a reactor with a volume of V, with the same feed rate $(F_{A0}/2)$, operating at a higher conversion $\bar{x}$.

The shaded area to the right of the figure, with diagonals running from upper left to lower right, is equal to $2\{(V_2/F_{A0}) - (V/F_{A0})\}$. This area is proportional to $(V_2 - V)$, the volume "penalty" associated with the larger reactor, operating at a conversion $x_{A,2}$, relative to a reactor with a volume of V, with the same feed rate $(F_{A0}/2)$, operating at a lower conversion $\bar{x}$.

Clearly, the shaded area on the right is larger than the shaded area on the left, so that,

$$V_1 + V_2 > 2V$$

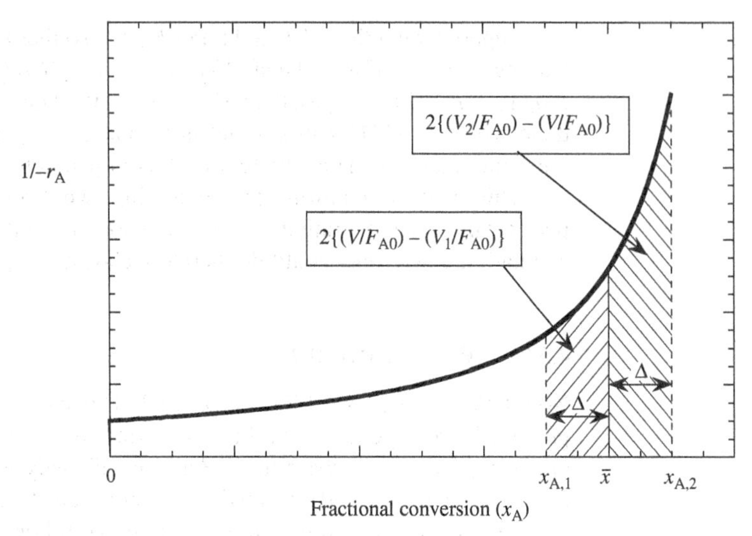

Figure 4-12 Graphical representation of the design equations for two PFRs in parallel.

For two PFRs in parallel, with the same feed to each reactor, the required total volume is greater if the reactors have different volumes and operate at different conversions than if they have the same volume and operate at the same conversion. This is the same result that we obtained for two CSTRs in parallel.

EXERCISE 4-14

Suppose that the feed rates to the parallel PFRs are adjusted so that both reactors operate with an outlet conversion of $\bar{x}$. Show that $V_1 + V_2 = 2V$.

4.3.3 Generalizations

The examples in the preceding sections were quite specific and were constructed to facilitate analysis. With either CSTRs or PFRs in parallel, it was best to have both reactors operating at the same conversion $\bar{x}$. The performance of parallel reactors operating at different conversions that averaged to $\bar{x}$ was inferior.

The results of these analyses can be generalized:

If a single reaction takes place in a network of reactors that has parallel branches, the optimum performance will result when the conversion is the same in any streams that merge.

This generalization applies to any combination of CSTRs and PFRs in series and parallel, for "normal" kinetics.

The above analysis also supports a second generalization:

A series arrangement of reactors is always as efficient or more efficient than a parallel arrangement.

This generalization also applies for any combination of CSTRs and PFRs, and for "normal" kinetics.

Despite the last generalization, there are occasions where it is necessary or desirable to use PFRs in parallel. For example, exothermic reactions sometimes are run in reactors that resemble shell-and-tube heat exchangers. A single reactor may have hundreds of tubes in parallel. The tubes are filled with catalyst pellets and a heat-transfer fluid is circulated through the shell side. Heat is removed through the walls of the tubes to maintain the temperature inside the tubes below some predetermined limit. The limit may be set, for example, by the need to avoid side reactions that occur at high temperatures, or by the need to control the rate of catalyst deactivation, which generally increases with increasing temperature.

According to the first generalization, the conversion leaving each tube must be the same if the overall reactor is to have optimum performance. Even if each individual tube behaves as an ideal PFR, the performance of the *overall* reactor will be less than that of an ideal PFR unless each and every tube produces the same fractional conversion.

This is a challenging requirement! It means, according to the design equation for an ideal PFR, that each and every tube must operate at the same W/F_{A0}. In order for that to happen, the catalyst must be added so that each tube contains the same weight. Moreover, the fluid that flows through the tubes must be fed so that the pressure drop across each tube is the same. Otherwise, F_{A0} will not be the same for each tube.

If a tube contains less than the required amount of catalyst, the resistance to fluid flow in that tube will be lower than the average because of the greater void volume in that tube. The lower resistance will result in a higher F_{A0}, so that the value of W/F_{A0} for this tube might be substantially below average.

As a final illustration of reactors in parallel, flip ahead to Chapter 9 and have a look at Figure 9-5. This figure shows a form of catalyst known as a "honeycomb" or "monolith." These catalysts consist of several hundred parallel channels per square inch of frontal area. Each channel is an independent reactor (although not necessarily an ideal PFR) since there is no flow of fluid between channels. One of the challenges of using this form of catalyst is to ensure that the value of W/F_{A0} is the same, or as close as possible, from channel to channel.

4.4 RECYCLE

Under some circumstances, it may be desirable to recycle some of the effluent stream that leaves a reactor back to the reactor inlet. Recycle can be used to control the temperature in the reactor and to adjust the product distribution if more than one reaction is taking place. The issues of multiple reactions and temperature control are addressed in Chapters 7 and 8, respectively.

A reactor with recycle is depicted in the following figure.

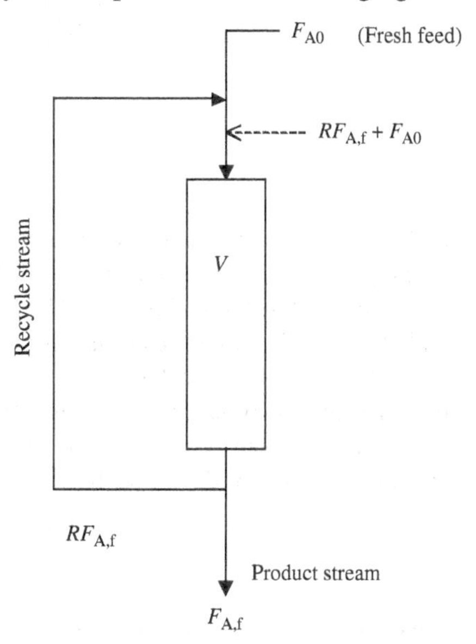

Species A is a reactant. The reactor is an ideal PFR. The molar ratio of the recycle stream to the product stream is R. This ratio is referred to as the "recycle ratio." For the time being, we will assume that the compositions of the product and recycle streams are identical.

Suppose that we want to apply the design equation to the PFR. One way to do this is to recognize that the total molar flow rate of A into the reactor is $RF_{A,f} + F_{A0}$, and to set the fractional conversion of A, x_A, equal to zero for the combined stream that enters the reactor. On this basis the design equation becomes

$$\frac{V}{F_{A0} + RF_{A,f}} = \int_0^{x_{out}} \frac{dx_A}{-r_A} \tag{4-38}$$

where

$$x_{out} = [(F_{A0} + RF_{A,f}) - (R+1)F_{A,f}]/(RF_{A,f} + F_{A0})$$
$$x_{out} = (F_{A0} - F_{A,f})/(RF_{A,f} + F_{A0}) \tag{4-39}$$

This conversion x_{out} is referred to as the "per pass" conversion. It is the fraction of the *total* feed of reactant A that is converted in a single pass through the reactor.

Equation (4-38) is difficult to use because $F_{A,f}$ (or x_{out}) appears on both sides of the equation. Therefore, we seek a simpler basis from which to analyze the problem.

Consider the following flowsheet.

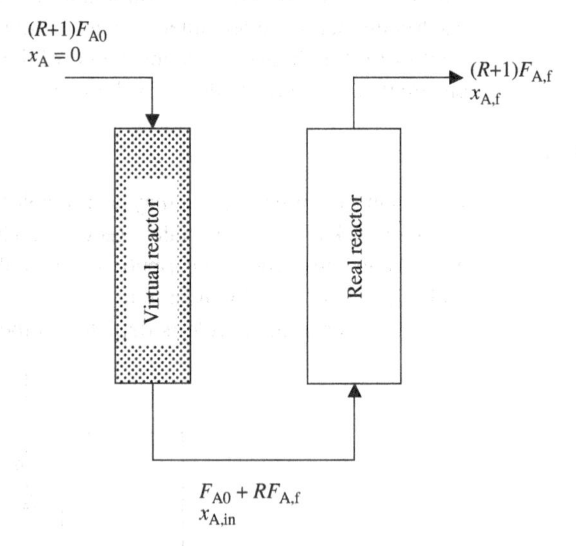

In this flowsheet, the first reactor is a "virtual" reactor. It takes the place of the mixing point in the previous figure, where the fresh feed and the recycle stream combine to form the stream that enters the real reactor. The function of the virtual reactor is to generate a stream with the same flow rate and composition as the stream that enters the real reactor. The molar flow rate of A to the virtual reactor is $(R+1)F_{A0}$. Just enough reaction takes place so that the flow rate of A leaving the virtual reactor is $F_{A0} + RF_{A,f}$.

Since the conversion of A in the feed entering the virtual reactor is 0, then the conversion in the outlet from the virtual reactor, and in the feed to the real reactor, is

$$x_{A,in} = \frac{(R+1)F_{A0} - (RF_{A,f} + F_{A0})}{(R+1)F_{A0}}$$

This conversion is based on the feed to the first (virtual) reactor, $(R+1)F_{A0}$.

Let $x_{A,f}$ be the fractional conversion of A in the outlet from the second (real) reactor. The basis for $x_{A,f}$ will be the same as for $x_{A,in}$, i.e., both conversions are based on the feed to the first (virtual) reactor. We then can write $F_{A,f} = F_{A0}(1 - x_{A,f})$. Substituting this relationship into the above equation gives

$$x_{A,in} = \left(\frac{R}{R+1}\right) x_{A,f} \tag{4-40}$$

The design equation for the real reactor now can be written as

$$\frac{V}{(R+1)F_{A0}} = \int_{x_{A,in}}^{x_{A,f}} \frac{dx_A}{-r_A} = \int_{\left(\frac{R}{R+1}\right)x_{A,f}}^{x_{A,f}} \frac{dx_A}{-r_A}$$

or

Design equation for recycle reactor

$$\frac{V}{F_{A0}} = (R+1) \int_{\left(\frac{R}{R+1}\right)x_{A,f}}^{x_{A,f}} \frac{dx_A}{-r_A} \tag{4-41}$$

Equation (4-41) is the design equation for a recycle reactor. This equation shows that the behavior of a recycle reactor can be varied continuously between an ideal PFR ($R = 0$) and an ideal CSTR ($R \to \infty$) by changing the recycle ratio. As the recycle ratio increases, the reactor behaves more like a CSTR.

EXAMPLE 4-11
Recycle Reactor Sizing

The first-order, homogeneous, liquid-phase reaction: A $\to$ products, takes place in an isothermal, ideal PFR with recycle. The recycle ratio is 1.0. The rate constant is $k = 0.15\ \text{min}^{-1}$. The concentration of A in the fresh feed is 1.5 mol/l, and the molar feed rate of A is 100 g·mol/min.

What reactor volume is required to achieve a final fractional conversion of 0.90? What is the "per pass" conversion for this case?

APPROACH

The required volume can be calculated by solving the design equation for a recycle reactor, Eqn. (4-41), recognizing that $-r_A = kC_A = kC_{A0}(1 - x_A)$. The "per pass" conversion can be calculated from Eqn. (4-39).

SOLUTION

For $R = 1$ and $x_{A,f} = 0.90$, Eqn. (4-41) becomes

$$\frac{V}{F_{A0}} = (1+1) \int_{\left(\frac{1}{1+1}\right)(0.90)}^{0.90} \frac{dx_A}{kC_{A0}(1 - x_A)}$$

$$V = \frac{2F_{A0}}{kC_{A0}} \int_{0.45}^{0.90} \frac{dx_A}{(1 - x_A)} = \frac{-2F_{A0}}{kC_{A0}} \ln(1 - x_A)\Big|_{0.45}^{0.90}$$

Substituting numbers

$$V = \frac{-2 \times 100(\text{mol/min})}{0.15(1/\text{min}) \times 1.0(\text{mol/l})} \{-2.303 + 0.598\}$$

$$V = 2273\ \text{l}$$

The "per pass" conversion x_{out} is given by Eqn. (4-39).

$$x_{out} = \frac{F_{A0} - F_{A0}(1 - x_{A,f})}{F_{A0} + RF_{A0}(1 - x_{A,f})} = \frac{x_{A,f}}{1 + R(1 - x_{A,f})}$$

Substituting numbers gives $x_{out} = 0.82$

EXERCISE 4-15

For the same F_{A0}, C_{A0}, k, and $x_{A,f}$, what volume of reactor would be required if there was no recycle and the reactor was (a) an ideal PFR? (Answer—1535 l); (b) an ideal CSTR? (Answer—6000 l).

SUMMARY OF IMPORTANT CONCEPTS

- The design equations for the three ideal reactors can be solved for unknown quantities such as the final (outlet) conversion, the weight of catalyst (or volume of reactor) required, or the feed rate of a component. There only can be one unknown in the design equation. All of the other parameters, including the complete rate equation, must be known.
- A stoichiometric table should be used to express all of the concentrations in the reactor as a function of a single composition variable.
- "Levenspiel plots" can be used to analyze the behavior of single reactors and systems of reactors. For "normal kinetics.
  - A PFR always requires less volume (or catalyst weight) than a CSTR to do a specified "job".
  - PFRs or CSTRs in parallel have, *at best*, the same performance as a single CSTR or PFR, respectively.
- A series of PFRs has the same performance as a single PFR.
- A series of CSTRs has a better performance than a single CSTR.
- If the concentrations and the temperature in a catalyst particle are the same as in the bulk fluid stream that surrounds the catalyst particle, then transport effects can be ignored and the reaction is controlled by intrinsic kinetics. When this is the case, the design equation can be solved without introducing additional equations to describe mass and heat transport. Intrinsic kinetic control is an important limiting case of reactor behavior.
- Always understand the chemical equilibrium of a reaction before attempting to solve problems involving reaction kinetics.

PROBLEMS

Single Reactor Problems

Problem 4-1 (Level 3)[6] When the isomerization of normal pentane

$$n\text{-}C_5H_{12} \rightleftarrows i\text{-}C_5H_{12}$$

is carried out in the presence of hydrogen at 750 °F over a catalyst composed of a "metal deposited on a refractory support," the kinetics are adequately described by the rate equation:

$$-r_n\left(\frac{\text{g·mol } C_5H_{12}}{\text{g·cat-h}}\right) = \frac{k\left(p_n - \dfrac{p_i}{K}\right)}{(1 + K_{H_2}p_{H_2} + K_i p_i + K_n p_n)^2}$$

where the subscript "i" refers to isopentane and, "n" refers to normal pentane, and "p" denotes partial pressure. At 750 °F, the values of the constants in the above expression are

$$K = 1.632$$
$$k = 2.08 \times 10^{-3}(\text{g·mol/g·cat-h-psia})$$
$$K_{H_2} = 2.24 \times 10^{-3}(\text{psia})^{-1}$$
$$K_n = 3.50 \times 10^{-4}(\text{psia})^{-1}$$
$$K_i = 5.94 \times 10^{-3}(\text{psia})^{-1}$$

It is desired to size a steady-state reactor that will operate isothermally at 750 °F and a total pressure of 500 psia. The fractional conversion of n-pentane to *iso*-pentane in the reactor effluent must be 95% of the maximum possible conversion. The feed to the reactor will be a mixture of hydrogen and n-pentane in the ratio 1.5 mol H_2/1.0 mol n-pentane. No other compounds will be present.

What value of the space time, $\tau = (WC_{A0}/F_{A0})$, will be required if

1. the reactor is an ideal, plug-flow reactor?
2. the reactor is an ideal, backmix reactor (CSTR)?

Be sure to specify the units of τ. Pressure drop through the reactor may be neglected, transport effects may be neglected, and the ideal gas laws may be assumed.

[6] Carr, N. L., *Ind. Eng. Chem.*, 52 (5) 391–396 (1960).

Problem 4-2 (Level 2) The following e-mail is in your box at 8 AM on Monday:

To: U. R. Loehmann

From: I. M. DeBosse

Subject: Sizing of LP Reactor

The kinetics of the liquid-phase reaction

$$\text{R.M.\#11} \rightarrow \text{L.P.\#7} + \text{W.P.\#31}$$

have been studied on a bench scale. (Unfortunately, because of network security concerns, I cannot be more specific about the chemicals involved. As you know, L.P.#7 is the desired product of this reaction.) The rate of disappearance of R.M.#11 (A) is adequately described by the zero-order rate equation

$$-r_A = k$$

The value of $k = 0.035$ lb·mol A/gallon-min at a certain temperature T.

A plug-flow reactor to convert R.M.#11 to L.P.#7 has been sized by Cauldron Chemicals' Applied Research Department, but there is some controversy about the result. I would like your help in analyzing the situation.

The reactor is intended to operate isothermally at temperature T. According to the Applied Research Department, the reactor volume will be 120 gallons, the volumetric flow rate to the reactor will be 2 gallon/min and the concentration of A in the feed to the reactor will be 1.0 lb·mol A/gallon. What will the concentration of A be in the effluent from the reactor?

Please report your answer to me in a short memo. Attach your calculations to the memo to support your conclusions.

Problem 4-3 (Level 2) The irreversible reaction A → B takes place in a solvent. The kinetics of the reaction have been studied at concentrations of A from 2.0 g·mol/l to 0.25 g·mol/l. Over this range, the kinetic data are well correlated by the rate equation

$$-r_A = kC_A^{0.5}$$

At 25 °C, in an isothermal, ideal batch reactor, the concentration of A drops from an initial C_{A0} of 2.0 g·mol/l to 1.0 g·mol/l in 15 min. At 50 °C, it takes 20 s for this same change to occur.

What will the concentration of A be in an ideal batch reactor operating isothermally at 40 °C after 10 min if the initial concentration C_{A0} is 2.0 g·mol/l? Explain your answer to whatever extent is required to make it plausible.

Problem 4-4 (Level 1) The irreversible, homogenous, gas-phase reaction A + B → R, is taking place isothermally in an ideal, constant-volume batch reactor. The temperature is 200 °C and the initial total pressure is 3 atm. The initial composition is 40 mol% A, 40 mol% B, and 20 mol% N_2. The ideal gas laws are valid.

The reaction obeys the rate equation

$$-r_A = kC_AC_B$$

At 100 °C, $k = 0.0188$ l/mol-min. The activation energy of the reaction is 85 kJ/mol.

1. What is the value of the rate constant at 200 °C?
2. What is the fractional conversion of A after 30 min?
3. What is the total pressure in the reactor after 30 min?

Problem 4-5 (Level 2) A reactor is to be sized to carry out the heterogeneous catalytic reaction

$$A \rightarrow R + S$$

The reactor will operate at 200 °C and 1 atm pressure. At these conditions, A, R, and S are ideal gases. The reaction is essentially irreversible, the heat of reaction is essentially zero and the intrinsic rate of reaction is given by

$$-r_A(\text{lb·mol A/lb·cat-h}) = kC_A$$

$$k = 275\ (\text{ft}^3/\text{lb·cat-h})$$

The feed to the reactor consists of A, R, and N_2 in a 4:1:5 molar ratio. The feed gas flow rate is 5.0×10^6 ft^3/h at 200 °C and 1 atm. The reactor must be sized to give 95% conversion of A.

A radial-flow, fixed-bed reactor, as shown in Problem 3-1 at the end of Chapter 3, will be used. The feed enters through a central pipe with uniformly spaced holes, flows radially outward through the catalyst bed, through a screen, and into an annulus, from which it flows out of the reactor. This direction of flow is opposite to that shown in Problem 3-1. Pressure drop through the catalyst bed can be neglected, and transport resistances are negligible.

For the purpose of sizing the reactor, assume that there is no mixing of fluid elements in the direction of flow (radial direction) and no concentration gradients in the axial or angular dimensions.

1. Calculate the weight of catalyst required to achieve at least 95% conversion of A.
2. The reactor has been brought on-stream and the conversion of A is only 83%. The temperature is measured at several positions in the catalyst bed, and in the inlet and outlet streams. The measured temperatures are all 200 °C. List as many reasons as you can that might explain the low conversion.

Problem 4-6 (Level 2) The catalytic cracking reaction

$$A \rightarrow B + C + D$$

is taking place in a fluidized-bed reactor. At reaction conditions, the reaction is essentially irreversible and is second order in A.

For the purpose of a preliminary analysis, the reactor can be assumed to operate as an ideal CSTR. The reactor contains 100,000 kg of catalyst. The feed to the reactor is pure A at a molar feed rate of 200,000 g·mol/min. The concentration of A in the feed to the reactor is 15.8 g·mol/m^3, at the temperature and pressure of the reactor.

When the reactor was started up with "fresh" catalyst, the fractional conversion of A was 71%. However, the catalyst deactivated with continued operation until the conversion had declined to 47%. At this point, 40% of the original catalyst was removed from the reactor and was replaced with an equal weight of "fresh" catalyst.

You may assume that the form of the rate equation does not change as a result of catalyst deactivation, i.e., the decrease in conversion is solely the result of a decrease in the rate constant. You may also assume that the ideal gas laws are applicable, and that transport resistances are negligible.

1. What was the value of the rate constant for the "fresh" catalyst, when the conversion was 71%?
2. By what percentage did the rate constant decrease as the conversion of A decreased from 71 to 47%? Does this answer seem reasonable?
3. What conversion of A would you expect when the reactor reached steady state after replacing 40% of the original catalyst with "fresh" catalyst?

Problem 4-7 (Level 1) Hydrodealkylation is a reaction that can be used to convert toluene (C_7H_8) into benzene (C_6H_6), which is historically more valuable than toluene. The reaction is

$$C_7H_8 + H_2 \rightarrow C_6H_6 + CH_4$$

Zimmerman and York[7] have studied this reaction between 700 and 950 °C in the absence of any catalyst. They found that the rate of toluene disappearance was well correlated by

$$-r_T = k_T[H_2]^{1/2}[C_7H_8]$$
$$k_T = 3.5 \times 10^{10} \exp(-E/RT)(\text{l/mol})^{1/2}/\text{s}$$
$$E = 50{,}900 \text{ cal/mol}$$

The reaction is essentially irreversible at the conditions of the study, and the ideal gas laws are valid.

Consider a feed stream that consists of 1 mol of H_2 per mole of toluene. A reactor is to be designed that operates at atmospheric pressure and a temperature of 850 °C. What volume (in liters) of reactor is required to achieve a fractional conversion of toluene of 0.50 with a toluene feed rate of 1000 mol/h?

First assume that the reactor is an ideal, plug-flow reactor. Then repeat the calculation for an ideal CSTR.

Problem 4-8 (Level 2)[8] The following e-mail is in your box at 8 AM on Monday morning:

To: U. R. Loehmann

From: I. M. DeBosse

Subject: Analysis of Patent Data

As you know, one of Cauldron Chemicals' strategic thrusts is the manufacture of pharmaceutical intermediates. The first step in the synthesis of a proprietary cardiovascular drug is the liquid-phase reaction of 2 mol of 4-cyanobenzaldehyde (A) with 1 mol of hydroxylamine sulfate (B) to give 2 mol of 4-cyanobenzaldoxime, 1 mol of sulfuric acid, and 2 mol of water. We are engaged in a race with Pheelgoode Pharmaceutical to scale up and optimize this reaction. A patent has just been issued to Pheelgoode. Our Patent Department does not believe that Pheelgoode's patent is pertinent to our own efforts. However, the patent contains data that may tell us something about what Pheelgoode is doing.

First, the patent states that the rate equation for the reaction is $-r_A = kC_AC_B$. Second, the expression for k is given as $k = 74,900 \exp(-8050/RT)$, where the units of the activation energy are cal/mole and the units of the rate constant are l/mol-min. Finally, the patent contains the following data, taken in an isothermal batch reactor.

Reaction temperature (°C)	Fractional conversion of A		
	$t = 15$ min	$t = 30$ min	$t = 120$ min
22	—	0.788	0.964
40	0.966	—	—

Unfortunately, the initial concentrations of A and B (C_{A0} and C_{B0}) are not given in the patent. It is stated that C_{A0} was the same for the experiment at 22 °C as for the experiment at 40 °C. However, the initial concentration of B was higher at 40 °C than at 22 °C.

Please see if you can figure out what the initial concentrations of A and B were in Pheelgoode's experiments. Report your findings to me in a memo that does not exceed one page in length. Attach your calculations in case someone wants to review them.

Problem 4-9 (Level 1) The irreversible, gas-phase trimerization reaction

$$3A \rightarrow B$$

is taking place at steady state in an ideal CSTR that has a volume of 10,000 l. The feed to the reactor is a 1/1 molar mixture of A and N_2 at 5 atm total pressure and a temperature of 50 °C. The reactor operates at 350 °C and 5 atm total pressure. The volumetric feed rate is 8000 l/h, at feed conditions. The gas mixture is ideal at all conditions.

The reaction is homogeneous and the rate of disappearance of A is given by

$$-r_A = kC_A$$

The value of k is 4.0×10^{-5} h^{-1} at 100 °C and the activation energy is 90.0 kJ/mol.

What is the fractional conversion of A in the stream leaving the reactor?

Problem 4-10 (Level 1) The homogeneous liquid-phase reaction

$$A + B \rightarrow C + 2D$$

[7] Zimmerman, C. C., and York, R., *I&EC Process Design Dev.*, 3(1), 254–258 (1962).

[8] Adapted from Chung, J., "Co-op student contribution to chemical process development at DuPont Merck," *Chem. Eng. Educ.*, 31(1), 68–72 (1992).

is taking place in an ideal CSTR. The reaction obeys the rate equation

$$-r_A = \frac{k C_A C_B}{(1 + K_B C_B)^2}$$

At 200 °C,

$$k = 0.12 \text{ l/mol-s}$$
$$K_B = 1.0 \text{ l/mol}$$

The feed to the reactor is an equimolar mixture of A and B, with $C_{A0} = C_{B0} = 2.0$ mol/l. The reactor operates at 200 °C. The molar feed rate of A, F_{A0}, is 20 mol/s.

How large a reactor is required if the final conversion of A must be greater than 90%?

Problem 4-11 (Level 2) The organic acid, ACOOH, reacts reversibly with the alcohol, BOH, to form the ester ACOOB according to the stoichiometric equation

$$\text{ACOOH} + \text{BOH} \rightleftarrows \text{ACOOB} + H_2O$$

The reaction will be carried out in an ideal batch reactor and the water will be removed rapidly by stripping with an inert gas as the reaction proceeds. Therefore, the reverse reaction can be neglected. The rate equation for the forward reaction is

$$-r_{\text{ACOOH}} = k\,[\text{ACOOH}]^2[\text{BOH}]$$

The value of the rate constant k is

$$0.16 \text{ l}^2/\text{mol}^2\text{-h}$$

at 373 K. The activation energy is 63.9 kJ/mol and the heat of reaction ΔH_R at 373 K is −126.3 kJ/mol.

The reaction is carried out in solution in a 2000-l reactor. The initial concentration of A is 2.0 mol/l and the initial concentration of B is 3.0 mol/l. The reactor is to be operated isothermally at 373 K.

1. How long must the reaction be run in order to obtain a fractional conversion of A of 0.90?
2. If the reactor is operated adiabatically, will it require more time, the same time, or less time to reach a final conversion of 0.90, if the initial temperature is 373 K. Explain your reasoning.
3. Suppose that water was not removed as the reaction proceeded. If the equilibrium constant (based on concentration) for the reaction were 10, what is the *maximum* fractional conversion of A that could be achieved? You may assume that all of the water that is produced remains dissolved in the solution.

Problem 4-12 (Level 2) A schematic diagram of a bubble-column reactor is shown in the following diagram. Both liquid and gas are fed to and withdrawn from the reactor continuously.

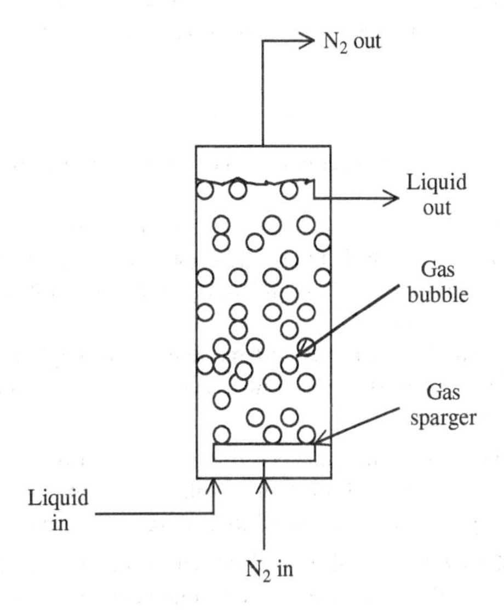

The volume fraction of gas in the column is referred to as the "gas holdup" ε. Values of the gas holdup for bubble columns can be obtained from correlations in the literature. The gas holdup is a function of the liquid velocity, the gas velocity (u_g), and the physical properties of the gas and liquid. Many of the available holdup correlations show that holdup is proportional to approximately the 0.6 power of the superficial gas velocity

$$\varepsilon \propto (u_g)^{0.60}$$

Consider a situation where the irreversible reaction

$$A \rightarrow R$$

is taking place in the *liquid* (no reaction takes place in the gas). Nitrogen is being sparged (bubbled) through the liquid. The liquid in the column is mixed intensely by the gas jets at the sparger, and by the rising gas bubbles. Therefore, the reactor approximates an ideal CSTR. The concentration of A in the liquid that is fed to the column is C_{A0}. The reaction is second-order in A.

At normal operating conditions, the fractional conversion of A in the effluent from the reactor is 0.80 and the value of the gas holdup is 0.30. However, on the midnight shift last night, one of the operators was careless and set the N_2 flow rate to triple the normal value.

Once a new steady state had been reached, what was the fractional conversion of A? As a first approximation, assume that the temperature of the reactor did not change when the N_2 flow rate was tripled.

Problem 4-13 (Level 1)

1. The cracking reaction

$$A \rightarrow B + C + D$$

is being carried out isothermally in an ideal plug-flow reactor (PFR) packed with a heterogeneous catalyst. The reactor operates at essentially atmospheric pressure. At the conditions of operation, the ideal gas laws are obeyed. The reaction follows the rate equation

$$-r_A = kC_A^2$$

At the operating temperature of the reactor, $k = 0.40$ (m^6/g mol-kg-min)

The feed to the reactor is a mixture of 2 mol of A and 1 mol of steam. Neither B, C, nor D is H_2O (steam). The molar feed rate of A is 100,000 g·mol/min and the concentration of A in the feed is 10.3 g·mol/m^3.

Assuming that transport resistances are negligible, what weight of catalyst is required to achieve 85% fractional conversion of A?

2. Suppose that the same reaction was carried out in a CSTR instead of a PFR, with all other conditions the same. What weight of catalyst would be required to achieve 85% conversion?

Problem 4-14 (Level 3)[9] Tungsten (W) is used as an interconnect in the manufacture of integrated circuits. Low-pressure chemical vapor deposition of tungsten can be carried out via the reaction of tungsten hexafluoride with hydrogen:

$$WF_6 + 3H_2 \rightarrow W\downarrow + 6HF \qquad (1)$$

Ideally, this reaction takes place only on the solid surface onto which W is being deposited. WF_6, H_2, and HF are gases; the metallic tungsten that is formed remains on the solid surface, and the thickness of the tungsten layer increases with time.

The rate of W deposition is given by

$$r_W = \frac{1.0 \exp[-8300/T]\, p_{H_2}^{0.5}\, p_{WF_6}}{1 + 450 p_{WF_6}}$$

where r_w = rate of W deposition (mol/cm^2-s), p_{WF_6} = partial pressure of WF_6 (mmHg), p_{H_2} = partial pressure of H_2 (mmHg), and T = temperature (K)

A disk of silicon with a diameter of 5.5 cm is placed inside a reactor. The temperature of the silicon disk is kept at 673 K. The feed to the reactor is continuous and consists of a mixture of WF_6, H_2, and argon (Ar), which is inert. The total gas rate is 660 standard cm^3/min. The inlet mole fractions are

$$y_{WF} = 0.045$$
$$y_{H_2} = 0.864$$
$$y_{Ar} = 0.091$$

The total absolute pressure in the reactor is 1 mmHg (1 Torr).

Assume that W deposits only on one side (the top side) of the silicon wafer. Assume that mixing of the gas in the reactor is so vigorous that the gas composition is the same everywhere in the reactor. Further assume that the gas composition in the reactor has reached steady state, and that transport resistances are negligible.

1. Derive expressions for p_{WF_6} and p_{H_2} in terms of the fractional conversion of WF_6 (x_A).
2. Let WF_6 be denoted "A." Show that the material balance on WF_6 for the whole reactor is

$$\frac{A}{F_{A0}} = \frac{x_A}{r_W}$$

where A is the exposed area of the silicon wafer ($A = \pi(5.5)^2/4\,\text{cm}^2$).

3. Calculate a numerical value of r_W for the conditions given.
4. Calculate the linear growth rate of the tungsten layer, in Å/min. (The density of metallic tungsten is 19.4 g/cc). Compare your answer to Figure 5 in the referenced article.

Problem 4-15 (Level 2) Altiokka and Çitak[10] have studied the esterification of acetic acid (A) with isobutanol (B) to form isobutyl acetate (E) and water (W). Amberlite IR-120 ion exchange resin was used as the catalyst. The ion exchange resin was in the form of small solid particles that were suspended in the liquid mixture of reactants and products. The rate equation was found to be

$$-r_A(\text{mol A/g·cat-h}) = \frac{k[C_A C_B - (C_E C_W/K_{eq})]}{(1 + K_B C_B + K_W C_W)}$$

At 333K, $k = 0.00384$ (l^2/g·cat-mol-h), $K_{eq} = 4$, $K_B = 0.460$ (l/mol B), and $K_W = 3.20$ (l/mol W).

If an ideal batch reactor is used, and if transport effects are negligible, what concentration of ion exchange resin (g/l) is required to reach a fractional conversion of acetic acid of 0.50 in 100 h at 333 K? The initial concentrations of acetic acid and isobutanol are both 1.50 mol/l, and there is no water or isobutyl acetate initially.

Problem 4-16 (Level 1) The reversible liquid-phase reaction

$$A + B \rightleftarrows C$$

is being carried out in an ideal batch reactor operating isothermally at 150 °C. The initial concentrations of A and B, C_{A0} and C_{B0}, are 2.0 and 3.0 g·mol/l, respectively. The initial concentration of C is 0.

The rate equation is

$$-r_A = k_f C_A C_B - k_r C_c$$

The value of k_f at 150 °C is 0.20 l/mol-h. The value of the equilibrium constant (based on concentration) for the reaction as written is 10 l/mol.

1. How long does it take for the conversion of A to reach 50%?
2. How long does it take for the conversion of A to reach 95%?

[9] Park, J.-H., *Korean J. Chem. Eng.*, 19(3), 391–399 (2002).

[10] Altiokka, M. L. and Çitak, A., *App. Catal A: General*, 239, 141–148 (2003).

Problem 4-17 (Level 2) The Cauldron Chemical Company currently supplies a commercial catalyst for the gas-phase isomerization reaction

$$A \rightleftarrows R$$

With Cauldron's catalyst, the forward reaction is first order in A and the reverse reaction is first order in R. Thermodynamic data for A and R are given in the table below.

A set of quality control tests is run on each batch of catalyst. One of these tests, designed to measure catalyst activity, is carried out as follows. Exactly 50 g of catalyst are charged to a small tubular reactor. The reactor operates isothermally at 300 °C and can be characterized as an ideal, plug-flow reactor. A mixture of A and N_2 is fed to the reactor at 300 °C and 1 atm total pressure. At these conditions, the inlet concentration of A, C_{A0}, is 0.00858 g·mol/l and the total volumetric feed rate is 500 l/h. Transport resistances are negligible. In order to pass the activity test, a catalyst sample must produce a fractional conversion x_A of 0.50 ± 0.01.

In response to pressure from competitors, Cauldron has started a research program to develop a catalyst with higher activity. One catalyst, EXP-37A, looks promising. In the quality control test, this catalyst reproducibly gives a conversion, $x_A = 0.68$.

The Marketing Department has started to provide samples of EXP-37A to potential customers along with some promotional literature. This literature claims that EXP-37A is 36% [(0.68 – 0.50) × 100/0.50] more active on a weight basis than Cauldron's standard commercial catalyst.

Comment on the validity of the Marketing Department's claim for improved catalyst activity. If you believe that 36% is an accurate quantification of the activity difference between EXP-37A and the standard catalyst, provide a rigorous justification of this number. If you feel that 36% is not an accurate quantification of the activity difference, generate a quantitative comparison of the activity difference. You may assume that mixtures of A, R, and N_2 obey the ideal gas laws at experimental conditions.

Thermodynamic data (at 298 K)			
Species	ΔG_f^0 (kcal/mol)	ΔH_f^0 (kcal/mol)	C_p (cat/mol) ^{0}K
A	−19.130	−7.380	11.18
R	−18.299	−3.880	11.18

Problem 4-18 (Level 2) The irreversible, homogeneous, liquid-phase reaction

$$A + B \rightarrow X + Y$$

is being carried out in the system shown below.

The PFR operates isothermally and has a volume of 5000 l. The volumetric flow rate of the fresh feed is $\upsilon = 10,000$ l/h, and the concentrations of A and B in the feed are 3.0 and 5.0 mol/l, respectively. The reaction is first order in A and first order in B, i.e., $-r_A = kC_AC_B$. At the temperature of operation, $k =$ 0.60 l/mol-h.

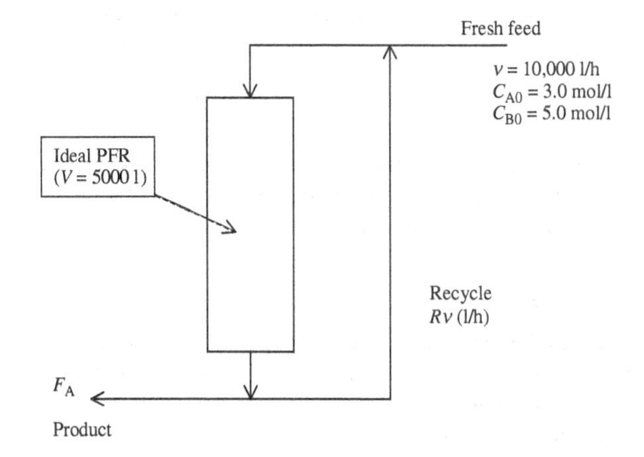

1. Suppose that the value of R (R = volumetric flow rate of recycle stream/volumetric flow rate of fresh feed) is unknown.

   (a) What is the highest possible conversion of A in the product stream?

   (b) What is the lowest possible conversion of A in the product stream?

Define "conversion of A" as $(F_{A0} - F_A)/F_{A0}$.

2. If $R = 2$, what conversion of A would you expect in the product stream?
3. Suppose that the actual conversion of A in a real facility was significantly less than you calculated in part 2 above. List at least three possible explanations.

Problem 4-19 (Level 1) Beltrame et al.[11] studied the oxidation of glucose to gluconic acid in aqueous solution at pH 7 using hyderase (a commercial enzyme catalyst system that is soluble in the reaction medium). The overall reaction is

$$C_5H_{10}O_6 + O_2 + H_2O \rightarrow C_5H_{10}O_7 + H_2O_2$$

Let E^0 designate the total amount of enzyme charged to the system, let G designate D-glucose, and let L designate D-gluconic acid.

The authors found that the rate of gluconic acid formation at 293 K was well described by the rate equation

$$r_L = \frac{k_c C_B C_G}{(k_c/k_1) + C_G}$$

provided that the concentration of dissolved O_2 was maintained at 1.18×10^{-3} mol/l. At 293 K, the values of k_c and k_1 were determined to be 0.465 mol/g-h and 18.8 l/g-h, respectively.

The reaction will be carried out in a batch reactor at 293 K, using an initial enzyme concentration of 0.0118 g/l and an initial glucose concentration of 0.025 mol/l. Assume that the O_2

[11] Beltrame, P., Comotti, M., Della Pina, C., and Rossi, M., "Aerobic oxidation of glucose L. Enzymatic catalysis," *J. Catal.*, 228, 282 (2004).

concentration can be maintained at 1.18×10^{-3} mol/l throughout the batch, and that it does not vary from point-to-point in the reactor.

1. How much time is required to reach 95% conversion of glucose?
2. An available reactor has a working volume of 1000 l. How much gluconic acid can be produced annually in this reactor, if a final glucose conversion of 95% is acceptable?

Multiple Reactor Problems

Problem 4-20 (Level 3) One of your friends, a chemist who does extracurricular research in his garage, has stumbled upon a new homogeneous catalyst that greatly increases the rate of the liquid-phase reaction:

$$\underset{(\mathrm{MW}=95)}{\mathrm{A}} + \underset{(\mathrm{MW}=134)}{\mathrm{B}} \xrightarrow{\text{catalyst}} \underset{(\mathrm{MW}=229)}{\mathrm{R}}$$

The reaction is essentially irreversible at the experimental conditions, and the catalyst is completely soluble in the reaction mixture.

As a chemical engineer, you know that R is a valuable product. With the idea of forming a company to manufacture it, you consult with your friend Harry, a marketing specialist. Harry estimates that 10,000,000 pounds per year of R (100% basis) could be sold without disturbing the current selling price. According to economic analyses performed by a third friend, Dick, a financier, this would be a highly profitable operation. Thus, the Embryonic Chemical Company is born.

Dick manages to negotiate a lease on the property of a defunct chemical company. The building contains only a single 100-gallon CSTR, but Dick is certain this reactor will be sufficient because it "... looked pretty ... big."

Meanwhile, back in the garage, your chemist friend has been studying the reaction in greater detail. He has found that the reaction is "clean," i.e., there are no side reactions, as long as the catalyst concentration, C_C, does not exceed 1.0×10^{-4} lb·mol/gal. You therefore choose this concentration for use in the plant. A stoichiometric mixture of A and B at reaction temperature contains 0.035 lb·mol/gal of each component and has a density of 8.00 lb/gal. In doing the research, attention has been confined to stoichiometric mixtures of A and B because Harry has determined that the product can be sold directly as it comes from the reactor (without a final purification) if the feed is stoichiometric and the final conversion is 93% or more. This is a big advantage because your building contains no separation equipment. In analyzing the kinetic data, you find that a sufficient rate equation is

$$-r_{\mathrm{A}}(\mathrm{lb\cdot mol/h\text{-}gal}) = \frac{kC_cC_{\mathrm{A}}C_{\mathrm{B}}}{1+KC_{\mathrm{B}}}$$

$$k = 1.51 \times 10^6 \ \mathrm{gal^2/lb\,mol^2\text{-}h}$$

$$K = 85.0 \ \mathrm{gal/lb\cdot mol}$$

1. Determine whether the existing 100-gallon CSTR is sufficient to produce 10,000,000 lb·R/year. The feed will be a stoichiometric mixture of A and B, and the conversion must be at least 93%. Allow for 10% downtime, i.e., 7890 operating hours per year. If the existing reactor is not sufficient, design a reactor *system* that will do the required job. Obviously, this system should include the existing CSTR. Minimizing the reactor volume is a sufficient design criterion.
2. Just as you are about to start production with the reactor system that you designed in Part 1, you get a call from Harry. After 18 holes of golf and 5 martinis, he has been persuaded by your largest (and only) customer that the product must have a final conversion of at least 97%. How much will your annual production rate be decreased?
3. What is the chemist's name?
4. What is the moral of the story?

Problem 4-21 (Level 2) Cauldron Chemical Company has just acquired Battsears Chemical Company! Executives at both companies have praised the transaction, saying that the two companies provide a "... great, long-term, strategic fit."

Cauldron has a plant in Salem, MA that makes L.P. #8 by the irreversible, homogenous, liquid-phase reaction of one mole of L.P. #7 to form 1 mol of L.P. #8:

$$\mathrm{L.P.\ \#7} \rightarrow \mathrm{L.P.\ \#8}$$

(Cauldron is very secretive, and all of its products and reactants have nondescriptive designations.)

The above reaction is first order in L.P. #7. Cauldron carries out this reaction in a plug-flow reactor, operating isothermally at 200 °C. The volume of the reactor is 2000 l. The feed to the reactor is a solution containing L.P. #7 at a concentration of 3.0 mol/l. The volumetric flow rate of the solution is 5000 l/h and the fractional conversion of L.P. #7 is 0.95.

1. What is the rate constant for the reaction at 200 °C?

   L.P. #8 is a very valuable intermediate, used for the production of L.P. #9. Prior to the acquisition, Battsears had been planning to start up a new facility in Eastwicke, Great Britain to produce L.P. #8, in competition with Cauldron. Battsears had contracted to buy the equivalent of 5000 l/h of the same solution of L.P. #7 that Cauldron uses as a feed. Battsears was ready to start up a PFR with a volume of 1000 l. This reactor was to have operated isothermally, also at 200 °C.
2. What would the fractional conversion of L.P. #7 have been in the outlet from Battsears' reactor?
3. If the Cauldron and Battsears plants both had operated as designed, what would be the combined production rate of L.P. #8 (mol/h)

   Cauldron needs more L.P. #8 very badly. However, the supply of the feed solution of L.P. #7 is limited. For the foreseeable future, the most that can be obtained is 10,000 l/h, i.e., the combined supply to Cauldron and Battsears.
4. What is the most L.P. #8 (mol/h) that can be obtained from the two reactors (Cauldron's plus Battsears')? The

operating temperature and the reactor volumes may not be changed.

Problem 4-22 (Level 2) The following e-mail is in your inbox at 8 AM on Monday:

To: U. R. Loehmann

From: I. M. DeBosse

Subject: L.P. #9 Expansion

Executives at Cauldron Chemical Company are very pleased with the success of our product, L.P. #9. (Cauldron is a very secretive company, so that all of our products have nondescriptive designations; please adhere to this convention in all of your work.)

Initially, we produced L.P. #9 in a batch reactor on an as-needed basis. Now, the largest batch reactor that Cauldron owns is dedicated to L.P. #9 production. In the near future, it will be necessary to produce the product continuously.

The reaction is

$$\text{L.P. \#8} \rightarrow \text{L.P. \#9}$$

This is a homogeneous reaction that takes place in solution. The reaction is second order in L.P. #8. The maximum operating temperature is 150 °C because L.P. #8 degrades at higher temperatures. The rate constant at 150 °C is 0.133 l/mol-h.

Cauldron has two continuous reactors that are not in use. One is an agitated, 2000-l reactor that behaves as a CSTR. The second is a 100-l tubular reactor that behaves as a PFR. Cauldron would like to produce 345 mol of L.P. #9 per hour from a feed that contains 4.0 mol L.P. #8/l. The fractional conversion of L.P. #8 must be 0.95 or higher because of constraints in the downstream product separation system.

Skip Tickle, Production Manager for L.P. Products, is in charge of the overall project. He has asked the following questions:

1. How should the CSTR and PFR be configured? (series or parallel? If in series, which reactor first?)
2. What will the final fractional conversion of L.P. #8 be for the reactor configuration in Part 1, assuming that the L.P. #8 feed rate is 363 mol/h (345 mol L.P. #9/0.95)?

Please report your answers to these questions to Skip in a short memo. Attach your calculations to the memo to support your conclusions.

Problem 4-23 (Level 3) The homogeneous, reversible, exothermic, liquid-phase reaction

$$\text{A} \rightleftarrows \text{R}$$

is being carried out in a reactor system consisting of an ideal PFR followed by an ideal CSTR. The volume of the PFR is 75,000 l, and the volume of the CSTR is 150,000 l. The PFR operates at 150 °C, and the CSTR operates at 125 °C. The volumetric flow rate entering the PFR is 55,000 l/h, the concentration of A in this stream is 6.5 g·mol/l, and the concentration of R is zero.

The reaction is first order in both directions. The rate constant for the forward reaction at 150 °C is 1.28 h^{-1} and the equilibrium constant based on concentration at 150 °C is 2.3. The rate constant at 125 °C is 0.280 h^{-1} and the equilibrium constant is 3.51.

1. What is the rate of R leaving the CSTR (g·mol/h)?
2. The operating temperature of the CSTR is *lower* than that of the PFR. Does this make sense? Explain your answer.
3. The flow rate to the PFR is increased so that the fractional conversion of A is 0.50 in the effluent from the CSTR. All other parameters remain unchanged. What is the new production rate of R?

Problem 4-24 (Level 2) The homogeneous, reversible, exothermic, liquid-phase reaction

$$\text{A} \rightleftarrows \text{R}$$

is being carried out in a reactor system consisting of two ideal CSTRs in series. Both reactors operate at 150 °C. The molar flow rate of A entering the first CSTR is 55,000 mol/h, the concentration of A in this stream is 6.5 g·mol/l, and the concentration of R is zero. The fractional conversion of A in the outlet stream from the *second* CSTR is 0.75. This fractional conversion is based on the molar flow rate entering the *first* CSTR.

The reaction is first order in both directions. The rate constant for the forward reaction at 150 °C is 1.28 h^{-1} and the equilibrium constant based on concentration at 150 °C is 10.0.

If the volume of the second CSTR is 10,000 l, what is the required volume of the first CSTR?

Problem 4-25 (Level 2) The reversible, liquid-phase isomerization reaction

$$\text{A} \rightleftarrows \text{R}$$

is first order in both directions. The equilibrium constant based on concentration for the reaction as written is 2.0 at temperature T_1.

An ideal CSTR with a volume of 1000 l is being operated at T_1. The molar flow rate of A to the CSTR is 1500 mol/min. The concentration of A in the feed is 2.5 mol/l; there is no R in the feed.

1. What is the *lowest possible* outlet concentration of A that can be obtained at T_1, for a feed containing 2.5 mol/l of A and no R?
2. The actual concentration of A leaving the CSTR is 1.5 mol/l. What is the value of the forward rate constant at T_1?
3. A second CSTR with a volume of 1000 l is added *in series* with the first. The molar flow rate of A to the first reactor is increased so that the concentration of A leaving the *second* CSTR is 1.5 mol/l. The feed composition is unchanged. What is the new flow rate of A to the first reactor?

APPENDIX 4 SOLUTION TO EXAMPLE 4-10: THREE EQUAL-VOLUME CSTRs IN SERIES

The following spreadsheet illustrates the use of GOALSEEK, a "Tool" in MS Excel, to solve Example 4-10: Three Equal-Volume CSTRs in Series.

	A	B	C	D	E	F	G
1	**Solution to Example 4-10**						
2							
3	$\upsilon =$	10000	l/h				
4	$k =$	3.5	l/mol-h				
5	$C_{A0} =$	1.2	mol/l				
6	$x_3 =$	0.75					
7							
8	V (l)	$k{*}C_{A0}{*}tau$	x_2	x_1	*Difference*		
9							
10	3754	1.577	0.651	0.460	−0.0005124		

Definitions:

$\upsilon =$ volumetric flow rate

$k =$ rate constant

$C_{A0} =$ concentration of A in feed

$V =$ volume of one reactor

$x_1 =$ fractional conversion of A leaving reactor 1

$x_2 =$ fractional conversion of A leaving reactor 2

$x_3 =$ fractional conversion of A leaving reactor 3

Explanation of spreadsheet:

Cell A10 contains the value of V
Cell B10 contains the formula "= B4 ∗ B5 ∗ A10/B3" (B10 contains $k{*}C_{A0}{*}$tau)
Cell C10 contains the formula "= B6 − B10 ∗ (1 − B6) ^ 2" (x_2 is calculated from Eqn. (4-37))
Cell D10 contains the formula "= C10 − B10 ∗ (1 − C10) ^ 2" (x_1 is calculated from Eqn. (4-36))
Cell E10 contains the formula "= B10 − (D10/(1 − D10) ^ 2)" (Eqn. (4-35))

Enter a value of V into cell A10. Manually change the value of V until the value in cell E10 becomes close to 0. Then use GOALSEEK to set the value of cell E10 to 0 by adjusting the value in cell A10.

Chapter 5

Reaction Rate Fundamentals (Chemical Kinetics)

LEARNING OBJECTIVES

After completing this chapter, you should be able to

1. determine the likelihood that a reaction is "elementary," i.e., that it proceeds on a molecular level *exactly* as written in a given balanced stoichiometric equation;
2. derive the form of the rate equation for the disappearance of a reactant or the formation of a product, given the sequence of elementary reactions by which an overall reaction proceeds;
3. analyze rate equations to determine their limiting forms.

In Chapter 2, the subject of reaction kinetics was approached from an empirical standpoint. The five rules of thumb that were developed in that chapter could be applied "blind" without any knowledge of the molecular details of the reaction taking place. In this chapter, we will take a more fundamental look at reaction kinetics and learn how rate equations can be derived from a knowledge of the *reaction mechanism*. This approach can lead to rate equations exhibiting behavior that cannot be captured by the empirical approaches of Chapter 2.

Let's begin by discussing what is meant by a "reaction mechanism."

5.1 ELEMENTARY REACTIONS

5.1.1 Significance

Elementary reactions are a critical building block in reaction kinetics. They are the *only* type of reaction for which the *form* of the rate equation can be written *a priori*, i.e., without analyzing experimental data. Moreover, for some of the simpler elementary reactions, the rate constant of the reaction can be predicted to within about an order of magnitude, and the activation energy can be predicted even more accurately.

For an elementary reaction, and *only* for an elementary reaction, the order of the *forward* reaction with respect to species "i" is equal to the number of molecules of species "i" that participate in that reaction. If a species does not participate in the forward reaction, its order in that reaction is zero. The same is true for the reverse reaction.

This property of elementary reactions sometimes is stated in a shorthand fashion as "order = molecularity." It can be derived from either collision theory or transition-state theory.[1]

If the reversible reaction

$$2A \rightleftarrows B$$

is elementary as written, then the rate of the forward reaction is given by $-r_{A,f} = k_f C_A^2$, where k_f is the rate constant for the forward reaction. By similar reasoning, the rate of the reverse reaction is given by $r_{A,r} = k_r C_B$, where k_r is the rate constant for the reverse reaction. Combining the two, the rate equation for the *net* disappearance of A is $-r_A = (-r_{A,f}) - (r_{A,r}) = k_f C_A^2 - k_r C_B$. The equilibrium constant, *based on concentration*, is $K_{eq}^C = k_f/k_r$.

In this illustration, both k_f and k_r have been defined based on species A. The units of k_f are volume/mole A-time, so that the units of the product $k_f C_A^2$ are mole A/volume-time. Similarly, the units of k_r are mole A/mole B time, so that the units of the product $k_f\, C_B$ are mole A/volume-time, as they must be. Unfortunately, the units of a first-order rate constant such as k_r invariably are written as time^{-1}. The mole A/mole B part of the rate constant almost never is shown explicitly, i.e., moles A is canceled against moles B.

Suppose that we wanted to write an expression for the net rate of formation of B (r_B), *after* we had already used k_f and k_r in the sense discussed above. Two approaches are available. The simplest is to apply Eqn. (1-16) to obtain

$$\frac{r_A}{\nu_A} = \frac{r_B}{\nu_B}; \quad \frac{r_A}{-2} = \frac{r_B}{1}$$

$$r_B = \tfrac{1}{2}(-r_A) = \tfrac{1}{2}(k_f C_A^2 - k_r C_B)$$

An alternative, which will produce the same result, is to redefine the rate constants. The units of r_B are moles B/volume-time, so we need rate constants, k_f' and k_r', such that the units of $k_f'\, C_A^2$ and $k_r'\, C_B$ are moles B/volume-time, i.e.,

$$r_B = k_f' C_A^2 - k_r' C_B$$

The units of k_f' must be volume-mole B/(mole A)^2-time. Therefore,

$$k_f'(\text{volume-mole B}/(\text{mole A})^2\text{-time}) = k_f(\text{volume/mole A-time})$$

$$\times\ (\text{mole A/mole B})$$

$$k_f' = k_f(\nu_B/-\nu_A) = k_f/2$$

By similar reasoning,

$$k_r'\,(1/\text{time}) = k_r(\text{mole A/mole B-time}) \times (\text{mole B/mole A})$$

$$k_r' = k_r(\nu_B/-\nu_A) = k_r/2$$

$$r_B = k_f'\, C_A^2 - k_r' C_B = \tfrac{1}{2}(k_f C_A^2 - k_r C_B)$$

[1] Space limitations preclude a serious treatment of either of these theories of chemical kinetics. Transition-state theory is developing rapidly as a result of the availability of inexpensive computing power and is a component of many graduate courses in chemical kinetics and chemical reaction engineering. The interested student can learn more about transition-state and collision theory from references such as Masel, R. L., *Chemical Kinetics and Catalysis*, Wiley-Interscience (2001), Benson, S. W., *Thermochemical Kinetics*, 2nd edition, Wiley-Interscience (1976), and Moelwyn-Hughes, E. A., *Physical Chemistry*, 2nd revised edition, Pergamon Press (1961).

EXAMPLE 5-1
Rate Equation for Reversible Elementary Reaction

If the reversible reaction

$$A + B \rightleftarrows 2R$$

is elementary as written, what is the form of the rate equation for the disappearance of A?

APPROACH

The "order = molecularity" principle will be applied separately to the forward and reverse reactions. The rate equation for the net rate of disappearance of A will be the difference between the rate at which A is consumed in the forward reaction and the rate at which A is formed in the reverse reaction.

SOLUTION

Forward reaction: $-r_{A,f} = k_f C_A C_B$
Reverse reaction: $r_{A,r} = k_r C_R^2$
Net: $-r_A(\text{net}) = k_f C_A C_B - k_r C_R^2$
Here, the rate constants have been defined based on reactant A.

EXERCISE 5-1

Using the expression for $-r_A$ (net) from the preceding example, write an expression for the net rate of formation of R.

EXERCISE 5-2

If k_f is the rate constant based on A, what is the relationship between k_f and k_f', the rate constant based on R?

The "order = molecularity" property of elementary reactions is *not* a two-way street. If a reaction is elementary, then the form of the rate equation can be written directly, as in the above example. However, we cannot *conclude* that a reaction is elementary just because the form of its rate equation, as determined from experimental data, is identical to the form that results from assuming an elementary reaction. For example, if experimental data show that the rate equation for ethylene hydrogenation

$$C_2H_4 + H_2 \rightarrow C_2H_6$$

is $-r_A = k[C_2H_4][H_2]$, this does not *prove* that the reaction is elementary.

5.1.2 Definition

An elementary reaction is a reaction that proceeds in a *single step, on a molecular level, exactly* as written in a balanced stoichiometric equation. Consider the following stoichiometric equation for the decomposition of ozone into oxygen and an oxygen free radical, $O^\bullet$

$$O_3 \rightarrow O_2 + O^\bullet \tag{5-A}$$

In order for this reaction to be elementary, it must occur *on a molecular level exactly* as written. This means that one molecule of ozone must spontaneously decompose into an oxygen molecule and an oxygen free radical. If Reaction (5-A) were elementary, the rate equation for the forward reaction would be $-r_{O_3} = k_f[O_3]$.

How else might this reaction proceed? Suppose that a molecule of oxygen collided with a molecule of ozone, transferring some energy to the ozone molecule and causing its decomposition. At the *molecular* level, this process would be represented as

$$O_2 + O_3 \rightarrow O_2 + O_2 + O^\bullet \tag{5-B}$$

If this reaction were elementary as written, the rate equation for the forward reaction would be $-r_{O_3} = k_f[O_3][O_2]$. From a purely stoichiometric point of view, a balanced equation for this second case could be written as

$$O_3 \rightarrow O_2 + O^\bullet \tag{5-A}$$

Although this equation would describe the *stoichiometry* correctly, it would *not* represent the molecular-level event for this second case. Therefore, it would not provide a valid basis for writing a rate equation, if the reaction actually proceeded on a molecular level as shown in Reaction (5-B).

5.1.3 Screening Criteria

Since it is virtually impossible to observe molecular-level events directly, how can we know whether a given reaction is elementary? The answer is that we never can know with *absolute certainty*. However, we can make some reasonable judgments based on *rules of simplicity*.

Above all, a single-step, molecular-level event must be simple. It must involve a small number of molecules, preferably only one or two, and it must involve the breaking and/or forming of a relatively small number of bonds, preferably only one or two. If too many molecules and/or too many bonds are involved, then the reaction probably will occur as a *series* of simpler elementary reactions, rather than as a single, molecular-level event.

Elementary reactions cannot involve fractional molecules. We may choose to write a balanced stoichiometric equation so that it contains fractional molecules, e.g.,

$$1/2N_2 + 3/2H_2 \rightarrow NH_3$$

However, at the molecular level, there is no such thing as half of a N_2 molecule or half of a H_2 molecule. Therefore, the above reaction, which describes the stoichiometry of the commercial synthesis of ammonia, cannot be elementary as written.

Suppose this reaction were written as

$$N_2 + 3H_2 \rightarrow 2NH_3$$

Is it reasonable to suppose that the reaction now is elementary as written? To answer this question, the rules of simplicity must be applied. The reaction as written requires a four-body collision, where one molecule of N_2 and three of H_2 collide *simultaneously*. This is extremely unlikely. Moreover, in order for the reaction to be elementary as written, one N–N bond and three H–H bonds would have to be broken and six N–H bonds would have to be formed, *simultaneously*, in a single, molecular-level event. This also is extremely unlikely. Both criteria, the molecularity of the event, i.e., the number of molecules colliding, and the number of bonds being broken and formed, lead to the conclusion that the reaction, as written, is not elementary.

One other factor that must be considered in analyzing reactions is the principle of *microscopic reversibility*. This principle states that an elementary reaction must follow the *same* path in both the forward and reverse directions. The practical implication of microscopic reversibility is that a reaction must pass the simplicity tests and the fractional molecule test *in both directions*. A reaction cannot be elementary in one direction and not in the other. Consider the reaction

$$2NOBr \rightleftarrows 2NO + Br_2$$

Table 5-1 Screening Criteria for Elementary Reactions

Criterion	Forward reaction
Fractional molecules?	Not permitted
Molecularity (number of molecules colliding)	One or two (unimolecular or bimolecular) Elementary termolecular processes are rare, and must be regarded with suspicion[2]
Total of bonds broken and formed	One or two, preferably. Three is acceptable. Some elementary reactions may involve as many as four. However, when the total is three or higher, look for simpler pathways, i.e., a sequence of simpler reactions rather than a single, complex reaction
Criterion	**Reverse reaction**
Microscopic reversibility	Repeat the above three analyses for the reverse reaction, even when the overall reaction is essentially irreversible at the conditions of interest

In the forward direction, this reaction involves the collision of two molecules of NOBr, a perfectly acceptable bimolecular process. Two N–Br bonds are broken and one Br–Br bond is formed. Although we would prefer that only one or two bonds be broken and formed, elementary reactions can involve the breaking and formation of more than two bonds. We are perhaps a little suspicious about the forward reaction, but it is reasonable to presume that it is elementary.

The principle of microscopic reversibility, however, tells us that the reverse reaction must be elementary if the forward reaction is elementary. The reverse reaction is termolecular, i.e., it requires a three-body collision. In addition, one Br–Br bond is broken and two N–Br bonds are formed. As discussed in Chapter 2, a three-body collision is much less probable than a two-body collision. The termolecular process in the reverse reaction reinforces our suspicion and makes it doubtful that the reaction is elementary as written.

Table 5-1 summarizes a methodology for evaluating whether a given reaction is likely to be elementary.

With heterogeneous catalytic reactions, i.e., reactions that take place on the surface of a solid catalyst, a more liberal interpretation of these criteria is permissible. This is especially true of the molecularity criterion. Consider the following process by which H_2 can adsorb on the surface of a solid catalyst:

$$H_2 + 2S^* \rightleftarrows 2H\text{–}S^*$$

The symbol S^* represents a vacant site on the surface of the catalyst and H–S^* represents a hydrogen atom bonded to a surface site. For example, S^* could be an atom on the surface of a

[2] Some reactions can *only* proceed via a termolecular mechanism. The recombination of two gas-phase hydrogen free radicals, $H^\bullet + H^\bullet \rightarrow H_2$, does not occur on a molecular level. Collisions can take place between two hydrogen free radicals. However, if an H–H bond formed, it would have to contain a great deal of the kinetic energy that the two $H^\bullet$ free radicals had before the collision. This would make the bond unstable, and it would break essentially as soon as it was formed. Recombination reactions, including the recombination of two $H^\bullet$ radicals, are believed to proceed according to a termolecular reaction, e.g., $H^\bullet + H^\bullet + M \rightarrow H_2 + M$, where M is any molecule that can absorb a substantial portion of the original kinetic energy of the two $H^\bullet$ radicals.

Pt nanoparticle. The above reaction is known as *disassociative chemisorption*, since it involves the disassociation of the H_2 molecule into two adsorbed H atoms.

At first glance, this process does not appear to be simple enough to be elementary. The forward reaction requires that one H–H bond be broken and two H–S^* bonds be formed. More importantly, the forward reaction appears to be termolecular, since three chemical entities are involved, one H_2 molecule and two vacant sites on the surface of the catalyst. However, the probability of this disassociative chemisorption process is much higher than the probability of a three-body collision in a fluid, since the two surface sites are adjacent and are not moving with respect to each other. Therefore, in evaluating the likelihood that a given reaction on the surface of a catalyst is elementary, we can be more tolerant of termolecular processes than we would be for homogeneous, fluid-phase reactions. However, only one of the chemical entities involved in an elementary, termolecular surface reaction should be a fluid species. For example, it is unlikely that the reaction

$$H_2 + C_2H_4 + S^* \rightleftarrows C_2H_6\text{–}S^*$$

could be elementary since it requires the simultaneous collision of two gas molecules with a single surface site.

Even with reactions involving the surface of a heterogeneous catalyst, processes involving four or more species, i.e., reactions with molecularities of four or more, are unlikely to be elementary.

EXAMPLE 5-2
Formation of Carbon Monoxide from Carbon Dioxide and Carbon

Is it reasonable to assume that the reaction

$$CO_2 + C\text{–}S^* \rightarrow 2CO + S^*$$

is elementary?

APPROACH

The screening criteria in Table 5-1 will be applied, allowing for the fact that the reaction involves a solid surface.

SOLUTION

Let's systematically analyze first the forward and then the reverse reaction.

Forward reaction:

No fractional molecules—OK

Two bonds broken (C–O, C–S^*), one bond formed (C–O)—Perhaps OK

Bimolecular—OK

Reverse reaction:

No fractional molecules—OK

One bond broken (C–O), two bonds formed (C–O, C–S^*)—Perhaps OK

Termolecular, but two of the molecules are in the gas phase—unlikely

Overall conclusion—probably not elementary.

In counting bonds broken and formed, bond order has not been considered. For example, in the earlier discussion of the ammonia synthesis reaction, we counted the N–N bond in N_2 as one bond, even though it is considered to be a triple bond. Similarly, in the above example we counted the C–O bond in CO_2 as one bond, even though it is a double bond.

Ignoring bond order is a significant approximation, since the energy required to break a given type of bond increases as the bond order increases. However, the screening methodology outlined here involves other approximations and is easier to apply if bond order and bond strength are not taken into account explicitly.

5.2 SEQUENCES OF ELEMENTARY REACTIONS

Most stoichiometrically simple reactions proceed via a *sequence* of elementary reactions, which is referred to as the *reaction mechanism*. For example, at about 1100 °C, the gas-phase reaction between nitric oxide (NO) and hydrogen is stoichiometrically simple and obeys the stoichiometric equation

$$2NO + 2H_2 \rightarrow N_2 + 2H_2O \tag{5-C}$$

A catalytic version of this reaction takes place in automotive exhaust catalysts at somewhat lower temperatures and is responsible for the removal of oxides of nitrogen from automobile exhaust gases. More generally, the reduction of oxides of nitrogen to N_2 is of immense practical importance in the field of air pollution control.

The mechanism of the uncatalyzed, gas-phase reaction has been of great interest for more than seven decades because it follows a third-order rate equation ($-r_{NO} = k[NO]^2[H_2]$), leading to speculation that a termolecular collision might be involved. However, from the beginning of research on this reaction, it has been hypothesized that the overall reaction proceeds in "stages,"[3] i.e., as a sequence of simpler reactions. One of the possibilities considered was

$$\begin{array}{l} 2NO \rightleftarrows N_2O_2 \\ N_2O_2 + H_2 \rightarrow N_2 + H_2O_2 \\ H_2O_2 + H_2 \rightarrow 2H_2O \\ \hline 2NO + 2H_2 \rightarrow N_2 + 2H_2O \end{array}$$

None of the reactions that make up this sequence can be considered elementary. The first probably requires a collision with an "inert" molecule to activate the N_2O_2 molecule and to absorb some of the energy associated with the combination of the two NO molecules. The second reaction involves the breaking of three bonds and the formation of three, and the third involves the breaking of two bonds accompanied by the formation of two. Each of the last two reactions probably can be broken into a few simpler reactions that meet the simplicity criteria discussed previously. Nevertheless, this scheme can be used to illustrate some important points about reaction sequences. However, we will not use it to derive a rate equation.[4]

How is it possible that Reaction (5-C) is stoichiometrically simple and obeys the Law of Definite Proportions, if N_2O_2 and hydrogen peroxide (H_2O_2) are produced in the first and second steps? Neither hydrogen peroxide nor N_2O_2 appears in the stoichiometric equation for the reaction of NO with H_2. If some H and O atoms are tied up in H_2O_2 and some N and O atoms are tied up in N_2O_2, how can reaction (5-C) be stoichiometrically simple?

The answer is that both N_2O_2 and H_2O_2 are *highly reactive* at the conditions of the study. Consequently, their concentrations are always negligibly small, so small that they do not affect the stoichiometry of the reaction. For all practical purposes, all of the H atoms are found in either H_2 or H_2O, all of the O atoms are found in either NO or H_2O, and all of the N atoms are in NO or N_2.

---

[3] Hinshelwood, C. N. and Green, T. F., *J. Chem. Soc.*, 730 (1926).

[4] As an interesting aside, one attempt to understand the kinetics and mechanism of Reaction (5-C) has involved computer modeling, via transition-state theory, of 38 simultaneous elementary reactions. (Diau, E. W., Halbgewachs, M. J., Smith, A. R., and Lin, M. C., Thermal reduction of NO by H_2: kinetic measurement and computer modeling of the HNO + NO reaction, *Int. J. Chem. Kinet.*, 27, 867 (1995)).

Species such as N_2O_2 and H_2O_2 that appear in the sequence of steps that make up an overall reaction, but are so reactive that their concentrations always are negligibly small are called *active centers*. The presence of active centers can be ignored in writing the stoichiometry of the overall reaction. However, as we shall see in a moment, active centers are a critical part of reaction kinetics. A sequence of elementary steps may contain any number of active centers.

5.2.1 Open Sequences

The sequence of reactions shown above is an "open" sequence. An "open" sequence is one in which the active centers are formed and consumed within the sequence of reactions that comprise the overall reaction. Hydrogen peroxide is formed in the second reaction and consumed in the third, and N_2O_2 is formed in the first and consumed in the second.

5.2.2 Closed Sequences

Consider the following sequence of elementary reactions:

$$\begin{array}{l} Br^{\bullet} + H_2 \rightarrow HBr + H^{\bullet} \\ \underline{H^{\bullet} + Br_2 \rightarrow HBr + Br^{\bullet}} \\ H_2 + Br_2 \rightarrow 2HBr \end{array}$$

There are two "active centers" in this sequence, the hydrogen and bromine free radicals, $H^{\bullet}$ and $Br^{\bullet}$ respectively. The overall, stoichiometrically-simple reaction is between H_2 and Br_2 to give two molecules of HBr.

A cursory look at this sequence gives rise to some disquieting questions: where does the $Br^{\bullet}$ come from in the first place? What eventually happens to it? Moreover, from a strictly stoichiometric standpoint, the sequence also could be written as

$$\begin{array}{l} H^{\bullet} + Br_2 \rightarrow HBr + Br^{\bullet} \\ \underline{Br^{\bullet} + H_2 \rightarrow HBr + H^{\bullet}} \\ H_2 + Br_2 \rightarrow 2HBr \end{array}$$

The stoichiometry is the same, but now we ask where does the $H^{\bullet}$ come from in the first place and what eventually happens to it? In this reaction, whichever way it is written, one of the active centers is *not* formed and consumed within the sequence of elementary reactions that make up the overall reaction.

This sequence of elementary reactions is called a "closed" sequence because at least one active center is created and consumed *outside* the sequence of elementary steps that make up the overall reaction. With a closed sequence, additional reactions are necessary to explain where the active centers come from in the first place and what eventually happens to them.

For the reaction of H_2 with Br_2, active centers are created by the decomposition of Br_2 into two $Br^{\bullet}$, and active centers are destroyed by the reverse of this reaction. Therefore, the complete sequence of elementary steps that are believed to be kinetically important is

$$\begin{array}{rll} Br_2 & \rightarrow 2Br^{\bullet} & \text{(initiation)} \\ Br^{\bullet} + H_2 & \rightarrow HBr + H^{\bullet} & \text{(propagation)} \\ H^{\bullet} + Br_2 & \rightarrow HBr + Br^{\bullet} & \text{(propagation)} \\ 2Br^{\bullet} & \rightarrow Br_2 & \text{(termination)} \end{array}$$

The elementary reactions that make up the closed sequence that gives rise to the overall reaction are called *propagation* reactions. The reaction(s) in which active centers are created are called *initiation* reactions, and the reaction(s) in which active centers are consumed are called *termination* reactions. In order for a reaction to be an initiation reaction, there must be a *net* creation of active centers. Similarly, with a termination reaction, there must be a *net* destruction of active centers.

Obviously, the termination reaction in this case is just the reverse of the initiation reaction. This is not always the case. Also, the initiation and termination reactions may not be elementary as written. As discussed in connection with Table 5-1, an "inert" molecule (M) may be required to serve as an energy source or sink. It may be more accurate to write the initiation and termination reactions as

$$M + Br_2 \rightleftarrows 2Br^{\bullet} + M$$

However, this issue does not affect the above discussion.

5.3 THE STEADY-STATE APPROXIMATION (SSA)

Consider two, first-order reactions occurring in series, i.e.,

$$A \xrightarrow{k_1} B$$

$$B \xrightarrow{k_2} C$$

or, more succinctly,

$$A \xrightarrow{k_1} B \xrightarrow{k_2} C$$

The parameters k_1 and k_2 are the first-order rate constants for the reactions $A \rightarrow B$ and $B \rightarrow C$, respectively.

Suppose these reactions occur in an ideal, batch reactor. The initial concentration of A at time, $t = 0$ will be designated C_{A0}, and we will assume that there is no B or C in the reactor at $t = 0$.

EXERCISE 5-3

1. Sketch the concentrations of A, B and C as a function of time.
2. What happens to the concentration of B as k_2 increases relative to k_1?
3. Use the design equation for an ideal, isothermal batch reactor to show that $C_A = C_{A0}e^{-k_1 t}$.
4. Set up the material balance for B in an ideal, isothermal batch reactor and show that $C_B = [k_1 C_{A0}/(k_2 - k_1)] \times [e^{-k_1 t} - e^{-k_2 t}]$.
5. Show that the time required for B to reach its maximum concentration, $C_{B,max}$, is given by $t_{max} = \ln(k_2/k_1)/(k_2 - k_1)$.
6. Show that the maximum value of C_B is given by $C_{B,max} = C_{A0}(k_1/k_2)^{(k_2/(k_2-k_1))}$.

Suggestion: Remember how you answered these questions. We will return to this sequence of reactions in Chapter 7, when systems of several stoichiometrically-simple reactions are considered.

Now suppose that B is an active center, in which case $A \rightarrow C$ must be a stoichiometrically-simple reaction. For this case, $C_{B,max}$ must be a very small fraction of C_{A0}. Let's arbitrarily require that $(C_{B,max}/C_{A0})$ be 10^{-6}. Practically, this ensures that the concentration of B is so low that the reaction $A \rightarrow C$ is stoichiometrically simple.

We will keep the ratio $(C_{B,max}/C_{A0})$ at or below 10^{-6} by making k_2 very large compared to k_1. Physically, a high value of k_2 relative to k_1 means that B is very reactive. Even a tiny concentration of B will be sufficient to keep the rate of the reaction $B \rightarrow C$ essentially equal to the rate of the reaction $A \rightarrow B$. In other words, B will react to form C essentially as fast as B is formed from A, even when the concentration of B is very low.

The relationship in Exercise 5-3, Part 6 simplifies to $C_{B,max}/C_{A0} \cong (k_1/k_2)$ when $k_2 \ggg k_1$. Therefore, the ratio (k_2/k_1) must be 10^6 or greater if the ratio $C_{B,max}/C_{A0}$ is to be

10^{-6} or less. The high value of k_2 relative to k_1 is consistent with the idea that an active center must be highly reactive.

Next, let's determine the fractional conversion of A when $C_B = C_{B,max}$. From Exercise 5-3, Parts 3 and 5, when $t = t_{max}$, the value of C_A/C_{A0} is given by $\exp\{-k_1 \ln(k_2/k_1)/(k_2 - k_1)\} \cong \exp\{-(k_1/k_2)\ln(k_2/k_1)\}$. For $(k_2/k_1) = 10^6$, $C_A/C_{A0} = (1 - x_A) = 0.999986$, so that $x_A = 1.4 \times 10^{-5}$. This calculation shows that the concentration of B builds up to its maximum value *very early* in the reaction before any significant quantity of A has reacted.

Finally, let's examine the rate of formation (or disappearance) of B relative to the rate of disappearance of A for times longer than t_{max}. The rate of disappearance of A is

$$-r_A = k_1 C_A = k_1 C_{A0} e^{-k_1 t}$$

The *net* rate of formation of B is

$$r_B = k_1 C_A - k_2 C_B = k_1 C_{A0} \left[e^{-k_1 t} - \frac{k_2}{(k_2 - k_1)} (e^{-k_1 t} - e^{-k_2 t}) \right]$$

For $k_2 \ggg k_1$,

$$\frac{r_B}{-r_A} \cong -\left(\frac{k_1}{k_2}\right) + \exp\{-(t/t_{max}) \times \ln(k_2/k_1)\}$$

EXERCISE 5-4

Prove the preceding relationship.

Values of $r_B/-r_A$ for various values of t/t_{max} are shown in the following table for $k_2/k_1 = 10^6$ and for $t \geq t_{max}$.

Dimensionless time, t/t_{max}	Fractional conversion of A (x_A)	$(r_B/-r_A)$
1	1.4×10^{-5}	0
1.1	1.5×10^{-5}	-7.5×10^{-7}
2	2.8×10^{-5}	-1.0×10^{-6}
1000	0.014	-1.0×10^{-6}
10,000	0.13	-1.0×10^{-6}
50,000	0.50	-1.0×10^{-6}
100,000	0.75	-1.0×10^{-6}
500,000	0.999	-1.0×10^{-6}

These calculations show that r_B is *very* small relative to $(-r_A)$ when t is greater than t_{max} and $k_2 \ggg k_1$. For all practical purposes, $-r_A = r_C$ when these conditions are met. This is just another way of saying that the reaction $A \rightarrow C$ is stoichiometrically simple, despite the existence of the active center, B.

Note that r_B is never *exactly* equal to zero, except when $C_B = C_{B,max}(t = t_{max})$. Note also that the rate of formation of B from A, $-r_A$, is *much* larger than r_B. The reason that r_B is so small relative to $-r_A$ is that B is so reactive that it reacts to form C essentially as fast as it is formed from A. In the above table, the *net* rate of formation of the active center, B, always is negligibly small compared to the rate of disappearance of A, and therefore to the rate of formation of C.

This simple analysis leads to an important and useful relationship known as the psuedo-steady-state approximation, or the Bodenstein steady-state approximation, or simply the steady-state approximation (SSA). As an approximation,

Mathematical expression of SSA

$$\boxed{r\,(\text{active center}) = 0} \tag{5-1}$$

The SSA is a generalization that is supported by two important features of the behavior of the active center B in the preceding example:

1. the concentration of B increased to a maximum value *very* rapidly at the start of a reaction, before any significant quantity of reactant was consumed;
2. the *net* rate of formation of B (r_B) was very small relative to the rate of disappearance of A, and the rate of formation of C.

Equation (5-1) simply is a mathematical expression of the second point. The first point suggests that Eqn. (5-1) is valid over the *complete duration* of a reaction, not just over some small period of time.

Equation (5-1) applies to each and every active center in a sequence of elementary reactions. Physically, this equation means that "rate of formation of active center ≅ rate of disappearance of active center." The *net* rate of formation of an active center is very small compared to the rate at which that active center is formed from other species and the rate at which it disappears by reacting to form other species. In fact, as illustrated in the next section, we will *use* Eqn. (5-1) by summing the rates of formation and disappearance of an active center and then setting the result to zero.

5.4 USE OF THE STEADY-STATE APPROXIMATION

Once a stoichiometrically simple reaction has been broken down into a sequence of elementary reactions, a rate expression for the overall reaction can be derived, at least in principle, by using the steady-state approximation. Whether we are dealing with a closed sequence or an open sequence, the procedure consists of three steps:

Procedure for using the SSA to derive a rate equation

1. *Pick a species* (reactant or product) for which the rate equation will be developed. Decide whether the rate equation will be for the *disappearance* or the *production* of this species.
2. *Write an expression for the net rate of formation or disappearance of the chosen species by summing up the contributions of every reaction in the sequence in which the species appears*. The rate of each elementary reaction in the sequence can be written using the "order = molecularity" property of elementary reactions.

   In general, this expression will contain the concentrations of some or all of the active centers that appear in the various elementary reactions. The active center concentrations must be treated as unknowns, since these concentrations cannot be related to the concentrations of the reactants and products through stoichiometry.

3. *Eliminate the active center concentrations* from the expression for the net reaction rate by writing the steady-state approximation (Eqn. (5-1)) for each active center.

EXAMPLE 5-3
Ozone Decomposition

Consider the thermal (homogeneous) decomposition of ozone to oxygen, as described by the balanced stoichiometric equation

$$2O_3 \rightarrow 3O_2$$

Let's assume that this reaction proceeds via the open sequence

$$O_3 \xrightarrow{k_1} O_2 + O^\bullet$$
$$O_2 + O^\bullet \xrightarrow{k_2} O_3$$
$$O^\bullet + O_3 \xrightarrow{k_3} 2O_2$$

Further, let's assume that each reaction is elementary as written. The symbols k_1, k_2, and k_3 are the rate constants for these reactions. Obviously, the second reaction is just the reverse of the first. Derive a rate equation for ozone decomposition.

APPROACH

We will follow the three steps discussed above.

SOLUTION

Step 1: We will write a rate equation for ozone disappearance.

Step 2: $-r_{O_3} = k_1[O_3] - k_2[O_2][O^\bullet] + k_3[O_3][O^\bullet]$

The right-hand side of this equation is simply the sum of the rates at which ozone is consumed in the three elementary reactions. The sign of the second term is negative because ozone is *formed* in the second reaction, but the rate equation is for ozone *disappearance*.

Step 3: $r_{O^\bullet} \cong 0 = k_1[O_3] - k_2[O_2][O^\bullet] - k_3[O_3][O^\bullet]$

The second and third terms are negative because $r_{O^\bullet}$ is the rate of *formation* of $O^\bullet$, and $O^\bullet$ *disappears* in the second and third reactions. Solving for $[O^\bullet]$ gives

$$[O^\bullet] = k_1[O_3]/(k_2[O_2] + k_3[O_3])$$

Substituting this expression into the rate equation in Step 2 and simplifying gives

$$-r_{O_3} = \frac{2k_1k_3[O_3]^2}{(k_2[O_2] + k_3[O_3])}$$

Only the concentrations of reactants and products appear in the final rate equation. The concentrations of active centers were eliminated in Step 3. All of the concentrations in the final rate equation are related through stoichiometry. For a single reaction, each concentration in the rate equation can be expressed in terms of one stoichiometric variable, e.g., extent of reaction, fractional conversion of a reactant, or the concentration of a single species, e.g., the concentration of the limiting reactant.

This rate equation contains some important information. First, it tells us that a simple, power-law rate equation will not provide an adequate description of the reaction kinetics over a wide range of oxygen and ozone concentrations. The apparent orders of the reaction with respect to ozone and oxygen will vary as the relative magnitude of the two terms in the denominator varies. If $k_2[O_2] \ggg k_3[O_3]$, the reaction will appear to be second order in ozone and negative first order in oxygen. However, if $k_2[O_2] \ll k_3[O_3]$, the reaction will appear to be first order in ozone and zero order in oxygen. The relative magnitude of the two terms in the denominator will depend on the ozone and oxygen concentrations and on the reaction temperature, since temperature determines the values of k_2 and k_3.

EXAMPLE 5-4
Hydrogenation of Ethylene

Consider the hydrogenation of ethylene

$$C_2H_4 + H_2 \rightarrow C_2H_6$$

This reaction is of no commercial interest since ethylene is more valuable than ethane. In fact, a mixture of ethane and propane is the feedstock in the major commercial process for manufacturing ethylene.

The overall reaction will be assumed to proceed homogeneously, according the following sequence of irreversible reactions, which we will assume to be elementary:

$$C_2H_4 + H_2 \xrightarrow{k_1} C_2H_5^{\bullet} + H^{\bullet} \tag{5-D}$$

$$H^{\bullet} + C_2H_4 \xrightarrow{k_2} C_2H_5^{\bullet} \tag{5-E}$$

$$C_2H_5^{\bullet} + H_2 \xrightarrow{k_3} C_2H_6 + H^{\bullet} \tag{5-F}$$

$$C_2H_5^{\bullet} + H^{\bullet} \xrightarrow{k_4} C_2H_6 \tag{5-G}$$

The rate constants for these four reactions are designated k_1, k_2, k_3, and k_4, respectively. There are two active centers, $C_2H_5^{\bullet}$ and $H^{\bullet}$, in this sequence. Derive a rate equation for this reaction mechanism.

APPROACH

We will follow the three steps for using the SSA.

SOLUTION

Step 1: Ethylene (C_2H_4) disappearance

Step 2: $-r_{C_2H_4} = k_1[C_2H_4][H_2] + k_2[H^{\bullet}][C_2H_4]$ (5-2)

This expression contains the concentration of hydrogen free radicals, $H^{\bullet}$, which is an unknown. To express this concentration in terms of the reactant and/or product concentrations, the steady-state approximation is used.

Step 3: Write the steady-state approximation for $H^{\bullet}$

$$r_{H^{\bullet}} \cong 0 = k_1[C_2H_4][H_2] - k_2[H^{\bullet}][C_2H_4] + k_3[C_2H_5^{\bullet}][H_2] - k_4[C_2H_5^{\bullet}][H^{\bullet}]$$

This equation contains the concentration of the second active center, the ethyl radical, $C_2H_5^{\bullet}$. To eliminate this concentration, we need to apply the steady-state approximation a second time, to $C_2H_5^{\bullet}$.

$$r_{C_2H_5^{\bullet}} \cong 0 = k_1[C_2H_4][H_2] + k_2[H^{\bullet}][C_2H_4] - k_3[C_2H_5^{\bullet}][H_2] - k_4[C_2H_5^{\bullet}][H^{\bullet}]$$

These two equations can be solved simultaneously to give

$$[H^{\bullet}] = \sqrt{\frac{k_1 k_3}{k_2 k_4}}[H_2]$$

which can be substituted into Eqn. (5-2) to give

$$-r_{C_2H_4} = \left(k_1 + k_2\sqrt{\frac{k_1 k_3}{k_2 k_4}}\right)[C_2H_4][H_2] \tag{5-3}$$

Since all of the rate constants are unknown and eventually have to be determined from experimental data, we can lump the quantity $\left(k_1 + k_2\sqrt{k_1 k_3/k_2 k_4}\right)$ into a single constant k and rewrite the rate equation as $-r_{C_2H_4} = k[C_2H_4][H_2]$.

This is exactly the form that would result if the overall reaction were elementary. However, in this example, the reaction proceeds via a sequence of four elementary reactions. This illustrates the danger of concluding that a reaction is elementary just because its rate equation happens to have the proper form.

EXERCISE 5-5

(a) Categorize each of these four reactions (Reactions (5-D), (5-E), (5-F), and (5-G)) according to the classifications discussed earlier.

(b) Analyze each reaction to determine whether it is reasonable to presume that the reaction is elementary as written.

(c) If the overall reaction takes place at 300 °C and 1 atm total pressure, and if the initial mixture composition is 75 mol% H_2 and 25 mol% C_2H_4, can the overall reaction be treated as irreversible?

5.4.1 Kinetics and Mechanism

Equation (5-3) is a very general expression for the rate of ethylene consumption, given the mechanism of Reactions (5-D)–(5-G). The first term of this rate equation ($k_1[C_2H_4][H_2]$) is the rate at which ethylene disappears in Reaction (5-D), and the second term ($k_2\sqrt{k_1k_3/k_2k_4}[C_2H_4][H_2]$) is the rate at which ethylene disappears in Reaction (5-E).

If $k_1 \ggg k_2\sqrt{k_1k_3/k_2k_4}$, then Reaction (5-D) accounts for essentially all of the ethylene that reacts, and essentially all of the ethane is formed by Reaction (5-G). Reactions (5-E) and (5-F) are not kinetically significant. In essence, if $k_1 \ggg k_2\sqrt{k_1k_3/k_2k_4}$, the overall reaction proceeds by the *open sequence* given by Reactions (5-D) and (5-G).

On the contrary, if $k_2\sqrt{k_1k_3/k_2k_4} \ggg k_1$, then essentially all of the ethylene that reacts is consumed in Reaction (5-E); the amount consumed in Reaction (5-D) is inconsequential. Similarly, essentially all of the ethane is formed in Reaction (5-F). The amount formed in Reaction (5-G) is negligible. Stated differently, if $k_2\sqrt{k_1k_3/k_2k_4} \ggg k_1$, the amount of ethylene consumed in the initiation reaction, Reaction (5-D), and the amount of ethane produced in the termination reaction, Reaction (5-G), are insignificant compared to the *total* amounts of ethylene consumed and ethane formed. The *closed sequence* of Reactions (5-E) and (5-F) accounts for essentially all of the reactants consumed and products formed.

For this example, the *form* of the rate equation is the same, independent of whether most of the ethylene is consumed via the open sequence:

$$C_2H_4 + H_2 \xrightarrow{k_1} C_2H_5^\bullet + H^\bullet \qquad \text{(5-D)}$$

$$C_2H_5^\bullet + H^\bullet \xrightarrow{k_4} C_2H_6 \qquad \text{(5-G)}$$

or via the closed sequence:

$$H^\bullet + C_2H_4 \xrightarrow{k_2} C_2H_5^\bullet \qquad \text{(5-E)}$$

$$C_2H_5^\bullet + H_2 \xrightarrow{k_3} C_2H_6 + H^\bullet \qquad \text{(5-F)}$$

or via a combination of the two. If experimental data showed that the best rate equation was $-r_{C_2H_4} = k[C_2H_4][H_2]$, we would not be able to discriminate between these three possibilities. This is a very simple illustration of the fact that *a reaction mechanism cannot be proven based solely on the form of the overall rate equation.* Different mechanisms can give rise to the same rate equation.

If the rate equation obtained from the experimental data did *not* match the form that was derived from a hypothetical reaction mechanism, this would be very strong evidence that the proposed mechanism was incorrect. New sequences of elementary reactions would have to be proposed and tested until a mechanism was discovered that led to a rate equation that was consistent with the experimental data.

EXERCISE 5-6

Suppose that the rate equation for the homogeneous hydrogenation of ethylene were found to be $-r_{C_2H_4} = k[C_2H_4][H_2]^{1/2}$. Find a sequence of elementary reactions that leads to this rate equation, using the SSA.

The emphasis in this text is on using a knowledge of reaction mechanisms to derive rate equations that will capture the kinetic behavior of a reaction as accurately and comprehensively as possible. The complementary issue of using rate equations obtained from experimental data to explore reaction mechanisms is not treated in detail. Nevertheless, reaction kinetics is one of the most powerful tools available to researchers that are intent on obtaining a molecular-level understanding of a particular reaction.

5.4.2 The Long-Chain Approximation

When we know (or are willing to assume) that the rates of reactant disappearance and product formation in the initiation and termination reactions are negligible, compared to their counterparts in the propagation reactions, we can invoke what is called the "long-chain approximation." This simplifies the algebra of the steady-state approximation to some extent. The simplification arises, in large part, because reactant consumption and product formation in the initiation and termination reactions are ignored. The long-chain approximation applies only to closed sequences.

To apply the long-chain approximation to the preceding ethylene hydrogenation mechanism, we write a rate expression that includes only the rate of ethylene consumption in the propagation steps, i.e., in the steps that carry the chain.

$$-r_{C_2H_4} = k_2[H^\bullet][C_2H_4]$$

To eliminate the concentrations of the active centers, we first apply the steady-state approximation to the *total concentration of all active centers*. Since there is no *net* creation or destruction of active centers in either propagation step, the resulting equation is

$$r_{AC} \cong 0 = 2k_1[C_2H_4][H_2] - 2k_4[C_2H_5^\bullet][H^\bullet]$$

The factors of 2 in front of both terms in this equation reflect the fact that two active centers are created or destroyed each time one of these reactions proceeds. Stated differently, in formulating this equation, it was assumed that the rate constant k_1 was based on C_2H_4 or H_2 and that the rate constant k_4 was based on $C_2H_5^\bullet$ or $H^\bullet$. This expression for r_{AC} is valid whether or not the long-chain approximation is applied. Except in certain "pathological" cases,[5] the rate of creation of active centers is balanced by the rate of destruction of active centers. In fact, this balance determines the total concentration of active centers, in the same way that the SSA on a specific active center determines the concentration of that center.

Next, we recognize that the rates of the two propagation steps must be the same if the amount of ethylene consumed is to be the same as the amount of hydrogen consumed and ethane produced.

$$k_2[C_2H_4][H^\bullet] = k_3[C_2H_5^\bullet][H_2]$$

This equation is a direct consequence of the long-chain approximation.

Using the last two equations, the concentrations of the two active centers can be eliminated and the resulting rate equation is

$$-r_{C_2H_4} = \left(k_2\sqrt{\frac{k_1k_3}{k_2k_4}}\right)[C_2H_4][H_2]$$

[5] One of the "pathological" cases is "chain branching." Chain-branching reactions are important in determining the region of composition and temperature in which a combustible mixture can explode. From the standpoint of safety, it is critical to know the location of the explosion limits of a combustible mixture *before* beginning design or experimentation. In many cases, processes are engineered to operate well outside of these limits, in order to avoid the possibility of an explosion.

5.5 CLOSED SEQUENCES WITH A CATALYST

There are many different kinds of catalysts. *Heterogeneous* catalysts are solid materials, usually with a high specific surface area. The heterogeneous catalysts that are used in various industrial processes have specific surface areas that range from about 10 to 1000 m^2/g. One or more fluid phases are in contact with the solid catalyst. Reactant molecules in the fluid phase adsorb on the surface of the solid catalyst, rearrange or react with another adsorbed molecule, and then the product(s) desorb back into the fluid(s). Chapter 9 treats the subject of heterogeneous catalysis in much greater detail.

Homogeneous catalysts, on the other hand, are dissolved in the fluid phase. Nevertheless, homogeneous catalysts function in much the same way as heterogeneous catalysts. A reactant molecule binds to the homogeneous catalyst, rearranges or reacts with another molecule, and the product(s) return to the fluid. *Enzyme* catalysts can be either homogeneous or heterogeneous.

All of these types of catalyst can be treated with the same kinetic tools. Catalytic reactions always proceed via a closed sequence of elementary reactions, and the steady-state approximation is the starting point for many analyses of catalytic kinetics. Let's illustrate the function of a catalyst using a greatly over-simplified example based on the water-gas-shift (WGS) reaction

$$CO + H_2O \rightleftarrows CO_2 + H_2$$

This is an important industrial reaction that is used, for example, in the manufacture of hydrogen, the manufacture of ammonia, and the manufacture of methanol from coal.

The shift reaction is carried out industrially at conditions where it is reversible. One of the important issues in the design of shift reactors and processes is sound "management" of the reaction equilibrium. Strictly for purposes of illustration, we will treat the reaction as though it were irreversible.

We will assume that the water-gas-shift reaction proceeds via the *hypothetical* sequence of elementary reactions shown below. This sequence is a major simplification of what actually occurs on a commercial WGS catalyst. However, this sequence will illustrate the important principles of catalytic kinetics, without requiring much algebra.

In the following reactions, the symbol S^* denotes an empty or unoccupied site on the catalyst surface, and the symbol $O{-}S^*$ denotes an oxygen atom bound to a site on the catalyst.

$$S^* + H_2O \xrightarrow{k_1} H_2 + O{-}S^*$$
$$O{-}S^* + CO \xrightarrow{k_2} CO_2 + S^*$$

These two reactions form a closed sequence with two active centers: S^* and $O{-}S^*$. Theoretically, the overall reaction can proceed an infinite number of times with a single unoccupied site, S^*, since the original site is regenerated on completion of the sequence. Note that both of these reactions are irreversible, so that the *overall* reaction is irreversible. If an overall reaction is reversible, then *all* of the elementary reactions leading from the reactants to the products must be written in reversible form.

In the first of these reactions, two bonds are broken and two are formed in a single, molecular-level event. This raises a question as to whether this reaction is elementary as written. However, for purposes of illustration, we will disregard this concern.

A rate equation for the *disappearance* of CO will be derived using the steady-state approximation. Carbon monoxide does not participate in the first reaction and it is consumed in the second. Therefore,

$$-r_{CO} = k_2[CO][O{-}S^*] \tag{5-4}$$

The quantity $[\mathrm{O{-}S^*}]$ is a surface concentration if the catalyst is heterogeneous, with units of, e.g., mol/m^2. The steady-state approximation must now be applied to eliminate this unknown concentration from the rate expression.

The steady-state approximation for $\mathrm{O{-}S^*}$ is

$$r_{\mathrm{O-S^*}} \cong 0 = k_1[\mathrm{S^*}][\mathrm{H_2O}] - k_2[\mathrm{CO}][\mathrm{O{-}S^*}] \tag{5-5}$$

This expression contains the concentration of the second active center, $\mathrm{S^*}$. The steady-state approximation for $\mathrm{S^*}$ is

$$r_{\mathrm{S^*}} \cong 0 = -k_1[\mathrm{S^*}][\mathrm{H_2O}] + k_2[\mathrm{CO}][\mathrm{O{-}S^*}] \tag{5-6}$$

Unfortunately, Eqn. (5-6) is just Eqn. (5-5) multiplied by -1. The two equations are not independent; they cannot be solved for *both* $[\mathrm{S^*}]$ and $[\mathrm{O{-}S^*}]$. Why did the steady-state approximation "fail" in this case?

The problem is as follows. The total concentration of sites on the catalyst is fixed. Sites are either empty ($\mathrm{S^*}$) or they are occupied by a bound O atom ($\mathrm{O{-}S^*}$). New sites are not created in the reaction sequence, nor are existing sites destroyed. The first SSA equation (Eqn. (5-5)) tells us that the rate of disappearance of $\mathrm{S^*}$ is equal to the rate of formation of $\mathrm{O{-}S^*}$. The second SSA equation (Eqn. (5-6)) tells us that the rate of disappearance of $\mathrm{O{-}S^*}$ is equal to the rate of formation of $\mathrm{S^*}$. This is not a new piece of information; it follows directly from the first SSA equation, plus the fact that the total number of sites, i.e., the total of $\mathrm{S^*}$ and $\mathrm{O{-}S^*}$, is constant.

The failure of the two equations for $r_{\mathrm{S^*}}$ and $r_{\mathrm{O-S^*}}$ to produce expressions for the concentrations of these two species can be looked at from another viewpoint. In earlier applications of the SSA, there was a *net* creation of active centers in some reactions (the initiation reactions) and a *net* destruction of active centers in other reactions (the termination reactions). This is not the case with the two reactions above. These reactions merely involve the transformation of one kind of active center into a different kind of active center. *There is nothing in the given reaction mechanism that allows us to calculate the total concentration of active centers.*

We resolve this dilemma by writing a conservation equation called a *site balance*, which expresses the fact that the number of occupied sites plus the number of unoccupied sites is constant. In other branches of catalysis, a site balance is referred to as a *catalyst balance* or an *enzyme balance*. If $\mathrm{S_T^*}$ is the total number of sites in the system, the site balance for this example is

Site balance for WGS reaction

$$[\mathrm{S_T^*}] = [\mathrm{S^*}] + [\mathrm{O{-}S^*}]$$

This is the second independent equation that is required to eliminate the active center concentrations from the rate equation. If $[\mathrm{S^*}] = [\mathrm{S_T^*}] - [\mathrm{O{-}S^*}]$ is substituted into the steady-state approximation for $\mathrm{O{-}S^*}$ (Eqn. (5-5)) and the resulting equation is solved for $[\mathrm{O{-}S^*}]$, the result is

$$[\mathrm{O{-}S^*}] = \frac{k_1[\mathrm{S_T^*}][\mathrm{H_2O}]}{k_1[\mathrm{H_2O}] + k_2[\mathrm{CO}]}$$

Substituting this into the rate expression (Eqn. (5-4)) gives

$$-r_{\mathrm{CO}} = \frac{k_1 k_2[\mathrm{S_T^*}][\mathrm{CO}][\mathrm{H_2O}]}{k_1[\mathrm{H_2O}] + k_2[\mathrm{CO}]}$$

The value of $[\mathrm{S_T^*}]$ may or may not be known *a priori*. With homogeneous catalysts and simple enzyme catalysts, $[\mathrm{S_T^*}]$ usually is known, e.g., it is the concentration of catalyst that is charged to the reactor initially. However, with solid heterogeneous catalysts, the total number of sites per unit surface area that *actually contribute* to catalyzing a reaction is extremely hard to determine.

If desired, the quantity $k_1 k_2 [S_T^*]$ can be "lumped" into a single constant k.

$$-r_{CO} = \frac{k[CO][H_2O]}{k_1[H_2O] + k_2[CO]} \tag{5-7}$$

This hyperbolic form of rate equation is known as a "Langmuir–Hinshelwood" type of rate equation in the field of heterogeneous catalysis, and as a "Michaelis–Menten" type of rate equation in biochemistry.

It is important to recognize that this rate equation was derived without specifying the kind of catalyst involved. The catalyst could have been heterogeneous or homogeneous, metallic or organometallic or enzyme. The tools required to develop the rate equation are common to all types of catalysis.

If the two terms in the denominator of Eqn. (5-7) are comparable in magnitude, this rate equation is *not* in a simple power-law form. When the water concentration is high, or more accurately, when $k_1[H_2O] \gg k_2[CO]$, then the rate equation simplifies to a form where the reaction is first order in CO and zero order in H_2O. However, when the carbon monoxide concentration is high, i.e., $k_2[CO] \gg k_1[H_2O]$, then the rate equation reduces to a different form, such that the reaction is first order in H_2O and zero order in CO.

A rate equation that is fractional order in both CO and H_2O might provide an adequate description of the reaction kinetics over a limited range of CO and H_2O concentrations. However, the use of fractional-order rate equations for reactor design can be dangerous.

In view of the preceding discussion of "catalyst balances," Step 3 of the procedure for using the steady-state approximation on page 314 requires modification as follows:

Modification of Step 3 in procedure for using the SSA to derive a rate equation

Step 3: For a catalytic reaction, eliminate the active center concentrations from the equation for the net reaction rate by writing a combination of SSAs and catalyst balances.[6] The number of such expressions must equal the number of active centers.

5.6 THE RATE-LIMITING STEP (RLS) APPROXIMATION

Consider a reversible isomerization reaction catalyzed by an enzyme E. The reaction will be represented as

$$S \rightleftarrows P$$

In the biochemical literature, a reactant frequently is referred to as a "substrate." The symbol "S" is used in the above reaction, in deference to this tradition. This reaction might represent, for example, the reversible isomerization of glucose to fructose, which is central to the process for producing the sweetener known as "high-fructose corn syrup." Glucose and fructose contain approximately the same number of calories. However, fructose is about five times sweeter to the taste. Therefore, high-fructose corn syrup is widely used as a sweetener in, e.g., soft drinks.

[6] More that one "catalyst balance" will be required if there is more than one distinct catalyst species in the reaction mechanism. Several of the end-of-chapter problems contain this extension.

Let's assume that the overall reaction proceeds according to the sequence of elementary reactions

$$S + E \underset{k_{-1}}{\overset{k_1}{\rightleftarrows}} E\text{–}S \tag{5-H}$$

$$E\text{–}S \underset{k_{-2}}{\overset{k_2}{\rightleftarrows}} E\text{–}P \tag{5-I}$$

$$E\text{–}P \underset{k_{-3}}{\overset{k_3}{\rightleftarrows}} P + E \tag{5-J}$$

In this sequence, E represents the free enzyme, E–S represents an enzyme that is bound to a molecule of substrate (an enzyme–substrate complex), and E–P represents an enzyme that is bound to a molecule of product (an enzyme–product complex). Reactions (5-H), (5-I), and (5-J) must be written as reversible because the overall reaction is reversible. If any one of the reactions leading from reactants to products were irreversible, there would be no pathway leading from the products back to the reactants, and the overall reaction would not be able to proceed in the reverse direction, i.e., it would be irreversible.

The sequence of elementary reactions shown above is closed, and there are three active centers E, E–S, and E–P.

A rate equation for this reaction could be developed using the SSA. However, the algebra would be tedious and the final expression would be complex. The final rate equation must have two terms because of reversibility. The SSA would have to be written on E–S and E–P, and each of these equations would have four terms. An enzyme (site) balance containing four terms also would be required.

5.6.1 Vector Representation

Suppose that both the forward and reverse rates of Reaction (5-I) were known to be *very* slow compared to the forward and reverse rates of Reactions (5-H) and (5-J). We might represent this situation as shown in Figure 5-1. In this figure, the length of each vector is proportional to the reaction rate, and the direction of the vector indicates whether the reaction is in the forward direction (arrow pointing to the right) or the reverse direction (arrow pointing to the left). The numbers and letters to the left of the vector show which reaction is being represented, e.g., 1F is the forward component of Reaction (5-H) and 2R is the reverse component of Reaction (5-I).

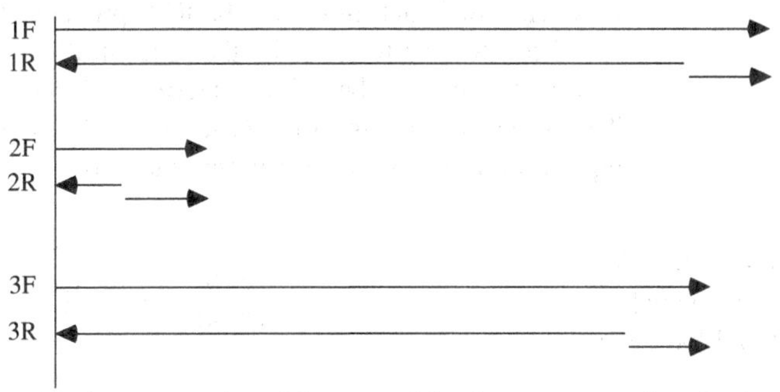

Figure 5-1 Graphical representation of the rates of the elementary reactions for the overall reaction $S \rightleftarrows P$. Reactions (5-H), (5-I), and (5-J) are labeled 1, 2, and 3, respectively. In the figure, "F" denotes the forward reaction and "R" denotes the reverse reaction.

The small vector that is located to the right of and slightly below the vector for the reverse reaction represents the *net forward* rate of reaction, i.e., the difference between the rate of the forward and reverse reactions. This net rate of reaction is identical for Reactions (5-H), (5-I), and (5-J) and is the *net* rate of the overall reaction S $\rightleftarrows$ P.

The equality of the three net rates is a direct consequence of the steady-state approximation. If the SSA is valid, these rates are necessarily equal.

EXERCISE 5-7

Prove that the net forward rates of Reactions (5-H), (5-I), and (5-J) are equal if the SSA is valid.

This picture leads to a tool, called the *rate-limiting step (RLS) approximation*, that is very useful in chemical kinetics. The figure suggests that Reactions 1 (5-H) and 3 (5-J) are essentially in chemical equilibrium because the rates of the forward and reverse reactions are *almost* equal. The *equilibrium expressions* for these fast reactions then can be used to solve for the concentrations of the active centers, instead of using the more cumbersome SSA. The rate of the overall reaction can be written in terms of the slow reaction, which is referred to as the *rate-limiting step* (or rate-determining step or rate-controlling step).

5.6.2 Use of the RLS Approximation

Let's illustrate how the RLS approximation can be used by deriving a rate equation for the disappearance of S, $-r_S$, in the above reaction. We might be tempted to begin by writing

$$-r_S = k_1[S][E] - k_{-1}[E\text{–}S]$$

as we did when we used the SSA. However, eventually we are going to assume that Reaction 1 (5-H) is in equilibrium, which will lead to $-r_S = 0$ if we use the above equation. Instead, *the expression for the overall rate of reaction must be written in terms of the rate of the RLS*, which can never be in equilibrium.

Starting point for use of RLS approximation

$$-r_S = (\text{rate of RLS}) \times (\text{molecules of S/RLS})$$

$$-r_S = \{k_2[E\text{–}S]\text{–}k_{-2}[E\text{–}P]\}(1) \tag{5-8}$$

The term (1) on the right-hand side of this equation results from the fact that 1 molecule of S is consumed each time that the RLS proceeds to the right, on a net basis.

The rate equation contains the concentrations of two active centers, E–S and E–P. These concentrations must be written in terms of the concentrations of the reactants and products. To do this, we assume that Reactions (5-H) and (5-J) are in equilibrium and write the equilibrium expressions for these reactions.

Use equilibrium expression to relate [active centers] to [reactants] and [products]

$$\frac{[E\text{–}S]}{[E][S]} = K_1; \quad [E\text{–}S] = K_1[E][S] \tag{5-9}$$

Here, K_1 is the equilibrium constant for Reaction (5-H), based on concentration. As an aside, K_1 is referred to as a "binding constant" in biochemistry nomenclature. In the field of heterogeneous catalysis, where Reaction (5-H) would represent the adsorption of molecule S onto the surface of a solid catalyst, K_1 is referred to as an "adsorption constant." More rigorously, K_1 in Eqn. (5-8) is the equilibrium constant for formation of the enzyme/substrate complex from the free enzyme and the substrate.

The equilibrium expression for Reaction (5-J) is

$$\frac{[\mathrm{E}][\mathrm{P}]}{[\mathrm{E{-}P}]} = K_3'; \quad [\mathrm{E{-}P}] = [\mathrm{E}][\mathrm{P}]/K_3' \tag{5-10}$$

Equations (5-9) and (5-10) contain the concentration of the *free* enzyme [E], which is an unknown. This concentration is eliminated by writing an *enzyme balance*, which is exactly analogous to the site balances that we write for heterogeneous catalysts.

Enzyme (catalyst) balance

$$[\mathrm{E}] + [\mathrm{E{-}S}] + [\mathrm{E{-}P}] = [\mathrm{E_0}] \tag{5-11}$$

In Eqn. (5-11), $[\mathrm{E_0}]$ is the initial concentration of enzyme charged to the reactor, or contained in the feed to the reactor. Equations (5-9) and (5-10) now are substituted into Eqn. (5-11) to give

$$[\mathrm{E}] = \frac{[\mathrm{E_0}]}{1 + K_1[\mathrm{S}] + \frac{1}{K_3'}[\mathrm{P}]} \tag{5-12}$$

Finally, substitution of Equations (5-9), (5-10), and (5-12) into the expression for $-r_\mathrm{S}$, Eqn. (5-8), gives

$$-r_\mathrm{S} = \frac{[\mathrm{E_0}]\{k_2K_1[\mathrm{S}] - k_{-2}[\mathrm{P}]/K_3'\}}{1 + K_1[\mathrm{S}] + \frac{1}{K_3'}[\mathrm{P}]}$$

The procedure that was used to derive a rate equation using the RLS approximation can be summarized as follows:

Procedure for using the RLS approximation to derive a rate equation

1. Decide whether the rate equation will describe the disappearance of a reactant or the formation of a product.
2. Write a rate equation *for the RLS* by employing the "order = molecularity" property of elementary reactions. Multiply the rate of the RLS by the number of molecules of reactant or product that are formed or consumed each time the RLS proceeds.
3. Eliminate the concentrations of the active centers from the rate equation using *equilibrium expressions* for the reactions that are not the RLS. If the reaction is catalytic, one or more "catalyst balances" also will be required to eliminate the concentrations of the active centers.

5.6.3 Physical Interpretation of the Rate Equation

The preceding expression is acceptable in a formal sense, since it contains only concentrations of reactants and products, plus the initial concentration of the enzyme. However, it can be put into a more understandable form by factoring k_2K_1 out of the bracketed term in the numerator to give

$$-r_S = \frac{k_2K_1[E_0]\{[S] - k_{-2}[P]/k_2K_1K_3'\}}{1 + K_1[S] + \frac{1}{K_3'}[P]}$$

Now, $k_{-2}/k_2K_1K_3' = 1/K_1K_2K_3'$ and $K_1K_2K_3' = K_{eq}^C$, where K_{eq}^C is the equilibrium constant (based on concentration) for the overall reaction, $S \rightleftarrows P$. Using these relationships, the rate equation becomes

$$-r_S = \frac{k_2K_1[E_0][S]}{1 + K_1[S] + \frac{1}{K_3'}[P]}\left\{1 - \frac{[P]/[S]}{K_{eq}^C}\right\} \tag{5-13}$$

The term $\{1 - ([P]/[S])/K_{eq}^C\}$ is a measure of how far the reaction is from equilibrium, i.e., the extent to which the rate of the reverse reaction influences the net rate. This term has a value between 0 and 1. If $[P]/[S]\ K_{eq}^C \lll 1$, the bracketed term is very close to 1, and the reverse reaction has no significant influence on the overall rate. On the other hand, if $[P]/[S] = K_{eq}^C$, the bracketed term is zero, the reaction is in equilibrium and $-r_S = 0$. The influence of the reverse reaction on the net rate becomes more and more significant as the value of the bracketed term decreases towards zero.

It is useful to put the rate equation for *every* reversible reaction into a form similar to Eqn. (5-13). The kinetic influence of the reverse reaction is quite easy to evaluate when the rate expression is in this form.

Note that the equilibria for Reactions (5-H) and (5-J), leading to Eqns. (5-9) and (5-10), were written in opposite senses. The constant K_1 is the equilibrium constant for *formation* of the enzyme–substrate complex E–S, from the substrate S, and the free enzyme E. The constant K_3' is the equilibrium constant for the *decomposition* of the enzyme–product complex into E and the product P. A common convention in catalysis is to use equilibrium constants based on the *formation* of the complex. The constant K_1 is consistent with this convention, but K_3' is not. If the equilibrium constant for the *formation* of E–P from E and P is K_3, then $K_3 = 1/K_3'$. Using this relationship in Eqn. (5-13) leads to

$$-r_S = \frac{k_2K_1[E_0][S]}{1 + K_1[S] + K_3[P]}\left\{1 - \frac{[P]/[S]}{K_{eq}^C}\right\} \tag{5-14}$$

The formation of catalyst-substrate and catalyst–product complexes is usually exothermic. Therefore, if the Ks are equilibrium constants for the formation of the complexes, the values of K generally will decrease with increasing temperature.

Equation (5-14) is worth some additional discussion. The relationship $K_3 = 1/K_3'$ can be substituted into Eqn. (5-12), which then can be rearranged to give

$$\frac{[E]}{[E_0]} = \frac{1}{1 + K_1[S] + K_3[P]}$$

The ratio $[E]/[E_0]$ is just the fraction of *unoccupied* binding sites, i.e., the fraction of sites that are *not* complexed with either S or P. Clearly, this fraction depends on the concentrations of S and P. If $(K_1[S] + K_3[P]) \gg 1$, very few of the binding sites will be free (unoccupied). Either P or S or both will be bound to the great majority of sites. On the other hand, if both $K_1[S]$ and $K_3[P] \lll 1$, the fraction of unoccupied sites is close to 1.

Equations (5-9) and (5-12) can be combined to give

$$\frac{[E-S]}{[E_0]} = \frac{K_1[S]}{1 + K_1[S] + K_3[P]} \tag{5-15}$$

The ratio [E–S]/[E$_0$] is the fraction of binding sites that are complexed with the reactant, S. If $K_1[\text{S}] \gg (1 + K_3[\text{P}])$, this fraction is close to 1. Conversely, if $K_1[\text{S}] \ll (1 + K_3[\text{P}])$, the fraction of sites with S bound to them is very small.

Similarly, using $K_3 = 1/K_3'$, Eqns. (5-10) and (5-12) can be combined to give

$$\frac{[\text{E}-\text{P}]}{[\text{E}_0]} = \frac{K_3[\text{P}]}{1 + K_1[\text{S}] + K_3[\text{P}]} \tag{5-16}$$

Depending on the value of $K_3[\text{P}]$ relative to $(1 + K_1[\text{S}])$, the fraction of sites complexed with P can range from essentially zero to almost 1.

Now, for purposes of illustration, let's examine the rate of the *forward* reaction. From Eqns. (5-8) and (5-14),

$$-r_{\text{S,f}} = k_2[\text{E–S}] = \frac{k_2 K_1[\text{E}_0][\text{S}]}{1 + K_1[\text{S}] + K_3[\text{P}]} \tag{5-17}$$

The rate of the forward reaction increases as [E–S] increases. The forward rate will have its largest possible value when all of the available binding sites are occupied by the substrate, i.e., when $[\text{E–S}] \cong [\text{E}_0]$. This occurs when $K_1[\text{S}] \gg (1 + K_3[\text{P}])$. When this condition is satisfied, $-r_{\text{S,f}} \cong k_2[\text{E}_0]$. The reaction is zero order in S. Physically, the rate is not sensitive to the reactant concentration because [E–S] has reached its maximum possible value, [E$_0$]. Increasing the concentration of S cannot increase [E–S] anymore. The enzyme catalyst is "saturated" with the substrate S.

When $K_1[\text{S}] \ll (1 + K_3[\text{P}])$, the fraction of sites with S bound to them is small. This fraction ([E–S]/[E$_0$] now increases linearly with [S], as shown by Eqn. (5-15). For this case, Eqn. (5-17) shows that $-r_{\text{s,f}}$ is first order in S.

Finally, let's examine the term K_3[P] in the denominator of Eqn. (5-17). As a result of this term, the rate of the forward reaction will decrease as the product concentration increases. This phenomenon is known as "product inhibition," and it is not uncommon in either enzyme catalysis or heterogeneous catalysis. Equations (5-15) and (5-16) provide the explanation for this behavior. As [P] increases, the fraction of sites occupied by P increases and the fraction occupied by S decreases. Since $-r_{\text{S,f}} = k_2[\text{E–S}]$, the rate also decreases.

5.6.4 Irreversibility

Suppose that Reaction (5-J) were essentially irreversible, i.e.,

$$\text{E–P} \rightarrow \text{E} + \text{P}$$

The equilibrium constant for this reaction, K_3', then would be essentially infinite. Moreover, the equilibrium constant for the overall reaction, K_{eq}^{C}, also would be infinite, since $K_{\text{eq}}^{\text{C}} = K_1 K_2 K_3'$. For this situation, Eqns. (5-13) and (5-14) reduce to

$$-r_{\text{S}} = \frac{k_2 K_1[\text{E}_0][\text{S}]}{1 + K_1[\text{S}]} \tag{5-18}$$

This illustrates one way to derive a rate equation when one or more of the elementary reactions in the overall sequence is irreversible. All of the reactions can be treated as though they were reversible and a rate equation can be derived. The resulting rate equation then can be simplified by setting the equilibrium constants for the irreversible steps equal to infinity. The advantage of this approach is that it is mechanical. The disadvantage is that requires more algebra and obscures some relevant conceptual issues.

Let's illustrate a simpler approach, one that underlines some of the implications of a rate-limiting step. We'll begin by drawing a vector picture of the rates of the individual

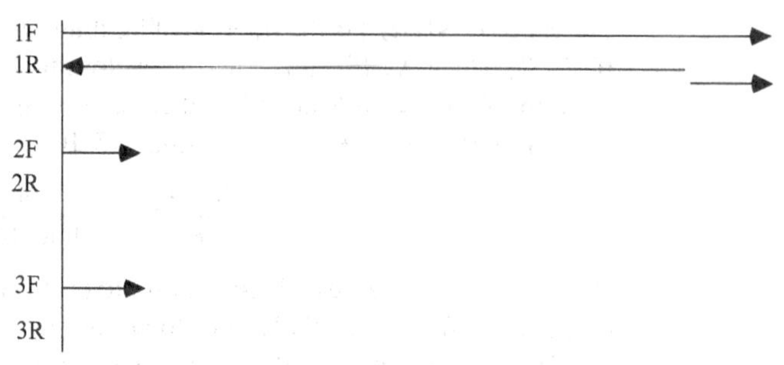

Figure 5-2 Graphical representation of the rates of the elementary reactions for the overall reaction S → P. Reaction 2 (5-I) is assumed to be rate-limiting and reaction 3 (5-J) is irreversible. Reactions (5-H), (5-I), and (5-J) are labeled 1, 2, and 3, respectively. In the figure, "F" denotes the forward reaction and "R" denotes the reverse reaction.

reactions (5-H), (5-I), and (5–J), assuming that (5-I) is the RLS and that (5-J) is irreversible. This picture, which will serve as a guide in deriving a rate equation, is shown in Figure 5-2.

Reaction (5-H) (i.e., 1 in the above figure) looks as it did previously, in Figure 5-1. This reversible reaction still is essentially in equilibrium because the *net* rate of reaction is very small compared to the individual forward and reverse rates. However, the vectors for Reactions (5-I) (2 in the above figure) and (5-J) (3 in the above figure) have changed significantly, relative to how they looked in Figure 5-1.

Conceptually, we have a situation where the overall reaction rate is determined by how fast the E–P complex is formed in Reaction (5-I). As soon as one of these complexes is formed, it immediately reacts to form E and P in Reaction (5-J). Reaction (5-J) is not reversible, so E–P is not reformed by reaction of E and P. Reaction (5-J) would "like" to go a lot faster, but its rate is limited by how fast E–P is formed in Reaction (5-I). Reaction (5-J) *cannot* go any faster than the rate at which E–P is formed. Therefore, the length of the vector 3F is exactly equal to the net rate of reaction. The length of the 3R vector is 0 since Reaction (5-J) is irreversible.

The length of the 2R vector also is 0 because this reaction, the reverse of the slow, rate-limiting step, also is very slow compared to Reaction (5-J). In essence, all of the E–P complexes that are formed by 2F are consumed in 3F. Essentially none are transformed back to E–S by the reverse of Reaction (5-I).

The derivation of the rate equation begins the same way it did previously

$$\begin{aligned} -r_S &= (\text{rate of RLS}) \times (\text{molecules of S/RLS}) \\ -r_S &= k_2[\text{E–S}](1) \end{aligned} \tag{5-19}$$

A term for the reverse of Reaction (5-I) is not required since drawing the vector picture convinced us that the rate of 2R was insignificant compared to that of 2F (and 3F).

The concentration of the active centers E–S can be eliminated using the equilibrium expression for Reaction (5-H)

$$\frac{[\text{E–S}]}{[\text{E}][\text{S}]} = K_1; \quad [\text{E–S}] = K_1[\text{E}][\text{S}] \tag{5-9}$$

Following the previous procedure, we use the enzyme balance to eliminate E.

$$[\text{E}] + [\text{E–S}] + [\text{E–P}] = [\text{E}_0] \tag{5-11}$$

However, we no longer have an equation that can be used to eliminate [E-P]! In the previous example, the equilibrium expression for Reaction (5-J) was used for this purpose. Now, Reaction (5-J) is irreversible.

In discussing the vector picture for this case, we recognized that the formation of the E–P complex limited the rate of Reaction (5-J) and that this complex was reacted away in Reaction (5-J) essentially as soon as the complex was formed by Reaction (5-I). Therefore, the concentration of E–P is very low, so that

$$[\text{E–P}] \cong 0^7$$

Substituting [E–P] = 0 and Eqn. (5-9) into Eqn. (5-11) reduces the enzyme balance to

$$[\text{E}] + K_1[\text{S}][\text{E}] = [\text{E}_0]$$

$$[\text{E}] = \frac{[\text{E}_0]}{1 + K_1[\text{S}]}; \quad [\text{E}-\text{S}] = \frac{K_1[\text{E}_0][\text{S}]}{1 + K_1[\text{S}]}$$

The rate equation, Eqn. (5-19), becomes

$$-r_\text{S} = \frac{k_2 K_1[\text{E}_0][\text{S}]}{1 + K_1[\text{S}]}$$

This is exactly the expression that was derived previously (Eqn. (5-18)) by assuming that all of the elementary reactions were reversible and then setting K_3' and $K_{eq} = \infty$.

5.7 CLOSING COMMENTS

Both the steady-state approximation and the rate-limiting step approximation require the use of the "order = molecularity" property of elementary reactions. For either of these tools to be useful, each reaction in the mechanism must be elementary and the reaction mechanism must be correct. Use of the screening criteria discussed in Section 5.1.3 of this chapter can avoid wasting time and energy deriving a rate equation for a proposed mechanism that contains nonelementary reactions. It is important to apply these screening criteria to *each* reaction in the presumed mechanism, *before* derivation of the rate equation is begun.

Even when the screening criteria are rigorously applied, *the derived rate equation must be tested against experimental data.* After all, we can never be certain that the assumed mechanism is correct, and we can never be certain that each reaction in the mechanism is elementary as written. Moreover, if the RLS approximation is used, we may or may not have identified the RLS correctly, if indeed a single RLS does exist.

Testing of rate equations against data is the subject of the next chapter.

SUMMARY OF IMPORTANT CONCEPTS

- For an elementary reaction (and only for an elementary reaction), the form of the rate equation can be written *a priori*, using the "order = molecularity" principle.
- Elementary reactions occur in a single step, on the molecular level, *exactly* as written in the balanced stoichiometric equation.
- To be considered elementary, a reaction must be simple. Screening tools, based on the principles of simplicity, can be used to assess the likelihood that a given reaction is elementary.
- The steady-state approximation (SSA) can be used to derive the form of a rate equation for a reaction whose mechanism is constructed of elementary reactions.
- The rate-limiting step (RLS) approximation also can be used to derive the form of the rate equation. The RLS Approximation is based on the assumption that a single RLS exists, and that it has been identified correctly. The RLS approximation contains all of the assumptions of the SSA, and is less general than the SSA.
- For catalytic reactions, one or more catalyst (site, enzyme) balances are required to derive a rate equation, with either the RLS approximation or the SSA.

---

[7] This relationship also could have been obtained from Eqn. (5-10) by recognizing that [E–P] must be essentially zero if K_3' is essentially infinite.

PROBLEMS

Problem 5-1 (Level 2) Derive a rate equation for the disappearance of A in a heterogeneous catalytic reaction, given the following sequence of elementary reactions:

$$A + S_1^* \rightleftarrows A\text{–}S_1^* \quad (1)$$

$$A\text{–}S_1^* + S_2^* \rightleftarrows S_1^* + B\text{–}S_2^* \quad (2)$$

$$B\text{–}S_2^* \rightleftarrows B + S_2^* \quad (3)$$

$$A \rightleftarrows B$$

There are two different types of site on the catalyst surface. S_1^* is an empty "Type 1" site and $A\text{–}S_1^*$ represents A adsorbed on a "Type 1" site. S_2^* is an empty "Type 2" site and $B\text{–}S_2^*$ represents B adsorbed on a "Type 2" site.

Assume that Reaction (2) is the rate-limiting step.

Problem 5-2 (Level 1) The methanol synthesis reaction

$$CO + 2H_2 \rightleftarrows CH_3OH$$

is reversible at typical operating conditions. With certain heterogeneous catalysts, the reaction is thought to proceed according to the following sequence of elementary reactions:

$$CO + S^* \rightleftarrows CO\text{–}S^* \quad (1)$$

$$H_2 + S^* \rightleftarrows H_2\text{–}S^* \quad (2)$$

$$H_2\text{–}S^* + CO\text{–}S^* \rightleftarrows CH_2O\text{–}S^* + S^* \quad (3)$$

$$CH_2O\text{–}S^* + H_2\text{–}S^* \rightleftarrows CH_3OH\text{–}S^* + S^* \quad (4)$$

$$CH_3OH\text{–}S^* \rightleftarrows CH_3OH + S^* \quad (5)$$

where S^* is an empty site on the catalyst surface, and $A\text{–}S^*$ is species "A" adsorbed on a site.

Reaction (3) is believed to be the rate-limiting step.

1. If K_i is the equilibrium constant for Reaction (*i*), show that the equilibrium constant for the overall reaction is

$$K_{eq} = K_1 K_2^2 K_3 K_4 K_5$$

2. Derive the form of the rate equation for methanol formation for this mechanism.

Problem 5-3 (Level 2) The following e-mail is in your inbox at 8 AM on Monday morning:

To: U. R. Loehmann

From: I. M. DeBosse

Subject: Rate Equation for Methanol Synthesis

Cauldron Chemical Company's Central Research Department has developed a new heterogeneous catalyst for methanol synthesis. The reaction is

$$CO + 2H_2 \rightleftarrows CH_3OH$$

The overall reaction is reversible at typical operating conditions.

The research team believes that the overall reaction proceeds according to the following sequence of elementary reactions

$$CO + S^* \rightleftarrows CO\text{–}S^* \quad (1)$$

$$H_2 + S^* \rightleftarrows H_2\text{–}S^* \quad (2)$$

$$H_2\text{–}S^* + CO\text{–}S^* \rightleftarrows CH_2O\text{–}S^* + S^* \quad (3)$$

$$CH_2O\text{–}S^* + H_2\text{–}S^* \rightleftarrows CH_3OH\text{–}S^* + S^* \quad (4)$$

$$CH_3OH\text{–}S^* \rightleftarrows CH_3OH + S^* \quad (5)$$

where S^* is an empty site on the catalyst surface and $A\text{–}S^*$ is species "A" adsorbed on a site.

Reactions (1), (2), and (5) are believed to be in equilibrium at normal operating conditions. The rates of Reactions (3) and (4) are believed to be comparable, i.e., neither one can be considered to be rate limiting.

The Central Research Department has collected a substantial quantity of kinetic data on this new catalyst, but they don't seem to know what rate equations to test against the data. Your assignment is to derive the form of the rate equation for methanol formation. Please report your results to me in a one-page memo, pointing out any important features of your rate equation. Please attach your derivation to the memorandum in case someone wants to review your work.

Problem 5-4 (Level 1) The oxidation of carbon monoxide on a heterogeneous, platinum-containing catalyst is believed to proceed according to the sequence of elementary reactions

$$CO + S^* \rightleftarrows CO\text{–}S^* \quad (1)$$

$$O_2 + 2S^* \rightleftarrows 2O\text{–}S^* \quad (2)$$

$$O\text{–}S^* + CO\text{–}S^* \rightarrow CO_2\text{–}S^* + S^* \quad (3)$$

$$CO_2\text{–}S^* \rightleftarrows CO_2 + S^* \quad (4)$$

The third reaction is irreversible, and is considered to be the rate-limiting step.

Questions

1. Derive a rate equation for the disappearance of CO.
2. Point out the important features of the rate equation, i.e., what happens to the rate as each of the partial pressures is varied?
3. How would your answer to part (a) change if Reaction (4) were irreversible, but Reaction (3) remained the rate-limiting step?

Problem 5-5 (Level 1) The overall reaction for the formation of HBr from H_2 and Br_2 in the gas phase is

$$H_2 + Br_2 \rightarrow 2HBr$$

This reaction may proceed via the following sequence of reactions

$$Br_2 \rightleftarrows 2Br^\bullet \quad (1)$$

$$Br^\bullet + H_2 \rightleftarrows HBr + H^\bullet \quad (2)$$

$$H^\bullet + Br_2 \rightarrow HBr + Br^\bullet \quad (3)$$

1. Identify the active centers in the above sequence.
2. Comment on the probability that Reaction (1) is elementary as written.
3. Derive a rate equation for the formation of HBr using the steady-state approximation, assuming that all of the above reactions are elementary as written.

Problem 5-6 (Level 1) The overall reaction for the formation of diethyl ether from ethanol is

$$2C_2H_5OH \rightleftarrows (C_2H_5)_2O + H_2O$$
$$(2A \rightleftarrows E + W)$$

This reaction takes place on the surface of a heterogeneous catalyst (a sufonated copolymer of styrene and divinylbenzene). Suppose that the reaction proceeds through the following sequence of elementary steps:

$$C_2H_5OH + S^* \rightleftarrows C_2H_5OH{-}S^* \quad (1)$$
$$2C_2H_5OH{-}S^* \rightleftarrows (C_2H_5)_2O{-}S^* + H_2O{-}S^* \quad (2)$$
$$(C_2H_5O)_2O{-}S^* \rightleftarrows (C_2H_5)_2O + S^* \quad (3)$$
$$H_2O{-}S^* \rightleftarrows H_2O + S^* \quad (4)$$

Derive a rate equation for the disappearance of ethanol assuming that Reaction (2) is the rate-limiting step.

Problem 5-7 (Level 1) Phosgene ($COCl_2$) is formed by the reaction of carbon monoxide (CO) and chlorine (Cl_2).

$$CO + Cl_2 \rightarrow COCl_2$$

Suppose that this reaction proceeds by the following sequence of reactions:

$$Cl_2 \rightleftarrows 2Cl^\bullet \quad (1)$$
$$Cl^\bullet + CO \rightarrow COCl^\bullet \quad (2)$$
$$COCl^\bullet + Cl_2 \rightarrow COCl_2 + Cl^\bullet \quad (3)$$

1. Classify each of these *four* reactions as either an initiation, propagation, or termination reaction.
2. Identify the active centers in the above sequence.
3. Is the sequence open or closed?
4. Comment on the probability that Reaction (1) is elementary as written.
5. Derive a rate equation for the formation of phosgene using the steady-state approximation, assuming that all of the reactions above are elementary.

Problem 5-8 (Level 1) The polymerization initiator I–I decomposes to give two free radicals, $I^\bullet$, via the elementary reaction

$$I{-}I \rightarrow 2I^\bullet \quad (1)$$

The free radical, $I^\bullet$, decomposes to another radical, $R^\bullet$, by the elementary-reaction

$$I^\bullet \rightarrow R^\bullet + C \quad (2)$$

The radical, $R^\bullet$, can attack I-I via the elementary reaction

$$R^\bullet + I{-}I \rightarrow R{-}I + I^\bullet \quad (3)$$

and two $R^\bullet$ radicals can combine by the elementary reaction

$$2R^\bullet \rightarrow R{-}R \quad (4)$$

Derive a rate equation for the disappearance of I–I.

Problem 5-9 (Level 2) In aqueous solution, the overall reaction

$$A \rightarrow P$$

takes place via the following sequence of elementary reactions:

$$A + H^+ \rightleftarrows AH^+ \quad (1)$$
$$A \rightarrow P \quad (2)$$

The last reaction (2) is rate controlling. Reaction (1) can be assumed to be in equilibrium. Note that the *total* concentration of A, A_T, is given by

$$[A_T] = [A] + [AH^+]$$

Sketch a curve of r_P versus $[H^+]$ at constant $[A_T]$. Make sure that the curve reflects the correct quantitative dependence of r_{AP} on $[H^+]$ at "high" and "low" $[H^+]$.

Hint: Derive a rate equation for the production of P. The rate equation should contain $[A_T]$, not [A] and/or $[AH^+]$. It also should contain $[H^+]$ and K_1, the equilibrium constant for Reaction (1).

Problem 5-10 (Level 3) The isomerization of 2,5-dihydrofuran (2,5-DHF) to 2,3-dihdyrofuran (2,3-DHF) has been studied over a Pd-containing catalyst at about 100 °C.[8] Suppose that the reaction proceeds via the following sequence of elementary reactions:

(2,5-DHF) + S* ⇌ (K_1) adsorbed 2,5-DHF (ring bonded to S) (1)

adsorbed 2,5-DHF + S* → (K_2) adsorbed intermediate (ring bonded to S) + H–S (2)

adsorbed intermediate + H–S → (k_3) adsorbed 2,3-DHF + S* (3)

adsorbed 2,3-DHF ⇌ ($1/K_4$) S* + (2,3-DHF) (4)

[8] Monnier, J. R., Medlin, J. W., and Kuo, Y.-J., Selective isomerization of 2,5-dihydrofuran to 2,3-dihydrofuran using CO-modified, supported Pd catalysts, *Appl. Catal. A: Gen.*, 194–195, 463–474 (2000).

Assume that the overall reaction is irreversible. Further assume that Steps 1 and 4 (the adsorption of 2,5-DHF and the desorption of 2,3-DHF, respectively) are very fast and are essentially in equilibrium. Finally, assume that Steps 2 and 3 are essentially irreversible.

Let K_1 be the equilibrium constant for Reaction (1) and let K_4 be the equilibrium constant for the *reverse* of Reaction (4), i.e., for the adsorption of 2,3-DHF onto the surface of the catalyst. Let k_2 and k_3 be the rate constants for Reactions (2) and (3), respectively.

1. What is the form of the rate equation for the disappearance of 2,5-DHF?
2. If Step 3 is the rate-limiting step, what is the form of the rate equation?
3. The value of the free energy change for the overall reaction at 100 °C is −8.2 kcal/mol ($\Delta G_R = -8.2$ kcal/mol). Consider a feed that is 100% 2,5-DHF. Is it justified to assume that the overall reaction is irreversible over the range of fractional conversion of 2,5-DHF from 0% to 99%?

Problem 5-11 (Level 3) In aqueous solution, peroxybenzoic acid decomposes to benzoic acid and molecular oxygen.[9]

$$\underset{\substack{\text{(peroxybenzoic acid)}\\ \text{(PBA)}}}{C_6H_5CO_3H} \rightarrow \underset{\substack{\text{(benzoic acid)}\\ \text{(BA)}}}{C_6H_5CO_2H} + 1/2\,O_2$$

The overall reaction may proceed according to the following sequence of elementary reactions:

$$PBA \rightleftarrows PBA^- + H^+ \quad (1)$$

$$PBA^- + PBA \rightarrow (PBA)_2^- \quad (2)$$

$$(PBA)_2^- \rightarrow BA + PBAO^- \quad (3)$$

$$PBAO^- \rightarrow BA^- + O_2 \quad (4)$$

$$BA^- + H^+ \rightleftarrows BA \quad (5)$$

1. Assume that Reaction (2) is the rate-limiting step. Derive a rate equation for the formation of O_2. In the rate equation, let [P] denote the *total* concentration of PBA, i.e., $[P] = [PBA] + [PBA^-] + 2[(PBA)_2^-] + [PBAO^-]$. Only [P] and $[H^+]$ should appear in the final answer; [PBA], $[PBA^-]$, $[(PBA)_2{}^-]$, $[PBAO^-]$, [BA], and $[BA^-]$ should not appear.
2. Consider a situation where [P] is constant, but the pH of the solution is varied. Use your answer to Question 1 to make a sketch of how the reaction rate (at constant [P]) varies with $[H^+]$.
3. At what value of $[H^+]$ does the maximum rate occur, at constant [P]?

Problem 5-12 (Level 2) At very high temperatures, acetone (C_3H_6O) decomposes to methane (CH_4) and ketene (CH_2CO)

$$C_3H_6O \rightarrow CH_2CO + CH_4$$

The reaction is believed to take place through the following series of elementary reactions:

$$CH_3COCH_3 \xrightarrow{k_1} CH_3^{\bullet} + CH_3CO^{\bullet} \qquad E_a = 84\ \text{kcal/mol}$$

$$CH_3CO^{\bullet} \xrightarrow{k_2} CH_3^{\bullet} + CO \qquad E_a = 10\ \text{kcal/mol}$$

$$CH_3^{\bullet} + CH_3COCH_3 \xrightarrow{k_3} CH_4 + CH_2COCH_3^{\bullet} \qquad E_a = 15\ \text{kcal/mol}$$

$$CH_2COCH_3^{\bullet} \xrightarrow{k_4} CH_3^{\bullet} + CH_2CO \qquad E_a = 48\ \text{kcal/mol}$$

$$CH_3^{\bullet} + CH_2COCH_3^{\bullet} \xrightarrow{k_5} C_2H_5COCH_3 \qquad E_a = 5\ \text{kcal/mol}$$

The amounts of CO and methyl ethyl ketone ($C_2H_5COCH_3$) that are formed are very small compared to the amounts of methane and ketene.

1. Derive a rate equation for the disappearance of acetone. In the final rate equation, neglect any terms that are very small.
2. Calculate the value of the activation energy that will be observed experimentally for the overall reaction. The values of the activation energies for the five elementary reactions are shown to the right of each reaction.

Problem 5-13 (Level 3)[10] Acetic acid (HOAc) is produced by carbonylation of methanol (MeOH).

$$CO + \underset{\text{(MeOH)}}{CH_3OH} \rightarrow \underset{\text{(HOAc)}}{C_2H_4O_2}$$

The reaction takes place in the liquid phase and is catalyzed by a soluble Rh or Ir organometallic compound. In addition to the soluble Rh or Ir compound, methyl iodide (CH_3I) is used as a cocatalyst. Methanol carbonylation is one of the most important commercial processes that is based on a homogeneous catalyst.

The Rh-catalyzed reaction proceeds according to the following sequence of elementary reactions:

$$CH_3OH + HI \rightleftarrows CH_3I + H_2O \quad (1)$$

$$RhL_m + CH_3I \rightarrow CH_3Rh(I)L_m \quad (2)$$

$$CH_3Rh(I)L_m + CO \rightleftarrows CH_3Rh(I)(CO)L_m \quad (3)$$

$$CH_3Rh(I)(CO)L_m \rightleftarrows CH_3CO{-}Rh(I)L_m \quad (4)$$

$$CH_3CO{-}Rh(I)L_m + MeOH \rightleftarrows RhL_m + HOAc + CH_3I \quad (5)$$

Reaction (2) is the rate-limiting step; let its rate constant be denoted k_2. Reactions (1), (3), (4), and (5) are in equilibrium; let their equilibrium constants be denoted K_1, K_3, K_4, and K_5. Let Rh_0 be the total atomic concentration of Rh in the system and let I_0 be the total atomic concentration of I in the system. RhL_m is an organometallic compound of Rh. Reaction (5) probably proceeds through a series of simpler steps. However, the exact sequence is not kinetically important in this case since Reaction (5) is in equilibrium.

[9] Goodman, J. F., Robson, P., and Wilson, E. R., *Trans. Farad. Soc.*, 58, 1846–1851 (1962).

[10] Adapted from Hjortkjaer, J. and Jensen, V. W., Rhodium complex catalyzed methanol carbonylation. *Ind. Eng. Chem. Prod. Res. Dev.*, 15(1), 46–49 (1976).

Derive a rate equation for the formation of HOAc. You may assume that $I_o \gg Rh_o$, so that the amount of I bound to Rh is very small compared to the amount in CH_3I and HI. You may also assume that the concentration of water is known, so that its concentration may appear in the final rate equation.

1. Assume that the equilibrium constant for Reaction (1) is very large, and that Reactions (3), (4), and (5) are irreversible.
2. Derive a more general expression without making the assumptions in part (1).

Problem 5-14 (Level 1) Which of the following chemical reactions can reasonably be assumed to be elementary? Explain your answers.

1. $O_3 \rightarrow O_2 + O^\bullet$
2. $C_4H_4S + 3H_2 \rightarrow C_4H_8 + H_2S$
3. $H_2 + \frac{1}{2}O_2 \rightarrow H_2O$
4. $H^\bullet + I_2 \rightarrow HI + I^\bullet$
5. $O_2 + 2S^* \rightarrow 2O{-}S^*$

(S^* = vacant site on catalyst surface; $O{-}S^*$ = oxygen atom adsorbed on site)

6. $3H^+ + PO_4^{\equiv} \rightarrow H_3PO_4$
7. C_6H_{12}(cyclohexane) $\rightarrow C_6H_6 + 3H_2$
8. $(CH_3)_3C{-}O{-}O{-}C(CH_3)_3 \longrightarrow 2\,(CH_3)_3C{-}O^\bullet$

Problem 5-15 (Level 2) The hydrogenation of butene to butane on certain oxide catalysts is believed to proceed according to the sequence of elementary reactions:

$$C_4H_8 + S_1 \underset{k_1'}{\overset{k_1}{\rightleftarrows}} C_4H_8S_1 \quad (1)$$

$$H_2 + S_2 \underset{k_2'}{\overset{k_2}{\rightleftarrows}} H_2{-}S_2 \quad (2)$$

$$H_2{-}S_2 + C_4H_8{-}S_1 \underset{k_3'}{\overset{k_3}{\rightleftarrows}} C_4H_{10}{-}S_1 + S_2 \quad (3)$$

$$C_4H_{10}{-}S_1 \underset{k_4'}{\overset{k_4}{\rightleftarrows}} C_4H_{10} + S_1 \quad (4)$$

In the above mechanism, S_1 and S_2 are two *distinctly different* types of sites. Only H_2 adsorbs on S_2; both C_4H_8 and C_4H_{10} adsorb on S_1 but H_2 does not.

Assume that Step 3 is the rate-limiting step. Derive a rate equation for the rate of disappearance of butene. Consider the overall reaction to be reversible.

Problem 5-16 (Level 2) Cyclopropane (CP) isomerizes to propylene (P) at temperatures in the range of 470 to 520 °C and at pressures between about 10 and 700 mm Hg.[11] Only cyclopropane is present initially. The overall reaction is

$$\text{cyclo-}C_3H_6 \longrightarrow CH_2{=}CH{-}CH_3 \quad (A)$$

This reaction is believed to proceed according to the following sequence of elementary reactions:[12]

$$CP + CP \rightleftarrows CP^* + CP \quad (1)$$

$$P + CP \rightleftarrows CP^* + P \quad (2)$$

$$CP^* \rightarrow P \quad (3)$$

Here, CP^* is an activated (highly energetic) molecule of cyclopropane.

1. Comment on the probability that Reaction (A) is elementary as written.
2. Is the sequence of Reactions (1) → (3) closed or open? Why?
3. Derive a rate equation for the disappearance of CP. You may assume that the forward rate constants for Reactions (1) and (2) are the same and that the reverse rate constants for these reactions also are the same: You also may assume that the overall reaction is irreversible.
4. What is the form of the rate equation at very high total pressure? What is the form at very low total pressure?
5. The following thermochemical data are available:[13]

Species	$\Delta H_f(298)$(kcal/mol)	$\Delta G_f(298)$(kcal/mol)
CP	12.74	24.95
Propylene	4.88	15.02

Comment on the validity of assuming that the overall reaction is irreversible,

Problem 5-17 (Level 2) Under certain conditions, the gas-phase thermal dehydrogenation of ethane is believed to proceed according to the sequence of reactions:

$$C_2H_6 \xrightarrow{k_1} 2CH_3^\bullet \quad (1)$$

$$CH_3^\bullet + C_2H_6 \xrightarrow{k_2} CH_4 + C_2H_5^\bullet \quad (2)$$

[11] Chambers, T. S. and Kistiakowsky, G. B., Kinetics of the thermal isomerization of cyclopropane, *J. Am. Chem. Soc.*, 56, 399 (1934).

[12] This mechanism often is referred to as the Lindemann mechanism.

[13] Dean, J. A., (ed.), *Lange's Handbook of Chemistry* (13th edition), Table 9-2, 9-70, McGraw-Hill, (1985).

$$C_2H_5^\bullet \xrightarrow{k_3} C_2H_4 + H^\bullet \quad (3)$$

$$H^\bullet + C_2H_6 \xrightarrow{k_4} H_2 + C_2H_5^\bullet \quad (4)$$

$$2\ C_2H_5^\bullet \xrightarrow{k_5} C_4H_{10} \quad (5)$$

$$H^\bullet + C_2H_5^\bullet \xrightarrow{k_6} C_2H_6 \quad (6)$$

Analysis of the gas mixture leaving an ethane dehydrogenation reactor shows that hydrogen and ethylene are the only significant products. The amounts of methane and butane formed are detectable, but negligible.

1. Are there any reactions in the above sequence that you suspect are not elementary. Justify your answer.
2. Classify the above reactions according to the following categories: initiation, termination, propagation.
3. Using the steady-state approximation, derive a rate equation for the disappearance of ethane. Simplify the rate expression as much as possible by neglecting the rates of any steps that are comparatively insignificant.

Problem 5-18 (Level 3) Parent et al.[14] have studied the selective hydrogenation of C=C bonds in nitrile–butadiene rubber (NBR) to produce hydrogenated nitrile–butadiene rubber (HNBR), which has superior resistance to thermal and chemical degradation. To obtain the desired product, the rate of hydrogenation of C=C bonds must be much higher than the rate at which the nitrile group (–C≡N) is hydrogenated.

Parent et al. used monochlorobenzene as a solvent and an osmium complex [$OsHCl(CO)(O_2)(P)_2$] as a homogeneous catalyst. In this formula, P represents a triphenyl phosphine group. Parent et al. studied NBR hydrogenation in a batch reactor over a range of temperatures (120–140 °C), H_2 pressures (21–80 bar), free P concentrations (0.37–1.38 ns.), catalyst concentrations (20–250 μm), and polymer concentrations (75–250 mM nitrile). Based on their research, the authors speculated that the overall reaction proceeds according to the following sequence of elementary reactions. In the following, OsCl(CO) is abbreviated Os, and RCN is the nitrile group in the polymer.

$$OsH(O_2)P_2 \rightarrow OsHP_2 + O_2 \quad (1)$$

$$OsHP_2 + RCN \rightleftarrows OsH(RCN)P_2 \quad (2)$$

$$OsHP_2 + H_2 \rightleftarrows OsH(H_2)P_2 \quad (3)$$

$$OsH(H_2)P_2 \rightleftarrows OsH(H_2)P + P \quad (4)$$

$$OsH(H_2)P + H_2 \rightleftarrows OsH_3(H_2)P \quad (5)$$

$$OsH_3(H_2)P + C{=}C \rightleftarrows OsH(H_2)(C{-}C)P \quad (6)$$

$$OsH(H_2)(C{-}C)P \rightarrow -C{-}C- + OsH(H_2)P \quad (7)$$

Reaction (7) is believed to be the rate-limiting step.

1. Derive an equation for the rate of disappearance of C=C groups in the polymer.
2. Use your rate equation to predict how the rate depends on
   i. Catalyst concentration
   ii. H_2 pressure
   iii. P concentration
   iv. RCN concentration
   v. C=C concentration

Do this by analyzing how the rate changes as the specified variable is changed with all of the other variables held constant.

3. Check your predictions against the results in the referenced article.

Problem 5-19 (Level 3) Chemical vapor deposition (CVD) is used to deposit thin films of polycrystalline silicon for electronic devices such as semiconductors.[15] One of the reactions that can be employed is the decomposition of silane (SiH_4). The overall reaction is

$$SiH_4(g) \rightarrow Si(s) + 2H_2$$

This reaction may proceed according to the following sequence of elementary steps:

$$SiH_4 \rightleftarrows SiH_2 + H_2 \quad (1)$$

$$SiH_2 + S \rightleftarrows SiH_2{-}S \quad (2)$$

$$SiH_2{-}S \rightarrow Si + H_2 + S \text{ (irreversible)} \quad (3)$$

Here, S is a site on the surface of solid Si, SiH_2 (silylene) is a reactive, gas-phase, active center, and $SiH_2{-}S$ is a complex of SiH_2 with a site on the silicon surface.

Let C_i = concentration of species i, K_j = equilibrium constant for Reaction j, C_V = concentration of vacant sites and C_T = total concentration of sites. Further, let I denote the intermediate SiH_2, I–S denote the $SiH_2{-}S$ surface complex, A denote SiH_4, and H denote H_2.

It is known that the reaction is first order in SiH_4 at low SiH_4 concentration and zero order at high SiH_4 concentrations.[16] It is also known that the reaction is inhibited by H_2.

1. An attempt has been made to explain the observed kinetics, as follows:

   Assume that Reaction (1) is rate limiting:
   $-r_A = k_1 p_A - k_{-1} p_I p_H$
   Equilibrium (Step 2): $C_{I-S} = K_2 p_I C_v$
   Site balance: $C_T = C_V + C_{I-S}$
   Combining the last two expressions: $C_V = C_T/(1 + K_2 p_I)$
   Assume that Step 1 is in equilibrium.

[14] Parent, J. S., McManus, N. T., Rempel, G. L., $OsHCl(CO)(O_2)(PCy_3)_2$-catalyzed hydrogenation of acrylonitrile–butadiene copolymers, *Ind. Eng. Chem. Res.*, 37, 4253–4261, (1998).

[15] For example, see Middleman, S. and Hochberg, A. K., *Process Engineering Analysis in Semiconductor Device Fabrication,* McGraw-Hill, New York (1993) and Lee, H. H., *Fundamentals of Microelectronics Processing,* McGraw-Hill, New York (1990).

[16] Roenigk, K. F. and Jensen, K. F., Analysis of multicomponent LPCVD processes, *J. Electrochem. Soc.,* 132, 448 (1985).

Equilibrium (Step 1): $p_H p_I / p_A = K_1$; $p_I = K_1 p_A / p_H$

Returning to rate equation: $-r_A = k_1 p_A - k_{-1} p_H (K_1 p_A / p_H) = k_1 p_A - k_1 p_A = 0$ (*Note*: $K_1 = k_1 / k_{-1}$)

Conclusion: Kinetics of this reaction cannot be analyzed by using the rate limiting step approximation.

Identify the error in this solution. (*Hint:* The error is *not* strictly mathematical. It is a fundamental error in reaction kinetics.)

2. Derive a rate equation for the disappearance of silane, assuming that Reaction (1) is the rate-limiting step.

3. Derive the rate equation, assuming that Reaction (2) is the rate-limiting step.

4. Derive the rate equation, assuming that Reaction (3) is the rate-limiting step.

5. Which of the three assumed rate-limiting steps leads to the rate equation that best describes the experimental observations?

Chapter 6

Analysis and Correlation of Kinetic Data

LEARNING OBJECTIVES

After completing this chapter, you should be able to

1. linearize rate equations, i.e., put them into a straight-line form;
2. test linearized rate equations against experimental data graphically, and obtain preliminary estimates of the unknown parameters in the rate equation;
3. obtain "best fit" values of the unknown parameters in a linearized rate equation using linear least-squares analysis;
4. obtain "best fit" values of the parameters in a nonlinear rate equation using nonlinear least squares analysis;
5. visually evaluate the overall fit of a rate equation to a set of experimental data;
6. test for systematic errors in the fit of a rate equation to a set of experimental data using graphical techniques.

The form of a rate equation *always* must be tested against experimental data. This assertion is easy to understand if the rate equation was postulated arbitrarily, e.g., a power-law form, as discussed in Chapter 2. However, the form of the rate equation must be tested against data, even if the kinetic expression was developed from a hypothetical sequence of elementary steps. First, the hypothesis may or may not be valid. Second, if the rate-limiting step approximation was used, we may or may not have guessed the correct rate-limiting step, or it may be that no single step is rate limiting.

If the rate equation being tested *does* fit the experimental data, then the unknown constants in the rate equation can be estimated from the data. If the rate equation does not fit, then a new kinetic expression must be postulated and tested.

Procedures to test rate equations against experimental data are discussed in this chapter. However, we first must deal with the question of how useful kinetic data can be obtained.

6.1 EXPERIMENTAL DATA FROM IDEAL REACTORS

Experimental kinetic data *always* should be taken in a reactor that behaves as one of the three ideal reactors. It is relatively straightforward to analyze the data from an ideal batch reactor, an ideal plug-flow reactor, or an ideal stirred-tank reactor. This is not the case if the reactor is nonideal, e.g., somewhere between a PFR and a CSTR. Characterizing the behavior of nonideal reactors is difficult and imprecise, as we shall see in Chapter 10. This can lead to major uncertainties in the analysis of data taken in nonideal reactors.

Many kinetic studies will involve heterogeneous catalysts, since they are so widely used commercially. *The kinetics of heterogeneously catalyzed reactions always must be studied under conditions where the reaction is controlled by intrinsic kinetics.* If either internal or external transport influences the rate of a reaction, the form of its rate equation may be distorted, and the parameters obtained from the data analysis will have little or no fundamental significance. Methods to eliminate transport effects from kinetic studies are discussed in detail in Chapter 9.

6.1.1 Stirred-Tank Reactors (CSTRs)

Reactors that behave as ideal CSTRs are sometimes referred to as "gradientless" reactors, especially when they are used for kinetic studies. This is because there are no spatial variations of concentration or temperature, and the rate is the same at every point inside the reactor.

The design equation for an ideal CSTR (either Eqn. (3-17) or (3-17a)) can be rearranged to give

$$-r_A = x_A F_{A0}/V(\text{or } W) \tag{6-1}$$

If the reaction is homogeneous, the volume V is used in Eqn. (6-1). If the reaction is a heterogeneous catalytic reaction, the catalyst weight W is used.

Equation (6-1) shows that the reaction rate can be obtained *directly* in a CSTR if the fractional conversion x_A is measured, and if the molar feed rate F_{A0} and the reactor volume V (or the catalyst weight W) are known. However, it is good practice to measure the complete composition of the effluent stream, even though the concentration of every species can be calculated from x_A. Measuring every concentration provides a check on the quality of the data, and allows the Law of definite proportions to be used to ensure that only one reaction takes place at the conditions of the experiment. The temperature of the reactor must also be measured and carefully controlled.

In order to obtain data that will provide a rigorous test of the assumed rate equation, the composition in the CSTR must be varied over a wide range in order to determine how the reaction rate, $-r_A$, responds to changes in the various species concentrations. This can be done by varying the composition of the inlet stream and by varying the space time τ_0 (either V/υ_0 or W/υ_0). For example, suppose that the difference between the highest and lowest concentrations of reactant A in a given set of data is 10%. If the reaction is first order in A, the rate at the highest C_A will be 10% higher than at the lowest C_A. If the reaction is second order in A, the difference in rates is 21%. Given normal errors in experimental data (small temperature fluctuations, analytical errors, etc.), it may be difficult to discriminate between these two possibilities. The solution is to take data over a much wider range of concentration. If the difference between the highest and lowest values of C_A is a factor of 10, the ratio of the rate at the highest concentration to that at the lowest is 10 for $n = 1$ and 100 for $n = 2$.

Stoichiometry is another issue that must be considered in designing kinetic experiments. Suppose that the reaction taking place is A + B → products, and it is necessary to determine the individual orders, α_A and α_B in the proposed rate equation:

$$-r_A = kC_A^{\alpha_A} C_B^{\alpha_B}$$

If A and B are always in the stoichiometric ratio, 1/1 in this case, then C_A will always equal C_B and the rate equation can be written as

$$-r_A = kC_A^{(\alpha_A+\alpha_B)}$$

It will be possible to determine the *overall* order, $\alpha_A + \alpha_B$, from the experimental data. However, there is no way that the *individual* orders, α_A and α_B, can be determined. To obtain values for both α_A and α_B, some experiments will have to be carried out where the ratios of the concentrations of A and B in the feed to the CSTR are considerably different from the stoichiometric ratio.

Usually, a number of data points are taken at a fixed temperature so that an isothermal version of the rate equation can be tested. Additional data then are taken at several different temperatures in order to determine the activation energy of the reaction and to determine the temperature dependence of any other constants in the rate equation.

Table 6-1 shows the type of data that are obtained from kinetic studies in an ideal CSTR.

Table 6-1 Typical Data from Experiments in a CSTR

Experiment number	Temperature	Reaction rate	Outlet concentration			
			A	B	C	etc.
1	T_1	$-r_A(1)$	$C_A(1)$	$C_B(1)$	$C_C(1)$	...
2	T_1	$-r_A(2)$	$C_A(2)$	$C_B(2)$	$C_C(2)$	...
3	T_1	$-r_A(3)$	$C_A(3)$	$C_B(3)$	$C_C(3)$	...
↓	↓	↓	↓	↓	↓	↓
N	T_1	$-r_A(N)$	$C_A(N)$	$C_B(N)$	$C_C(N)$	...
$N+1$	T_2	$-r_A(N+1)$	$C_A(N+1)$	$C_B(N+1)$	$C_C(N+1)$	...
$N+2$	T_2	$-r_A(N+2)$	$C_A(N+2)$	$C_B(N+2)$	$C_C(N+2)$	...
↓	↓	↓	↓	↓	↓	↓

The first N data points in Table 6-1 were taken at a constant temperature T_1. The feed composition and/or the space time τ_0 were varied in order to vary the outlet concentrations. The most convenient way to vary τ_0 is to vary υ_0, rather than changing V or W.

Subsequent data were taken at a different temperature, T_2. Ideally, additional data would be obtained at several more temperatures.

The major advantage of using a CSTR to study the kinetics of a reaction is that values of the reaction rate $-r_A$ can be obtained directly from the data via Eqn. (6-1). As we shall soon see, this makes data analysis easier. A disadvantage is that it is difficult to control the *outlet* concentrations, which determine the reaction rate.

6.1.2 Plug-Flow Reactors

6.1.2.1 Differential Plug-Flow Reactors

A differential plug-flow reactor is a plug-flow reactor, as described in Chapter 3, that is operated at very low (differential) conversion. Differential PFRs are widely used for studying the kinetics of heterogeneous catalytic reactions.

The fractional conversion of a reactant in a differential PFR can be kept low (typically less than 10%) by operating the reactor at a very low space time τ_0. For a heterogeneous catalyst, $\tau_0(=W/\upsilon_0)$ can be made low by using a very small amount of catalyst. This can be an advantage when experimental catalysts are being studied, as the amount of catalyst available may be limited.

The design equation for an ideal PFR in differential form is

$$\frac{dV(\text{or}\, dW)}{F_{A0}} = \frac{dx_A}{-r_A} \tag{3-27/27a}$$

Suppose that the PFR is operated so that the fractional conversion x_A is very small, of the order of a few percent, and so that the reactor is isothermal. For these conditions, the reaction rate will not vary substantially between the reactor inlet and the reactor outlet. Equation (3-27/27a) then can be integrated by assuming that $-r_A$ is constant at some average value, $-\bar{r}_A$. Thus,

$$\frac{V(\text{or } W)}{F_{A0}} = \frac{x_A}{-\bar{r}_A} \tag{6-2}$$

$$-\bar{r}_A = x_A F_{A0}/V(\text{or } W)$$

A value of the average reaction rate can be calculated from the measured outlet conversion x_A using Eqn. (6-2). The concentrations that are associated with this value of $-\bar{r}_A$ usually are taken to be the average of the inlet and outlet concentrations.

By carrying out experiments at different inlet concentrations and temperatures, a differential PFR can be used to generate the type of data shown in Table 6-1. In fact, it is easier to generate data in the format of Table 6-1 with a differential PFR than with a CSTR. With a CSTR, the *outlet* concentration must be controlled. This involves manipulating the inlet concentration and/or the space time in a trial-and-error manner, since the kinetics are not known *a priori*. This type of manipulation can be especially difficult when one outlet concentration is being varied while the others are being held constant. In a differential PFR, it is only necessary to set the inlet concentrations.

6.1.2.2 Integral Plug-Flow Reactors

In an *integral* PFR, the reactant conversion is significant. Therefore, it is not valid to assume that the reaction rate is essentially constant in the direction of flow. Suppose that an ideal, isothermal PFR is operated with a constant feed composition at several different values of V/F_{A0} (or W/F_{A0}), and the fractional conversion, x_A, is measured at each value of V/F_{A0}. The resulting data will have the form shown in the following figure.

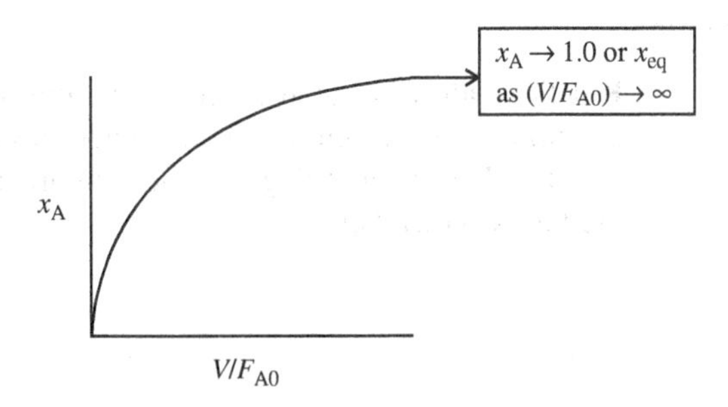

We can work directly with this kind of data to test a postulated rate equation. This approach will be discussed later, in Section 6.3 of this chapter. However, it is also possible to calculate values of the reaction rate $-r_A$ at various values of x_A, to obtain the type of data shown in Table 6-1.

The integral form of the PFR design equation

$$\frac{V}{F_{A0}} = \int_0^{x_A} \frac{dx_A}{-r_A} \tag{3-33}$$

can be differentiated to give

$$-r_A = \frac{dx_A}{d(V/F_{A0})}$$

This equation shows that the value of the reaction rate at any value of x_A is equal to the slope of the curve in the figure above, taken at the specified value of x_A. This relationship is represented in the following figure.

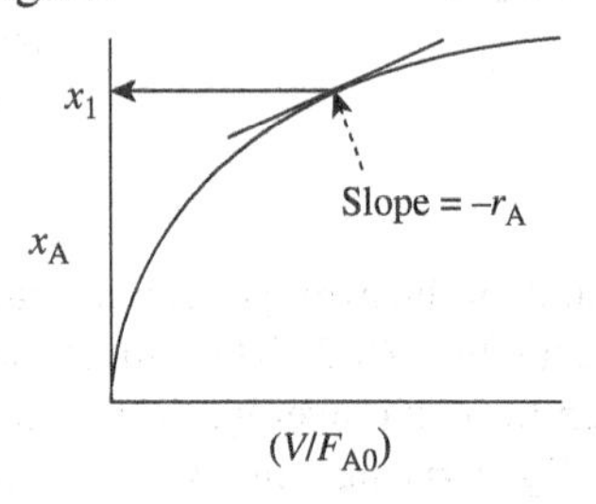

In other words, a value of the reaction rate at x_1 can be obtained by taking the derivative of the x_A versus V/F_{A0} curve at x_1.

If the x_A versus V/F_{A0} curve was obtained at a constant temperature, a subset of data similar to the first part of Table 6-1 can be generated by taking slopes at various values of x_A. The values of C_A, C_B, etc. in Table 6-1 can be calculated from x_A and the known feed composition.

If x_A varies over a significant range, then the reactant and product concentrations will vary significantly, and it will be possible to test the concentration dependencies of the rate equation. However, a single set of experiments at one feed composition, especially if the feed composition is stoichiometric, may not allow the kinetic effects of the individual species to be separated. This is because all of the concentrations will be related through stoichiometry. Desirably, data such as these shown above should be taken with several different feed compositions at the same temperature.

Finally, additional experiments should be carried out at different temperatures. This will lead to a set of data like the one in Table 6-1.

6.1.3 Batch Reactors

Kinetic studies in batch reactors are almost always performed under constant-volume conditions. This discussion is restricted to such systems.

The data taken during a single experiment in an ideal batch reactor might have one of the forms shown below.

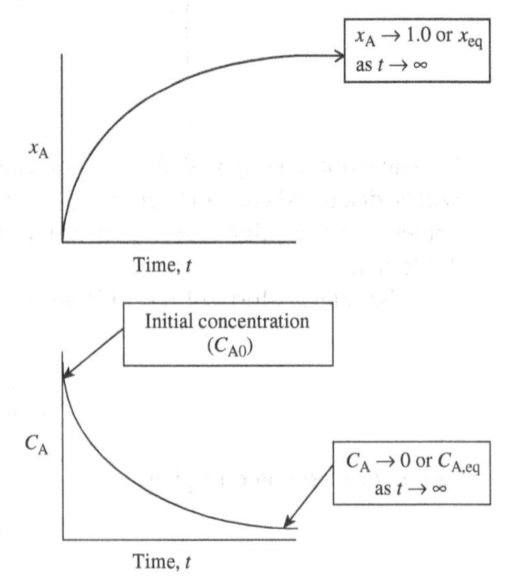

As with the PFR, we can work directly with this kind of data, as discussed later, in Section 6.3 of this chapter. We can also obtain the reaction rate at any value of $C_A(x_A)$ by differentiating the above curves. For example, Eqn. (3-8)

$$-r_A = -\frac{dC_A}{dt} \tag{3-8}$$

shows that the rate of disappearance of reactant A at any value of C_A is the negative of the slope of a plot of C_A versus t, taken at the specified value of C_A. This relationship is illustrated below.

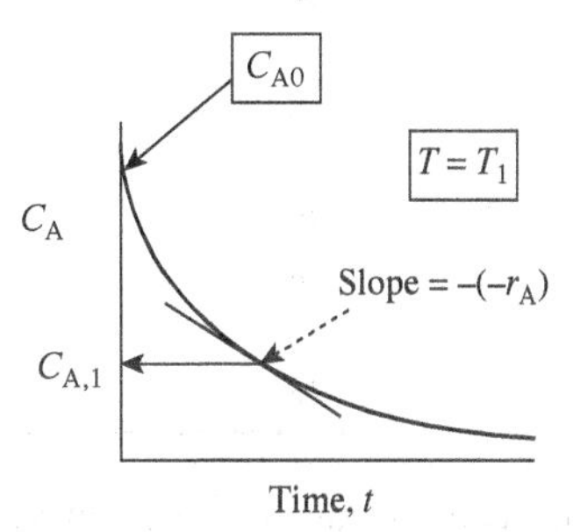

By taking the slope at various points on the curve, a subset of data similar to the first part of Table 6-1 can be generated. Once again, it is necessary to cover a wide range of $C_A(x_A)$, and to use initial mixtures with different stoichiometries to obtain data that will provide a rigorous test of a rate equation. This will require some more experiments at the same temperature T_1.

Table 6-1 then can be completed by carrying out additional experiments at different temperatures, T_2, T_3, etc.

The "method of initial rates" is another way to obtain the type of data shown in Table 6-1 using a batch reactor. In the method of initial rates, it is only necessary to obtain the slope of the C_A versus time curve at $t = 0$. This procedure requires more experiments, since only one data point is obtained per experiment. The initial compositions will have to cover a wide range of species concentrations. Moreover, the start of the experiment ($t = 0$) must be precisely defined, and enough data must be taken close to $t = 0$ to permit the initial rate to be calculated accurately. However, the experiments can be shorter, and the initial concentrations are known accurately and are under the control of the experimenter.

6.1.4 Differentiation of Data: An Illustration

In order to analyze the data obtained from a batch reactor or from an integral PFR, the data may have to be differentiated, as schematically illustrated above. Several techniques can be used to differentiate data, but each introduces some error as illustrated below.

Suppose that the data in Table 6-2 were taken in an isothermal, constant-volume batch reactor. Let's differentiate these data and prepare a table of the reaction rate $-r_A$ as a function of the concentration C_A.

There are several procedures that can be used to differentiate a set of data such as the one above. One (very tedious) way is to plot the data in Table 6-2 on a graph, manually construct tangents to the curve at various points, and measure the slope of the tangents. Two simpler methods are illustrated below.

Procedure 1 (Numerical differentiation): The average reaction rate between two points in time, say t_1 and t_2, can be approximated as

$$-\bar{r}_A = \frac{C_A(t_1) - C_A(t_2)}{t_2 - t_1}$$

Table 6-2 Concentration of Reactant A as a Function of Time in an Ideal Batch Reactor

Time (min)	Concentration of A (mol/l)
0	0.850
2	0.606
4	0.471
6	0.385
12	0.249
18	0.184
24	0.146
30	0.121
36	0.103
42	0.0899
48	0.0797

This reaction rate is associated with the arithmetic average concentration of A over the time interval, i.e.,

$$\overline{C}_A = \frac{C_A(t_1) + C_A(t_2)}{2}$$

The results of these calculations for the data in Table 6-2 are shown in Table 6-2a.

Table 6-2a Results of Numerical Differentiation of Data in Table 6-2

Time interval (min)	Average reaction rate, $-\overline{r}_A$ (mol A/l-min)	Average concentration of A, $\overline{C}_A$ (mol/l)
0–2	0.122	0.728
2–4	0.0675	0.539
4–6	0.0430	0.428
6–12	0.0227	0.317
12–18	0.0108	0.217
18–24	0.00633	0.165
24–30	0.00417	0.134
30–36	0.00300	0.112
36–42	0.00218	0.0965
42–48	0.00170	0.0848

Procedure 2 (Polynomial fit followed by analytical differentiation): A polynomial can be fit to the data in Table 6-2 using any standard fitting program. For a sixth-order polynomial, the result is

$$C_A = 0.84458 - 0.13835t + 1.4017 \times 10^{-2}t^2 - 7.9311 \times 10^{-4}t^3 + 2.4312 \times 10^{-5}t^4 - 3.7655 \times 10^{-7}t^5 + 2.3042 \times 10^{-9}t^6$$

This polynomial can be differentiated analytically to yield

$$\begin{aligned} -r_A &= -\mathrm{d}C_A/\mathrm{d}t \\ &= 0.13835 - 2.8034 \times 10^{-2}t + 2.3793 \times 10^{-3}t^2 - 9.7248 \times 10^{-5}t^3 + 1.8828 \times 10^{-6}t^4 - 1.3825 \times 10^{-8}t^5 \end{aligned}$$

The above equation then can be used to calculate values of the reaction rate at each of the times in Table 6-2. The results are given in Table 6-2b.

Table 6-2b Calculation of Reaction Rates from Data in Table 6-2 by Fitting and Differentiating a Sixth-Order Polynomial

Time (min)	Reaction rate, $-r_A$ (mol A/l-min)	Concentration of A (mol/l)[a]
0	0.138	0.850
2	0.0911	0.606
4	0.0585	0.471
6	0.0371	0.385
12	0.0121	0.249
18	0.00901	0.184
24	0.00624	0.146
30	0.00211	0.121
36	0.00188	0.103
42	0.00486	0.0899
48	−0.00845	0.0797

[a]From Table 6-2.

EXERCISE 6-1

Fit a fifth-order or a seventh-order polynomial to the data in Table 6-2. Then calculate $-r_A$ and C_A at each of the times in the above table. How well do your reaction rates agree with those shown in Table 6-2b?

Figure 6-1 compares the rates obtained with the two methods of differentiation discussed above. The data in Table 6-2 were generated using the second-order rate equation:

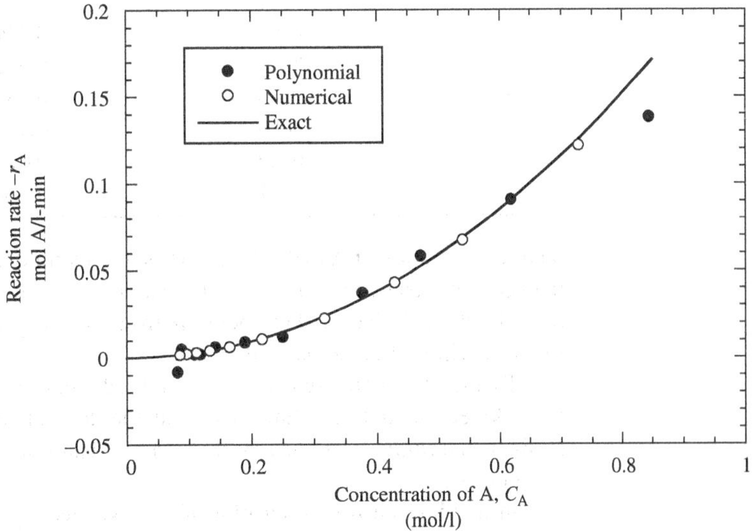

Figure 6-1 Comparison of the values of the reaction rate $-r_A$ obtained by different methods of differentiation. The unfilled points are the results from numerical differentiation (Procedure 1). The filled points are the results from fitting a polynomial to the concentration versus time data in Table 6-2 and differentiating the polynomial analytically (Procedure 2). The solid line is the second-order rate equation: $-r_A = 0.2368 \times C_A^2$, which was used to generate the data in Table 6-2.

$-r_A = 0.2368 \times C_A^2$. This equation is shown as the solid line in Figure 6-1; it provides a means to evaluate the accuracy of the two methods of differentiation.

For this example, the numerical differentiation technique provides more accurate estimates of $-r_A$ than the "polynomial" technique. The latter procedure is particularly inaccurate at high and low values of C_A. In fact, at the lowest value of C_A in Table 6-2b, the value of $-r_A$ is negative.

In general, the values of $-r_A$ obtained by differentiation of experimental data will contain significant errors unless the data are quite accurate and closely spaced. The data in this example are essentially error free. For this reason, the close correspondence between the exact results and those from the numerical differentiation is atypical.

6.2 THE DIFFERENTIAL METHOD OF DATA ANALYSIS

The differential method of analysis can be used when numerical values of the reaction rate have been obtained at various concentrations and at a constant temperature. The kind of data that are required is represented in Table 6-1.

6.2.1 Rate Equations Containing Only One Concentration

6.2.1.1 Testing a Rate Equation

Let's illustrate the differential method with a simple example. Suppose that the reaction A → B + C + D is being studied and that you have been asked to determine whether the rate equation

$$-r_A = kC_A^2 \tag{6-3}$$

fits the data in Table 6-3.

Table 6-3 Rate of Disappearance of A as a Function of Concentration of Reactant A

Experiment number	Temperature (K)	Reaction rate, $-r_A$ (mol/l-s)	Concentration of A (mol/l)
1	397.8	0.034	0.050
2	398.1	0.046	0.100
3	398.0	0.060	0.150
4	398.0	0.075	0.250
5	397.9	0.099	0.500
6	398.1	0.150	1.00

The concentration of A in this table varies by a factor of 20, making it possible to distinguish between various dependencies of the reaction rate on this concentration. In analyzing the data, the slight variation of temperature from experiment to experiment will be ignored, and the data will be treated as isothermal.

To test whether the reaction is second order in A, we will make a plot of $-r_A$ versus C_A^2. If the kinetic model, i.e., Eqn. (6-3), fits the data: (1) the points on this plot will fall on a straight line *through the origin*, and (2) the data points will scatter *randomly* around the straight line.

In making the following plot of $-r_A$ versus C_A^2, the method of linear least squares (linear regression) was used to fit a straight line through the data. This technique is a standard feature of many graphics and data analysis applications. The method of linear least squares produces the line that minimizes the sum of the squares of the deviations between the actual data points and their predicted values. Therefore, this line is referred to as the "best fit" line. The dashed line in Figure 6-2 shows the "best fit" of a straight line to the data in Table 6-3.

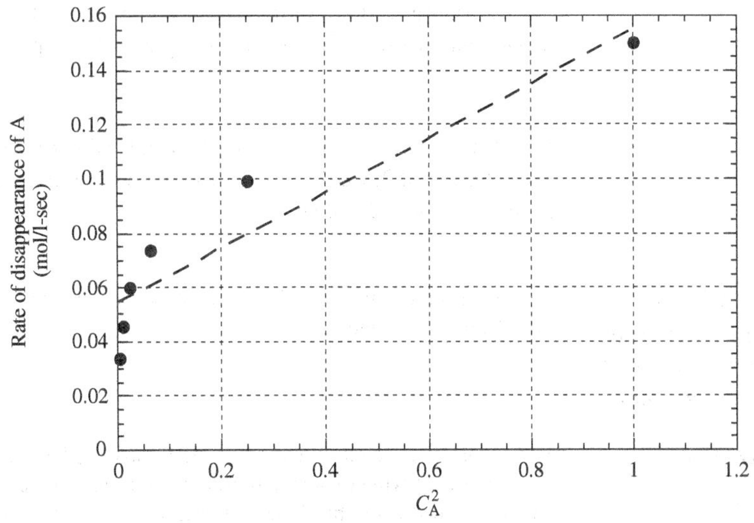

Figure 6-2 Test of a second-order rate equation against the data in Table 6-3. The dashed line is the "best fit" straight line, obtained by linear least-squares analysis.

Obviously, the data points do not fall on a straight line. There is pronounced curvature and the scatter about the "best fit" straight line is *not* random. The data points at the two lowest values of C_A fall below the line, the three data points at intermediate values of C_A fall above the line, and the data point at the highest value of C_A again is below the line. The deviations between the data and the model are *systematic*, not random. Finally, the "best fit" straight line does not come close to going through the origin.

Even though the second-order rate equation does not fit the data, two important points should be emphasized concerning the *procedure* that was used to carry out the analysis.

Procedure for graphical data analysis using the differential method

1. *We linearized the rate equation.* In other words, we put the rate equation into the straight-line form, $y = mx + b$. In this case, $y = -r_A$, $m = k$, $x = C_A^2$, and $b = 0$. The eye can recognize a straight line. However, the eye cannot distinguish between various functions that exhibit curvature.
2. *We made a graph.* A graph allows us to *see* the fit (or lack thereof) between the data and the postulated rate equation. Visual observation tells us many things that are hard to glean from a purely statistical analysis. For example, it is easy to see that the data points in Figure 6-2 form a curve, not a straight line. We also can *see* that the scatter of the data points around the straight line is not random and that the intercept of the straight line is nowhere near zero.

The correlation coefficient for the "best fit" straight line in Figure 6-2 turns out to be 0.936. Sounds pretty good, doesn't it? Does that mean that the model *does* fit the data? A quick look at the graph tells you that the fit is not acceptable, despite the value of the correlation coefficient.

The data in Figure 6-2 form a relatively smooth curve that is concave down. This suggests that the order of the reaction with respect to A is less than 2. The downward curvature indicates that $-r_A$ is a *weaker* function of C_A than assumed. If the curvature of the data on the same plot had been concave up, a *stronger* dependency of $-r_A$ on C_A would be suggested.

Returning to the data in Table 6-3, we would like to find a kinetic model that does provide an adequate fit. Perhaps the power-law rate equation

$$-r_A = kC_A^n \tag{6-4}$$

will fit these data, if we can find the right value of the order n. However, even though we suspect that $n < 2$, we don't want to assume different values of n and test them individually as we tested $n = 2$ above.

EXAMPLE 6-1
Testing an nth-Order Rate Equation Against Experimental Data

Test the rate equation $-r_A = kC_A^n$ against the data in Table 6-3. If this rate law fits the data, find the "best" values of n and k.

APPROACH

We will *linearize* the rate equation, and then *make a plot*. If the linearized rate equation fits the data, values of n and k will be estimated from the slope and intercept of the straight line. (Actually, n and k will be obtained from the slope and intercept of the "best fit" straight line through the data.)

SOLUTION

To linearize a rate equation of this form, take the logarithm of both sides

$$\ln(-r_A) = \ln(k) + n\ln(C_A) \tag{6-5}$$

If Eqn. (6-4) fits the experimental data, a plot of $\ln(-r_A)$ against $\ln(C_A)$ should be a straight line with a slope "n" and an intercept $\ln(k)$. This plot is shown in Figure 6-3.

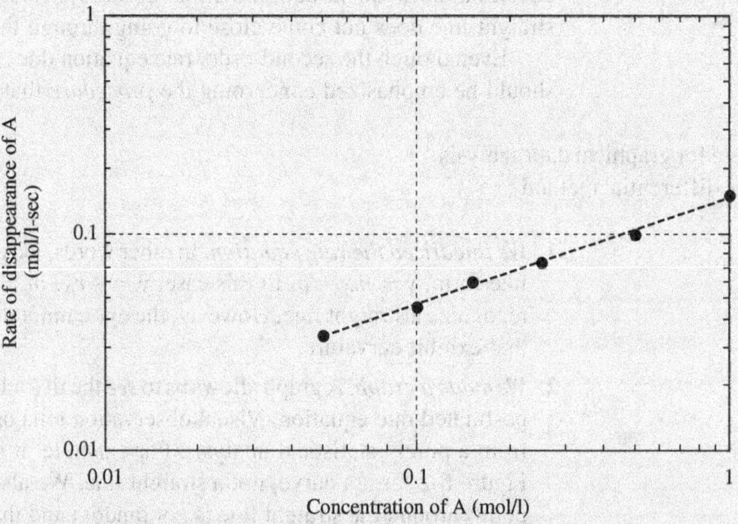

Figure 6-3 Test of an nth order rate equation against the data in Table 6-3. The dashed line is the "best fit" straight line obtained by linear least-squares analysis.

Visually, this fit appears to be much better than the one in Figure 6-2. The data form a good straight line and the scatter is random. The slope of the line is 0.49, which is the value of the order "n". The order of 0.49 is consistent with our earlier analysis of Figure 6-2 that concluded that $n = 2$ was too high. The "intercept" k is the value of $-r_A$ when $\ln C_A = 0$, i.e., when $C_A = 1$. This value is $k = 0.146\ (\text{mol/l})^{0.51}/\text{s}$. The values of "$n$" and "$k$" for this example were determined from the equation for the "best fit" straight line, which is shown as the dashed line in Figure 6-3.

6.2.1.2 Linearization of Langmuir–Hinshelwood/Michaelis–Menten Rate Equations

Fractional orders sometimes are observed when power-law rate equations are used in place of more fundamental forms, for example, Langmuir–Hinshelwood or Michaelis–Menten kinetic expressions. Consider the rate equation

$$-r_A = \frac{kC_A}{1 + K_A C_A} \tag{6-6}$$

When the value of $K_A C_A$ is small compared to 1, the reaction is nearly first order in A. On the other hand, when the value of $K_A C_A$ is large compared to 1, the reaction is close to zero order in A. At concentrations between these extremes, the reaction might appear to have a fractional order, perhaps 0.5 or so. However, the use of fractional-order rate equations can lead to difficulty in reactor design and analysis. Therefore, let's test Eqn. (6-6) to determine whether it provides an adequate fit of the data in Table 6-3.

EXAMPLE 6-2
Testing a Langmuir–Hinshelwood or Michaelis–Menten Rate Equation Against Experimental Data

Test the rate equation given by Eqn. (6-6) against the data in Table 6-3. If this rate law fits the data, determine the "best" values of k and K_A.

APPROACH

We will linearize the rate equation and then make a graph. If the linearized rate equation fits the data, values of k and K_A will be estimated from the slope and intercept of the "best fit" straight line.

SOLUTION

One way to linearize Eqn. (6-6) is to divide through by C_A and invert both sides to obtain

$$\frac{C_A}{-r_A} = \frac{1}{k} + \frac{K_A}{k} C_A \tag{6-7}$$

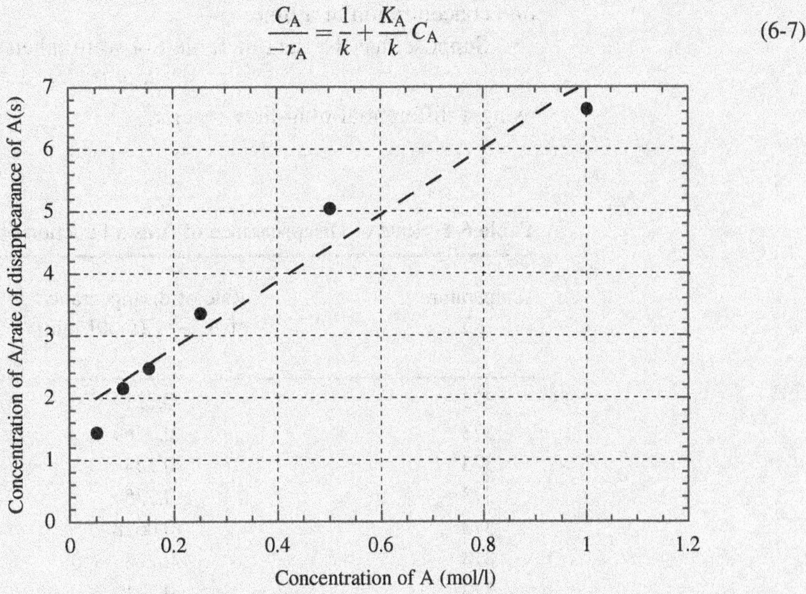

Figure 6-4 Test of the rate equation given by Eqn. (6-6). The dashed line is the "best fit" of Eqn. (6-7), obtained by linear least-squares analysis.

The parameter $C_A/-r_A$ can be plotted against C_A. If the model fits the data, the result will be a straight line with random scatter. The intercept of the line will be $1/k$ and the slope will be K_A/k, allowing values of both k and K_A to be calculated. This plot is shown as Figure 6-4.

This graph shows distinct curvature and systematic deviations between the data and the "best fit" straight line. Equation (6-4), with $n = 0.49$ and $k = 0.146\ (\text{mol/l})^{0.51}/\text{s}$, provides a better fit of the data than Eqn. (6-6).

To illustrate the calculation of k and K_A, the intercept of the line in Figure 6-4 is about 1.75 s. Since the intercept is $1/k$, the value of $k = 0.57\ \text{s}^{-1}$. The slope of the line is about 5.8 s-l/mol A. The slope is equal to K_A/k, so that $K_A = 5.8\ (\text{s-l/mol A}) \times 0.57\ \text{s}^{-1} = 3.3$ l/mol A.

EXERCISE 6-2

Linearize the rate equation

$$-r_A = \frac{kC_A}{(1 + K_A C_A)^2}$$

Explain how to plot a set of experimental data, with all of the experiments at the same temperature, to test this rate equation. If the model fits the data, how can values of k and K_A be obtained?

6.2.2 Rate Equations Containing More Than One Concentration

Consider the rate equation

$$-r_A = kC_A^{\alpha} C_B^{\beta} \tag{6-8}$$

The reaction rate now depends on two concentrations, those of species A and B. Moreover, there are three arbitrary constants in the rate equation, k, α, and β, that must be determined from the experimental data. Obviously, the dependence of $-r_A$ on both C_A and C_B cannot be determined using a single plot. Moreover, we cannot extract values of the three unknowns k, α, and β from two parameters, a slope and an intercept.

To test rate equations containing more than one concentration graphically, the experiments leading to the kinetic data must be planned carefully, so as to isolate the effect of the individual concentrations. The analysis of the kinetic data then must be carried out in stages, one concentration at a time.

Suppose that the data in Table 6-4 were taken during a study of the reaction

$$A + B \rightarrow C + D$$

using a differential plug-flow reactor.

Table 6-4 Rate of Disappearance of A as a Function of the Concentrations of A and B

Temperature (K)	Rate of disappearance of A, $-r_A$ (mol/l-min)	Concentrations (mol/l)	
		A	B
373	0.0214	0.10	0.20
373	0.0569	0.25	0.20
373	0.144	0.65	0.20
373	0.235	1.00	0.20
373	0.0618	0.40	0.10
373	0.228	0.90	0.25
373	0.211	0.55	0.60
373	0.0975	0.20	0.95

Does the power-law rate equation given by Eqn. (6-8) fit these data? If so, what are the approximate values of α, β, and k?

When we face a problem where the reaction rate may depend on more than one concentration, the data must be examined to determine if there are any experiments where all of the concentrations have been held constant except for one. If so, that subset of data can be used to determine the effect of the varying concentration. In the present case, the first four experiments in Table 6-4 were carried out with different concentrations of A, but with a constant concentration of B.

The rate equation can be linearized, as was done with Eqn. (6-4), by taking the logarithm of both sides.

Linearize the rate equation

$$\ln(-r_A) = \ln(kC_B^\beta) + \alpha \ln C_A$$

If Eqn. (6-8) fits the subset of data comprised of the first four experiments in Table 6-4, a log-log plot of $-r_A$ versus C_A will give a straight line with a slope of α. For each of these four experiments, kC_B^β is constant since both the temperature and C_B are constant. This plot is shown in Figure 6-5a.

Make a graph with the first four data points

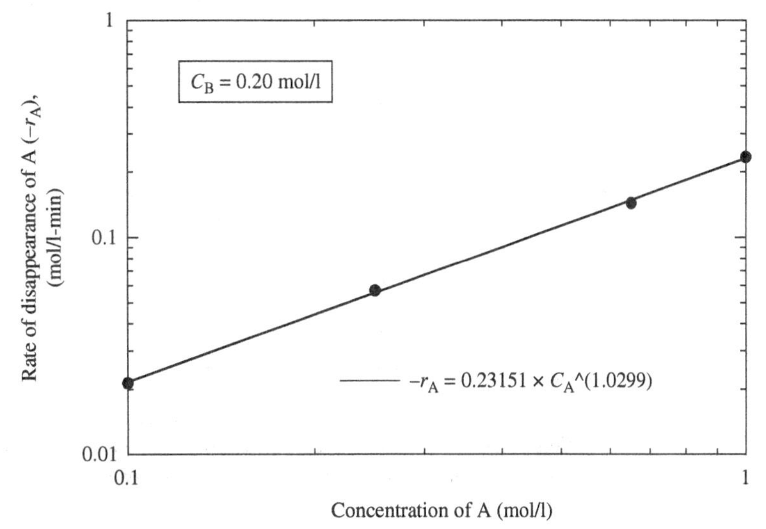

Figure 6-5a Test of the rate equation given by Eqn. (6-8) for the first four data points in Table 6-4. The solid line is the linear least-squares "best fit" to the data.

The model of Eqn. (6-8) fits this subset of data very well. From the equation for the "best fit" line, the value of α is 1.03.

This value of α now can be used to remove the effect of C_A from the data, so that the effect of C_B can be determined. Divide both sides of the rate equation by C_A^α to obtain

Use the value of α to define a new variable

$$\frac{-r_A}{C_A^\alpha} = kC_B^\beta$$

A numerical value of the left-hand side of this equation can be calculated for every data point in Table 6-4, if the value of α is known.

The above equation can be linearized by taking the logarithm of both sides.

Linearize again

$$\ln\left(\frac{-r_A}{C_A^\alpha}\right) = \ln k + \beta \ln C_B$$

If the model fits the data, a plot of $\ln(-r_A/C_A^\alpha)$ versus $\ln(C_B)$ will be a straight line with a slope of β and an intercept of ln k.

Make a graph using all of the data

This plot is shown in Figure 6-5b. All eight of the data points in Table 6-4 have been used in constructing this graph. Four points are tightly clustered at $C_B = 0.20$ mol/l. These are the four data points that were used to determine α from Figure 6-5a.

The data fits the model quite well. The "best fit" straight line gives $\beta = 0.51$, and $k = 0.52\ (1/\text{mol})^{0.54}/\text{min}$.

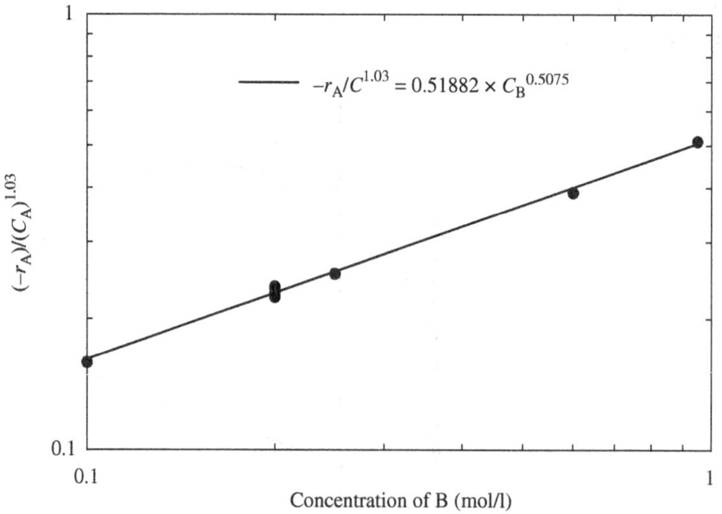

Figure 6-5b Test of rate equation (Eqn. (6-8)) against the data of Table 6-4. The solid line is the "best fit" obtained by linear least-squares analysis.

When kinetic data have been analyzed in a stagewise manner, as we have done here, errors in the early stages of analysis can propagate and distort the results obtained in later stages. Therefore, it is advisable to test the *final* rate equation, using *all* of the data. This can be done with a parity plot, as illustrated later in Section 6.4. of this chapter. However, for isothermal data and a power-law rate equation, the final kinetic expression can also be tested by plotting the measured reaction rate against the concentration-dependent term in the rate equation. This method is shown in Figure 6-5c, which is a plot of $-r_A$ against $C_A^{1.03} C_C^{0.51}$. If the model of Eqn. (6-8) fits the data, and if the values of α and β are correct, the experimental points should form a straight line *through the origin*, with a slope of k, the rate constant.

Visual test of final rate equation

The proposed rate equation fits the experimental data quite well. The scatter is random, the correlation coefficient is high, and the value of the rate constant k obtained from Figure 6-5c (i.e., the slope m_1) agrees closely with the value obtained from Figure 6-5b.

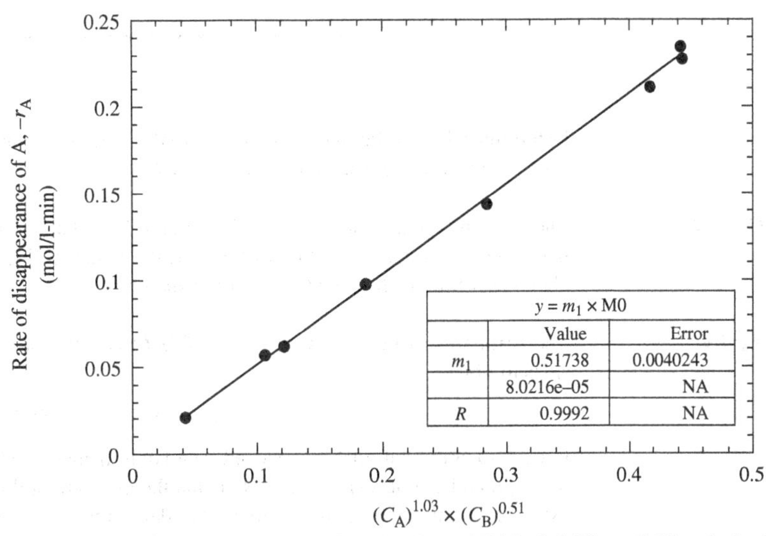

Figure 6-5c Test of final rate equation against the data of Table 6-4. The solid line is the "best fit" obtained by linear least-squares analysis.

6.2.3 Testing the Arrhenius Relationship

Suppose that the rate constant for a reaction has been determined at several different temperatures, using the techniques described above. We then can determine whether the Arrhenius relationship is obeyed. If it is, the value of the activation energy can be obtained from the data.

First, the Arrhenius relationship

$$k = A \exp(-E/RT) \tag{2-2}$$

is linearized:

$$\ln k = \ln A - E/RT$$

If the measured rate constants obey the Arrhenius relationship, a plot of $\ln k$ against $1/T$ will be a straight line. The value of the slope will be $(-E/R)$ and the activation energy, E, can be calculated from the slope.

EXAMPLE 6-3
Decomposition of AIBN

The compound 2,2′-azobis(isobutyronitrile) (AIBN) commonly is used to initiate polymerization reactions because it decomposes spontaneously into two free radicals. Each free radical then can react with a monomer to start a growing polymer chain. Not surprisingly, compounds such as AIBN are called "initiators."

The use of supercritical carbon dioxide ($scCO_2$) as a polymerization medium has attracted widespread attention because it can eliminate the need for organic solvents and/or eliminate the need to treat large quantities of wastewater produced in the polymerization reactor. Therefore, the decomposition of AIBN has been studied in $scCO_2$, with the following results.

Table 6-5 First-Order Rate Constant (k_D) for AIBN Decomposition as a Function of Temperature

Temperature, (°C)	k_D ($\times 10^5$) (s^{-1})
80	7.6
90	15
95	52
100	62

Determine whether the rate constant for AIBN decomposition obeys the Arrhenius relationship and estimate the value of the activation energy E.

APPROACH

The Arrhenius expression will be linearized. The data in Table 6-5 will be plotted to test the linearized equation. The values of the activation energy E and the preexponential factor A will be obtained from the slope and intercept of the graph.

SOLUTION

The Arrhenius expression, $k = A\exp(-E/RT)$ can be linearized by taking the logarithm of both sides to obtain

$$\ln(k) = \ln(A) - (E/R)(1/T)$$

If the Arrhenius relationship is obeyed, a plot of $\ln k_D$ against $1/T$ should be a straight line with a slope of $-E/R$ and an intercept of $\ln A$. Note that the intercept of this plot is at $T = \infty$. Therefore, a large extrapolation in temperature is required to determine A, so that a small error in E can have a large effect on the value of A. The values of A and E will be obtained from the equation for the "best fit" straight line through the data.

SOLUTION

A plot of $\ln k_D$ against $1/T$ is shown in Figure 6-6.

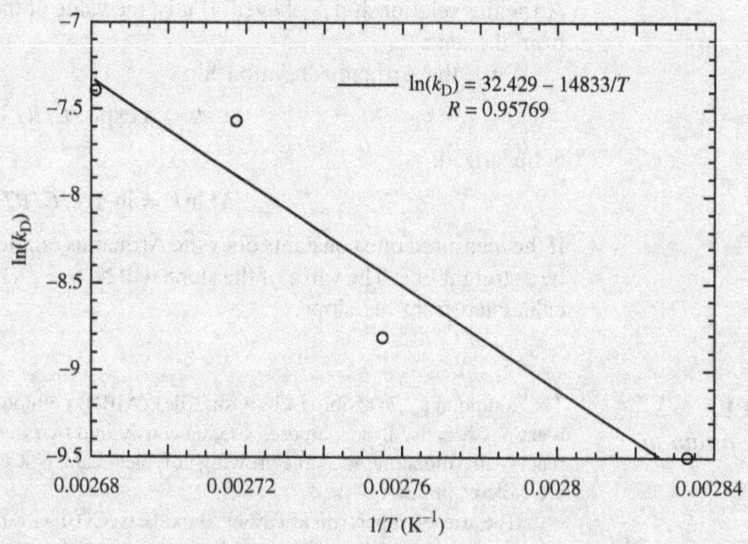

Figure 6-6 Arrhenius plot for AIBN decomposition rate constants (Example 6-3). The solid line is the "best fit" obtained by linear least-squares analysis.

There is scatter in the data in Figure 6-6, although the scatter appears to be random. The value of the activation energy, determined from the slope of the "best fit" straight line, is $E = 14{,}833(\text{K}) \times 0.008314(\text{kJ/mol-K}) = 123.3\,\text{kJ/mol}$. The value of the preexponential factor is $A = \exp(32.429) = 1.21 \times 10^{14}\,\text{s}^{-1}$.

Figure 6-6 illustrates the difficulty is estimating a value of E from two points, instead of using all of the data. If E were estimated from the first two points in Table 6-5 (80 and 90 °C), or from the last two points (95 and 100 °C), the resulting values of E would be much lower than 123 kJ/mol. On the contrary, if E were estimated from the middle two points (90 and 95 °C), the calculated value of E would be much higher than 123 kJ/mol.

EXERCISE 6-3

What value of E is obtained from a calculation using only the first and last points in Table 6-5?

6.2.4 Nonlinear Regression

In some of the previous examples, the independent variable, y, obtained by linearizing the rate equation, was not the reaction rate, $-r_A$. Therefore, when the "best" values of the slope and the intercept were determined by linear regression, we were minimizing the sum of the squares of the deviations between the calculated and experimental values of *some variable other than* $-r_A$. For example, in the first stage of the analysis of the data in Table 6-4, the value of α was determined by minimizing the sum of the squares of the deviations in $y = \ln(-r_A)$. In the second stage of the analysis, the values of β and k were determined by minimizing the sum of the squares of the deviations in $y = \ln(-r_A/C_A^{\alpha})$. These values of α, β, and k are not necessarily the same as the ones that would have been obtained if we had minimized the sum of the squares of the deviations in $-r_A$ itself. A new set of tools is required to find the "best" values of the parameters when $-r_A$ is not linear in the various concentrations.

Fortunately, powerful *nonlinear regression* programs now are available. These programs allow us to minimize the sum of the squares of the deviations in any variable we choose, linear or not. Moreover, some of the easier nonlinear regression problems can be solved with a simple spreadsheet. Let's illustrate the use of a spreadsheet to carry out nonlinear regression by reanalyzing the AIBN decomposition data in Table 6-5.

To begin, the Arrhenius relationship will be written in the equivalent form

$$k(T) = k(T_0)\exp\left[\frac{-E}{R}\left(\frac{1}{T} - \frac{1}{T_0}\right)\right] \tag{6-9}$$

This transformation usually improves the convergence and stability of the numerical techniques that are used in nonlinear regression programs. Let's choose T_0 as the midpoint of the range of temperatures in Table 6-5, i.e., $T_0 = 90\,°\text{C} = 363\,\text{K}$. We will use nonlinear regression to find the values of $k(363)$ and E.

There are two common approaches to determining parameter values by nonlinear regression. The first is to minimize the sum of the squares of the *absolute* deviations in the objective function, i.e., the rate constant, k, for the present problem. This involves finding the values of $k(363)$ and E that produce a minimum value of $\sum_{i=1}^{N}(k_{i,\text{theo}} - k_{i,\text{exp}})^2$, where $k_{i,\text{exp}}$ is the measured (experimental) value of k for the ith data point, $k_{i,\text{theo}}$ is the value of k calculated from the Arrhenius equation for the ith data point, and N is the total number of data points.

The second approach is to minimize the *fractional* deviations in k, i.e., to minimize $\sum_{i=1}^{N}\{(k_{i,\text{theo}} - k_{i,\text{exp}})/k_{i,\text{exp}}\}^2$. The second approach is a variation of the first, in which a weighting factor of $(1/k_{i,\text{exp}})$ is applied to each of the data points. Therefore, these two approaches will not necessarily produce the same answer.

Let's begin by applying the second approach to the data for AIBN decomposition in Table 6-5. Appendix 6-A shows how the problem can be solved with an EXCEL spreadsheet. The subroutine "SOLVER" was used to carry out the mathematical operations that determine the values of $k(363)$ and E that produced a minimum value of $\sum_{i=1}^{N}\{(k_{i,\text{theo}} - k_{i,\text{exp}})/k_{i,\text{exp}}\}^2$. *Initial estimates of k(363) and E are required to begin the calculations. The values determined by linear regression usually provide a good starting point for nonlinear regression problems.* These values were used as initial estimates in Appendix 6-A.

The values of $k(363)$ and E determined by nonlinear regression (Approach 2) are $k(363) = 1.96 \times 10^{-4}\ \text{s}^{-1}$ and $E = 123$ kJ/mol. The value of A is calculated from $k(363)$ and E to be $1.10 \times 10^{14}\ \text{s}^{-1}$. The minimum value of $\sum_{i=1}^{N}\{(k_{i,\text{theo}} - k_{i,\text{exp}})/k_{i,\text{exp}}\}^2$ is 0.25057.

With all nonlinear regression problems, it is advisable to check the final solution to be sure that a true minimum has been achieved. This is done by varying the values of $k(363)$ and E by a small amount in both directions around the values determined by nonlinear regression. The results in Appendix 6-A show that $\sum_{i=1}^{N}\{(k_{i,\text{theo}} - k_{i,\text{exp}})/k_{i,\text{exp}}\}^2$ increases when $k(363)$ and E are either increased or decreased slightly from the values determined by nonlinear regression. This is the behavior that would be expected if a minimum in $\sum_{i=1}^{N}\{(k_{i,\text{theo}} - k_{i,\text{exp}})/k_{i,\text{exp}}\}^2$ had, indeed, been found.

Although this exercise does not *prove* that SOLVER found a true minimum, it does give us some confidence. Even if the values determined by this calculation do represent a true minimum, they do not necessarily represent an absolute minimum, as opposed to a local minimum. In order to determine whether the point $k(363) = 1.96 \times 10^{-4}\ \text{s}^{-1}$, $E =$ 123 kJ/mol is a local or an absolute minimum, it would be necessary to vary the initial estimates over a wide range, and determine whether the calculation converges to a value of $\sum_{i=1}^{N}\{(k_{i,\text{theo}} - k_{i,\text{exp}})/k_{i,\text{exp}}\}^2$ that is lower than 0.25057. We will not attempt to perform these calculations here.

When we try the first approach, minimizing the absolute deviations in k, our attempts are much less successful, as shown in Appendix 6-B1. The problem is that SOLVER thinks that it has found a minimum value of $\sum_{i=1}^{N}(k_{i,\text{theo}} - k_{i,\text{exp}})^2$ and stops working. However, when E is manually decreased to check for a minimum, the value of $\sum_{i=1}^{N}(k_{i,\text{theo}} - k_{i,\text{exp}})^2$ declines, showing that SOLVER had not really reached a minimum.

This problem can arise when the value of the quantity being minimized ($\sum_{i=1}^{N}(k_{i,\text{theo}} - k_{i,\text{exp}})^2$ in this problem) is very small relative to the parameters that are being optimized ($k(363)$ and E in this problem). Let $\sum_{i=1}^{N}(k_{i,\text{theo}} - k_{i,\text{exp}})^2$ be represented by Σ. When SOLVER calculates values of $\partial\Sigma/\partial E$ and $\partial\Sigma/\partial k(363)$ to determine whether a minimum has been reached, the value of $\partial\Sigma/\partial E$ is so small that the program is "fooled" into thinking that it has found the minimum.

One way to solve this problem is to arbitrarily increase Σ by multiplying it by a constant factor. Appendix 6-B2 shows the results that are obtained when SOLVER is used to find the values of $k(363)$ and E that minimize the value of $10^8 \times \Sigma$. When the problem is reformulated in this way, SOLVER appears to reach a true minimum. The resulting values are $k(363) = 2.42 \times 10^{-4}\ \text{s}^{-1}$ and $E = 113$ kJ/mol, giving $A = 4.05 \times 10^{12}\ \text{s}^{-1}$.

The results from the three approaches to finding the "best" values of A and E are summarized in Table 6-6 and Figure 6-7.

Table 6-6 Values of A, E, and k_D for AIBN Decomposition in Supercritical Carbon Dioxide

Technique	A (s^{-1})	E (kJ/mol)	k_D (363) (s^{-1})
Linear regression	1.21×10^{14}	123.3	2.19×10^{-4}
Nonlinear regression (Approach 2)	1.10×10^{14}	123.3	1.98×10^{-4}
Nonlinear regression (Approach 1)	4.05×10^{12}	112.7	2.42×10^{-4}

Figure 6-7 shows that there is not much difference between the three procedures for fitting the data in Table 6-5, at least relative to the scatter in the experimental data. However, the differences between the three methods would have been much greater if the data had covered a wider range of temperature and k_D values.

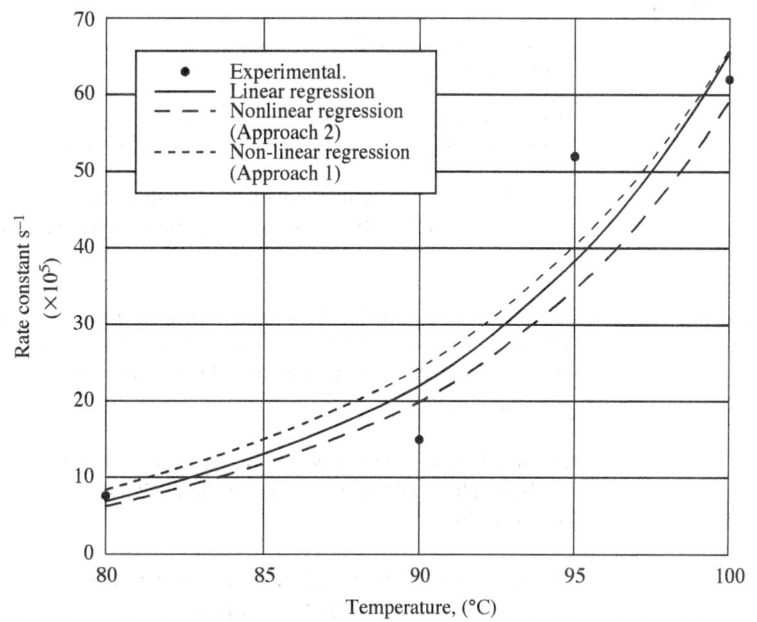

Figure 6-7 Values of k_D for AIBN decomposition in supercritical CO_2, calculated from the results given in Table 6-6.

6.3 THE INTEGRAL METHOD OF DATA ANALYSIS

6.3.1 Using the Integral Method

The integral method of analysis can be used when the available data are in the form of concentration (or fractional conversion) versus time or space time (or V/F_{A0} or W/F_{A0}). As pointed out earlier in this chapter, this kind of data are obtained when an ideal batch reactor or an ideal plug-flow reactor is used. For these two reactors, use of the integral method avoids the need for numerical or graphical differentiation.

Procedure for integral method

The steps in the integral method are

1. A rate equation is assumed.
2. The appropriate design equation is integrated to generate a relationship between concentration (or conversion) and time (or space time).

3. The relationship is linearized.
4. The data are plotted so as to test the linearized equation.
5. If the equation fits the data, the values of the slope and the intercept are used to estimate the unknown parameters in the rate equation.

Let's illustrate this procedure with an example.

EXAMPLE 6-4
Decomposition of Aqueous Bromine

A small quantity of liquid bromine was dissolved in water in a glass container. The liquid was well stirred and the temperature was 25 °C throughout the experiment. The following data were obtained:

Time (min)	0	10	20	30	40	50
$[Br_2]$ (μmol/ml)	2.45	1.74	1.23	0.88	0.62	0.44

Use the integral method of data analysis to test whether the reaction is zero, first, or second order in Br_2. If one of these kinetic models fits the data, determine the value of the rate constant.

APPROACH

The design equation for a batch reactor will be integrated for each of the three specified rate equations. The resulting expressions will be linearized and the data will be plotted to test the linearized equation. If appropriate, the rate constant will be estimated from the slope of the resulting line.

SOLUTION

Assume that the container behaves as an ideal, isothermal, batch reactor. Since the reaction occurs in the liquid phase, constant density (constant volume) can be assumed. The subscript "A" will be used to denote Br_2.

A. Assume that the reaction is zero order. The design equation for an ideal, isothermal, constant-volume batch reactor is

$$-dC_A/dt = k_0, \quad C_A > 0$$
$$-dC_A/dt = 0, \quad C_A = 0$$

Integrating from $t = 0$, $C_A = C_{A0}$ to $t = t$, $C_A = C_A$:

$$C_A = C_{A0} - k_0 t, \quad C_A > 0\,(t < C_{A0}/k_0)$$
$$C_A = 0, \quad C_A = 0\,(t \geq C_{A0}/k_0) \tag{6-10}$$

Equation (6-10) shows that C_A should be linear in time (t) if the reaction is zero order. The zero-order model can be tested directly with this equation since it is already linear; no further manipulations are required. We simply plot C_A against time, as shown in Figure 6-8a.

This plot has distinct curvature. The zero-order rate equation does not fit the data very well. The fact that the curvature is upward suggests that the reaction order is greater than 0.

B. Assume that the reaction is first order. The design equation is

$$-dC_A/dt = k_1 C_A$$

Integrating from $t = 0$, $C_A = C_{A0}$ to $t = t$, $C_A = C_A$:

$$\ln(C_A/C_{A0}) = -k_1 t \tag{6-11}$$

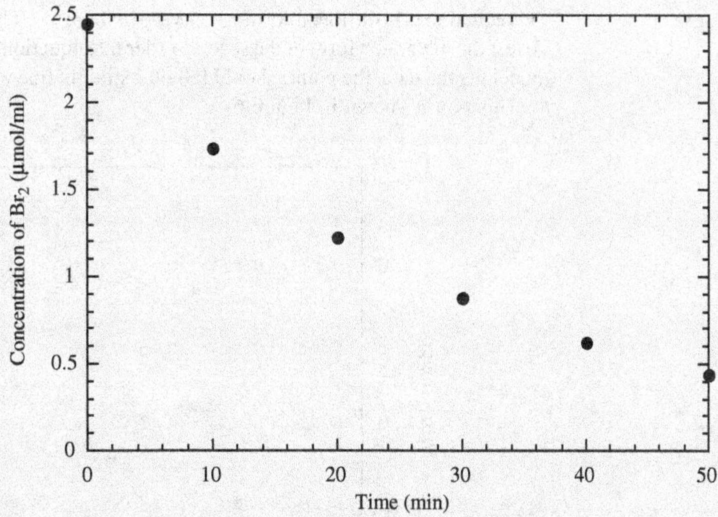

Figure 6-8a Test of zero-order rate equation for bromine decomposition.

Equation (6-11) is in linear form if the independent variable is taken to be $\ln(C_A/C_{A0})$. A plot of $\ln(C_A/C_{A0})$ against time (t) should be a straight line if the reaction is first order. A graphical test of the first-order model is shown in the Figure 6-8b.

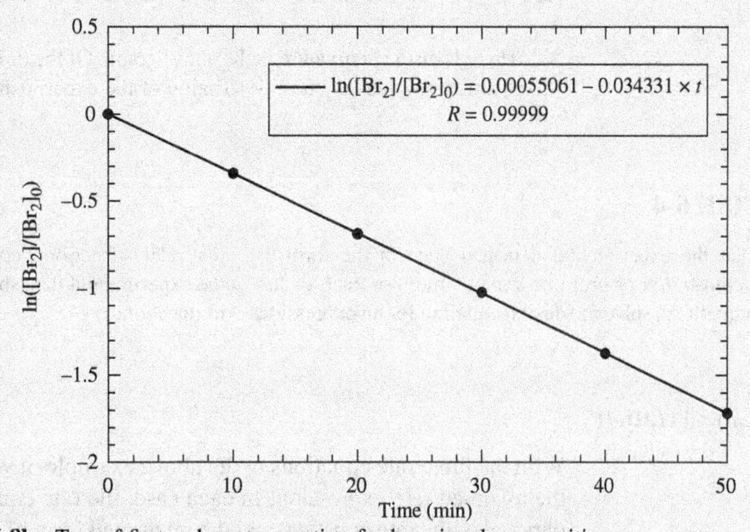

Figure 6-8b Test of first-order rate equation for bromine decomposition.

The integrated form of the first-order rate equation fits the data very well. According to Eqn. (6-11), the slope of the line in the above graph is $-k_1$. Therefore, $k_1 = 0.0343 \text{ min}^{-1}$.

At this point, the question of the reaction order and the rate constant appears to be answered. However, to be absolutely sure, the second-order rate equation will be tested.

C. Assume that the reaction is second order. The design equation is

$$-\mathrm{d}C_A/\mathrm{d}t = k_2(C_A)^2$$

Integrating from $t = 0$, $C_A = C_{A0}$ to $t = t$, $C_A = C_A$:

$$(1/C_A) - (1/C_{A0}) = k_2 t \tag{6-12}$$

This equation is in the linear form $y = mx + b$, where $y = (1/C_A)$; $b = (1/C_{A0})$, $x = t$, and $m = k_2$. To test the integrated form of the second-order rate equation, $1/C_A$ can be plotted against time. If the model fits the data, the points should fall on a straight line with a y-intercept at $1/C_{A0}$ and a slope of k_2. This plot is shown in Figure 6-8c.

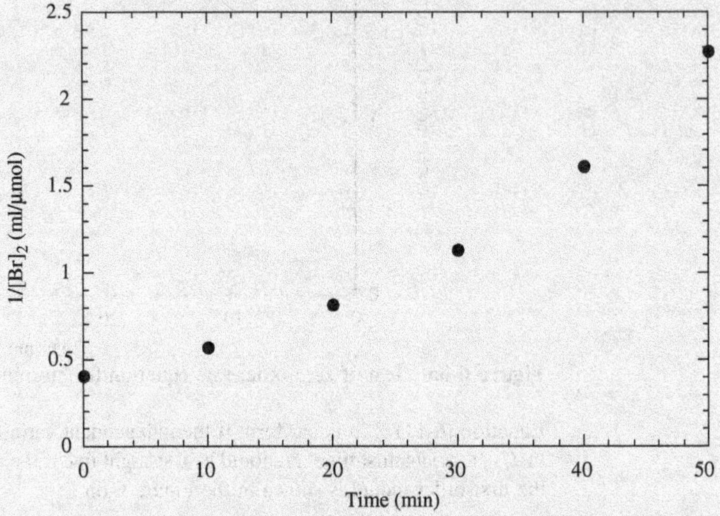

Figure 6-8c Test of second-order rate equation for bromine decomposition.

There is distinct curvature in the above graph. Of the three rate equations tested, the first-order rate equation provides the best description of the experimental data.

EXERCISE 6-4

Suppose that the experimental data had been in the form of *fractional conversion* of bromine versus time. For each of the three rate equations, solve the design equation for an expression that relates fractional conversion to time, and then explain how the experimental data should be plotted to test the rate equation in question.

6.3.2 Linearization

With the three rate equations in the above example, it was relatively easy to see how to test the assumed rate expression. In each case, the rate equation contained only one unknown parameter, the rate constant, and the integrated form of the design equation always took the form:

$$f(C_A) = k_n t + \text{constant}$$

Therefore, to test the rate equation, we simply had to plot $f(C_A)$ against t.

Life is not always so straightforward. Suppose we wished to test the rate equation

$$-r_A = kC_A/(1 + K_A C_A) \tag{6-13}$$

against the bromine decomposition data using the integral method. For this rate equation, the integrated form of the design equation for an ideal, isothermal batch reactor is

$$\ln(C_A/C_{A0}) + K_A(C_A - C_{A0}) = -kt \tag{6-14}$$

What should be plotted against what? We can't plot $\{\ln(C_A/C_{A0}) + K_A(C_A - C_{A0})\}$ against time because we don't know the value of K_A. Therefore, we can't calculate a value of $\{\ln(C_A/C_{A0}) + K_A(C_A - C_{A0})\}$ for every data point. We can't plot $\ln(C_A/C_{A0})$ against time because $K_A(C_A/C_{A0})$ is not a constant, since C_A varies with time.

Clearly, we have to "linearize" the above equation by performing some algebraic manipulations to transform the equation into the form $y = mx + b$. In doing this, we must remember that values of x and y have to be calculated for every data point, i.e., x and y can't contain the unknowns k and K_A. Moreover, m and b must be true constants; they can't contain either C_A or t.

There are several ways to linearize Eqn. (6-14). One is to divide through by $(C_A - C_{A0})$ to obtain

$$\frac{-\ln(C_A/C_{A0})}{(C_{A0} - C_A)} = \frac{kt}{(C_{A0} - C_A)} - K_A$$

We can plot $-\ln(C_A/C_{A0})/(C_{A0} - C_A)$ against $t/(C_A - C_{A0})$. If the model fits the data, the data points should fall on a straight line with an intercept of $-K_A$ and a slope of k. This graph is shown in Figure 6-8d.

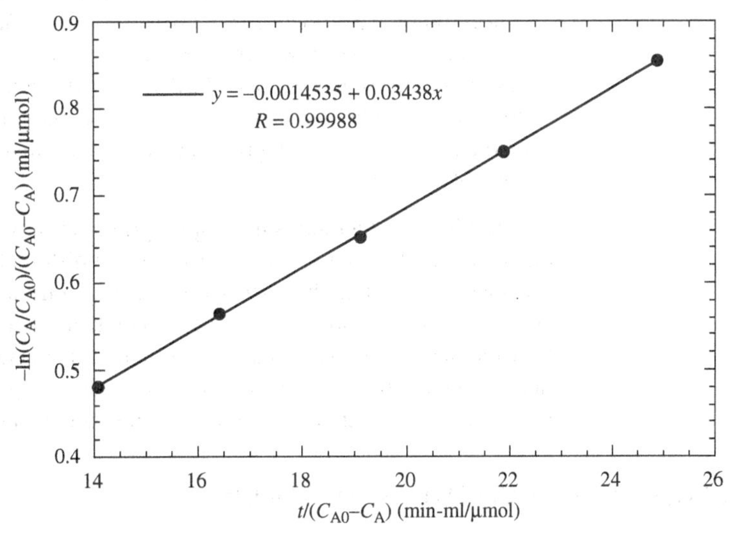

Figure 6-8d Test of hyperbolic rate equation (Eqn. (6-13)) against the bromine decomposition data.

The model appears to fit the data quite well. The equation for the "best fit" straight line gives $K_A = 0.00145$ ml/μmol and $k = 0.0344\ \text{min}^{-1}$.

How does this good fit reconcile with our previous conclusion that the reaction is first-order? Let's examine the magnitude of the term K_AC_A in Eqn. (6-13). The *largest* value of K_AC_A occurs when $C_A = C_{A0} = 2.45$ μmol/ml. Therefore, the *largest* value of K_AC_A is $2.45 \times 0.00145 = 0.00355$. This is negligible compared to 1, so that the term K_AC_A in the denominator of Eqn. (6-13) can be ignored. This reduces Eqn. (6-13) to a simple first-order rate equation. Note that the rate constant that was obtained from the "pure" first-order analysis was $k_1 = 0.0343\ \text{min}^{-1}$, and that the rate constant obtained from the preceding graph was $k = 0.0344\ \text{min}^{-1}$.

6.3.3 Comparison of Methods of Data Analysis

Two important points have emerged from the discussions of the differential and integral methods of data analysis:

1. Differentiation of concentration (or conversion) versus time (or space time) data is inherently inaccurate.
2. The differential method is somewhat more flexible than the integral method. For example, in the case of a power-law rate equation, the reaction order(s) can be determined directly from the data if the differential method is used. If the integral method is used, an order must be assumed, the design equation must be integrated, the integrated design equation must be linearized, and the data must be plotted. If the model does not fit the data, a new order must be assumed and the process repeated.

This "trial-and-error" approach is not so bad if the analysis is limited to integral orders, as with the bromine decomposition example (Example 6-4). However, the process can be tedious if fractional orders must be considered.

These points lead to the following approach:

- Use the differential method unless differentiation of experimental data would be required in order to obtain numerical values of the reaction rate;
- Use the integral method when the available experimental data are in the form of concentration (or fractional conversion) as a function of time (or space time or V/F_{A0} or W/F_{A0}). However, if several "guesses" of the rate equation lead to integrated design equations that do not fit the data, then differentiate the data numerically and use the differential method. Finally, test the rate equation that you obtain via the differential method by using the integral method, and estimate the unknown parameters from the integral analysis.

The integral method is most appropriate when we have a good idea of the form of the rate equation, but need to test that form in a different context. For example, suppose that previous analyses had shown that a certain reaction was second order in A and first order in B at temperatures T_1 and T_2. Now, a new set of data (say C_A versus time) becomes available at a third temperature T_3. We need to confirm the existing rate equation, and obtain a value of the rate constant at the new temperature. The most straightforward way to do this would be to use the integral method, assuming that the rate equation $-r_A = kC_A^2 C_B$ remained valid.

6.4 ELEMENTARY STATISTICAL METHODS

In addition to the graphical techniques that have been illustrated in previous sections, some basic statistical tools should be brought to bear in the analysis of kinetic data. In fact, in most cases, graphical analyses merely set the stage for the *efficient* use of statistical analysis. Some of the most useful statistical tools are illustrated in the following example.

6.4.1 Fructose Isomerization

Vieille et al.[1] studied the activity of the enzyme xylose isomerase, which was derived from the thermophilic organism *T. neapolitana*, for the isomerization of fructose (F) to glucose. This is the reverse of the reaction that is used to produce high-fructose corn syrup for the beverage industry. The reverse reaction was studied to help understand the biochemistry and the behavior of the enzyme.

---

[1] Vieille, C., Hess, J. M., Kelly, R. M., and Zeikus, J. G., *xylA* cloning and sequencing and biochemical characterization of xylose isomerase from thermotoga neapolitana, *Appl. Environ. Microbiol.*, 61(5) 1867–1875 (1995).

Conversion of fructose to glucose at 70 °C			
$-r_F$ (μmol/min-mg)[a]	[F] (mmol/l)	$-r_F$ (μmol/min-mg)[a]	[F] (mmol/l)
9.46	1000	6.21	100
7.95	500	5.86	90
7.57	325	5.79	80
7.80	250	5.37	70
7.87	200	5.14	60
7.04	175	4.73	50
7.04	160	4.12	40
6.82	140	3.48	30
6.74	120	2.77	20
6.52	110	1.60	10

[a]Of enzyme.

The table above shows the rate of disappearance of fructose as a function of its concentration.

An isothermal batch reactor was used at 70 °C and pH 7, with a sodium phosphate buffer. The initial concentration of glucose was zero in all cases. The rates in this table are initial rates, i.e., the rate at essentially zero fructose conversion.

As an aside, the rate of disappearance, $-r_F$, would usually be designated V (for reaction *velocity*) in the biochemistry and biochemical engineering literature. In addition, the reactant, fructose in this case, would be referred to as the substrate.

6.4.1.1 First Hypothesis: First-Order Rate Equation

To begin, let's determine whether an irreversible, first-order rate equation, $-r_F = k[F]$, fits the data. At 70 °C, fructose isomerization is quite reversible. However, in this study, the conversions of fructose were close to zero, and there was no glucose in the initial mixture. Therefore, the reverse reaction was not kinetically important for these particular experiments.

To test the first-order rate equation, we plot $-r_F$ against [F] as shown in Figure 6-9a.

If the first-order model fits the data, the points will scatter *randomly* around a straight line *through the origin*. The "best fit" straight line through the origin for the present data is shown in the graph. The first-order rate constant, the slope of this line, is 0.01701 μmol/mg enzyme-min-mmol.

Clearly, the first-order rate equation provides a very poor fit of the data. The points are not randomly scattered about the "best fit" straight line; the error is quite systematic. In fact, only 2 out of 20 points fall below the "best fit" line. These two points have the highest rates and fructose concentrations among the set of data points.

Residual Plots A more formal way to test the "randomness" of the error distribution is by means of a *residual plot*. A "residual" is the difference between an actual data point (an experimental value of $-r_F$ in this case) and the value predicted by a model. For example, the residual for the data point at $-r_F = 6.21$ μmol/min-mg, $[F] = 100$ mmol/l is

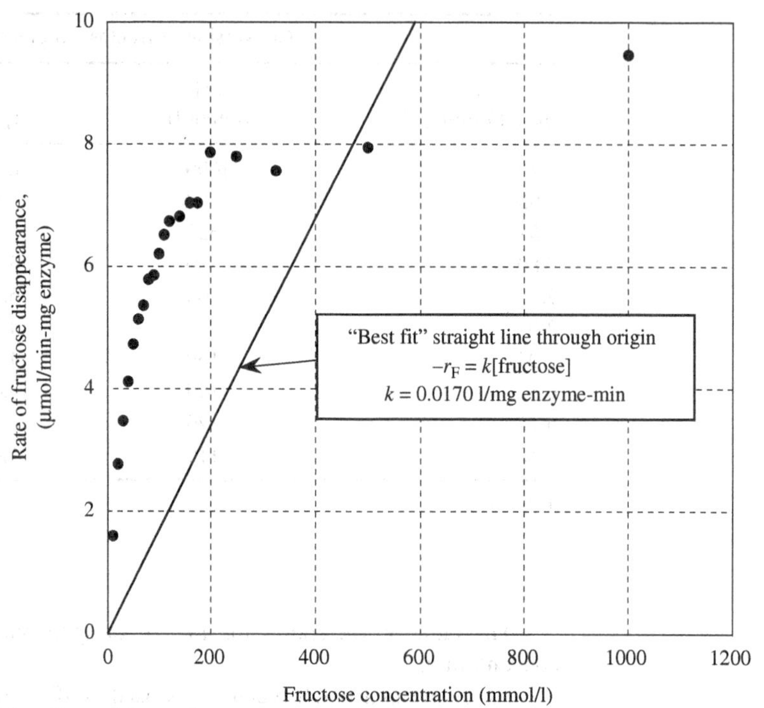

Figure 6-9a Test of first-order rate equation for fructose isomerization at 70 °C using xylose isomerase derived from *T. neapolitana.*

$(6.21 - 0.0170 \times 100) = 4.51$ μmol/min-mg. If a model fits a set of data, the residuals will be randomly scattered around zero when they are plotted against any significant variable.

A plot of the residuals in the reaction rate versus the fructose concentration is shown in Figure 6-9b, for the first-order rate equation.

This figure shows that the first-order rate equation is not adequate. First, most of the residuals are very large compared to the values of the experimental rates. Second, the residuals are not randomly distributed around zero. The residuals vary in a very systematic fashion with fructose concentration. All of the positive residuals occur at low fructose concentrations, <500 mmol/l. The only two negative residuals occur at fructose concentrations above this value.

We could construct residual plots for variables other than the fructose concentration. For example, it is common to plot the residuals against the measured values of the dependent variable, in this case the rate of fructose disappearance. We shall construct and discuss such a plot shortly, when we test a Michaelis–Menten rate equation against the data.

We might also construct a plot of the residuals against the technician who ran each experiment, to look for systematic "operator error." Another possibility is to examine the residuals against the source of a key raw material, e.g., fructose, if the material was obtained from more than one source.

Parity Plots A *parity plot* is used to present the overall results of an analysis visually. The parity plot is especially valuable when a complex model has been developed by piecewise analysis of various subsets of data.

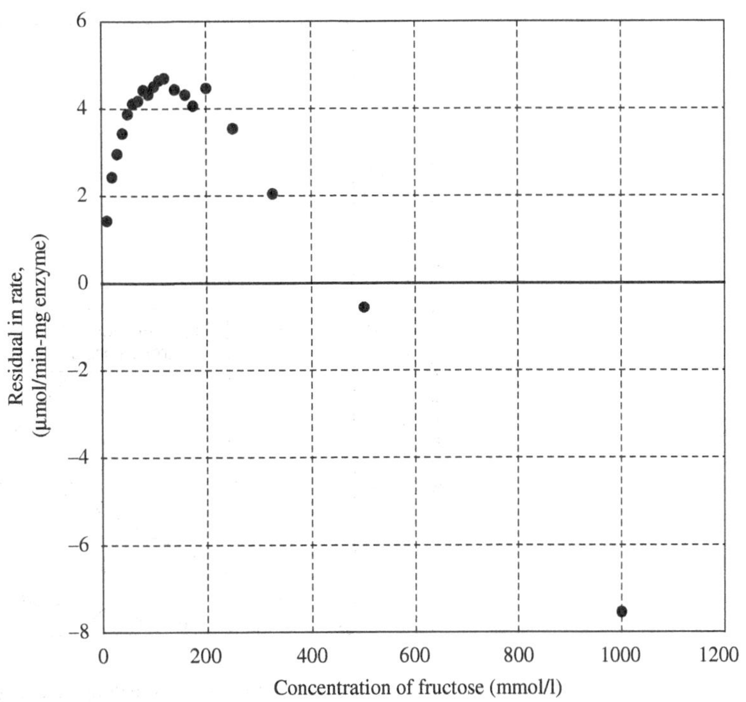

Figure 6-9b Residual plot to test first-order rate equation for fructose isomerization using xylose isomerase derived from *T. neapolitana*.

A parity plot is nothing more than a plot of the result calculated from the model (in this case, the first-order rate equation) against the experimental result. If the model were perfect and there was no error in the data, every point on a parity plot would fall on a line though the origin with a slope of 1. In reality, the data points will contain experimental error and will scatter around this line. However, if the model fits the data, the deviations will not be large, and the scatter will be random.

Figure 6-9c is a parity plot for the present example. Once again, the deficiency of the first order, model is apparent. The model overestimates the actual results when the rates are low to moderate, and underestimates the actual results when the rates are high. The deviations between the model and the data are generally large, and they vary systematically, not randomly, with the fructose concentration.

6.4.1.2 Second Hypothesis: Michaelis–Menten Rate Equation

The Michaelis–Menten rate equation

$$V = V_m[S]/(K_m + [S])$$

frequently provides a good description of the kinetics of simple enzymatic reactions. In this equation, S is the substrate (reactant) concentration, K_m is called the Michaelis constant, and V_m is the maximum reaction velocity. This rate equation can be derived from a simple reaction mechanism and can provide insight into the behavior of the

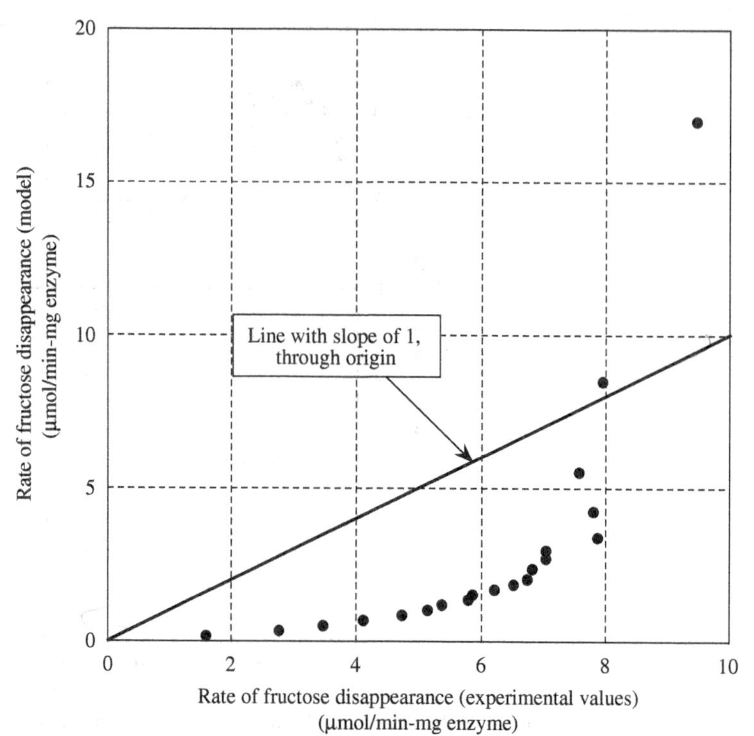

Figure 6-9c Parity plot to test first-order rate equation for fructose isomerization using xylose isomerase derived from *T. neapolitana*.

enzyme. For the present reaction, the Michaelis–Menten rate equation can be written as

$$-r_F = k[F]/(K_m + [F])$$

This rate equation can be linearized in several ways. One way is to simply invert both sides to obtain

$$\frac{1}{-r_F} = \left(\frac{K_m}{k}\right)\left(\frac{1}{[F]}\right) + \frac{1}{k}$$

If this model fits the data, a plot of $1/-r_F$ versus $1/[F]$ will be linear, with an intercept, I, of $1/k$ and a slope, S, of K_m/k. In biochemistry, this type of plot is known as a Lineweaver–Burke plot. Such a plot is shown in Figure 6-9d for the data of this example.

The Michaelis–Menten model appears to provide a much better description of the fructose isomerization data than the simple, first-order model. The points in the above graph fall very close to the "best fit" straight line, and the scatter seems to be random. The values of the slope and the intercept are shown on the graph; we will return to them shortly. These values were used to calculate the values of the residuals.

Figure 6-9e is one form of residual plot for the Michaelis–Menten rate equation. The values of the residuals generally are much smaller than they were for the first-order rate equation. Moreover, the scatter about the zero line is rather random. An equal number of points fall above and below this line.

Figure 6-9f shows the residuals plotted against the measured reaction rates. This kind of plot was not used in analyzing the first-order rate equation. However, it provides some useful insights here. First, the residuals appear to scatter randomly about zero, suggesting that there

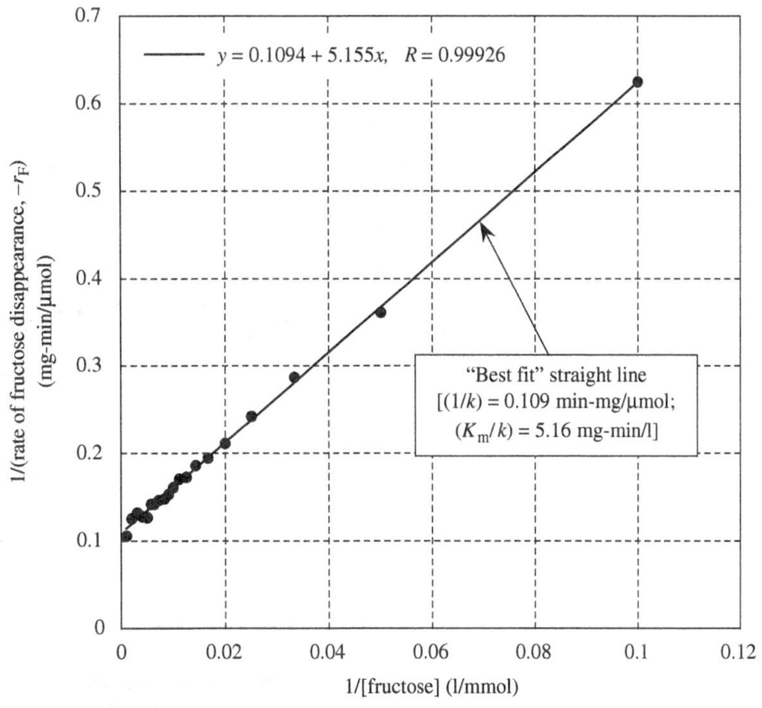

Figure 6-9d Test of Michaelis–Menten rate equation for fructose isomerization at 70 °C using xylose isomerase derived from *T. neapolitana*.

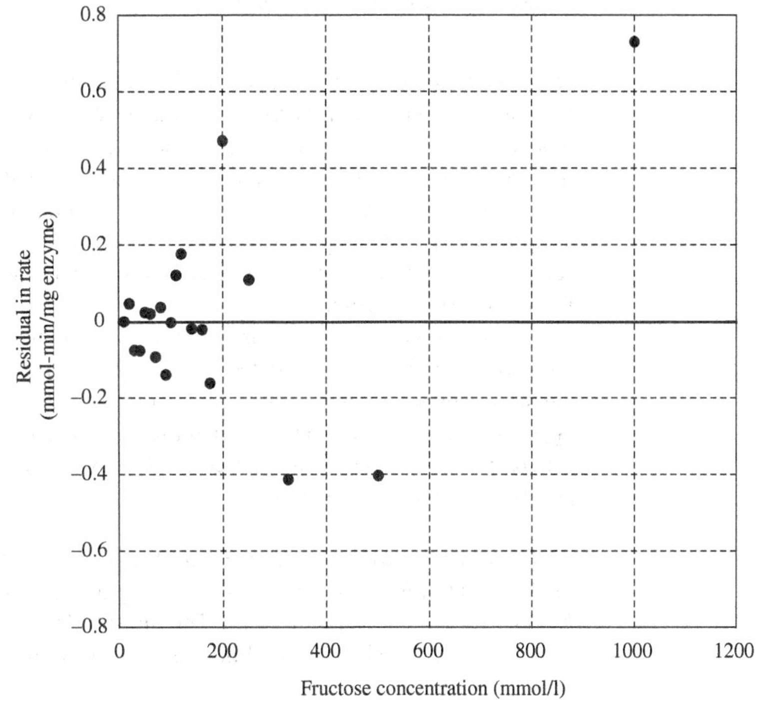

Figure 6-9e One form of residual plot to test Michaelis–Menten rate equation for fructose isomerization using xylose isomerase derived from *T. neapolitana*. The residuals in reaction rate are plotted against fructose concentration.

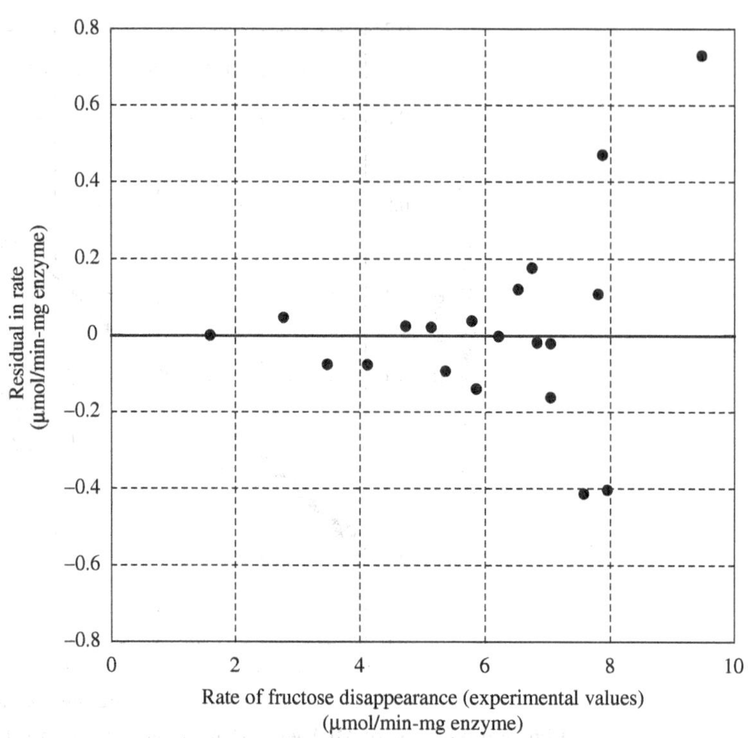

Figure 6-9f Another form of residual plot to test Michaelis–Menten rate equation for fructose isomerization using xylose isomerase derived from *T. neapolitana*. Residuals in reaction rate plotted against experimental values of reaction rate.

are no systematic errors in the model. However, the residuals appear to increase as the rate increases. The plot has a characteristic funnel shape. This does *not* indicate that the model is inadequate. However, it does suggest that the errors in the data are nonhomogeneous. One of the assumptions in the least squares analysis is that the errors are independent, random variables. Figure 6-9f suggests that this assumption is questionable, for the present data. Unfortunately, residual plots that resemble Figure 6-9f are not uncommon in the analysis of scientific data. A more complete discussion of the interpretation of residual plots can be found in various textbooks.[2]

Finally, Figure 6-9g is a parity plot for the Michaelis–Menten rate equation. This rate equation provides a much better fit of the experimental data than the first-order rate equation. The scatter in the parity plot is random, although the magnitude of the deviations appears to increase as the rate increases, consistent with the conclusions drawn from Figure 6-9f. Parity plots can be used to identify "outlying" data points. However, no such points are evident in Figure 6-9g.

Constants in the Rate Equation: Error Analysis As shown earlier, the "best fit" value of the intercept in Figure 6-9d is 0.109 min-mg/μmol. Since the intercept is $1/k$, the value of k is $1/0.109 = 9.17$ μmol/mg-min. The "best fit" value of the slope is 5.16 mg-min-mmol/μmol-l. Since the slope of the line is K_m/k, the value of K_m is 47.3 μmol/l.

[2] For example: Hines, W. W., Montgomery, D. C., Goldsman, D. M., and Borror, C. M., *Probability and Statistics in Engineering,* 4th edition, John Wiley & Sons, Inc. (2003); Walpole, R. E., Myers, R. H., Myers, S. L., and Ye, K., *Probability and Statistics for Engineers & Scientists,* 7th edition, Prentice-Hall, Inc. (2002).

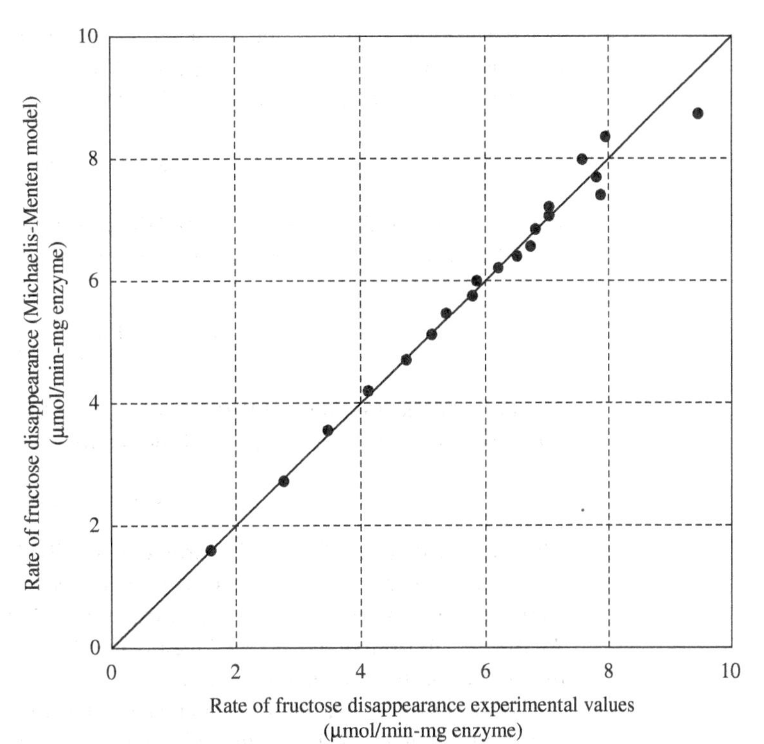

Figure 6-9g Parity plot to test Michaelis–Menten rate equation for fructose isomerization using xylose isomerase derived from *T. neapolitana.*

The uncertainty of the slope and intercept values can be calculated by standard statistical techniques. These calculations are straightforward in an EXCEL spreadsheet. An example is provided in Appendix 6-C. In fact, the calculations can be done "automatically" with the "Data Analysis" package in EXCEL.

Let y_i be an experimentally measured value of the dependent variable y, and let $\widehat{y}_i$ be the value of y predicted by the equation obtained by linear least squares analysis. The *error sum of squares* (SS_E) is given by

$$SS_E = \sum_{i=1}^{N} (y_i - \widehat{y}_i)^2$$

In this summation, N is the number of data points that were used to establish the "best fit" straight line. The value of SS_E is easily calculated in a spreadsheet, as shown in Appendix 6-C. The mean squared error (MS_E) is closely related to the error sum of squares:

$$MS_E = SS_E/(N - 2)$$

The values of the slope S and the intercept I that were estimated via the least-squares analysis contain some uncertainty because of errors in the data. Let $\bar{x}$ be the average of all of the values of x, and let S_{xx} be defined by

$$S_{xx} = \sum_{i=1}^{N} (x_i - \bar{x})^2$$

The *variance* of the slope, a measure of the uncertainty in the value of S obtained from the least squares analysis, is given by

$$V(S) = MS_E/S_{xx}$$

For the present example, the value of $V(S)$ is 0.00219 (mg-min-mmol/μmol-1)^2. The calculation of $V(S)$ is shown in Appendix 6-C.

The *estimated standard error* of the slope is given by

$$s_S = \sqrt{V(S)}$$

For this problem, the value of s_S is 0.0468 (mg-min-mmol/μmol-1).

The variance of the intercept is given by

$$V(I) = \mathrm{MS_E}\left[\frac{1}{\mathrm{N}} + \frac{\bar{x}^2}{S_{xx}}\right]$$

For this example, the value of $V(I)$ is 1.73×10^{-6} (mg-min/μmol)^2, as shown in Appendix 6-C. The estimated standard error of the intercept is

$$s_I = \sqrt{V(I)}$$

The numerical value of s_I for this problem is 0.00132 mg-min/μmol.

All of the parameters that have been calculated up to this point, plus a host of additional parameters, can be obtained directly by using the "Regression" tool that is under the "Data Analysis" submenu of the "Tools" menu in EXCEL. A printout of the "Data Analysis" for the present problem is given in Appendix 6-C.

In this example, we are more interested in the uncertainty of k and K_m than in that of the slope and the intercept. The estimated standard error in k can be approximated through some simple calculus.

Since $k = 1/I$, $|dk| = |dI|/I^2$. Therefore,

$$s_k \equiv s_I/I^2 = 0.00132/(0.109)^2 \ \mu\text{mol/mg-min} = 0.110 \ \mu\text{mol/mg-min}$$

The error in K_m can be developed in a similar manner. Since $K_m = S/I$, $|dK_m| = \{(|dS|/I) + (S|dI|/I^2)\}$. Therefore,

$$\begin{aligned} s_{K_m} &\cong (s_S/I) + (s_I S/I^2) = (0.0468)/(0.109) + (0.00132) \times 5.16/(0.109)^2 \\ &= 1.02 \ \text{mmol/l} \end{aligned}$$

The estimated standard errors in both k and K_m are small compared with the values of these constants, consistent with the excellent fit of the Michaelis–Menten rate equation shown in Figure 6-9d.

Nonlinear Least Squares Similar estimates of the accuracy and precision of parameter estimates can be made when calculations are performed with a nonlinear regression program. A discussion of how to develop these estimates is beyond the scope of this book. Many "canned" NLLS programs produce these estimates automatically.

6.4.2 Rate Equations Containing More Than One Concentration (Reprise)

Section 6.2.2 of this chapter dealt with the graphical analysis of data where the rate is affected by the concentrations of more than one species. If the experiments are properly designed, graphical techniques can be used for preliminary data analysis. However, in some cases, the experiments may not have been designed to facilitate stagewise graphical analysis of the data, so that it is not possible to isolate the effect of each concentration. In such a case, multiple linear regression can be used to begin the data analysis.

Consider the reaction

$$A + B \rightarrow C + D$$

which may (or may not) be described by the rate equation

$$-r_A = \frac{kC_AC_B}{(1 + K_AC_A + K_BC_B + K_CC_C + K_DC_D)^2}$$

This equation can be linearized to

$$\left(\frac{C_AC_B}{-r_A}\right)^{1/2} = \frac{1}{\sqrt{k}} + \frac{K_A}{\sqrt{k}}C_A + \frac{K_B}{\sqrt{k}}C_B + \frac{K_C}{\sqrt{k}}C_C + \frac{K_D}{\sqrt{k}}C_D$$

If a subset of data was available where the values of three concentrations, say C_A, C_B, and C_C, were constant, while the value of the fourth concentration C_D varied, then the data analysis could be started using the graphical techniques described earlier. In the absence of this kind of data, a multiple-linear-regression program, such as the one contained in the "Data Analysis" package in EXCEL, can be used to estimate values for $1/\sqrt{k}$, $K_A/\sqrt{k}$, $K_B/\sqrt{k}$, $K_C/\sqrt{k}$, and $K_D/\sqrt{k}$. Values of k, K_A, K_B, K_C, and K_D then can be calculated, along with estimates of their precision. Finally, values of $-r_A$ can be calculated from the model.

At this point, the fit of the model to the data must be evaluated carefully, since it was not possible to perform an initial graphical analysis. Residual plots should constructed to identify any systematic errors between the data and the model. For this example, at least four residual plots are appropriate, i.e., plots of the residuals in $-r_A$ versus C_A, C_B, C_C, and C_D. A parity plot should also prepared to allow a visual comparison of the model and the data, and to permit outlying points to be identified. Finally, the parameter estimates should be refined via nonlinear regression.

After one model has been thoroughly evaluated, alternatives must be explored, using the same approach. For the present example, it would certainly be desirable to test the rate equation:

$$-r_A = \frac{kC_AC_B}{(1 + K_AC_A + K_BC_B + K_CC_C + K_DC_D)}$$

SUMMARY OF IMPORTANT CONCEPTS

- Rate equations, and the integrated forms of rate equations, can be tested visually against experimental data. The rate equation is first linearized, and the data are then plotted in a straight-line form, as suggested by the linearized rate equation.
- Linear least-squares analysis can be used to obtain "best fit" values of the unknown parameters in a linearized rate equation.
- Nonlinear least-squares analysis can be used to obtain "best fit" values of the unknown parameters in a nonlinear rate model. Elementary nonlinear regressions can be performed using the SOLVER function in EXCEL.
- A parity plot is a direct, straightforward way to visually evaluate the fit of a model to experimental data.
- Residual plots can be used to detect systematic error between a model and one of the independent variables in the model.

PROBLEMS

Problem 6-1 (Level 1) Shreiber[3] has studied the dimerization of formaldehyde to methyl formate

$$2CH_2O \rightarrow HCOOCH_3$$

using a continuous slurry reactor that behaved as an ideal CSTR. The catalyst was Raney copper. Some of the data at 325 °C are given in the following table.

Selected kinetic data for the formation of methyl formate from formaldehyde ($T = 325$ °C; Raney Cu)

Partial pressure of formaldehyde (p_{HCHO}) (psi)	Rate of methyl formate formation (mol/g cat-h)
0.1797	3.63E-02
0.2916	8.51E-02
0.2495	8.12E-02
0.4976	2.64E-01
0.5031	2.51E-01
0.4630	2.46E-01
0.4077	1.93E-01
0.3894	1.52E-01
0.4217	1.78E-01
0.0613	3.75E-03

1. The following rate equation has been suggested:

$$r_{C_2H_4O_2} = k(p_{HCHO})^2$$

where $r_{C_2H_4O_2}$ is the rate of methyl formate formation. Test this rate equation against the data. Discuss the fit of the rate equation to the data.

2. Assuming that the rate equation is acceptable, what is the value of k?
3. What is the estimated standard error for the rate constant?

Problem 6-1A (Level 2) The following memo is in your inbox at 8 AM on Monday:

To: U. R. Loehmann
From: I. M. DeBosse
Subject: Methyl Formate

U. R.,

As you know, Cauldron Chemical Company has a strategic interest in methyl formate. Shreiber[3] has studied the dimerization of formaldehyde to methyl formate:

$$2CH_2O \rightarrow HCOOCH_3$$

using a continuous slurry reactor that behaved as an ideal CSTR. The catalyst was Raney copper. Some of Shreiber's data at 325 °C are given in the following table.

Selected kinetic data for the formation of methyl formate from formaldehyde ($T = 325$ °C; Raney Cu)

Partial pressure of formaldehyde (p_{HCHO}) (psi)	Rate of methyl formate formation (mol/g cat-h)
0.1797	3.63E-02
0.2916	8.51E-02
0.2495	8.12E-02
0.4976	2.64E-01
0.5031	2.51E-01
0.4630	2.46E-01
0.4077	1.93E-01
0.3894	1.52E-01
0.4217	1.78E-01
0.0613	3.75E-03

See if you can find a simple, power-law rate equation that fits these data. Work in terms of partial pressure rather than concentration. Find the value of the rate constant.

Please write me a short memo (not more than one page) giving the results of your analysis. Attach your calculations, graphs, etc. to the memo in case someone wants to review the details.

Thanks,
I. M.

Problem 6-2 (Level 3)[4] The vapor phase dehydration of ethanol to diethyl ether and water

$$2C_2H_5OH \rightleftarrows (C_2H_5)_2O + H_2O$$

was carried out at 120 °C over a solid catalyst (a sulfonated copolymer of styrene and divinylbenzene in acid form). The following table shows some experimental data for the rate of reaction as a function of composition. In this table, the subscript "A" denotes ethanol, "E" denotes diethyl ether, and "W" denotes water.

[3] Shreiber, E. H., *In situ* formaldehyde generation for environmentally benign chemical synthesis," *Ph.D. thesis*, North Carolina State University (1999).

[4] Adapted from Kabel, R. L. and Johanson, L. N., Reaction kinetics and adsorption equilibrium in the vapor-phase dehydrogenation of ethanol, *AIChE J.*, 8(5), 623 (1962).

Experimental data for ethanol dehydration

Experiment number	Reaction rate, $-r_A$ ($\times 10^4$) (mol/g·cat-min)	Partial pressures (atm) A	E	W
3–1	1.347	1.000	0.000	0.000
3–2	1.335	0.947	0.053	0.000
3–3	1.288	0.877	0.123	0.000
3–4	1.360	0.781	0.219	0.000
3–6	0.868	0.471	0.529	0.000
3–7	1.003	0.572	0.428	0.000
3–8	1.035	0.704	0.296	0.000
3–9	1.068	0.641	0.359	0.000
4–1	1.220	1.000	0.000	0.000
4–2	0.571	0.755	0.000	0.245
4–3	0.241	0.552	0.000	0.448
4–4	0.535	0.622	0.175	0.203
4–7	1.162	0.689	0.000	0.000

1. Using a graphical analysis, show that the rate equation
$$-r_A = \frac{k[p_A^2 - (p_E p_W/K_{eq})]}{(1 + K_A p_A + K_E p_E + K_W p_W)^2}$$
fits the above data reasonably well. The parameter K_{eq} is the equilibrium constant for the reaction at 120 °C, based on pressure. Calculate the value of K_{eq} using thermodynamic data.
2. From your graphical analysis, estimate values for the constants k, K_A, K_E, and K_w.
3. Determine the values of these four constants using multiple *linear* regression.
4. Determine the value of the four constants using *nonlinear* regression.
5. Compare the values of k, K_A, K_E, and K_w from Part 4 with those given in Table 3 of the referenced article.
6. The referenced article contains one additional data point at 120 °C that was obtained with a different reactor configuration. For this point; $p_A = 0.381$ atm, $p_E = 0$, and $p_w = 0.619$ atm, and $-r_A = 0.0866 \times 10^{-4}$ mol/g·cat-min. How well does the rate equation, with the constants determined in Part 4, describe this data point?

Problem 6-3 (Level 2) Yadwadkar[5] has studied the kinetics of the saponification of methyl acetate:

$$CH_3-\overset{O}{\overset{\|}{C}}-O-CH_3 + NaOH \longrightarrow CH_3-\overset{O}{\overset{\|}{C}}-O^-Na^+ + CH_3OH$$

in an isothermal, plug-flow reactor. The temperature was 25 °C, an acetone–water mixture was used as a solvent, and the initial concentrations of methyl acetate and sodium hydroxide were both 0.05 g·mol/l. The experimental data are given in the following table:

Kinetics of methyl acetate saponification

Experiment	Space time, (s)	Fractional Conversion of NaOH
1	34.4	0.208
2	69.0	0.321
3	139	0.498
4	69.5	0.342
5	139	0.466
6	281	0.670
7	94.8	0.415
8	189	0.580

1. Find a rate equation that adequately describes the data in the above table. Specify numerical values for all unknown constants in the rate equation.
2. Using *only* the information provided above, is it possible to determine the individual orders for the two reactants? If so, what are the appropriate values? If not, what additional experiments should be performed in order to obtain the individual reaction orders?

Problem 6-4[6] (Level 2) The reaction of trityl chloride with methanol (MeOH) at 25 °C in the presence of pyridine and phenol is stoichiometrically simple and essentially irreversible.

$$(C_6H_5)_3CCl + CH_3OH \rightarrow (C_6H_5)_3COCH_3 + HCl$$

Phenol does not react with trityl chloride as long as some methanol is present. The data in the table below were taken to study the kinetics of the reaction. An ideal, isothermal batch reactor was used.

Suppose that the reaction is second order overall and that the concentrations of the products do not enter into the rate equation.

1. If the reaction is second order overall, what values of the individual orders for trityl chloride and methanol provide the best description of the observed kinetics? Limit your analysis to integer orders, either 0, 1, or 2. What is the approximate value of the rate constant in the best kinetic model?

[5] Yadwadkar, S. R., The influence of shear stress on the kinetics of chemical reactions, M. S. thesis, Washington University (St. Louis) (1972).

[6] Adapted from Swain, C. G., Kinetic evidence for a termolecular mechanism in displacement reactions of triphenylmethyl halides in benzene solution, *J. Am. Chem. Soc.*, 70, 1119 (1948).

2. Does the rate equation $-r_{MeOH} = k[MeOH]^2[\text{trityl chloride}]$ provide a better description of the data than the rate equation you found in Part 1?

Data for trityl chloride/methanol reaction

Temperature: 25 °C

Solvent: dry benzene

Initial concentrations:

CH_3OH—0.054 g·mol/l

$(C_6H_5)_3\ CCl$—0.106 g·mol/l

Phenol—0.056 g·mol/l

Pyridine—0.108 g·mol/l

Time (min)	Fractional conversion of methanol
39	0.318
53	0.420
55	0.389
91	0.582
127	0.702
203	0.834
258	0.870
434	0.924
1460	0.976

Problem 6-5 (Level 1) The isomerization of *cis*-2-wolftene to *trans*-2-wolftene is believed to be elementary. The kinetics of the reaction have been studied in a constant-volume, isothermal, ideal batch reactor. The following table shows the results of a run at 30 °C, using ramsoil as a solvent. Wolftene forms an ideal solution in ramsoil.

Concentrations of *cis*- and *trans*-2-wolftene in ramsoil as a function of time at 30 °C

Time (h)	Concentration (mol/l)	
	cis-2-wolftene	*trans*-2-wolftene
0	1.00	0
1	0.821	—
2	0.700	—
4	0.522	—
8	0.380	—
96	0.310	—
168	0.310	0.690

In another experiment at 50 °C in the same system with the same starting solution, the concentration of *cis*-2-wolftene was 0.452 mol/l after 15 min and 0.310 mol/l after 168 h.

1. Do the data at 30 °C support the hypothesis that the reaction is elementary? Show how you arrived at your conclusion.
2. Is the reaction exothermic or endothermic? Explain your reasoning.
3. If the value of the forward rate constant is 0.200 h^{-1} at 30 °C and 4.38 h^{-1} at 50 °C, what is the value of the activation energy of the forward reaction?

Problem 6-6 (Level 1) The catalytic, gas-phase reaction $A \rightarrow B$ is taking place in an isothermal, ideal plug-flow reactor that is filled with catalyst particles. Some data on the performance of the reactor are shown in the following table.

Space time h-kg (catalyst)/l	Fractional conversion of A (x_A)
1	0.32
5	0.55
20	0.90
30	0.96

Are the above data consistent with the hypothesis that the reaction is first order and irreversible? Show your analysis and explain your reasoning.

Problem 6-7 (Level 1) The following data were obtained for the irreversible, gas-phase reaction

$$A + B \rightarrow R + S$$

using an isothermal, ideal plug-flow reactor. The feed contained a stoichiometric mixture of A and B in N_2. The feed concentration of A was 0.005 mol/l.

Space time, τ (min)	Concentration of A in outlet, mol/l
200	0.00250
600	0.00125

Are the above data consistent with the assumption that the reaction is second order overall? Support your answer quantitatively.

Problem 6-8 (Level 1) Toluene (T) can be produced by dehydrogenation of methylcyclohexane (M) over various transition metal catalysts:

$$M \rightarrow T + 3H_2$$

Sinfelt and coworkers[7] studied the kinetics of this reaction over a Pt/Al_2O_3 catalyst in a differential plug-flow reactor. Data showing the rate of reaction (r_T) as a function of the partial pressure of M (p_M) at 315 °C are summarized in the following table.

[7] Sinfelt, J. H., et al., *J. Phys. Chem.*, 64, 1559 (1990).

p_M (atm)	r_T (g T formed/h-g-cat)
0.36	0.012
0.36	0.012
0.07	0.0086
0.24	0.011
0.72	0.013

The rate of formation of T was found to be nearly independent of the partial pressure of M at higher pressures. This led the following rate equation to be proposed:

$$r_T = \frac{kbp_M}{1 + bp_M}$$

where k and b are constants that are functions of temperature.

1. Does the proposed rate equation fit the experimental data? Show how you arrived at your conclusion.
2. Determine values of the constants k and b using *all* of the available data. A graphical analysis is sufficient.

Problem 6-9 (Level 1) The kinetics of the irreversible, gas-phase reaction $A \rightarrow B$ were studied in an ideal CSTR. In one experiment, the following data was obtained in a reactor with a volume of 0.50 l, operating at a temperature of T_1 and a pressure of P_1.

Experiment number	Inlet volumetric flow rate at T_1 and P_1	Inlet concentrations, mol/l		Outlet concentrations, mol/l	
		A	B	A	B
4	0.25 l/h	0.020	0.0050	0.0035	0.0215

1. What was the rate of disappearance of A (mole A/l h) for Experiment #4?

The complete set of experimental results is shown in the following table.

Experiment number	Reaction rate $(-r_A)$(mol/l-h)	Inlet concentrations (mol/l)		Outlet concentrations (mol/l)	
		A	B	A	B
1	0.152	0.025	0	0.015	0.010
2	0.0674	0.015	0.010	0.010	0.015
3	0.0285	0.015	0.010	0.0065	0.0185
4		0.020	0.0050	0.0035	0.0215

2. Does the rate equation $-r_A = kC_A^2$ fit the experimental data? Justify your answer using a graphical analysis.

Problem 6-10 (Level 1) The kinetics of the liquid-phase, enzyme-catalyzed reaction

$$A \rightarrow P + Q$$

have been studied in an ideal, isothermal batch reactor. The rate equation is believed to be

$$-r_A(\text{mol/l-s}) = kC_A/(1 + K_P C_P)$$

In one particular experiment, the concentrations of A, P, and Q were measured as a function of time. The initial concentration of A in this experiment was C_{A0}. There was no P or Q present initially, i.e., $C_{P0} = C_{Q0} = 0$.

1. Demonstrate how you would use the *integral* method of data analysis to test (graphically) this rate equation against the experimental data. Carry out any mathematical operations that are required. Sketch the graph that you would make, showing what you would plot against what.
2. Assuming that this rate equation did fit the data, how would you obtain estimates of k and K_P from the graph that you constructed?

Problem 6-11 (Level 2) The reaction $A + B \rightarrow C + D$ has been studied in a differential plug-flow reactor, with the following results.

Temperature, (K)	Rate of disappearance of A $-r_A$, (mol/l-min)	Concentrations, (mol/l) A	B
373	0.0214	0.10	0.20
373	0.0569	0.25	0.20
373	0.144	0.65	0.20
373	0.235	1.00	0.20
373	0.0618	0.40	0.10
373	0.228	0.90	0.25
373	0.211	0.55	0.60
373	0.0975	0.20	0.95

Does the rate equation $-r_A = kC_A^{\alpha} C_B^{\beta}$ fit the data? If so, what are the approximate values of α, β, and k?

Problem 6-12 (Level 1) The kinetics of the reaction of hexamethylenetetramine bromine $[(CH_2)_6N_4Br_4]$ (A) with thiomalic acid $[HOOCCH_2CH(SH)COOH]$ (B) to form dithiomalic acid (C) has been studied by Gangwani and coworkers.[8] The reaction is stoichiometrically simple and may be represented as

$$A + 4B \rightarrow 2C + \text{other products}$$

The reaction was carried out in glacial acetic acid at 298 K. Some of the data are given in the following table.

$-r_A \times 10^6$ (mol/l-s)	$C_A \times 10^3$ (mol/l)	C_B (mol/l)
6.64	1.0	0.10
10.30	1.0	0.20
14.10	1.0	0.40
17.40	1.0	0.80
19.50	1.0	1.50
20.90	1.0	3.00
22.00	2.0	0.20
40.40	4.0	0.20
64.80	6.0	0.20
84.80	8.0	0.20

The reaction is believed to be first order in A and first order in B. Does this model fit the data? Show how you arrived at your conclusion. If the model fits the data, what is the value of the rate constant?

Problem 6-13 (Level 3) The dehydration of mixtures of methanol and isobutanol to form ethers and olefins has been studied over a sulfonic acid (Nafion H) catalyst.[9] At high pressures, dimethylether (DME) is the predominant product. The reaction of methanol to form DME is a dehydration reaction:

$$2CH_3OH \rightleftarrows \underset{\text{(DME)}}{CH_3OCH_3} + H_2O$$

The rate equation for the formation of DME is postulated to be

$$r_{DME} = \frac{k_1 K_M^2 p_M^2}{(1 + K_M p_M + K_B p_B)^2} \qquad (6\text{-}1)$$

where p_M is the partial pressure of methanol, p_B is the partial pressure of isobutanol, and K_M and K_B are the corresponding adsorption equilibrium constants.

To test the rate equation partially, kinetic data were taken using a differential plug-flow reactor operating at 375 K and a total pressure of 1.34×10^3 kPa. For a feed consisting only of methanol and nitrogen, the following data were obtained.

r_{DME} (mol/kg$_{cat}$-h)	p_M (kPa)
0.155	30
0.156	40
0.200	80
0.217	120
0.219	160
0.220	240

1. Carry out a graphical analysis to determine whether the postulated rate equation fits the above data. Determine the "best" values of the unknown rate and equilibrium adsorption constants, k_1 and K_M, from the results of the graphical analysis.
2. Carry out an nonlinear regression analysis to determine the "best" values of k_1 and K_M for the model of Eqn. 6-1. Compare these values with the ones obtained in Part 1.

Problem 6-14 (Level 1) Hexamethylene tetramine-bromine $[(CH_2)_6N_4Br_4)]$, abbreviated as HABR, can oxidize thioglycollic acid (TGA), abbreviated as RSH, to the corresponding disulfide, abbreviated as RSSR. The overall reaction can be represented as

$$4RSH + HABR \rightarrow 2RSSR + 4HBr + (CH_2)_6N_4$$

Gangwani et al.[8] studied the kinetics of this reaction at 298 K using glacial acetic acid as a solvent. Experiments were run in an ideal batch reactor using a large excess of TGA relative to HABR.

[8] Gangwani, H., Sharma, P. K., and Banerji, K. K., Kinetics and Mechanism of the oxidation of some thioacids by hexamethylenetetramine-bromine, *React. Kinet. Catal. Lett.*, 69(2) 369–374 (2002).

[9] Nunan, J. G., Klier, K. and Herman, R. G., Methanol and 2-methyl-1-propanol (isobutanol) coupling to ethers and dehydration over nafion H: selectivity, kinetics, and mechanism. *J. Catal.* 139, 406–420 (1993).

The initial rate of reaction was measured at various combinations of initial concentrations. The results are shown in the following table:

TGA (mol/l)	HABR (mol/l)	$-r_{HABR} \times 10^6$ (mol/l-s)
0.10	0.001	2.33
0.20	0.001	3.76
0.40	0.001	5.41
0.80	0.001	6.92
1.5	0.001	7.97
3.0	0.001	8.72
0.20	0.002	6.66
0.20	0.004	15.6
0.20	0.006	21.4
0.20	0.008	24.0

How well does the rate equation

$$-r_{HABR} = k[\text{HABR}][\text{TGA}]/(1 + K[\text{TGA}])$$

fit the data? Determine values of k and K via a graphical analysis.

Problem 6-15 (Level 1) Xu et al.[10] have studied the hydrogenation of polystyrene to poly(cyclohexylethylene) using a Pd/ $BaSO_4$ catalyst in an isothermal batch reactor. The reaction is as given below.

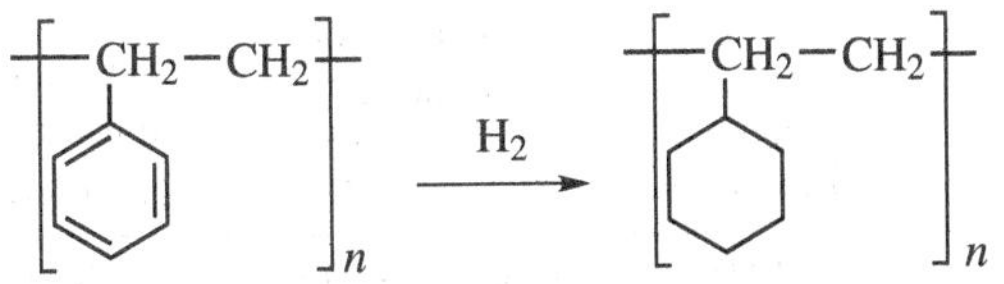

The polystyrene is dissolved in a solvent, decahydronaphthalene. The catalyst is in the form of small solid particles that are suspended in the polystyrene solution.

The fraction of aromatic rings that were hydrogenated to cyclohexane rings was measured as a function of time, catalyst concentration, and polymer concentration. The H_2 pressure was 750 psig and the temperature was 150 °C. At these conditions, the reaction rate was independent of H_2 pressure. Some of the experimental data are given in the following table.

[10] Xu, D., Carbonell, R. G., Kiserow, D. J., and Roberts, G. W., *Ind. Eng. Chem. Res.*, 42(15), 3509 (2003).

Polystyrene concentration (wt.%)	Reaction time (h)	Catalyst concentration (g/l)	Percent conversion of aromatic rings
1	10	9.05	40
2	10	18.3	65
3	10	8.13	24
3	10	45.2	89
3	10	27.7	83
3	10	27.7	83
3	10	27.7	80
3	10	4.50	13
2	10	18.3	72
3	3.0	27.7	48
3	6.0	27.7	69
3	10	27.7	82

1. Determine whether the reaction is first order in the concentration of aromatic rings. Justify your answer. [*Note:* At any time, the concentration of aromatic rings is directly proportional to $C_{PS}(1 - x_A)$, where C_{PS} is the initial concentration of polystyrene, as shown in the above table, and x_A is the fractional conversion of aromatic rings, as shown above.]
2. Independent of your answer to Part (a), estimate a value of the first-order rate constant.

Problem 6-16 (Level 2) The kinetics of the polymerization of methyl methacrylate monomer were studied at 77 °C using benzene as a solvent and azo-bisisobutyronitrile (AIBN) as the free radical initiator.[11] The following table contains some data from an ideal batch reactor that shows the initial rate of reaction as a function of the initial concentration of monomer M and the initial concentration of initiator I.

Kinetic Data for Methyl Methacrylate Polymerization ($T = 77$ °C, Benzene Solvent, AIBN Initiator)

Initial reaction rate (mol M/l-s) $\times 10^4$	Initial monomer [M] concentration (mol/l)	Initial initiator [I] concentration (mol/l)
1.93	9.04	0.235
1.70	8.63	0.206
1.65	7.19	0.255
1.29	6.13	0.228
1.22	4.96	0.313
0.94	4.75	0.192
0.87	4.22	0.230
1.30	4.17	0.581
0.72	3.26	0.245
0.42	2.07	0.211

[11] Arnett, L. M., *J. Am. Chem. Soc.*, 74, 2027 (1962).

Find a rate equation of the form

$$-r_M = k[M]^b[I]^a$$

that adequately fits the data. Estimate the values of the unknown constants in the rate equation.

Bridging Problems

Problem 6-17 (Level 2) The reaction

$$NaBH_4(aq) + 2H_2O \rightarrow 4H_2 + NaBO_2(aq)$$

has been considered as a means of generating H_2 to power small fuel cells.[12] The reaction is catalyzed by Ru supported on an ion exchange resin. No reaction takes place in the absence of the catalyst.

In an experiment in an ideal, isothermal batch reactor, 0.25 g of 5 wt.% Ru/IRA-400 resin was dispersed in an aqueous solution containing 6 g $NaBH_4$, 3 g NaOH, and 21 g H_2O. The following data were obtained at 25 °C.

Time (s)	Cumulative amount of H_2 generated (l@STP)
0	0.00
500	0.31
1000	0.62
1500	0.93

1. What is the fractional conversion of $NaBH_4$ at 1500 s?
2. The reaction is claimed to be zero order overall, and zero order with respect to each of the reactants. If this is true, what is the value of the rate constant at 25 °C?
3. Is it reasonable to conclude that the reaction is zero order based only on the data above? Explain your answer.
4. A CSTR that operates at 25 °C and generates 1 l of H_2/min is to be designed. The feed will be an aqueous solution of $NaBH_4$ and NaOH, with the same $NaBH_4$/NaOH/H_2O ratio shown above. If the conversion of $NaBH_4$ in the effluent is 75%, how much catalyst must be in the reactor? (You may use the zero-order assumption for this calculation.)

Problem 6-18 (Level 3) The following memo is in your inbox at 8 AM on Monday:

To: U. R. Loehmann

From: I. M. DeBosse

Subject: Preliminary Reactor Design

Corporate Research has run the experiments shown in Tables P6-18a–c to determine the kinetics of the irreversible, liquid-phase reaction:

$$A + 2B \xrightarrow{\text{solvent}} R + S$$

[12] Adapted from Amendola, S. C., Sharp-Goodman, S. L., Janjva, M. S., Kelly, M. T., Petillo, P. J., and Binder, M., An ultrasafe hydrogen generator: aqueous, alkaline borohydride solutions and Ru catalyst, *J. Power Sources*, 85, 186–189 (2000).

Table P6-18a

Reactor—Batch
Reactor temperature—25.0 °C (constant)
Initial concentration of A—0.50 g·mol/l
Initial concentration of B—1.00 g·mol/l

Time (min)	Fractional conversion of A
0	0.00
2	0.17
4	0.28
6	0.38
8	0.43
10	0.50
15	0.60
20	0.68
25	0.71
30	0.75
50	0.85
70	0.87
100	0.91
200	0.95
600	0.98
Overnight	≅1.0

Table P6-18b

Reactor—CSTR
Reactor temperature—25 °C

Run number	Space time (min)	Feed concentration (g·mol/l) A	Feed concentration (g·mol/l) B	Fractional conversion of A in effluent
1	40.0	0.50	1.00	0.61
2	10.0	0.50	1.00	0.38
3	2.5	0.50	2.00	0.17
4	10.0	1.00	2.50	0.50
5	10.0	0.25	1.00	0.27

Table P6-18c

Reactor—CSTR
Reactor temperature—75 °C

Run number	Space time (min)	Feed concentration (g·mol/l) A	Feed concentration (g·mol/l) B	Fractional conversion of A in effluent
6	0.20	0.50	1.00	0.38
7	1.00	0.50	1.00	0.64

Below 75 °C, no significant side reactions occur. Above this temperature, the desired product (R), decomposes rapidly to worthless byproducts.

Based on these data, a preliminary design for a reactor to manufacture R (MW = 130) at a rate of 50,000,000 pounds per year must be prepared. Because of limitations associated with the design of the purification system downstream of the reactor, the fractional conversion of A must be at least 0.80.

Would you please do the following:

1. Find a rate equation that describes the effect of temperature and the concentrations of A and B on the reaction rate. Evaluate all arbitrary constants in this rate equation.
2. Assuming that the concentrations of A and B in the feed to the reactor are 0.50 and 1.00 g mol/l, respectively:
   i. What is the smallest CSTR that can be used?
   ii. What is the smallest isothermal PFR that can be used?
   iii. Suppose that three CSTRs in series are used, each with the same volume and operating at the same temperature. What total reactor volume is required?

Please write me a short memo (not more than one page) giving the results of your analysis. Attach your calculations, graphs, etc. to the memo in case someone wants to review the details.

Thanks,

I. M.

Problem 6-19 (Level 3) Trimethyl gallium (Me_3Ga), triethylindium (Et_3In), and arsine (AsH_3) are gases that are used to grow films of gallium indium arsenide (GaInAs). However, the deposition of these films is complicated by a reaction between Et_3In and AsH_3 to form a complex of the two compounds:

$$Et_3In + AsH_3 \rightleftarrows (Et_3In)\text{–}(AsH_3)$$

The kinetics of this reaction have been studied by Agnello and Ghandi,[13] using an ideal, plug-flow reactor that operated isothermally at room temperature. Some of their data are given in the following table.

Steady-state Et_3In partial pressures as a function of axial position

Total flow: 3 slpm (standard liters per minute)

Tube diameter: 50 mm (inside diameter)

Inlet partial pressures: Et_3In—0.129 Torr

AsH_3—1.15 Torr

H_2—152 Torr

[13] Agnello, P. D. and Ghandi, S. K., A mass spectrometric study of the reaction of triethyl indium with arsine gas, *J. Electrochem. Soc.*, 135(6), 1530–1534 (1988).

Note: 1 Torr = 1 mmHg

Axial position (cm)	Et_3In partial pressure (Torr)
0 (inlet)	0.129
2	0.0667
3	0.0459
4.5	0.0264
7.0	0.0156
9.5	0.0156

The ideal gas law may be used at these conditions.

1. Is it reasonable to assume that the reaction between Et_3In and AsH_3 is elementary as written? Support your answer.
2. If this reaction is elementary, what is the form of the rate equation for the disappearance of Et_3In?
3. How do you explain the data at 7.0 and 9.5 cm?
4. Since AsH_3 is present in large excess, the rate of the forward reaction should be pseudo-first order in Et_3In. In other words, the concentration of AsH_3 varies so little as the reaction proceeds that the AsH_3 concentration can be considered constant, and lumped into the rate constant. Using this assumption, simplify the rate equation that you derived in Part 2 and determine whether this rate equation fits the data. Explain the reasons for your conclusion.
5. Estimate of the value of the forward rate constant. You may continue to keep the AsH_3 concentration as part of the rate constant.

Problem 6-20 (Level 2) The Cauldron Chemical Company was boiling with activity. The first production batch of L.P. #9 was scheduled to start in 10 days, based on the irreversible reaction

$$\text{L.P. \#8} \rightarrow \text{L.P. \#9}$$

(Cauldron is a very secretive company so all of the chemicals have nondescriptive designations.) This is a homogeneous reaction that takes place in solution. The reaction will be carried out in an isothermal batch reactor at 150 °C. The conversion of L.P. #8 must be at least 95%, starting from an initial concentration of 4.0 mol L.P. #8/l.

The Department of Blue Sky Research carried out some quick experiments to determine the reaction kinetics. They reported that the initial rate (the rate at 150 °C and at a concentration of 4.0 mol L.P. #8/l) was 2.11 mol/l-h. The Engineering Department then assumed that the reaction was first order in L.P. #8 and they calculated a rate constant from the measured initial rate. Then they used this rate constant to calculate the time required to reach 95% conversion.

1. What value of the rate constant did the Engineering Department calculate?

2. What value of the time required to reach 95% conversion did the Engineering Department calculate?

Skip Tickle, Area Production Manager for L.P. Products, has never had any use for the Department of Blue Sky Research, and he is not willing to bet his career on the Engineering Department's estimate of the required reaction time. Therefore, Skip asked Mal Ingerer, the plant chemist, to run some additional experiments to determine the reaction kinetics. Skip's final words as he was leaving Mal's office were "Those guys in Engineering think every reaction is first order!" Mal, who never came to work earlier than 8 AM and never left later than 5 PM, produced the data in the following table.

3. Is the reaction first order? Explain your reasoning.
4. How long is it going to take to reach 95% conversion in a batch reactor at 150 °C with an initial concentration of 4.0 mol/l of L.P. #8?

Kinetic data for conversion of L.P. #8 to L.P. #9 ("A" denotes L.P. #8)

Initial [A] (mol/l)	Time (h)	[A] (mol/l)
2	0	2.00
	2	1.31
	4	0.97
	6	0.77
4	0	4.00
	2	1.95
	4	1.29
	6	0.96
6	0	6.00
	2	2.32
	4	1.44
	6	1.04

Use the integral method of data analysis.

APPENDIX 6-A NONLINEAR REGRESSION FOR AIBN DECOMPOSITION

Approach 2: Minimize fractional error

$A =$	$1.21\,E+14$	(s^{-1})	*Note:* A and E are initial estimates from
$E =$	123.3	(kJ/mol)	linear regression
$k(363)$(init) $=$	$2.19\,E-04$	(s^{-1})	*Note:* Initial value of $k(363)$ calculated from A and E
$k(363) =$	$1.9759\,E-04$	(s^{-1})	*Note:* SOLVER finds the values of $k(363)$
$E =$	123.33	(kJ/mol)	and E that minimize the boxed Sum below
$A =$	$1.10446\,E+14$	(s^{-1})	*Note:* Calculated from $k(363)$ and E

Note: In the following, $\text{Del}(k) = k(\text{theo}) - k(\text{exp})$.

Temperature (K)	k(theo)	k(exp)	[Del(k)/k(exp)]^2	Del(k)^2
353	6.2088E−05	7.6000E−05	3.3510E−02	1.9356E−10
363	1.9759E−04	1.5000E−04	1.0066E−01	2.2647E−09
368	3.4427E−04	5.2000E−04	1.1421E−01	3.0882E−08
373	5.9097E−04	6.2000E−04	2.1925E−03	8.4279E−10
		Sum =	2.5057E–01	3.4183E−08

Check for minimum by changing values of $k(363)$ and E:

$k(363) =$	1.9600E−04	(s^{-1})
$E =$	123.33	(kJ/mol)

Temperature (K)	k(theo)	k(exp)	[Del(k)/k(exp)]^2	Del(k)^2
353	6.1588E−05	7.6000E−05	3.5961E−02	2.0771E−10
363	1.9600E−04	1.5000E−04	9.4044E−02	2.1160E−09
368	3.4150E−04	5.2000E−04	1.1783E−01	3.1862E−08
373	5.8622E−04	6.2000E−04	2.9688E−03	1.1412E−09
		Sum =	2.5081 E–01	3.5327E−08

Results:

$k(363)$	E	Sum
1.9759E−04	123.00	2.5058E−01
1.9759E−04	123.70	2.5058E−01
1.9900E−04	123.33	2.5076E−01
1.9600E−04	123.33	2.5081E−01

SOLVER appears to have converged with respect to both $k(363)$ and E

APPENDIX 6-B1 NONLINEAR REGRESSION FOR AIBN DECOMPOSITION

Approach 1: Minimize total error

$A =$	1.21 E + 14	(s^{-1})
$E =$	123.3	(kJ/mol)
k(363)(init) =	2.1913E−04	(s^{-1})
k(363) =	2.2438E−04	(s^{-1})
$E =$	123.3	(kJ/mol)
$A =$	1.24204 E + 14	(s^{-1})

Temp (K)	k(calc)	k(exp)	Del(k)^2	[Del(k)/k(exp)]^2
353	7.0525E−05	7.6000E−05	2.9973E−11	5.1892E−03
363	2.2438E−04	1.5000E−04	5.5323E−09	2.4588E−01
368	3.9089E−04	5.2000E−04	1.6668E−08	6.1643E−02
373	6.7092E−04	6.2000E−04	2.5929E−09	6.7453E−03
		Sum =	2.4824 E–08	3.1946E−01

Check for minimum by changing values of k(363) and E

Example: Decrease k(363)

k(363) =	2.2300E−04	(s^{-1})
$E =$	123.30	kJ/mol

Temperature (K)	k(theo)	k(exp)	Del(k)^2	[Del(k)/k(exp)]^2
353	7.0092E−05	7.6000E−05	3.4909E−11	6.0438E−03
363	2.2300E−04	1.5000E−04	5.3290E−09	2.3684E−01
368	3.8849E−04	5.2000E−04	1.7295E−08	6.3960E−02
373	6.6680E−04	6.2000E−04	2.1898E−09	5.6967E−03
		Sum =	2.4848 E–08	3.1254E−01

Results:

k(363)	E	Sum	
2.2438E−04	123.40	2.4840E−08	Changing E at constant k(363)
2.2438E−04	**123.30**	**2.4824E−08**	
2.2438E−04	123.00	2.4779E−08	
2.2438E−04	122.00	2.4684E−08	
2.2438E−04	121.00	2.4667E−08	
2.2438E−04	120.00	2.4728E−08	
2.2600E−04	123.30	2.4858E−08	Changing k(363) at constant E
2.2300E−04	123.30	2.4848E−08	

The numbers in the column labeled "Sum" show that $\Sigma_{i=1}^{N}(k_{i,\text{theo}} - k_{i,\text{exp}})^2$ increases slightly when k(363) is either increased or decreased slightly from the value of 2.2438×10^{-4} found by SOLVER, when E is fixed at 123.30 kJ/mol. Also, the "Sum" increases slightly when E is increased slightly above the value of 123.30 kJ/mol found by SOLVER, when k(363) is fixed at 2.2438×10^{-4}. However, when E is decreased below 123.30 kJ/mol, the "Sum" decreases. This shows that SOLVER had not reached a minimum with respect to E. In fact, the "Sum" continued to decrease with E at constant k(363) until E was in the region of 121–122 kJ/mol.

APPENDIX 6-B2

Approach 1: Minimize total error (scaled by 10^8)

$A =$	1.21 E + 14	(s^{-1})
$E =$	123.3	(kJ/mol)
$k(363)$(init) =	2.1913E−04	(s^{-1})
$k(363) =$	2.4160E−04	(s^{-1})
$E =$	112.745256	kJ/mol
$A =$	4.0494 E + 12	(s^{-1})

Temp (K)	k(calc)	k(exp)	Del(k)^2	[Del(k)/k(exp)]^2
353	8.3846E−05	7.6000E−05	6.1557E−11	1.0657E−02
363	2.4160E−04	1.5000E−04	8.3902E−09	3.7290E−01
368	4.0136E−04	5.2000E−04	1.4076E−08	5.2056E−02
373	6.5775E−04	6.2000E−04	1.4251E−09	3.7073E−03
		10 ^ 8 * Sum =	2.3953 E+00	4.3932E−01

Sum = 2.3953 E–08

Check for minimum by changing values of $k(363)$ and E

Example: Decrease both $k(363)$ and E

$k(363) =$	2.4100E-04	(s^{-1})
$E =$	111.00	(kJ/mol)

Temperature (K)	k(theo)	k(exp)	Del k ^ 2	[Del(k)/k(exp)^2]
353	8.5020E−05	7.6000E−05	8.1356E−11	1.4085E−02
363	2.4100E−04	1.5000E−04	8.2810E−09	3.6804E−01
368	3.9723E−04	5.2000E−04	1.5072E−08	5.5740E−02
373	6.4603E−04	6.2000E−04	6.7748E−10	1.7624E−03
		Sum =	2.4112 E–08	4.3963E−01

Results:

$k(363)$	E	Sum
2.4160E−04	114.00	2.4015E−08
2.4160E−04	111.00	2.4070E−08
2.4200E−04	112.75	2.4046E−08
2.4100E−04	112.75	2.4112E−08

SOLVER appears to have converged with respect to both $k(363)$ and E

APPENDIX 6-C ANALYSIS OF MICHAELIS–MENTEN RATE EQUATION VIA LINEWEAVER–BURKE PLOT BASIC CALCULATIONS

$-r_F$	[F]	$1/-r_F$	1/[F]	$-r_F$(M-M)	Resid.1(MM)	[x−x(bar)]^2	[y−y(hat)]^2
9.46	1000	0.10570825	0.001	8.72943128	0.73056872	0.000263259	7.82651E−05
7.95	500	0.12578616	0.002	8.35352101	−0.403521	0.000231808	3.69198E−05
7.57	325	0.1321004	0.00307692	7.98329649	−0.4132965	0.000200175	4.677E−05
7.8	250	0.12820513	0.004	7.69112444	0.10887556	0.000174907	3.29376E−06
7.87	200	0.1270648	0.005	7.39781764	0.47218236	0.000149457	6.57753E−05
7.04	175	0.14204545	0.00571429	7.20164609	−0.1616461	0.000132502	1.01653E−05

APPENDIX 6-C (Continued) ANALYSIS OF MICHAELIS–MENTEN RATE EQUATION VIA LINEWEAVER–BURKE PLOT BASIC CALCULATIONS

7.04	160	0.14204545	0.00625	7.06121188	−0.0212119	0.000120456	1.82077E−07
6.82	140	0.14662757	0.00714286	6.83894289	−0.0189429	0.000101655	1.64948E−07
6.74	120	0.14836795	0.00833333	6.56347427	0.17652573	7.90663E−05	1.59231E−05
6.52	110	0.15337423	0.00909091	6.3994415	0.1205585	6.61676E−05	8.34865E−06
6.21	100	0.1610306	0.01	6.21310966	−0.0031097	5.22043E−05	6.49569E−09
5.86	90	0.17064846	0.01111111	5.99960003	−0.1396	3.73828E−05	1.57664E−05
5.79	80	0.17271157	0.0125	5.75249874	0.03750126	2.23281E−05	1.26771E−06
5.37	70	0.18621974	0.01428571	5.46320144	−0.0932014	8.64091E−06	1.00926E−05
5.14	60	0.19455253	0.01666667	5.11989078	0.02010922	3.12023E−07	5.83906E−07
4.73	50	0.21141649	0.02	4.70588235	0.02411765	7.6992E−06	1.17399E−06
4.12	40	0.24271845	0.025	4.19683139	−0.0768314	6.04466E−05	1.97442E−05
3.48	30	0.28735632	0.03333333	3.55576627	−0.0757663	0.00025947	3.7491E−05
2.77	20	0.36101083	0.05	2.72368242	0.04631758	0.001074184	3.76894E−05
1.6	10	0.625	0.1	1.60025604	−0.000256	0.006851658	1E−08

Sum = 0.34450513
x(bar) = 0.01722526

S_{xx} = 0.00989378

SS_E = 0.000389634
MS_E = 2.16463E–05

V(S) = 0.002187871
s(S) = 0.046774682
V(I) = 1.73148E–06
s(I) = 0.001315856

Results from use of "Data Analysis" tool in EXCEL

Summary output

Regression statistics	
Multiple R	0.99925983
R square	0.99852022
Adjusted R square	0.99843801
Standard error	0.00465256
Observations	20

ANOVA

	df	SS	MS	F	Significance F
Regression	1	0.26291464	0.26291464	12145.9381	6.3146E−27
Residual	18	0.00038963	2.1646E−05		
Total	19	0.26330427			

	Coefficients	Standard error	t Stat	P-value	Lower 95%	Upper 95%
Intercept	0.10940382	0.00131586	83.1427031	9.9824E−25	0.1066393	0.112168328
X variable 1	5.1549713	0.04677467	110.208612	6.3146E−27	5.05670129	5.253241308

Chapter 7

Multiple Reactions

LEARNING OBJECTIVES

After completing this chapter, you should be able to

1. define selectivity and yield, and explain how they are related to conversion;
2. classify *systems* of reactions;
3. qualitatively analyze various design options for systems in which more than one reaction takes place;
4. calculate the complete composition in a batch reactor as a function of time for a system in which more than one reaction takes place;
5. calculate the complete composition in the effluent from a CSTR or PFR as a function of space time for a system in which more than one reaction takes place;
6. calculate the complete composition of a system in which multiple reactions take place, given the concentration of one component.

7.1 INTRODUCTION

In Chapters 2–6, we considered systems where only one stoichiometrically simple reaction took place. However, in most real situations, several reactions take place simultaneously. For example, light olefins such as ethylene and propylene are produced in steam crackers like the one shown in Figure 7-1.

The feed to a steam cracker is a mixture of steam with either LPG (liquefied petroleum gas; a mixture of primarily ethane, propane, and butane) or a heavier petroleum fraction, such as naphtha or gas oil. A large number of individual reactions take place. The feed hydrocarbons are "cracked" to smaller molecules and are dehydrogenated to produce olefins. The reactor operates at about 850 °C and the residence time typically is less than 1 s. Steam cracking is a homogeneous (non-catalytic) reaction. Steam crackers produce monomers that form the building blocks for important, high volume polymers such as polyethylene and polypropylene.

In industrial processes, reactions usually can be categorized as "desirable" or "undesirable." As trivial as these two categories might seem, valuable perspective can be obtained by assigning one of these labels to each known reaction. A reaction that leads to the formation of the intended product certainly would be "desirable." "Undesirable" reactions usually result in the formation of low-value by-products from either the reactant(s) or the desired product(s).

As an example, consider the partial oxidation of methanol (CH_3OH) to formaldehyde (CH_2O) using air

$$CH_3OH + \tfrac{1}{2}O_2 \rightarrow CH_2O + H_2O \tag{7-A}$$

Figure 7-1 Steam cracker at ExxonMobil's Singapore Refinery. This reactor is capable of producing almost 1 million tons per year of light olefins such as ethylene and propylene (Photo, ExxonMobil 2004 Summary Annual Report.)

Formaldehyde is an important chemical intermediate. It ranks approximately 25th in annual production among all chemicals. Most formaldehyde is used in plastics, including polyacetals, phenolic resins, urea resins, and melamine resins.

Two commercial processes have been built around Reaction (7-A). The first is based on an elemental silver catalyst. This process is operated at about 650 °C, atmospheric pressure, and with a large stoichiometric excess of methanol. Formaldehyde is formed by the dehydrogenation of methanol

$$CH_3OH \rightleftarrows CH_2O + H_2 \tag{7-B}$$

This reaction is "desirable" since it leads to the target product, formaldehyde. A valuable by-product, H_2, also is formed.

Reaction (7-B) is endothermic and highly reversible. The *equilibrium* fractional conversion of methanol to formaldehyde is substantially less than 1, even at 650 °C. However, a second reaction takes place over the silver catalyst

$$H_2 + \tfrac{1}{2}O_2 \rightarrow H_2O \tag{7-C}$$

We might be tempted to classify Reaction (7-C) as "undesirable" because it results in the conversion of valuable H_2 to H_2O. However, Reaction (7-C) serves two useful purposes. First, it drives the equilibrium of Reaction (7-B) to the right, increasing the amount of CH_2O that is produced. Second, it is exothermic, and it provides most of the heat that is required by Reaction (7-B).

If the silver catalyst did not catalyze Reaction (7-C), or if we attempted to operate without O_2, the conversion of CH_3OH would be much lower, and heat would have to be added to the reactor to maintain the necessary high temperature. As we shall see in the next chapter, heating or cooling complicates the mechanical design of a reactor. For these reasons, Reaction (7-C) is "desirable," in the context of formaldehyde production.

The second commercial formaldehyde process is based on an iron molybdate catalyst. This process operates at about 450 °C, atmospheric pressure, and with a large stoichiometric excess of oxygen. Reaction (7-A) is the principal source of CH_2O. The reaction temperature is so low that Reaction (7-B) is unimportant.

A number of undesirable reactions do occur in both commercial processes. An effective oxidation catalyst is present in both cases, the temperature is high, and O_2 (air) is present. Both CH_2O and CH_3OH are reactive, and we would expect them to oxidize to some extent via the reactions

$$CH_2O + \tfrac{1}{2}O_2 \rightarrow CO + H_2O \tag{7-D}$$

$$CH_2O + O_2 \rightarrow CO_2 + H_2O \tag{7-E}$$

$$CH_3OH + O_2 \rightarrow CO + 2H_2O \tag{7-F}$$

$$CH_3OH + \tfrac{3}{2}O_2 \rightarrow CO_2 + 2H_2O \tag{7-G}$$

These reactions clearly are "undesirable." Reactions (7-D) and (7-E) cause the product, CH_2O, to be degraded into other compounds, CO, CO_2, and H_2O, that have very little value. Reactions (7-F) and (7-G) cause the reactant, CH_3OH, to be degraded into the same set of low-value products.

Small amounts of carbon monoxide and/or carbon dioxide are formed in both commercial processes. The catalyst, the reactor, and the reactor operating conditions must be designed to minimize the formation of CO and CO_2.

As an aside, we might ask why both of the commercial formaldehyde processes operate with feeds where the ratio of air to methanol is far removed from the stoichiometric ratio. The answer is safety, specifically the need to avoid the possibility of an explosion. In both processes, the feed compositions are outside the flammability limits of methanol/air mixtures, so that an explosion is not possible, even if an ignition source is present. In the "silver catalyst" process, the methanol/air ratio is above the upper flammability limit. In the "iron molybdate" process, the methanol/air ratio is below the lower flammability limit.

7.2 CONVERSION, SELECTIVITY, AND YIELD

In previous chapters, we used the fractional conversion of a reactant to measure the progress of a single reaction. Reactant conversion still is a valid concept, even when more than one reaction takes place. However, the conversion of a single reactant is not sufficient to describe the progress of more than one reaction, and is not sufficient to define the complete composition of a system in which more than one reaction takes place. In fact, the concept of fractional conversion must be applied with some care when more than one reaction occurs, as illustrated later in this chapter.

The ability to convert a reactant into the *desired* product, without the formation of undesired products, is measured by the *selectivity*. Selectivity is defined as

Definition of overall selectivity

$$\begin{aligned}&\text{selectivity to product D with}\\&\text{respect to reactant A} = S(\text{D/A}) = (-\nu_\text{A})\\&\times \text{moles of D formed}/\nu_\text{D} \times \text{moles of A reacted}\end{aligned} \tag{7-1}$$

In Eqn. (7-1), ν_D and ν_A are the stoichiometric coefficients of D and A, respectively, in the balanced chemical equation for the reaction that leads to D. With this definition, the selectivity S(D/U) will be 1.0 (100%) when *all* of the A that reacts is converted into the desired product, D. If there are reactions that result in the conversion of A into compounds other than D, or reactions that result in the conversion of D into other products, then S(D/A) will be less than 1.0 (100%). Note that both the product (D) and the reactant (A) are specified in the definition of selectivity.

The definition of selectivity given by Eqn. (7-1) can be applied to a complete reactor by looking at the inlet and outlet streams if the reactor operates continuously or by looking at the initial and final compositions of a batch reactor. The selectivity for the reactor as a whole is called the *overall* selectivity.

There is a second type of selectivity, the *point* or *instantaneous* selectivity. The instantaneous selectivity describes the selectivity in a batch reactor at one particular time, and the point selectivity describes the selectivity at one point in a continuous reactor operating at steady state. The point (instantaneous) selectivity is defined by

Definition of point selectivity

$$\begin{aligned}&\text{point (instantaneous) selectivity to product D}\\&\text{with respect to reactant A} = s(\text{D/A}) = (-\nu_\text{A})\\&\times \text{rate of D formation}/\nu_\text{D} \times \text{rate of A consumption}\end{aligned} \tag{7-2}$$

The rates of the reactions that take place will vary with time in a batch reactor because the composition, and perhaps the temperature, will vary with time. Therefore, the *instantaneous* selectivity for a batch reactor will not necessarily be the same as the *overall* selectivity. Similarly, for a PFR, the composition and temperature will vary from point to point so that the *overall* selectivity will not necessarily be the same as the *instantaneous* selectivity. However, for a CSTR, the overall and point selectivities *are* the same, since the composition and temperature do not vary from point to point in a CSTR.

Suppose that "N" products can be formed, directly or indirectly, from reactant A. The sum of the selectivities for all products must be unity, i.e.,

$$\sum_{i=1}^{N} S(i/\text{A}) = 1; \quad \sum_{i=1}^{N} s(i/\text{A}) = 1 \tag{7-3}$$

There is a third parameter, "yield," that frequently is used to describe the behavior of systems in which more than one reaction takes place. The overall yield is defined as

Definition of overall yield

$$\begin{aligned}&\text{yield of D with respect to A} = Y(\text{D/A}) = (-\nu_\text{A})\\&\times \text{moles of D formed}/\nu_\text{D} \times \text{moles of A fed}\end{aligned} \tag{7-4}$$

The definitions of yield and selectivity are similar. However, selectivity is based on the amount of reactant *actually consumed*, whereas yield is based on the amount of reactant that is *fed*. These two definitions are not always used consistently in the literature. Always be sure how "yield" and "selectivity" are defined when working with a new source, such as an article from the literature.

The sum of all of the yields of all products will *not* be equal to unity as long as some A remains unconverted.

Yield and selectivity are related through the fractional conversion. In words,

$$\begin{aligned}&\text{(moles of D formed/moles of A fed)}\\&= \text{(moles of A reacted/moles of A fed)} \times \text{(moles of D formed/moles of A reacted)}\end{aligned}$$

Relationship between selectivity and yield

$$Y(\mathrm{D/A}) = x_{\mathrm{A}} \times S(\mathrm{D/A}) \tag{7-5}$$

In Eqn. (7-5), x_A is the fractional conversion of reactant A.

EXAMPLE 7-1
Partial Oxidation of Methanol to Formaldehyde

The partial oxidation of methanol to formaldehyde takes place in a batch reactor. The reactor initially contains 100 moles of CH_3OH, 28 moles of O_2, and 140 moles of N_2. Some time later, at time $= t$, the reactor contains 44 moles of CH_2O, 6 moles of CO, and 140 moles of N_2. A significant quantity of water (H_2O) also is present. However, there are no other chemical species present in measurable quantities.

Questions

A. How many moles of CH_3OH, O_2, and H_2O are present at time $= t$?

B. What is the selectivity of CH_2O based on CH_3OH?

C. What is the yield of CH_2O based on CH_3OH?

D. What is the fractional conversion of CH_3OH?

E. What is the selectivity of H_2O based on O_2?

F. What is the fractional conversion of CH_2O?

Part A: How many moles of CH_3OH, O_2, and H_2O are present at time = t?

APPROACH

To determine the number of moles of CH_3OH, O_2, and H_2O at time t, we must carry out some stoichiometric calculations using the techniques of Chapter 1. We will use the "extent of reaction" to describe the progress of each of the reactions taking place. However, the problem statement does not tell us which reactions to consider. In fact, we are not even told how *many* reactions to consider.

To begin, let's ask the question: What is the minimum number of variables that are required to completely describe the composition of a system in which R independent reactions take place? In Chapter 4, we saw that a single variable, e.g., fractional conversion, extent of reaction, or the concentration of one compound, is sufficient to describe the composition of a system, when only *one* reaction takes place. How many variables are required if R reactions occur?

To answer this question, we turn to Eqn. (1-9) in Chapter 1.

$$\Delta N_i = N_i - N_{i0} = \sum_{k=1}^{R} \nu_{ki}\xi_k \tag{1-9}$$

If the initial composition of the system is specified, i.e., if all of the N_{i0}'s are known, then all of the N_i's can be calculated, provided that the "extent" of every independent reaction is known, i.e., if all of the ξ_k's are known. *To completely define the composition of the system, one variable is required for each independent reaction.*

The next question is: How many stoichiometrically *independent* reactions take place? Suppose that a system contains "E" elements. In this example, $E = 4$ (C, H, O, N). Further, suppose that the system contains "S" chemical species *that are present in sufficient concentration to affect the stoichiometry*. Based on the remarks in Chapter 5, it is clear that active centers should *not* be considered in determining the value of S. For this example, $S = 6$ (CH_3OH, CH_2O, CO, H_2O, O_2, and N_2).

The rule for determining the number of independent reactions, R, is

$$R = S - E \qquad (7\text{-}6)$$

In this example, $R = 6 - 4 = 2$.

We can choose any two reactions as long as these reactions contain all of the species that are known to be present, and do not contain any species that are not present in a significant quantity. Moreover, one reaction cannot be a multiple or a linear combination of the others.

Let's choose Reactions (7-A) and (7-D) to describe the stoichiometry of the system of this example.

$$CH_3OH + \tfrac{1}{2}O_2 \rightarrow CH_2O + H_2O \qquad (7\text{-A})$$

$$CH_2O + \tfrac{1}{2}O_2 \rightarrow CO + H_2O \qquad (7\text{-D})$$

According to the problem statement, no CO_2 is present at time t. Therefore, reactions leading to CO_2, such as (7-E) and (7-G), should not be considered. Moreover, it is not necessary to include N_2 in the reactions because N_2 is an inert species in this example.

SOLUTION The extent of the first reaction (7-A) will be designated ξ_1 and the extent of the second reaction (7-D) will be designated ξ_2. Using these two variables, we construct the following stoichiometric table.

Species	Initial moles	Moles at time = t
CH_3OH	100	$100 - \xi_1$
O_2	28	$28 - 0.5\xi_1 - 0.5\xi_2$
N_2	140	140
CH_2O	0	$\xi_1 - \xi_2(=44)$
CO	0	$\xi_2(=6)$
H_2O	0	$\xi_1 + \xi_2$
Total	268	$268 + 0.5\xi_1 + 0.5\xi_2$

The number of moles of CH_2O and CO are known at time t, so that

$$\xi_2 = 6$$
$$\xi_1 - \xi_2 = 44$$
$$\xi_1 = 50$$

Therefore, at time t,

$$\text{moles } CH_3OH = 100 - \xi_1 = 50$$

$$\text{moles } O_2 = 28 - 0.5\xi_1 - 0.5\xi_2 = 0$$

$$\text{moles } H_2O = \xi_1 + \xi_2 = 56$$

Part B: What is the selectivity of CH_2O based on CH_3OH?

APPROACH The values of ξ_1 and ξ_2 calculated in Part A can be used in Eqn. (7-1) to calculate $S(CH_2O/CH_3OH)$.

SOLUTION $S(CH_2O/CH_3OH) = -(-1) \times 44/(1) \times (100 - 50) = 0.88$

Part C: What is the yield of CH_2O based on CH_3OH?

APPROACH The values of ξ_1 and ξ_2 calculated in Part A can be used in Eqn. (7-2) to calculate $Y(CH_2O/CH_3OH)$.

SOLUTION $Y(CH_2O/CH_3OH) = -(-1) \times 44/(1) \times 100 = 0.44$

Part D: What is the fractional conversion of CH_3OH?

APPROACH The value of ξ_1 calculated in Part A can be used to calculate the fractional conversion of methanol.

SOLUTION $x_{CH_3OH} = (100 - \xi_1)/100 = 0.50$. Therefore, according to Eqn. (7-5), $Y(CH_2O/CH_3OH) = x_{CH_3OH} \times S(CH_2O/CH_3OH) = 0.50 \times 0.88 = 0.44$. This checks the result for $Y(CH_2O/CH_3OH)$ obtained in Part C above.

Part E: What is the selectivity of H_2O based on O_2?

APPROACH The values of ξ_1 and ξ_2 calculated in Part A can be used in Eqn. (7-1) to calculate $S(H_2O/O_2)$. Oxygen is consumed, and H_2O is formed in both Reactions (7-A) and (7-D). However, their stoichiometric coefficients are the same in both reactions.

SOLUTION $S(H_2O/O_2) = -(-1/2) \times (\xi_1 + \xi_2)/(1) \times [(\xi_1/2) + (\xi_2/2)] = (56/2)/(56/2) = 1.0$. If the stoichiometric coefficients in the two reactions had been different, it would have been necessary to specify which reaction was used for the calculation.[1]

Part F: What is the fractional conversion of CH_2O?

APPROACH This question is difficult to understand. The conversion of a *reactant* is defined as

$$x = (\text{initial moles} - \text{moles at time } t)/(\text{initial moles})$$

Using this definition, the conversion of CH_2O is infinite and negative, since the initial number of moles is 0 and the number of moles at time t is finite. If some formaldehyde were present initially, the conversion would not be infinite. However, as long as there was a *net production* of CH_2O, the conversion would be negative, another counterintuitive result.

The problem is that CH_2O is not a reactant. It is an *intermediate product*, formed in Reaction (7-A), but consumed in Reaction (7-D). In order to preserve the characteristic of fractional conversion that we learned in previous chapters, i.e., $0 \leq x \leq 1$, the concept of conversion should not be applied to intermediate products, such as CH_2O in this example.

EXERCISE 7-1

We could have selected two reactions other than (7-A) and (7-D) to describe the stoichiometry of the system in this example. Suppose we had chosen

$$CH_3OH + O_2 \rightarrow CO + 2H_2O \quad (7\text{-H})$$

$$CO + CH_3OH \rightarrow 2CH_2O \quad (7\text{-I})$$

Show that the numerical answers to Parts A, B, C, D, and E do not change if the calculations are based on these reactions instead of Reactions (7-A) and (7-D).

---

[1] In very complex cases, e.g. where a product is formed via many different reactions, selectivity and yield are best defined based on an element, rather than a compound. For example, in the formation of ethylene (C_2H_4) by steam cracking of a complex mixture of hydrocarbons, the *carbon selectivity* would be defined as (moles C_2H_4 formed) $\times$ 2/(moles C in all reaction products).

7.3 CLASSIFICATION OF REACTIONS

Systems of multiple reactions usually fall into one of four categories. These classifications are parallel, independent, series, and mixed series/parallel. Recognizing the structure of a multiple-reaction system frequently helps to identify the best approach to design or analysis.

7.3.1 Parallel Reactions

A parallel reaction network can be represented schematically as

$$A \rightarrow B, \quad A \rightarrow C, \quad A \rightarrow D$$

In this example, three reactions, *originating from a common reactant, A*, are taking place.

Any number of reactions may occur in parallel. Moreover, it is not necessary that there be only one reactant. For example,

$$CO + \begin{cases} 2H_2 \rightarrow CH_3OH \\ 3H_2 \rightarrow CH_4 + H_2O \end{cases}$$

The top reaction, between carbon monoxide and hydrogen to form methanol, is the basis for all commercial methanol synthesis plants, and is "desirable." This reaction is carried out using a heterogeneous catalyst containing copper and zinc oxide, and is quite reversible at commercial reaction conditions. The bottom, "undesirable" reaction is referred to as "methanation." It is relatively slow with today's methanol synthesis processes and catalysts. However, methanation can be important if the catalyst becomes contaminated with elements such as nickel and iron, which catalyze the methanation reaction.

Another important example of parallel reactions is the selective catalytic oxidation of CO in the presence of H_2:

$$1/2O_2 + \begin{cases} CO \rightarrow CO_2 \\ H_2 \rightarrow H_2O \end{cases}$$

The top reaction is "desirable" since the objective of this process is to oxidize a relatively low concentration of CO in the presence of a high concentration of H_2, with little or no consumption of H_2. Therefore, the bottom reaction is "undesirable."

Designing a catalyst for the selective oxidation of CO in the presence of high concentrations of H_2 is a major scientific challenge. Nevertheless, the selective catalytic oxidation of CO has been used to increase the production rate of existing ammonia synthesis plants. This reaction network also is a critical element in the developing technology for H_2-powered fuel cells. In this context, the process is referred to as PROX (preferential oxidation).

7.3.2 Independent Reactions

An independent reaction network can be represented as

$$A \rightarrow P_1 + P_2 + \cdots$$

$$B \rightarrow P_I + P_{II} + \cdots$$

The symbol P_i represents a reaction product.

The independent reaction network is similar to the parallel network, with one important difference. The reactants are not the same. Reactants A and B above are different compounds. There can be more than one reactant in each reaction. However, there are no common reactants in the two reactions. In general, the products of the two reactions will be different. However, this is not a necessary condition.

An important example of independent reactions occurs in the fluid catalytic cracking (FCC) units that are found in every major petroleum refinery. The function of these reactors is to reduce the average molecular weight of a heavy petroleum fraction, producing a "lighter" product that has a higher value. In essence, high-molecular-weight hydrocarbons are "cracked" into smaller fragments using a zeolite catalyst. Since the typical feed to an FCC unit contains literally thousands of different molecules, there are literally thousands of independent reactions taking place in an FCC reactor.

7.3.3 Series (Consecutive) Reactions

The common representation for this category of reactions is

$$A \longrightarrow R \longrightarrow S$$

Series reaction networks are very important commercially. In many cases, the intermediate product, R, is desired and the terminal product, S, is undesired.

Series reaction networks are not limited to two reactions. Butadiene (C_4H_6) is an important monomer that goes into a large number of elastomeric products, including automobile tires. Butadiene can be produced by the catalytic dehydrogenation of butane (C_4H_{10}) as shown below.

$$C_4H_{10} \xrightarrow{-H_2} C_4H_8 \xrightarrow{-H_2} C_4H_6 \xrightarrow{-H_2} \text{“Coke”}$$

Butane is first dehydrogenated to butene (C_4H_8), which is further dehydrogenated to the desired product, butadiene. However, the reactions do not stop there. Butadiene can dehydrogenate further to a carbonaceous material, referred to as "coke", which collects on the heterogeneous catalyst, causing it to lose activity (i.e., "deactivate"). The coke eventually must be burned off the catalyst. Processes based on this chemistry once were an important source of butadiene. They largely have been supplanted by butadiene that is produced as a by-product in ethylene plants (i.e., steam crackers such as the one shown in Figure 7-1) and in fluid catalytic crackers (FCC units) in refineries.

Another example of a series reaction occurs in methanol synthesis, which we discussed in Section 7.3.1. Once CH_3OH is formed, it can react further to dimethyl ether (CH_3OCH_3),

$$CO + 2H_2 \rightarrow CH_3OH \xrightarrow{CH_3OH} CH_3OCH_3 + H_2O$$

The second reaction, the combination of two methanol molecules to form dimethyl ether plus water, actually is used for the commercial production of dimethyl ether. However, this reaction is undesirable when it occurs in a conventional methanol plant, as the presence of dimethyl ether and water complicates the design of the separation system.

7.3.4 Mixed Series and Parallel Reactions

This category of multiple reactions is represented as

$$A + B \longrightarrow R$$
$$R + B \longrightarrow S$$

If we focus on reactant A, the reactions appear to be in series,

$$A \longrightarrow R \longrightarrow S$$

However, if reactant B is the focus, the reactions appear to be occurring in parallel,

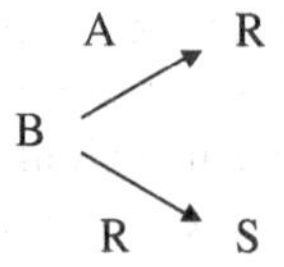

$$B \begin{cases} \xrightarrow{A} R \\ \xrightarrow{R} S \end{cases}$$

There are many important industrial examples of mixed series/parallel reaction networks. The partial oxidation of methanol to formaldehyde, as discussed above, is one. We can see this from Reactions (7-A) and (7-D):

$$CH_3OH + \tfrac{1}{2}O_2 \rightarrow CH_2O + H_2O \tag{7-A}$$

$$CH_2O + \tfrac{1}{2}O_2 \rightarrow CO + H_2O \tag{7-D}$$

Most partial oxidation reactions display the same kind of mixed series/parallel structure.

In addition, a large number of chlorinations, sulfonations, alkylations, and nitrations follow mixed series/parallel networks. Consider the nitration of benzene,

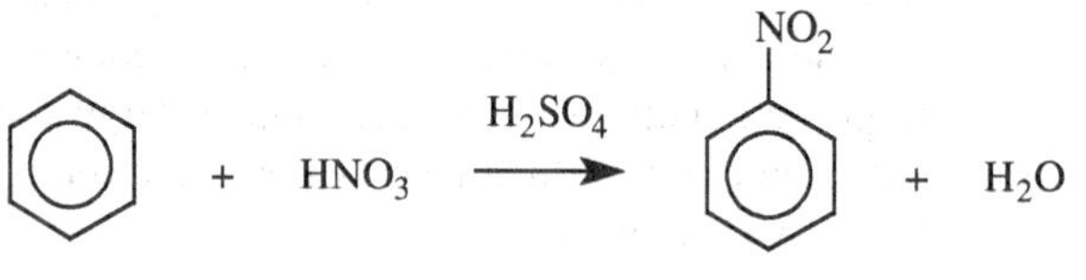

$$C_6H_6 + HNO_3 \xrightarrow{H_2SO_4} C_6H_5NO_2 + H_2O \tag{7-J}$$

$$C_6H_5NO_2 + HNO_3 \xrightarrow{H_2SO_4} m\text{-}C_6H_4(NO_2)_2 + H_2O \tag{7-K}$$

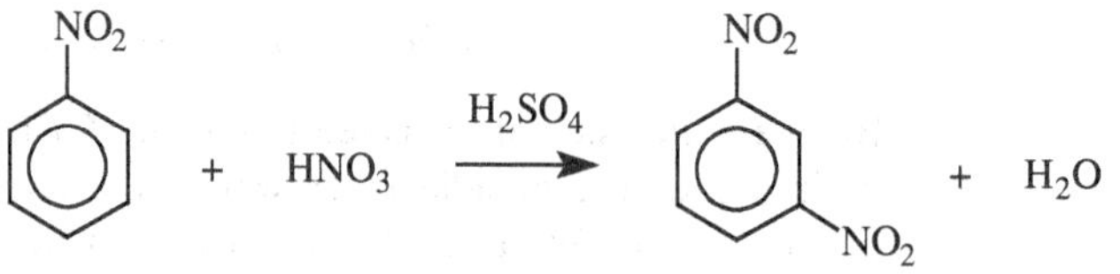

This reaction is carried out at about 50 °C. There are two phases in the reactor, an organic phase (benzene and the various nitrated benzenes) and an aqueous phase (originally a mixture of sulfuric and nitric acid, which becomes diluted by the water that is formed as the reactions proceed). Mononitrobenzene is the desired product; *m*-dinitrobenzene is an undesired by-product.

Chain-growth polymerization is perhaps the most important and complex example of a mixed series/parallel reaction. For example, polystyrene is formed by adding one styrene molecule at a time to the end of a growing polymer chain.

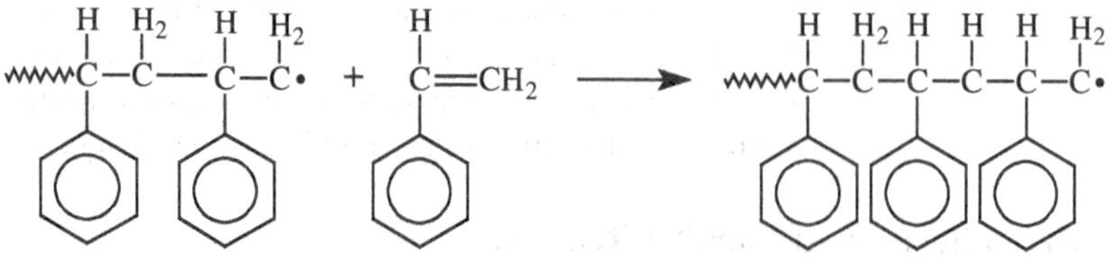

$$\sim\!\sim CH(C_6H_5)-CH_2-CH(C_6H_5)-CH_2\bullet + CH(C_6H_5)=CH_2 \longrightarrow \sim\!\sim CH(C_6H_5)-CH_2-CH(C_6H_5)-CH_2-CH(C_6H_5)-CH_2\bullet$$

A molecule of polystyrene may contain several thousand styrene molecules, indicating that several thousands reactions, such as that shown above, took place. The reaction is series from the standpoint of the polymer radical, but parallel from the standpoint of the styrene molecule.

Actually, chain-growth polymerization involves other reactions. The reaction shown above is a *propagation reaction*, as defined in Chapter 5. This reaction is accompanied by an *initiation* reaction, which provides a source of free radicals, and by a *termination* reaction, which consumes free radicals. For polystyrene polymerization, the termination reaction is the combination of two "live" polymer radicals, such as those shown above, to form a molecule of "dead" polymer.

7.4 REACTOR DESIGN AND ANALYSIS

7.4.1 Overview

Suppose that we are asked to design or analyze a reactor system in which multiple reactions are taking place. In general, our work would have two simultaneous objectives:

1. To produce the desired product at the specified production rate using the smallest reactor possible (or the smallest amount of catalyst possible), i.e., to maximize the reaction rate.
2. To minimize the formation of low-value by-products, i.e., to maximize the reaction selectivity.

The second objective makes the problem much more difficult than the single-reaction problems that we solved earlier, in Chapter 4. Usually, it is not possible to minimize the reactor volume *and* maximize the reaction selectivity *simultaneously*. Frequently, there is a trade-off between rate and selectivity. Ultimately, this trade-off requires an economic analysis. However, the final analysis often favors selectivity over rate because there is a large and *continuing* cost penalty associated with converting a valuable raw material into a low-value by-product.

There are several important questions that must be considered in designing/analyzing multiple-reactor systems:

1. What kind of reactor or system of reactors should be used?
   Backmixing usually lowers the overall reaction rate. Does it help or hurt selectivity? Is there any reason to consider reactors in series?
2. What feed concentration should be used?
   Should the feed be concentrated or dilute? Does the same answer apply to all of the feed components? Should one or more reactants be added as the reaction proceeds, as opposed to adding all of the reactants initially?
3. What temperature should be used?
   Should the temperature be changed as the reaction proceeds? How?

The answers to these questions will affect both the reaction rate and selectivity. There are interesting trade-offs involved, as we shall see shortly.

When we design or analyze a reactor with more than one reaction taking place, we need to calculate the concentrations of *all* species in the system. In a batch reactor, these concentrations must be calculated as functions of time. In a continuous reactor, the concentrations must be calculated as functions of the space time τ.

In order to perform these calculations, we need more tools than were required for a single reaction. In general, for multiple reactions

(a) we must have a rate equation for *each* independent reaction;

(b) we must write a component material balance (design equation) for each independent reaction;

(c) we must have an independent composition variable for each independent reaction.

As we shall see, in carrying out calculations for multiple-reaction systems, careful selection of the composition variables and the component material balances can decrease the difficulty of solving the problem.

Two types of problem arise in multiple-reaction systems. The first is analogous to the kind of problem associated with single reactions. In this category are questions such as: Given a system of reactions with known kinetics, what reaction time (or space time) will be required to obtain a specified concentration of reactant or product, and what concentrations of the other species will exist at this time? Another example is the converse of this question, i.e., what reactant and product concentrations will result for a specified reaction time (or space time)?

The second type of problem does not involve reaction time or space time. The question here is given a system of reactions with known kinetics and given the concentration of one component at some (unspecified) time (or space time), what are the concentrations of the other species at that time (or space time)? This type of problem is referred to as a "time-independent" problem. Time-independent problems can be solved by forming the ratio of various reaction rates to eliminate time (or space time) as an explicit variable. However, the solution to such problems provides no information about the time or reactor volume required to obtain a given composition.

The solution of various types of multiple-reaction problems is illustrated below.

7.4.2 Series (Consecutive) Reactions

7.4.2.1 Qualitative Analysis

The simplest and most-studied example of series reactions is

$$\mathrm{A} \xrightarrow{k_1} \mathrm{R} \xrightarrow{k_2} \mathrm{S}$$

where both reactions are first-order and irreversible. The rate constants for these reactions are k_1 and k_2, respectively. We dealt with this reaction network in Chapter 5, during the discussion of the steady-state approximation. In an ideal, constant-volume, isothermal batch reactor with $C_A = C_{A0}$ and $C_R = C_S = 0$ at $t = 0$, we showed that

$$C_A = C_{A0}e^{-k_1 t}$$
$$C_R = C_{A0}\left(\frac{k_1}{k_2 - k_1}\right)(e^{-k_1 t} - e^{-k_2 t}); \quad k_1 \neq k_2 \tag{7-7}$$

If both C_A and C_R are known, C_S can be calculated from stoichiometry.

EXERCISE 7-2

Show that when $k_1 = k_2$, $C_R = k_1 t C_{A0} e^{-k_1 t}$.

Figure 7-2 is a plot of the concentrations of A, R, and S versus the dimensionless time, $k_1 t$, for $k_2/k_1 = 0.5$.

Figure 7-2 illustrates a critical feature of series-reaction networks. The concentration of the intermediate product, R in this case, increases rapidly once the reaction has started. This concentration then goes through a maximum and declines. The value of the time at which R is maximum has been labeled "optimum time" or t_{opt} in the figure. If the reaction that consumes R is irreversible, C_R will approach zero at very long times.

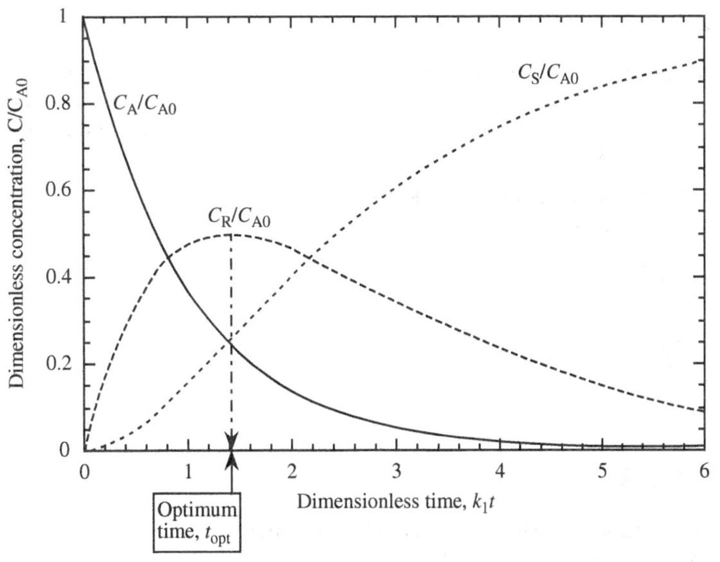

Figure 7-2 Dimensionless concentrations of A, R, and S versus dimensionless time $(k_1 t)$ for the irreversible, first-order reactions $A \xrightarrow{k_1} R \xrightarrow{k_2} S$ taking place in an ideal, isothermal, constant-volume batch reactor. The initial concentrations are $C_A = C_{A0}$, $C_R = C_S = 0$; $k_2/k_1 = 0.50$.

The behavior of C_R can be understood in terms of the rate equation for R. The *net* rate of formation of R is the difference between the rate at which R is formed from A and the rate at which R is converted to S, i.e.,

$$r_R = k_1 C_A - k_2 C_R$$

For this example, C_A is large and C_R is very small at short times. The formation term $(k_1 C_A)$ is larger than the consumption term $(k_2 C_R)$, so that $r_R > 0$. The concentration of R increases rapidly. However, as time increases, the concentration of A decreases and the concentration of R increases. At some time, the concentration of A has declined and the concentration of R has increased, to the point that the two terms in the rate equation are equal. At this point, the net rate of formation of R (r_R) is zero. This corresponds to the maximum in the C_R/C_{A0} curve in Figure 7-2, which occurs at t_{opt}. At times that are longer than t_{opt}, the second term in the rate expression is larger than the first, so that $r_R < 0$. The net rate of formation of R is negative, i.e., R is consumed. This corresponds to the portion of the C_R/C_{A0} curve to the right of t_{opt}, where the concentration of R is decreasing.

If R is the desired product, as is often the case, the exact location of the maximum in the C_R/C_{A0} curve is critical. This point corresponds to the highest possible value of the yield, $Y(R/A)$, since $Y(R/A) = C_R/C_{A0}$ when there is no R in the feed. If the reactor is operated at a time that is less than t_{opt}, the final concentration of R is lower than it could be because not enough A has been converted to R. The yield $Y(R/A)$ also is less than its maximum possible value. On the other hand, if the reactor is operated for a time that is longer than t_{opt}, then the final concentration of R again is less than it could be, this time because too much R has been converted to S. The yield $Y(R/A)$ again is less than its value at the optimum time.

EXERCISE 7-3

Show that the value of the optimum time t_{opt} is given by

$$t_{opt} = \ln(k_2/k_1)/(k_2 - k_1); \quad k_2 \neq k_1$$
$$t_{opt} = 1/k_1; \quad k_2 = k_1$$

Also, show that the maximum yield of R, i.e., the value of C_R/C_{A0} at t_{opt} is given by

$$\left(\frac{C_R}{C_{A0}}\right)_{max} = \left(\frac{k_1}{k_2}\right)^{[k_2/(k_2-k_1)]}; \quad k_2 \neq k_1$$
$$\left(\frac{C_R}{C_{A0}}\right)_{max} = 1/e; \quad k_2 = k_1 \tag{7-8}$$

The equations for $k_2 \neq k_1$ show that t_{opt} increases as k_2 decreases if k_1 is constant, and that $(C_R/C_{A0})_{max}$ increases as k_2 decreases if k_1 is constant. Explain these trends in physical terms.

EXERCISE 7-4

The discussion of Figure 7-2 makes no mention of the selectivity, $S(R/A)$. Derive an expression for $S(R/A)$ as a function of time. Make a plot of $S(R/A)$ versus k_1t for $k_2/k_1 = 0.50$.

EXERCISE 7-5

H. I. Pschuetter is Area Production Manager for Specialty Oxides at the Cauldron Chemical Company. "Hip" is well known for his very conservative approach to any and all new projects.

You have designed a small PFR to make a new product, whose code name is PARTOX. The reaction sequence can be represented as A → R → S, where PARTOX is species R. "Hip" has reviewed your design and has sent you the following e-mail:

Sonny Boy:

I looked over the calculations that you did to size the reactor. The R&D team did a great job (for once) in developing good rate equations. Your calculations are right on the money, except that you forgot to include a safety factor. Even though the feed rate and feed composition are fixed, you need to oversize the reactor to compensate for who knows what. Make the reactor 200 gallons instead of the 400 liters that you calculated and I'll sign off on the design.

Hip:

Compose a reply. *Hint:* A two-word response probably will not help your career.

EXERCISE 7-6

Examine the initial ($t = 0$) slopes of the curves of C_R and C_S in Figure 7-2. Suppose that you knew that products B and C were formed from reactant A, but you were not sure whether there were two parallel reactions, A → B and A → C, or whether the reactions were in series, i.e., A → B and B → C. Discuss how you might use the initial slopes to help resolve this question.

7.4.2.2 Time-Independent Analysis

The variable of time can be eliminated from the above problem by dividing the design equation for R by the design equation for A. Thus,

$$\frac{dC_R}{dt} = r_R = k_1C_A - k_2C_R \quad \text{(design equation for R)}$$
$$\frac{dC_A}{dt} = -k_1C_A \quad \text{(design equation for A)}$$
$$\frac{dC_R}{dC_A} = -1 + \left(\frac{k_2}{k_1}\right)\frac{C_R}{C_A} \tag{7-9}$$

Equation (7-9) is a linear, first-order, ordinary differential equation that can be solved by the "integrating factor" approach. The solution is

$$\frac{C_R}{C_{A0}} = \frac{1}{[1-(k_2/k_1)]}\left[\left(\frac{C_A}{C_{A0}}\right)^{(k_2/k_1)} - \frac{C_A}{C_{A0}}\right]; \quad k_2 \neq k_1$$

$$\frac{C_R}{C_{A0}} = -\left(\frac{C_A}{C_{A0}}\right)\ln(C_A/C_{A0}); \quad k_2 = k_1$$

The concentrations C_R and C_S can be calculated *as functions of* C_A with the above equations. However, none of the concentrations can be calculated as a function of time (or space time) using these equations alone. The time variable was removed when the two material balances were divided.

Figure 7-3 is a plot of C_R/C_{A0} versus C_A/C_{A0}, generated from the above equations, for several values of k_2/k_1.

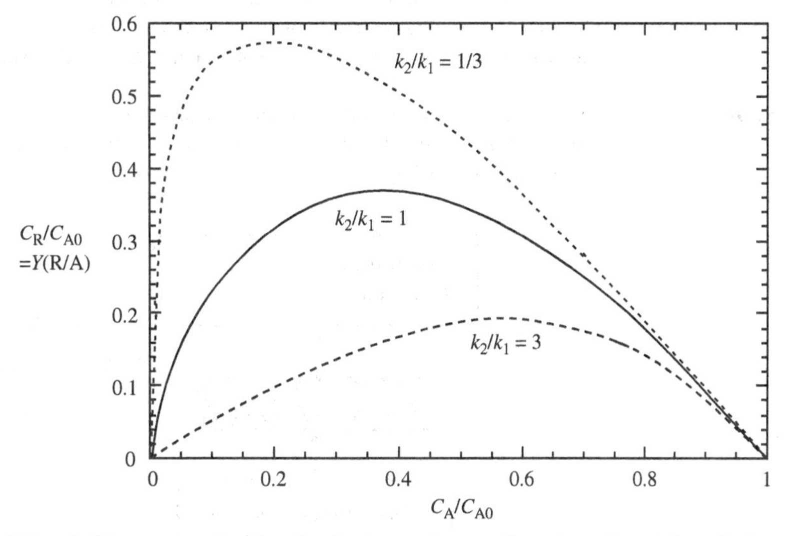

Figure 7-3 C_R/C_{A0} versus C_A/C_{A0} for the first-order reactions A → R and R → S, for several values of the rate constant ratio k_2/k_1. $C_R = C_S = 0$.

EXERCISE 7-7

You have carried out the series reactions A → R and R → S in an isothermal, constant-volume, batch reactor, starting at $t = 0$ with $C_A = C_{A0}$, and $C_R = C_S = 0$. When the fractional conversion of A was 0.50, $C_R/C_{A0} = 0.25$. Estimate the value of k_2/k_1.

7.4.2.3 Quantitative Analysis

EXAMPLE 7-2
Reactions in Series

The two, irreversible, homogeneous reactions

$$A \rightarrow B$$
$$2B \rightarrow C$$

are taking place in the liquid phase in an isothermal, ideal plug-flow reactor (PFR). The first reaction is first order in A with a rate constant (k_1) of 0.50 min^{-1}. The second reaction is second order in B with a rate constant (k_2) of 0.10 l/mol-min.

The concentration of A in the feed to the reactor, C_{A0}, is 1.0 mol/l. The concentration of B in the feed to the reactor, C_{B0}, is 0.10 mol/l. There is no C in the feed. The volume of the reactor is 400 l and the volumetric flow rate is 10 l/min so that the space time τ is 40 min.

What are the concentrations of A, B, and C in the reactor effluent?

APPROACH

There are two independent reactions, A → B and 2B → C. The complete rate equation for each reaction is given. Two material balances are required that, in general, will have to be solved simultaneously. The required material balances are nothing more than design equations for an ideal PFR, written for two different species. We can choose the two species for which to write the design equations.

For this problem, choosing A and R simplifies the mathematics a bit. As we shall see, this choice permits the two design equations to be solved sequentially, rather than simultaneously, and the design equation for A can be solved analytically.

The result of solving the two material balances is values for the concentrations of A and B at the specified space time of 40 min. The concentration of C then can be calculated from stoichiometry.

Rather than starting with the design equations, as developed in Chapter 3, we will begin this problem by writing material balances for A and B in a PFR, for a constant-density system. This approach provides a more fundamental, and perhaps safer, starting point.

SOLUTION

Let's begin with reactant A and perform a material balance over a differential element of the reactor, as shown in Figure 7-4. This is the same control volume that we used to derive the design equation for an ideal PFR in Chapter 3.

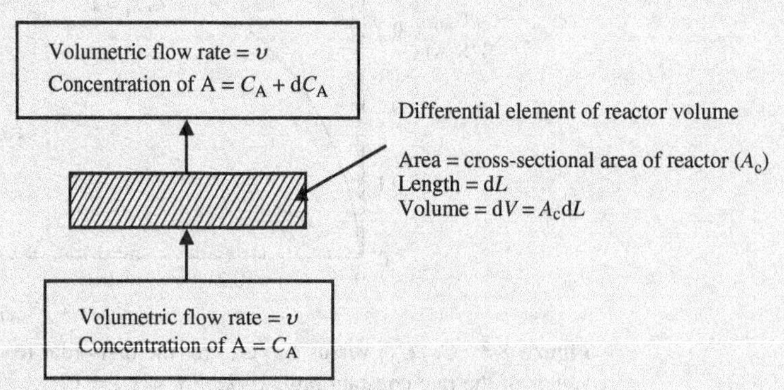

Figure 7-4 Control volume for material balances: Example 7-2.

Since constant density can be assumed for liquid-phase reactions, the volumetric flow rate into the element (υ) is the same as the volumetric flow rate out. However, the concentration of A leaving the element is different from the concentration entering by a differential amount, dC_A.

At steady state, a material balance on reactant A gives

$$\text{rate A in} - \text{rate A out} = \text{rate A disappearance}$$

$$(\upsilon C_A) - (\upsilon[C_A + dC_A]) = (-r_A\, dV)$$

$$\boxed{\frac{dV}{\upsilon} = d\tau = -\frac{dC_A}{-r_A}} \tag{7-10}$$

Equation (7-10) is the design equation for an ideal PFR, for a constant-density system, in differential form. It is the same as Eqn. (3-31).

When $-r_A = k_1 C_A$ is substituted into Eqn. (7-10) and the resulting equation is integrated from $\tau = 0$ to $\tau = \tau$, we obtain

$$C_A = C_{A0} \exp(-k_1 \tau) \tag{7-11}$$

Evaluating C_A at $\tau = 40$ min gives

$$C_A = 1.0\,(\text{mol/l}) \exp\left[(-0.5\,\text{min}^{-1}) \times 40\,(\text{min})\right] = 2.1 \times 10^{-9} (\text{mol/l})$$

This is the concentration of A in the stream that leaves the reactor. For all practical purposes, A has been completely converted.

Now, consider species B. At steady state, a material balance on B over the differential element of reactor volume shown in Figure 7-4 gives

$$\text{rate B in} - \text{rate B out} = \text{rate B disappearance}$$

$$(v C_B) - (v[C_B + dC_B]) = (-r_B\, dV)$$

$$\boxed{\frac{dV}{v} = d\tau = -\frac{dC_B}{-r_B}}$$

Again, this is just the differential form of the design equation for species B, for a constant-density system.

We must be careful to formulate $-r_B$ correctly. Species B participates in two reactions; it is formed in the first reaction at a rate of $k_1 C_A$, and it is consumed in the second reaction at a rate of $k_2 C_B^2$. Therefore, the *net* rate of *disappearance* is

$$-r_B = k_2 C_B^2 - k_1 C_A$$

Substituting this expression into the material balance on B, substituting Eqn. (7-11) for C_A, and simplifying gives the differential equation

$$\frac{dC_B}{d\tau} = k_1 C_A - k_2 C_B^2 = k_1 C_{A0} \exp(-k_1 \tau) - k_2 C_B^2 \tag{7-12}$$

The challenge now is to solve this equation. An analytical, closed-form solution to Eqn. (7-12) does not exist. However, differential equations like this can be solved numerically using programs such as Matlab, Maple, or Mathcad. Appendix 7-A shows how to solve Eqn. (7-12) using a spreadsheet, i.e., EXCEL. From Appendix 7-A, the value of C_B at $\tau = 40$ min is 0.22 mol/l.

The value of C_C can be calculated from the Law of Definite Proportions. Since the density is constant,

$$C_A - C_{A0} = \nu_{1,A}\xi_1 = (2.1 \times 10^{-9} - 1.0) = (-1)\xi_1$$

$$\xi_1 = 1.0\,\text{mol/l}$$

$$C_B - C_{B0} = \nu_{1,B}\xi_1 + \nu_{2,B}\xi_2 = (0.22 - 0.10) = 0.12\,\text{mol/l}$$

$$(1)(1) + (-2)\xi_2 = 0.12$$

$$\xi_2 = 0.44\,\text{mol/l}$$

$$C_C - C_{C0} = \nu_{2,C}\xi_2 = (1)(0.44\,\text{mol/l})$$

$$C_C = 0.44\,\text{mol/l}$$

Summarizing, there were two independent reactions in this problem: A → B and 2B → C. A rate equation was available for each reaction. To solve the problem, we used two material balances, one on species A and the other on species B. The concentrations C_A and C_B were chosen as the variables to describe the system composition. The material balances (design equations) for A and B were solved to determine C_A and C_B at $\tau = 40$ min. The concentration of C at this space time was calculated from stoichiometry.

EXERCISE 7-8

In the above solution, we performed material balances on A and B. Suppose that we had chosen to perform balances on B and C instead, and had used C_B and C_C to describe the system composition. Set up the material balances for B and C, and explain how to solve the problem when it is formulated in this way. Remember that C_B, C_C, and τ are the only variables that can be present in the material balances that will be solved; C_A may not be present.

Before leaving this example, let's look at the concentrations of A, B, and C versus τ. Values of C_A as a function of τ were calculated from Eqn. (7-11). Values of C_B were obtained from the table in Appendix 7-A, and C_C was calculated from C_A and C_B by stoichiometry. A graph of these concentrations as a function of τ is presented in Figure 7-5.

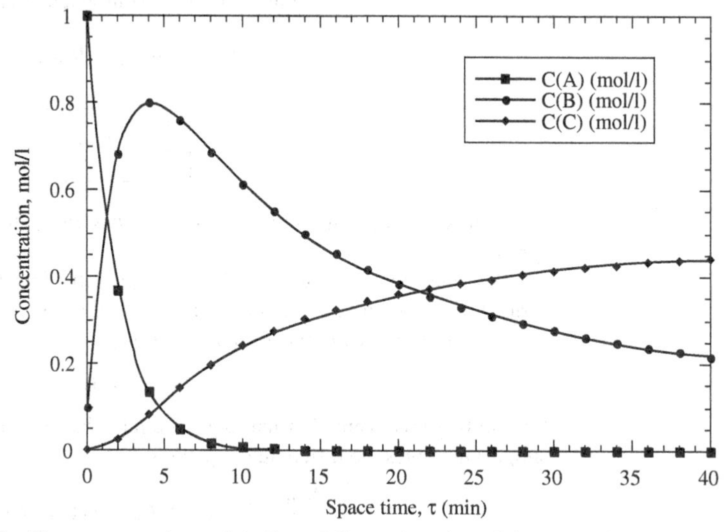

Figure 7-5 The concentrations of A, B, and C as a function of the space time τ for the reactions $A \rightarrow B$ and $2B \rightarrow C$ occurring in a liquid-phase, ideal plug-flow reactor. The reaction $A \rightarrow B$ is first order in A with a rate constant, $k_1 = 0.50\ \text{min}^{-1}$. The reaction $2B \rightarrow C$ is second order in B with a rate constant, $k_2 = 0.10$ l/mol-min. Inlet concentrations: A—1.0 mol/l; B—0.10 mol/l; C—0.

The trends in Figure 7-5 are similar to those in Figure 7-2. The concentration of A declines monotonically towards zero. The intermediate product B increases at low τ, goes through a maximum, and then declines towards zero as τ increases. The concentration of C increases monotonically towards an asymptote of $(C_{A0} + C_{B0})/2$. The fact that the second reaction is second order instead of first order does not affect the behavior of the reaction system, on a qualitative level.

7.4.2.4 Series Reactions in a CSTR

In a batch or plug-flow reactor, every fluid element spends exactly the same time in the reactor. If the series of reactions $A \rightarrow R \rightarrow S$ is occurring, it is straightforward, at least conceptually, to design a batch or plug-flow reactor that will operate at t_{opt} (or τ_{opt}), and produce a concentration of R equal to $C_{R,max}$.

What happens when these reactions are carried out in a CSTR, where mixing is intense, and not every molecule spends the same time in the reactor? In a CSTR, the feed mixes instantaneously into the contents of the reactor and the effluent is a random sample of the contents of the reactor. Some fluid elements remain in the reactor for a very short time, and some are in the reactor for a very long time. Very few are in the reactor for exactly τ_{opt}.

Let's analyze the performance of the two reactions:

$$A \xrightarrow{k_1} R \xrightarrow{k_2} S$$

in an ideal CSTR. Each reaction is first order and irreversible. Since there are two independent reactions, two material balances are required. Let's do balances on A and R, and use C_A and C_R as the composition variables.

Material Balance on A The material balance on A is just the CSTR design equation, written for reactant A. Since the total number of moles does not change as the reaction proceeds, and since a CSTR is isothermal by definition, the density of the system is constant. The constant-density form of the design equation is

$$\tau = \frac{C_{A0} - C_A}{-r_A} \tag{3-24}$$

Since $-r_A = k_1 C_A$,

$$C_A = \frac{C_{A0}}{1 + k_1 \tau} \tag{7-13}$$

Material Balance on R Using the whole reactor as the control volume,

$$\text{rate R in} - \text{rate R out} + \text{rate of R generation} = \text{rate of accumulation of R}$$

at steady state, and for the case of no R in the feed,

$$0 - \upsilon C_R + V(k_1 C_A - k_2 C_R) = 0$$

Again, we could have obtained this equation by starting with the design equation for a constant-density CSTR (Eqn. (3-24)), applying it to species B, and recognizing that $r_B = k_1 C_A - k_2 C_R$.

Substituting Eqn. (7-13) for C_A and rearranging gives

$$\frac{C_R}{C_{A0}} = \frac{k_1 \tau}{(1 + k_1 \tau)(1 + k_2 \tau)} \tag{7-14}$$

The maximum value of C_R can be found by differentiating the above equation with respect to τ and setting the derivative equal to zero.

$$\frac{dC_R}{d\tau} = 0 = \frac{k_1}{(1 + k_1 \tau)(1 + k_2 \tau)} - \frac{k_1^2 \tau}{(1 + k_2 \tau)(1 + k_1 \tau)^2} - \frac{k_1 k_2 \tau}{(1 + k_1 \tau)(1 + k_2 \tau)^2}$$

$$\tau_{\text{opt}} = \sqrt{1/k_1 k_2} \tag{7-15}$$

Substituting this relationship into Eqn. (7-14) gives

$$\frac{C_{R,\max}}{C_{A0}} = \frac{1}{\left(1 + \sqrt{k_2/k_1}\right)^2} \tag{7-16}$$

Equation (7-16) gives the maximum yield of R based on A ($Y(R/A)$) that can be obtained in a CSTR, for a specified value of k_2/k_1. Equation (7-15) gives the value of τ at which this maximum is obtained.

Figure 7-6 shows a comparison of $C_{R,\max}/C_{A0}$ for a CSTR and a PFR, as a function of the ratio k_2/k_1. The values of $C_{R,\max}/C_{A0}$ for the PFR were calculated from Eqn. (7-8).

The value of $C_{R,\max}/C_{A0}$ begins at 1 for $k_2/k_1 = 0$ and declines monotonically to 0 as the rate constant ratio increases to infinity. Only the range $0 \le k_2/k_1 \le 1$ is shown in

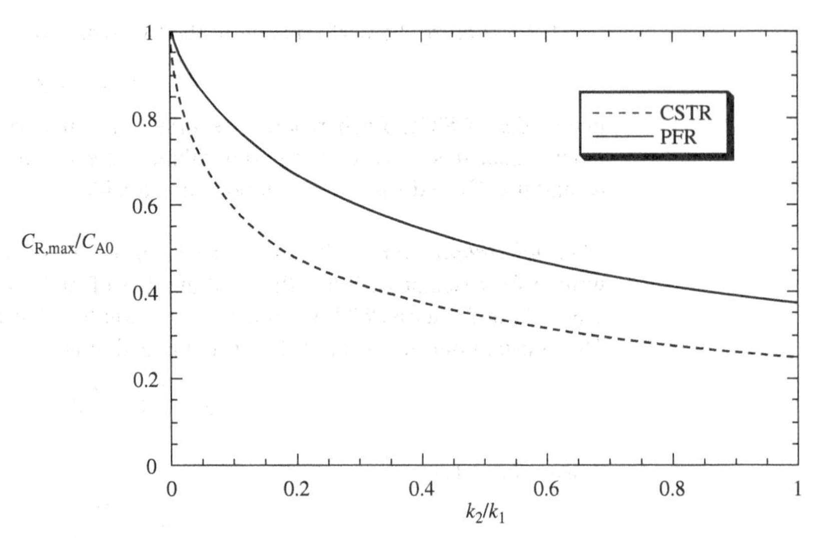

Figure 7-6 The maximum value of the concentration of the intermediate product, R, in the series reactions A → R → S, as a function of the ratio of the rate constant for the second reaction (k_2) to the rate constant for the first reaction (k_1). Both reactions are first order and irreversible. There is no R in the feed. The reactors are isothermal. Solid line—PFR; dashed line—CSTR.

Figure 7-6 because the yield of R is quite low when $k_2/k_1 > 1$. It is unlikely that an industrial reactor would be operated to produce R in the region $k_2/k_1 > 1$, unless there were special economic circumstances (e.g., S has a high value and a large market compared to R; A is very inexpensive and disposal of S is safe and inexpensive). At every k_1/k_2 ratio, the yield $Y(R/A)$ is always greater in the PFR than in the CSTR.

Figure 7-6 compares the performance of a PFR with that of a CSTR at conditions where the values of τ are different for the two reactors. In general, the maximum in C_R for a PFR occurs at a different value of τ than the maximum in C_R for a CSTR.

Figure 7-7 is a plot of $k_1\tau_{opt}$ versus k_1/k_2 for the two types of reactor. When $k_1/k_2 = 1$, the values of $k_1\tau_{opt}$ for the PFR and CSTR are the same. Otherwise, $k_1\tau_{opt}$ for the CSTR is greater than $k_1\tau_{opt}$ for the PFR. In other words, $C_{R,max}$ is lower for a CSTR than for a PFR, and it takes a longer space time to reach this lower value!

This discussion shows that the PFR is a better choice than a CSTR for series reactions where the desired product is an intermediate, such as R in this example, and where it is important to maximize the yield of the intermediate.

7.4.3 Parallel and Independent Reactions

7.4.3.1 Qualitative Analysis

The rate equations hold the key to sound design and operation for all multiple-reaction systems. To illustrate, let's consider the parallel system

$$A + B \begin{cases} \nearrow D \text{ (desired)} \\ \searrow U \text{ (undesired)} \end{cases}$$

Suppose that the rate equation for the desired reaction is

$$r_D = k_D C_A^{\alpha_D} C_B^{\beta_D}$$

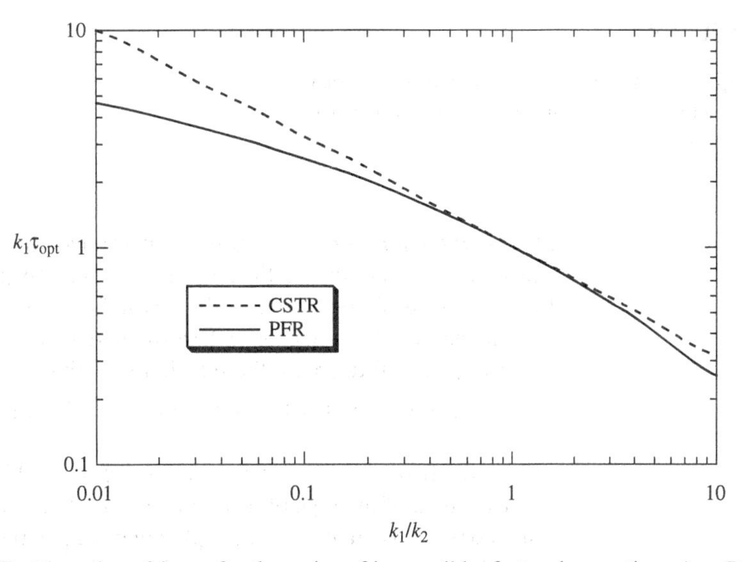

Figure 7-7 The value of $k_1\tau_{opt}$ for the series of irreversible, first-order reactions $A \rightarrow R \rightarrow S$, as a function of the ratio of the rate constant for the second reaction (k_2) to the rate constant for the first reaction (k_1). The parameter τ_{opt} is the value of τ at which the concentration of the intermediate product, R, has a maximum value ($C_{R,max}$). There is no R in the feed, and the reactors are isothermal. Solid line—PFR; dashed line—CSTR.

In this equation, α_D is the order of the desired reaction with respect to A, and β_D is the order of the desired reaction with respect to B. The rate of the undesired reaction is

$$r_U = k_U C_A^{\alpha_U} C_B^{\beta_U}$$

The order of the undesired reaction with respect to A is α_U and β_U is the order of the undesired reaction with respect to B.

To begin the analysis, let's form the ratio of the rate of the desired reaction to the rate of the undesired reaction

$$\frac{r_D}{r_U} = \left(\frac{k_D}{k_U}\right) C_A^{(\alpha_D-\alpha_U)} C_B^{(\beta_D-\beta_U)} = \left(\frac{A_D}{A_U}\right) e^{-(E_D-E_U)/RT} C_A^{(\alpha_D-\alpha_U)} C_B^{(\beta_D-\beta_U)} \tag{7-17}$$

Here, we substituted the Arrhenius relationship for each of the rate constants, and labeled the activation energy for the desired reaction E_D and the activation energy for the undesired reaction E_U.

Effect of Temperature First, let's look at temperature. If $E_D > E_U$, both r_D and r_D/r_U will increase as the temperature is increased. Therefore, both the *rate* of the desired reaction and the *selectivity* to the desired product will be maximized by operating at the highest possible temperature. The highest possible temperature might be set, for example, by limitations on the materials of construction of the reactor, by the onset of additional, undesired, side reactions, or by the onset of catalyst deactivation.

On the other hand, if $E_U > E_D$, the situation is more complicated. Although the *rate* of the desired reaction still increases with temperature, the ratio r_D/r_U *decreases*. If the reaction temperature is too high, the selectivity will be poor.

The question of what temperature to use ultimately must be answered through an economic analysis. Since selectivity usually is a dominant issue in the economics, the optimum temperature often is "low," but not so low that the reactor volume is excessive.

EXERCISE 7-9

The ratio r_D/r_U is not the same as the reaction selectivity, $s(D/A)$. Using the rate equations given above, derive an expression for $s(D/A)$.

Effect of Reactant Concentrations What concentrations of A and B should be used? From Eqn. (7-17), we can see that the ratio r_D/r_U is directly proportional to $C_A^{(\alpha_D-\alpha_U)}$. If $\alpha_D > \alpha_U$, the rate of the desired reaction will increase faster than the rate of the undesired reaction as C_A is increased. This will cause $s(D/A)$ to increase as C_A is increased. However, if $\alpha_U > \alpha_D$, increasing C_A will decrease the reaction selectivity.

How can we employ this kind of analysis in reactor and process design?

- **α_D>α_U** If $\alpha_D > \alpha_U$, we want to operate at the highest practical concentration of A. Independent of the type of reactor that is used, the composition of the feed to the reactor should be adjusted to make C_A high. For example, if the feed is a liquid, the concentration of solvent or any other diluent that is present might be reduced. If the feed is a gas, we might consider removing diluents or operating at higher pressure.

  Adjusting the feed composition may have consequences outside the realm of reactionselectivity. For example, we shall see in the next chapter that the feed composition can have a significant effect on the energy balance for the reactor. The energy balance might constrain the changes that can be made to the feed composition.

  Continuous Processes If we are considering a continuous process, the highest value of C_A is obtained by using a plug-flow reactor. If a CSTR is used, C_A immediately drops to the exit concentration, since the feed stream mixes instantaneously into the reactor contents.

  If agitation of the reactor is required, for example, to keep a catalyst suspended, then a series of CSTRs could be used to minimize dilution of the feed stream.

  Batch Processes If the reactions are to be carried out in a batch reactor, C_A can be kept high by charging all of the A at the beginning of the reaction.

- **α_D>α_U** If $\alpha_U > \alpha_D$, we want to keep C_A as low as possible. Following the line of reasoning above, $\alpha_U > \alpha_D$ will favor a dilute feed. It also favors the use of a CSTR rather than a PFR for continuous processes.

  So far, this analysis has included only the concentration of reactant A. However, the concentration of reactant B also appears in the expression for r_D/r_U. The analysis of the influence of C_B follows the same path as the analysis of C_A. If $\beta_D > \beta_U$, the design and process variables should be manipulated to keep C_B as high as possible. If $\beta_D < \beta_U$, the opposite is true.

  If $\alpha_D > \alpha_U$ *and* $\beta_D > \beta_U$, then *both* C_A and C_B should be kept as high as possible, using one or more of the techniques discussed above. However, suppose that $\alpha_D > \alpha_U$ but $\beta_D < \beta_U$. In this situation, we want to keep C_A as high as possible and simultaneously keep C_B as low as possible. *How can this be done?*

  First, consider the feed concentration. A large stoichiometric excess of A could be used. This keeps C_A high, and the large excess of A dilutes the B, keeping C_B low. Of course, a large quantity of unreacted A would have to be separated from the reactor effluent and recycled. This might be a worthwhile compromise if the excess of A had a significant beneficial effect on the reaction selectivity.

  Some reactor design options also must be considered.

Continuous Processes We can keep C_A high and C_B low by using reactors in series with an interstage feed. If $\alpha_D > \alpha_U$ and $\beta_U > \beta_D$, all of the A can be fed to the first reactor, along with some of the B. The remainder of the B is fed between reactors. The application of this concept to a series of PFRs and to a series of CSTRs is shown in the Figures 7-8a and 7-8b.

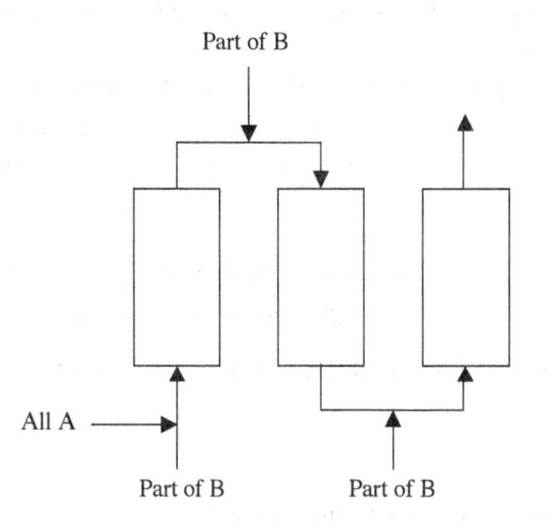

Figure 7-8a A series of PFRs in which the concentration of A is kept high by feeding all of the A to the first reactor, and the concentration of B is kept low by feeding B between reactors.

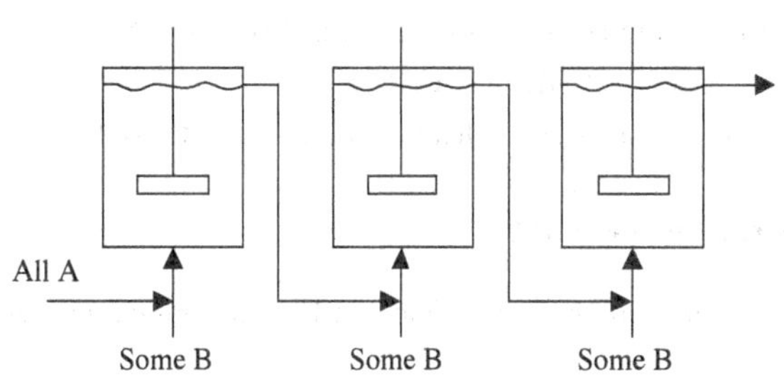

Figure 7-8b A series of CSTRs in which the concentration of A is kept high by feeding all of the A to the first reactor, and the concentration of B is kept low by feeding B between reactors.

Batch Processes In an ideal batch reactor, as defined in Chapter 3, all of the reactants are charged at once, after which there is no flow of mass across the system boundaries. With this definition, B cannot be added as the reaction proceeds, as in the continuous examples shown above. On the other hand, we could employ a modification of the batch reactor, known as a "semibatch" reactor. All of the A is added initially. Compound B is fed slowly as the reaction proceeds. This idea is represented in the Figure 7-9.

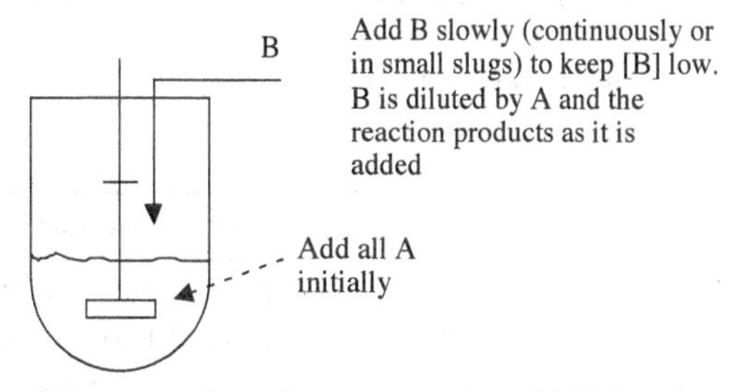

Figure 7-9 "Semibatch" reactor to keep the concentration of A high and the concentration of B low.

The design and analysis of semibatch reactors is more complicated than the design and analysis of batch reactors. Example 7-5 will illustrate the procedure for solving problems involving semibatch reactors.

7.4.3.2 Quantitative Analysis

EXAMPLE 7-3
Hydrogenation of Two Olefins

A stream containing two olefins, A and B, is fed to a catalytic, fluidized bed reactor, which operates at 250 °C and a total pressure of 1 atm. As a first approximation, the reactor may be treated as an ideal CSTR, and transport effects may be neglected. The weight of catalyst in the reactor, *W*, is 1000 kg. The feed to the reactor is 1000 kg · mol/h of A, 1000 kg · mol/h of B, and 2000 kg · mol/h of H_2.

The olefins are hydrogenated according to the irreversible reactions

$$A + H_2 \rightarrow C; \quad -r_A = k_1[A][H_2] \tag{1}$$

$$B + H_2 \rightarrow D; \quad -r_B = k_2[B][H_2] \tag{2}$$

At 250 °C,

$$k_1 = 4.80 \times 10^6 \text{ l}^2/\text{h-kg-(cat)-kg mol}$$

$$k_2 = 4.80 \times 10^5 \text{ l}^2/\text{h-kg-(cat)-kg mol}$$

The hydrogenation of A is the desired reaction. Hydrogenation of B is undesired. The ideal gas laws are valid.

Questions

A. How would you classify this reaction network?

B. What are the fractional conversions of A and B at the above conditions?

C. What is the overall selectivity, $S(C/H_2)$?

D. Do you have any suggestions that might increase $S(C/H_2)$?

Part A: How would you classify this reaction network?

SOLUTION

This is a parallel reaction network. Note that H_2 is a common reactant.

Part B: What are the fractional conversions of A and B at the above conditions?

APPROACH

First, a stoichiometric table will be constructed to keep track of the composition changes taking place. Let's use "extent of reaction" to characterize the composition of the system, and let's work in terms of molar flow rates. Two independent reactions are occurring, so two variables are required to describe the composition of the system. Let's designate the extent of Reaction (1) as ξ_1 and the extent of Reaction (2) as ξ_2, and use ξ_1 and ξ_2 to describe the system composition.

Next, the design equations for A and B will be written. These algebraic equations will contain the unknowns ξ_1 and ξ_2. Finally, the two equations will be solved simultaneously for ξ_1 and ξ_2. The fractional conversions of A and B will be calculated from these two extents of reaction.

SOLUTION

The stoichiometric table is

Species	Molar flow rate in	Molar flow rate out
A	F_{A0} (=1000 kg·mol/h)	$F_{A0} - \xi_1 (=F_A)$
B	F_{B0} (=1000 kg·mol/h)	$F_{B0} - \xi_2 (=F_B)$
H_2	$F_{H_2,0}$ (=2000 kg·mol/h)	$F_{H_2,0} - \xi_1 - \xi_2$
C	0	ξ_1
D	0	ξ_2
Total	$F_{A0} + F_{B0} + F_{H_2,0}$	$F_{A0} + F_{B0} + F_{H_2,0} - \xi_1 - \xi_2$

Now we can write the design equations for A and B:

$$\frac{W}{F_{A0}} = \frac{x_A}{-r_A} = \frac{(F_{A0} - F_A)/F_{A0}}{k_1 C_A C_{H_2}} \quad (3\text{-}17a)$$

$$\frac{W}{F_{B0}} = \frac{x_B}{-r_B} = \frac{(F_{B0} - F_B)/F_{B0}}{k_2 C_B C_{H_2}} \quad (3\text{-}17a)$$

From the stoichiometric table, $F_A = F_{A0} - \xi_1$ and $F_B = F_{B0} - \xi_2$. The concentrations C_A, C_B, and C_{H_2} can by expressed in terms of ξ_1 and ξ_2 using the ideal gas law and the stoichiometric table.

$$C_A = p_A/RT = (P/RT)y_A = (P/RT)[(F_{A0} - \xi_1)/(F_{A0} + F_{B0} + F_{H_2,0} - \xi_1 - \xi_2)]$$

$$C_B = (P/RT)[(F_{B0} - \xi_2)/(F_{A0} + F_{B0} + F_{H_2,0} - \xi_1 - \xi_2)]$$

$$C_{H_2} = (P/RT)[(F_{H_2,0} - \xi_1 - \xi_2)/(F_{A0} + F_{B0} + F_{H_2,0} - \xi_1 - \xi_2)]$$

Substituting the above expressions for F_A, F_B, C_A, C_B, and C_{H_2} into the two design equations gives

$$Wk_1\left(\frac{P}{RT}\right)^2 = \frac{\xi_1(F_{A0} + F_{B0} + F_{H_2,0} - \xi_1 - \xi_2)^2}{(F_{A0} - \xi_1)(F_{H_2,0} - \xi_1 - \xi_2)} \quad (7\text{-}18)$$

$$Wk_2\left(\frac{P}{RT}\right)^2 = \frac{\xi_2(F_{A0} + F_{B0} + F_{H_2,0} - \xi_1 - \xi_2)^2}{(F_{B0} - \xi_2)(F_{H_2,0} - \xi_1 - \xi_2)} \quad (7\text{-}19)$$

Equations (7-18) and (7-19) contain the two unknowns, ξ_1 and ξ_2. All other parameters in these equations are known. One way to solve the problem is to solve these two nonlinear algebraic equations simultaneously, to obtain values of ξ_1 and ξ_2.

Another way to solve the problem is to first relate ξ_1 and ξ_2. This will illustrate the "time-independent" method referred to earlier. First, divide Eqn. (7-19) by Eqn. (7-18) to obtain

$$\frac{\xi_2}{\xi_1} = \left(\frac{k_2}{k_1}\right)\left(\frac{F_{B0} - \xi_2}{F_{A0} - \xi_1}\right)$$

Solving for ξ_2 in terms of ξ_1,

$$\xi_2 = \frac{(k_2/k_1)F_{B0}\xi_1}{F_{A0} - \xi_1[1 - (k_2/k_1)]} \quad (7\text{-}20)$$

This type of equation can be very useful. If F_{A0}, F_{B0}, and k_2/k_1 are known, as they are in this problem, ξ_2 can be calculated for any value of ξ_1. For example, using *only* Eqn. (7-20), you could calculate that the conversion of B is 9.09% when the conversion of A is 50%.

Let's return to the problem as stated. Equations (7-18) and (7-20) can be solved using either the GOALSEEK or SOLVER routine in an EXCEL spreadsheet. The resulting values are $\xi_1 =$ 516 kg mol/h and $\xi_2 = 96.4$ kg mol/h. The fractional conversion of A, x_A, is $\xi_1/F_{A0} = 0.516$, and the fractional conversion of B, x_B, is $\xi_2/F_{B0} = 0.096$.

To summarize the solution to this problem, we had two independent reactions: $A + H_2 \rightarrow C$ and $B + H_2 \rightarrow D$. A rate equation was given for each reaction. To solve the problem, we introduced two variables, ξ_1 and ξ_2, to describe the composition of the system. We then used two material balances, one on species A (the design equation for a ideal CSTR based on A) and the other on species B (the design equation for an ideal CSTR based on B). The two material balances were solved simultaneously for the two unknowns, ξ_1 and ξ_2. In this case, we manipulated the two material balances to obtain a "time-independent" relationship between ξ_1 and ξ_2 before carrying out the simultaneous solution.

Part C: What is the overall selectivity, $S(C/H_2)$?

APPROACH The overall selectivity will be calculated from Eqn. (7-1), using the values of ξ_1 and ξ_2 obtained in Part B.

SOLUTION According to Eqn. (7-1),

$$S(C/H_2) = -(-1)(F_C - F_{CO})/(+1)(F_{H_2} - F_{H_{2,0}}) = \xi_1/(\xi_1 + \xi_2)$$
$$S(C/H_2) = 516/(516 + 96) = 0.84(84\%)$$

Part D: Do you have any suggestions that might increase $S(C/H_2)$?

APPROACH The ratio $r_C/-r_{H_2}$ will be formed and the effects of temperature and the concentrations will be analyzed for ways to make $r_C/-r_{H_2}$ as large as possible.

SOLUTION The ratio of the rate of C formation to the rate of H_2 consumption is

$$\frac{r_C}{-r_{H_2}} = \frac{k_1[A][H_2]}{k_1[A][H_2] + k_2[B][H_2]} = \frac{1}{1 + \left(\frac{k_1}{k_2}\right)\left(\frac{[B]}{[A]}\right)}$$

To improve the overall selectivity, reduce the [B] in the feed and/or increase the [A] in the feed, if possible. Operate in a PFR or a series of CSTRs since the conversion of B is so low that the [B] actually is higher in the effluent from the single CSTR of this example than it is in the feed. Since A is hydrogenated faster than B, the ratio [B]/[A] increases as the reaction proceeds. Examine the feasibility of operating with a series of PFRs or CSTRs with fresh feed added between reactors.

The ratio k_1/k_2 will influence the selectivity. However, the two activation energies must be known in order to analyze the effect of temperature.

EXAMPLE 7-4
Selective Oxidation of Carbon Monoxide

The selective oxidation of CO in the presence of H_2 is being carried out in a catalytic, ideal plug-flow reactor at atmospheric pressure and 100 °C. Although the reactions are exothermic, we will assume that the reactor is isothermal in order to develop a preliminary understanding of reaction behavior. We also will neglect pressure drop, assume that transport effects are negligible, and assume that the ideal gas laws are valid.

The feed to the reactor contains 1.0 mol% CO, 30 mol% H_2, "w" mol% O_2, and the balance N_2. The mole fraction of O_2 in the feed will have to be calculated based on the performance that is required from the reactor. The concentration of O_2 in the reactor effluent must be less than 10 ppm. The reaction

$$CO + 1/2\,O_2 \rightarrow CO_2$$

obeys the rate equation

$$-r_{CO} = k_1 p_{CO} p_{O_2}^{1/2}$$

The reaction

$$H_2 + 1/2\,O_2 \rightarrow H_2O$$

obeys the rate equation

$$-r_{H_2} = k_2 p_{H_2} p_{O_2}^{1/2}/(1 + K_{CO} p_{CO})^2$$

The values of the constants are $k_1 = 155$ mol/g cat-min-atm$^{1.5}$; $k_2 = 1.95$ mol/g cat-min-atm$^{1.5}$; $K_{CO} = 1000$ atm^{-1}. The symbol p_i denotes the partial pressure of species i in atmospheres.

A. What feed concentration of O_2 (w) is required to reduce the concentration of CO in the reactor effluent to 10 ppm or less?

B. What percentage of the H_2 in the feed is consumed?

C. What space time ($\tau = W/v$) is required to reach an effluent concentration of 10 ppm CO?

Part A: What feed concentration of O_2 (w) is required to reduce the concentration of CO in the reactor effluent to 10 ppm or less?

APPROACH The first and second questions can be answered without calculating the space time. These are "time-independent" questions. The material balances (design equations) for CO and H_2 will be formulated. The balance for H_2 will be divided by that for CO. The resulting differential equation will be solved to obtain the outlet concentration of H_2, since the outlet concentration of CO is known. The required inlet concentration of O_2 then can be calculated from stoichiometry.

SOLUTION If we assume that the mole fraction of O_2 in the feed (w) is small, the change in density on reaction can be neglected. Material balances on H_2 and CO over the differential volume element shown in Figure 7-4 are

$$H_2: \quad -\upsilon dC_{H_2} = -r_{H_2} dW \tag{7-21}$$

$$CO: \quad -\upsilon dC_{CO} = -r_{CO} dW \tag{7-22}$$

These balances are identical to the PFR design equation for a constant-density, catalytic reaction (Eqn. (3-31)). The problem solution could have begun with these design equations, one for CO and one for H_2.

Dividing Eqn. (7-22) by Eqn. (7-21) and using the ideal gas law:

$$\frac{dC_{CO}}{dC_{H_2}} = \frac{dp_{CO}}{dp_{H_2}} = \frac{-r_{CO}}{-r_{H_2}} = \frac{k_1 p_{CO} p_{O_2}^{1/2}}{k_2 p_{H_2} p_{O_2}^{1/2}/(1 + K_{CO} p_{CO})^2}$$

$$\frac{dp_{CO}}{dp_{H_2}} = \frac{k_1 p_{CO}(1 + K_{CO} p_{CO})^2}{k_2 p_{H_2}}$$

This equation does not contain W, υ, or the space time τ.

Separating variables and integrating from the reactor inlet to the outlet,

$$\int_{10^{-2}}^{10^{-5}} \frac{k_2 dp_{CO}}{p_{CO}(1 + K_{CO} p_{CO})^2} = \int_{0.30}^{Y} \frac{k_1 dp_{H_2}}{p_{H_2}}$$

The limits of integration in the above equation have units of atmosphere and the symbol Y designates the outlet partial pressure of H_2. Carrying out the indicated integration for an isothermal reactor yields

$$\left[\frac{1}{(1 + K_{CO} p_{CO})} - \ln\left(\frac{1 + K_{CO} p_{CO}}{p_{CO}}\right)\right]_{10^{-2}}^{10^{-5}} = \left(\frac{k_1}{k_2}\right)[\ln p_{H_2}]_{0.30}^{Y}$$

Substituting the specified values of K_{CO}, k_1, and k_2 gives $Y = 0.2866$ atm. This is the partial pressure of H_2 in the outlet from the reactor. An extra significant figure has been carried in the calculation of Y for accuracy in calculating the required inlet partial pressure of O_2, as shown below.

Since the inlet and outlet concentrations of both H_2 and CO now are known, the required inlet concentration of O_2 can be calculated from stoichiometry. If CO oxidation is designated as Reaction (1), then the extent of this reaction is

$$\xi_1 = (F_{CO}^{in} - F_{CO}^{out})/\nu_{1,CO} = \upsilon(p_{CO}^{in} - p_{CO}^{out})/\nu_{1,CO}RT$$

and the extent of Reaction (2), H_2 oxidation, is

$$\xi_2 = (F_{H_2}^{in} - F_{H_2}^{out})/\nu_{1,H_2} = \upsilon(p_{H_2}^{in} - p_{H_2}^{out})/\nu_{1,H_2}RT$$

According to the Law of Definite Proportions

$$F_{O_2}^{in} - F_{O_2}^{out} = \nu_{1,O_2}\xi_1 + \nu_{2,O_2}\xi_2 = \upsilon(p_{O_2}^{in} - p_{O_2}^{out})/RT$$

Substituting the expressions for ξ_1 and ξ_2 into this equation gives

$$\left(p_{O_2}^{in} - p_{O_2}^{out}\right) = \frac{\nu_{1,O_2}}{\nu_{1,CO}}\left(p_{CO}^{in} - p_{CO}^{out}\right) + \frac{\nu_{2,O_2}}{\nu_{2,H_2}}\left(p_{H_2}^{in} - p_{H_2}^{out}\right) \tag{7-23}$$

Finally, inserting the known values for the inlet and outlet partial pressures of CO and H_2, the known outlet partial pressure for O_2, and the three stoichiometric coefficients gives

$$p_{O_2}^{in} = 10^{-4} + (0.5/1)(10^{-2} - 10^{-5}) + (0.5/1)(0.300 - 0.2866) = 0.0118\ \text{atm}$$

This calculation shows that a little over half of the oxygen in the feed to the reactor is consumed in the undesired reaction, the oxidation of H_2. Nevertheless, the selectivity of the (hypothetical) catalyst of this example is remarkable, since the ratio of H_2 to CO in the feed is 30.

Part B: What percentage of the H_2 in the feed is consumed?

APPROACH

The partial pressure of H_2 in the reactor effluent (0.287 atm) was calculated in Part A, and the partial pressure of H_2 in the feed was specified as 0.300 atm. The percentage of H_2 reacted can be calculated from these numbers.

SOLUTION

Since the density is essentially constant, the percentage of the inlet H_2 that is consumed is $(0.300 - 0.287) \times 100/0.300 = 4.33\%$.

Part C: What space time ($\tau = W/v$) is required to reach an effluent CO concentration of 10 ppm?

APPROACH

The two ordinary differential equations, Eqns. (7-21) and (7-22), will be solved simultaneously, using the given rate equations and the ideal gas law. Both of these equations contain the partial pressure of O_2, which is related to the partial pressures of CO and H_2 through Eqn. (7-32). Due to the complexity of the rate equations, the simultaneous solution of Eqns. (7-21) and (7-22) will have to be accomplished numerically.

SOLUTION

Equation (7-22) can be combined with the rate equation for CO disappearance and with the ideal gas law to give

$$\frac{dC_{CO}}{d\tau} = \frac{1}{RT}\frac{dp_{CO}}{d\tau} = -(-r_{CO}) = -k_1 p_{CO} p_{O_2}^{1/2}$$
$$\frac{dp_{CO}}{d\tau} = -(k_1 RT) p_{CO} p_{O_2}^{1/2} \tag{7-24}$$

Similar operations with Eqn. (7-21) yield

$$\frac{dp_{H_2}}{d\tau} = -(k_2 RT) p_{H_2} p_{O_2}^{1/2}/(1 + K_{CO} p_{CO})^2 \tag{7-25}$$

These simultaneous ordinary differential equations, Eqns. (7-24) and (7-25), must be solved numerically subject to the initial conditions:

$$\tau = 0;\ p_{O_2} = 0.0118\ \text{atm};\ p_{CO} = 0.010\ \text{atm}$$

The right-hand side of both differential equations contains the partial pressure of O_2. This suggests that a third differential equation, i.e., a material balance on O_2, is required. However, since there are only two independent reactions, a third differential equation would be redundant. If the partial pressures of H_2 and CO both are known, the partial pressure of O_2 can be calculated from stoichiometry, i.e., from Eqn. (7-23).

The numerical solution of simultaneous, ordinary differential equations can be accomplished with a number a standard mathematical packages. Appendix 7-A.2 illustrates the use of a spreadsheet to solve these equations, via a fourth-order Runge–Kutta technique. The result is $\tau = 0.0246$ g cat-min/l.

In summary, there were two independent reactions in this problem. A rate equation was available for each reaction. Two material balances were required, one on CO and the other on H_2.

The partial pressures p_{CO} and p_{H_2} were used as the independent variables to describe the system composition. The partial pressure of oxygen was not an independent variable. It was calculated from stoichiometry using the known partial pressures of carbon monoxide and hydrogen.

EXAMPLE 7-5
Semibatch Reactor

The reaction of A with B to give C

$$A + B \rightarrow C$$

is accompanied by the undesired side reaction

$$2B \rightarrow D$$

These reactions take place in the liquid phase, with rates that are given by

$$-r_A = k_1 C_A C_B \tag{7-26}$$

$$r_D = k_2 C_B^2 \tag{7-27}$$

To minimize the amount of D that will be formed, an ideal, semibatch reactor will be used. The reactor will be operated isothermally at a temperature where the values of the rate constants are $k_1 = 0.50$ l/mol-h; $k_2 = 0.25$ l/mol-h.

The total volume of the reactor is 10,000 l. The initial charge is 2000 l, containing 2.0 mol/l of A, 0.25 mol/l of B, and no C or D. A solution containing 0.50 mol/l of B and no A, C, or D is fed to the reactor at a rate of 1000 l/h for 7.0 h, beginning immediately after the initial charge has been added.

A. What are the concentrations of A, B, C, and D in the reactor after 7.0 h?

B. What is the value of the selectivity $S(C/B)$ for the overall process?

Part A: What are the concentrations of A, B, C, and D in the reactor after 7.0 h?

APPROACH

There are two independent reactions, and the complete rate equation is specified for both reactions. Two independent material balances are required. We will choose to perform balances on A and B, and use the concentrations of these compounds to describe the system composition. The two ordinary differential equations that result from the balances will be solved simultaneously to obtain C_A and C_B after 7 h of reaction. Once the concentrations of A and B are known, the concentrations of C and D can be calculated from stoichiometry.

SOLUTION

Balance on A for whole reactor:

$$\text{rate in} - \text{rate out} + \text{rate generation} = \text{rate accumulation}$$

$$0 - 0 + r_A V = \mathrm{d}(VC_A)/\mathrm{d}t$$

In this equation, V is the volume of the *liquid* in the reactor, which increases with time. The volume at any time is given by

$$V = V_0 + \upsilon t \tag{7-28}$$

where $V_0 = 2000$ l and $\upsilon = 1000$ l/h. From Eqn. (7-28), $\mathrm{d}V/\mathrm{d}t = \upsilon$.

Returning to the balance on A,

$$r_A V = V\frac{\mathrm{d}C_A}{\mathrm{d}t} + C_A\frac{\mathrm{d}V}{\mathrm{d}t} = V\frac{\mathrm{d}C_A}{\mathrm{d}t} + C_A\upsilon$$

Substituting the rate equation for $-r_A$,

$$\frac{\mathrm{d}C_A}{\mathrm{d}t} = -\frac{\upsilon C_A}{V} - k_1 C_A C_B \tag{7-29}$$

Balance on B:

$$\text{rate in} - \text{rate out} + \text{rate generation} = \text{rate accumulation}$$

$$\upsilon C_{B,f} - 0 + V(r_A - 2r_D) = d(VC_B)/dt$$
$$\upsilon C_{B,f} + V(-k_1 C_A C_B - 2k_2 C_B^2) = V\frac{dC_B}{dt} + C_B\frac{dV}{dt}$$
$$\frac{dC_B}{dt} = \frac{\upsilon(C_{B,f} - C_B)}{V} - k_1 C_A C_B - 2k_2 C_B^2 \qquad (7\text{-}30)$$

In these equations, $C_{B,f}$ is the concentration of B in the solution that is fed to the reactor over the 7-h period, i.e., 0.50 mol/l.

Equations (7-29) and (7-30) are now solved numerically, using the technique described in Example 7-4. The initial conditions are

$$C_A = 2.0\,\text{mol/l}; \quad C_B = 0.25\,\text{mol/l}; \quad t = 0$$

The results are

$$C_A = 0.238\,\text{mol/l}$$
$$C_B = 0.173\,\text{mol/l}$$

The concentrations of C and D now can be calculated from stoichiometry, using the extent of reaction:

$$\Delta N_A = \nu_{1A}\xi_1 + \nu_{2A}\xi_2$$
$$9000 \times 0.238 - 2000 \times 2.0 = -1858 = (-1)\xi_1; \quad \xi_1 = 1858$$
$$\Delta N_C = \nu_{1C}\xi_1 + \nu_{2C}\xi_2$$
$$9000C_C - 0 = (1)\xi_1 = 1858; \quad C_C = 0.206\,\text{mol/l}$$
$$\Delta N_B = \nu_{1B}\xi_1 + \nu_{2B}\xi_2$$
$$9000 \times 0.173 - (2000 \times 0.25 + 7000 \times 0.50) = (-1)\xi_1 + (-2)\xi_2$$
$$-2443 = -1858 - 2\xi_2; \quad \xi_2 = 293$$
$$\Delta N_D = \nu_{1D}\xi_1 + \nu_{2D}\xi_2$$
$$9000C_D - 0 = \xi_2; \quad C_D = 0.033\,\text{mol/l}$$

Part B: What is the value of the selectivity $S(C/B)$ for the overall process?

APPROACH

Values of the moles of C formed and the moles of B reacted can be calculated from the results of Part A. These values then can be substituted into the definition of $S(C/B)$, i.e., Eqn. (7-1).

SOLUTION

$$S(C/B) = \text{moles C formed/moles B reacted}$$
$$S(C/B) = 9000 \times 0.206/(2000 \times 0.25 + 7000 \times 0.50 - 9000 \times 0.173) = 0.76$$

7.4.4 Mixed Series/Parallel Reactions

7.4.4.1 Qualitative Analysis

This classification covers reactions that combine the characteristics of both series and parallel reactions. The intermediate product that is formed in the series sequence will have the same general behavior as the intermediate product in a pure series sequence. The concentration and the yield of this intermediate product may go through a maximum with time or space time. Moreover, the kinetics of the parallel reactions can be used to analyze the effects of the concentrations and the temperature on the reaction selectivity, as was done for pure parallel reactions.

7.4.4.2 Quantitative Analysis

EXAMPLE 7-6
Mixed Series/Parallel Reactions in PFR

The first-order irreversible reactions

$$A \xrightarrow{k_1} R \xrightarrow{k_2} S$$
$$A \xrightarrow{k_3} D$$

are being carried out in an ideal, isothermal PFR. At the reactor temperature, $k_1 = k_2 = 1.0\ \text{s}^{-1}$ and $k_3 = 2\ \text{s}^{-1}$. The concentration of A in the feed to the reactor, C_{A0}, is 2.0 mol/l. There is no R, S, or D in the feed.

A. What is the composition of the stream leaving the reactor when the concentration of A in the exit stream is 0.40 mol/l?

B. What space time is required to achieve this effluent composition?

Part A: What is the composition of the stream leaving the reactor when the concentration of A in the exit stream is 0.40 mol/l?

APPROACH

This is a "time-independent" question. To begin, we will write the design equations for A and R, divide the design equation for R by that for A, and then solve the resulting differential equation to obtain C_R as a function of C_A. This equation will permit the outlet concentration of R to be calculated, since the outlet concentration of A is given.

The same thing then will be done for A and D to obtain an equation that permits the outlet concentration of D to be calculated.

Finally, the outlet concentration of S will be calculated from stoichiometry.

SOLUTION

Since there is no change in moles on reaction, and since the reactor is isothermal, the design equations for A and R are given by Eqn. (3-31),

$$d\tau = \frac{-dC_A}{-r_A}; \quad d\tau = \frac{dC_R}{r_R}$$

Dividing,

$$\frac{-dC_R}{dC_A} = \frac{r_R}{-r_A}$$

At any point in the PFR, the *net* rate of R formation is given by $k_1 C_A - k_2 C_R$ and the *total* rate of A consumption is given by $(k_1 + k_3)C_A$. Therefore,

$$-\frac{dC_R}{dC_A} = \frac{k_1}{k_1 + k_3} - \frac{k_2}{k_1 + k_3}\left(\frac{C_R}{C_A}\right)$$

This equation is similar to Eqn. (7-9) and also can be solved using the "integrating factor" approach. The solution is

$$\frac{C_R}{C_{A0}} = \frac{k_1}{k_1 + k_3 - k_2}\left[\left(\frac{C_A}{C_{A0}}\right)^{[k_2/(k_1+k_3)]} - \left(\frac{C_A}{C_{A0}}\right)\right]; \quad k_1 + k_3 \neq k_3 \qquad (7\text{-}31)$$

Substituting numbers into Eqn. (7-31) gives $C_R(\text{exit})/C_{A0} = 0.192$; $C_R(\text{exit}) = 0.385$.

We will follow the same approach to calculate C_D(exit).

$$-\frac{dC_D}{dC_A} = \frac{r_D}{-r_A} = \frac{k_3 C_A}{(k_1 + k_3)C_A} = \frac{k_3}{k_1 + k_3}$$

Integrating,

$$C_D = \left(\frac{k_3}{k_1 + k_3}\right)(C_{A0} - C_A)$$

Substituting numbers gives $C_D = 1.066$ mol/l.

The effluent concentration of S can be calculated from stoichiometry. The result is $C_S = 0.149$ mol/l.

The composition of the effluent stream is

$$C_A(\text{exit}) = 0.40 \text{ mol/l} \quad C_S(\text{exit}) = 0.15 \text{ mol/l}$$
$$C_R(\text{exit}) = 0.39 \text{ mol/l} \quad C_D(\text{exit}) = 1.07 \text{ mol/l}$$

Part B: What space time is required to achieve this effluent composition?

APPROACH Since the fractional conversion of A is known $[x_{Ae} = (C_{A0} - C_{Ae})/C_{A0} = 0.80]$, this question can be answered simply by solving the design equation.

SOLUTION

$$\frac{V}{F_{A0}} = \int_0^{x_{Ae}} \frac{dx_A}{-r_A} = \int_0^{x_{Ae}} \frac{dx_A}{(k_1 + k_3)C_A} = \frac{1}{(k_1 + k_3)C_{A0}} \int_0^{x_{Ae}} \frac{dx_A}{1 - x_A}$$

$$\tau = \frac{-\ln(1 - x_{Ae})}{k_1 + k_3} = \frac{-\ln(1 - 0.80)}{(1 + 2)\text{ s}^{-1}} = 0.54 \text{ s}$$

SUMMARY OF IMPORTANT CONCEPTS

- In a system where more than one reaction occurs, the product distribution is influenced by temperature, the various concentrations, and the choice of reactor system.
- Semibatch reactors are an extension of batch reactors that offer more control of the product distribution.
- Selectivity is a measure of how efficiently a reactant is converted into a specified product. Yield is a measure of how much of a specified product is formed from a given amount of feed.
- In order to describe the complete composition of a system in which multiple reactions take place, a separate composition variable (extent of reaction, concentration of a species, etc.) is required for *each* independent reaction.
- The "time-independent" method can be used to calculate the complete product distribution in a system where multiple reactions take place, if a rate equation is available for each independent reaction. However, the real time or space time required to reach the calculated product distribution cannot be obtained via the "time-independent" method.
- To solve for the complete product distribution as a function of time (or space time) in a multiple-reaction system, a design equation (component material balance) and a rate equation are required for *each* independent reaction. These design equations must be solved simultaneously.

PROBLEMS

Single-Reactor Problems

Problem 7-1 (Level 2) An ideal CSTR is to be sized for the polymerization of styrene monomer to polystyrene. Pure styrene monomer containing a small amount of the free-radical initiator 2,2′-azobisisobutyronitrile (AIBN) will be fed to the reactor as a liquid. The stream leaving the reactor will be a liquid containing dissolved polymer, unconverted monomer and unconverted AIBN. In the following, styrene monomer is denoted as M and AIBN is denoted as I.

The initiator decomposes by a first-order reaction to form two free radicals

$$I \rightarrow 2R^{\bullet}$$

The rate constant for this reaction (based on AIBN) is k_d. Each free radical starts one polymer chain.

The rate equation for the disappearance of styrene monomer is

$$-r_M = k_p[M][I]^{1/2}$$

The reactor will operate at 200 °C with an inlet AIBN concentration of 0.010 g-mol/l, an inlet monomer concentration of 8.23 g-mol/l, and a feed rate of 1900 l/h. At this temperature,

$$k_d = 9.25\,s^{-1}$$

$$k_p = 0.925\,l^{1/2}/mol^{1/2}\text{-s}$$

1. What reactor volume is required to achieve a monomer conversion of 60%?
2. Assume that "dead" polymer is formed by the combination of two growing polymer chains. In other words, each chain of "dead" polymer contains two AIBN fragments, one on each end of the polymer chain. What is the *average* number of monomer molecules contained in each "dead" polymer molecule?
3. What design features, other than size, might be important to the successful operation of this reactor?

Problem 7-2 (Level 1) The irreversible, liquid-phase reactions $A \rightarrow B$ and $2B \rightarrow C$ are taking place in an ideal CSTR at a temperature of 150 °C. The volume of the reactor is 1000 l. The volumetric flow rate into the CSTR is 167 l/min. The feed concentrations are $C_{A0} = 5$ mol/l and $C_{B0} = C_{C0} = 0$.

At 150 °C:

$$-r_A(\text{mol/l-min}) = k_1 C_A$$

$$k_1 = 0.50\,\text{min}^{-1}$$

$$r_C(\text{mol/l-min}) = k_2 (C_B)^2$$

$$k_2 = 1.0\,\text{l/mol-min}$$

What are the concentrations of A, B, and C leaving the reactor?

Problem 7-3 (Level 3) The homogeneous, gas-phase reactions

$$A \rightarrow B + C$$
$$B \rightarrow D + E$$

are taking place in an ideal, isothermal, plug-flow reactor. The temperature is 400 K and the total pressure is 0.20 MPa. The feed to the reactor is a 4/1 (molar) mixture of N_2 and A, and the feed rate is 360 l/min at inlet conditions. The reactor volume is 450 l, and pressure drop may be neglected.

The rate equation for the disappearance of A is

$$-r_A = k_1 C_A; \quad k_1 = 4.0\,\text{min}^{-1}$$

and the rate equation for the formation of D is

$$r_D = k_2 C_B; \quad k_2 = 2.0\,\text{min}^{-1}$$

1. What is the composition of the stream leaving the reactor?
2. Without changing the operating conditions or feed composition, what can be done to increase the production rate of B?

Problem 7-4 (Level 1) The homogeneous, gas-phase reactions

$$A \rightarrow B + C$$
$$B \rightarrow D + E$$

are taking place in an ideal, continuous stirred-tank reactor (CSTR). The temperature is 400 K and the total pressure is 0.20 MPa. The feed to the reactor is a 4/1 (molar) mixture of N_2 and A, and the feed rate is 360 l/min at inlet conditions. The reactor volume is 450 l, pressure drop may be neglected, and the ideal gas laws are valid.

The rate equation for the disappearance of A is

$$-r_A = k_1 C_A; \quad k_1 = 4.0\,\text{min}^{-1}$$

and the rate equation for the formation of D is

$$r_D = k_2 C_B; \quad k_2 = 2.0\,\text{min}^{-1}$$

1. What is the composition of the stream leaving the reactor?
2. Without changing the operating conditions or feed composition, what can be done to increase the production rate of B?

Problem 7-5 (Level 2) The homogeneous, gas-phase reactions

$$A \rightarrow B + C$$
$$B \rightarrow D + E$$

are taking place in an ideal, continuous stirred-tank reactor. The temperature is 400 K and the total pressure is 0.20 MPa. The feed to the reactor is a 4/1 (molar) mixture of N_2 and A, and the feed rate is 360 l/min at inlet conditions. The ideal gas laws are valid.

The rate equation for the disappearance of A is

$$-r_A = k_1 C_A; \quad k_1 = 4.0\,\text{min}^{-1}$$

and the rate equation for the formation of D is

$$r_D = k_2 C_B; \quad k_2 = 2.0\,\text{min}^{-1}$$

1. What volume of reactor is required to produce a fractional conversion of A equal to 0.75?
2. What is the effluent molar flow rate of B (mol/min) when the reactor volume is equal to what you calculated in Part 1?

Problem 7-6 (Level 2) The reactions

$$C_{16}H_{34} \rightarrow C_8H_{18} + C_8H_{16} \quad (1)$$

$$C_8H_{18} + H_2 \rightarrow C_7H_{16} + CH_4 \quad (2)$$

are being carried out in a fluidized bed reactor, which may be approximated as an ideal, continuous stirred-tank reactor. The reactor temperature is 400 K and the total pressure is 2.0 atm.

At these conditions, the ideal gas laws are valid and the reactions are essentially irreversible. The feed to the reactor is a 4/1 (molar) mixture of H_2 and $C_{16}H_{34}$. The feed rate is 360 l/min at reaction conditions. The weight of catalyst in the reactor is 200 kg.

The rate equation for the disappearance of $C_{16}H_{34}$ is

$$-r_A = k_1 C_A; \quad k_1 = 4.0 \text{ l/kg-min}$$

and the rate equation for the formation of C_7H_{16} is

$$r_D = k_2 C_B; \quad k_2 = 2.0 \text{ l/kg-min}$$

1. What is the fractional conversion of $C_{16}H_{34}$ in the stream leaving the reactor?
2. What is the production rate of C_8H_{18}, i.e., the molar flow rate of C_8H_{18} out of the reactor?
3. What is the production rate of CH_4?
4. If $k_2 = 0$, will the fractional conversion of $C_{16}H_{34}$ be higher, the same, or lower than the conversion you calculated in Question 1? Explain your answer *qualitatively*. No calculations are necessary.

Problem 7-7 (Level 1) In the hindsight of history, it was clear that the war had entered its final phase. The Confederate Army, under Robert E. Lee, and the Union Army, under Ulysses S. Grant, had fought two bloody battles in the spring of 1864, the first in the Wilderness and the second at Spotsylvania Court House. Grant broke off the battle at Spotsylvania and attempted to move around Lee's right flank. Lee anticipated the move and withdrew to a position near Hanover Junction, on the North Anna River, about 20 miles north of Richmond, Virginia.

Lee had a total of about 50,000 troops at Hanover Junction. The Confederates were well entrenched and protected by well-constructed barricades. Troops defending such a strong position were more effective on a man-for-man basis, i.e., they suffered fewer casualties per capita than troops attempting to attack such a position.

Grant's army had moved to Hanover Junction in three wings, one under Winfield Scott Hancock, containing about 30,000 troops, the second under Gouverner K. Warren, containing about 50,000 troops, and the third under Ambrose E. Burnside, containing about 20,000 troops. Warren's wing had crossed the North Anna at Jericho Mills and taken a position on Lee's left. Hancock's wing had crossed at the river at Chesterfield Bridge and had taken a position on Lee's right. Burnside's troops were in front of Lee, on the north side of the river, effectively pinned down by Lee's artillery. The situation is shown in the following sketch.

Neither Hancock nor Warren planned an immediate attack since Grant himself had not arrived. However, Lee was not going to wait for Grant. He could easily shift troops from the left to the right of his position (and vice versa) to oppose an attack or to launch an offensive. Lee's plan was to leave enough troops on his left to hold Warren, i.e., to fight a stalemate or better if Warren's wing attacked. He would shift the remainder of his troops to the right and attack Hancock. After disposing of Hancock's wing, he would shift the remaining troops back to the left, combine them with the troops that had been left on the defensive, and await an attack from Warren's wing. Lee was confident that Warren eventually would attack because Grant was known to favor the offensive.

There were several keys to Lee's plan: (1) Ox Ford was not usable by the Union troops since it was covered very effectively by the Confederate artillery; (2) Warren and Hancock were about 1 day's march apart, so that Warren's wing could not reinforce Hancock's very easily; (3) it would be very difficult for Hancock's troops to retreat across the river over Chesterfield Bridge if they were under attack by the Confederates; (4) neither of the Union wings had entrenched or constructed barricades; (5) Burnside was effectively pinned down by Lee's artillery. He could not reinforce either Hancock or Warren.

You may assume that the rate at which the soldiers of one army are killed or incapacitated is proportional to the number of soldiers in the *opposing* army. If both armies are out in the open, the proportionality constants are the same, i.e., the Union and Confederate soldiers were equally effective at killing/incapacitating each other. (This assumption would be disputed by the ghosts of both armies.) However, if one army is fighting behind an established defensive position and the other army is attacking, the proportionality constant for the defending army is approximately four times that of the attacking army. In mathematical terms,

$$-\frac{dS_{\text{attack}}}{dt} = 4\,kS_{\text{defend}}; \quad -\frac{dS_{\text{defend}}}{dt} = kS_{\text{attack}}$$

where S_{attack} is the number of attacking soldiers and S_{defend} is the number of defending soldiers.

Suppose that Lee left 15,000 troops facing Warren and attacked Hancock with 35,000 troops. Moreover, suppose that Burnside stayed in his position and made no attempt to cross Ox Ford or otherwise engage any of Lee's troops.

1. Assume that Warren's wing did not attack Lee's left during the course of the battle with Hancock's wing, but that Warren's wing did attack once the battle between Lee and Hancock was over, after Lee had consolidated all his remaining troops on the left. Predict the outcome of the battle.[2]
2. If Warren's wing attacked Lee's left *during* the battle with Hancock on the right, Lee would have to fight two independent battles, without shifting troops in either direction. Predict the outcome of the battle between Warren's 50,000 troops on the offensive and Lee's 15,000 troops on the defensive. Assume that Lee's troops would fight to the last man, instead of surrendering when their losses reached 50%. However, assume that Warren's troops would withdraw if their losses reached 50% before Lee had lost his last soldier.

[2] To predict the "outcome of the battle," assume that the first army to lose one half of its men will surrender in order to avoid further losses. The "outcome" then means (1) which army surrendered? and (2) how many troops were left in the victorious army?

Battle of Hanover Junction
Placement of Opposing Armies (May 24, 1864)

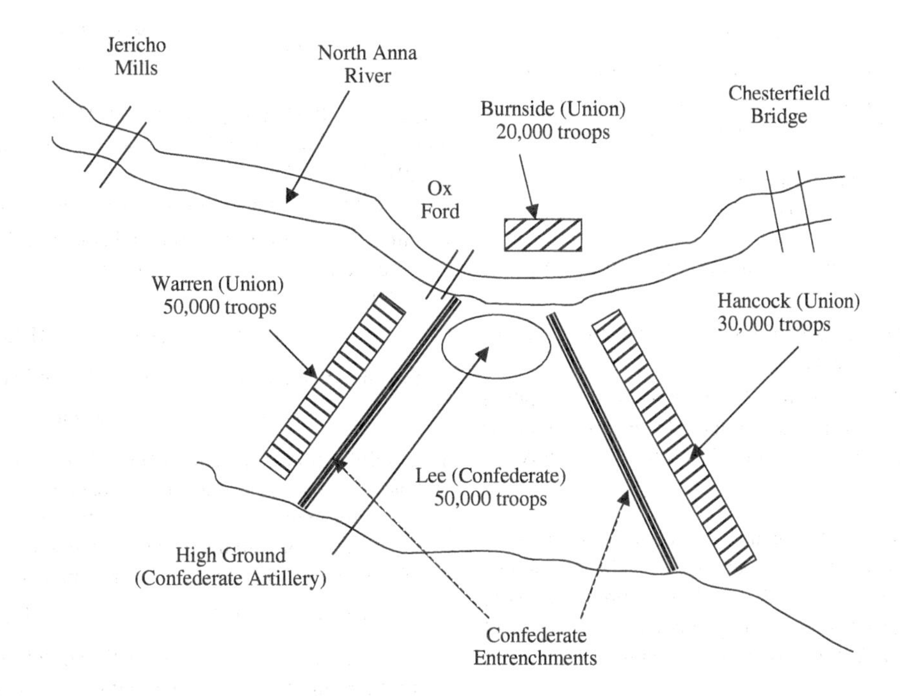

3. How many troops should Lee have left opposing Warren to ensure a stalemate or better if Warren attacked? Could Lee have attacked Hancock and defeated him if that many troops were left facing Warren?
4. What was the *actual* result of the battle?

Problem 7-8 (Level 2) The homogeneous first-order irreversible reactions

$$A \xrightarrow{k_1} R \xrightarrow{k_2} S$$

are taking place in the liquid phase in an ideal, isothermal PFR. The feed to the reactor contains 1.0 mol/l of A, 0.20 mol/l of R, and no S.

The rate constant $k_1 = 0.025\,\text{min}^{-1}$ and the rate constant $k_2 = 0.010\,\text{min}^{-1}$. The space time τ for the reactor is 100 min.

1. What are the outlet concentrations of A, R, and S?
2. What would the outlet concentrations of A, R, and S be if the inlet concentration of R was 0, but everything else remained the same?
3. What value of τ produces the maximum concentration of R leaving the reactor, for the actual inlet concentrations?

Problem 7-9 (Level 2) The irreversible, gas-phase reactions

$$A \xrightarrow{k_1} B + C$$

$$A \xrightarrow{k_2} D + E$$

are taking place in an isothermal, ideal plug-flow reactor. Both reactions are first order in A. The value of k_1 is $3.2 \times 10^3\,\text{h}^{-1}$ and the value of k_2 is $5.1 \times 10^3\,\text{h}^{-1}$.

The molar flow rate of A entering the reactor is 10,000 mol/h, the concentration of A in the feed to the reactor is $5.0 \times 10^{-5}\,\text{mol/cm}^3$, and the feed is a 1/1 (molar) mixture of A and N_2, which is inert. Pressure drop through the reactor can be neglected.

1. What volume of reactor is required to achieve a fractional conversion of A, x_A, equal to 0.80?
2. What are the mole fractions of A, B, D, and N_2 in the reactor effluent?

Problem 7-10 (Level 1) An ideal batch reactor is to be sized for the polymerization of styrene monomer to polystyrene. Pure styrene monomer containing a small amount of the free-radical initiator 2,2′-azobisisobutyronitrile (AIBN) will be charged to the reactor initially, as a liquid. At the end of the polymerization, the reactor will contain a liquid comprised of dissolved polymer, unconverted styrene, and unconverted AIBN. In the following, styrene monomer is denoted as M and AIBN is denoted as I.

The AIBN initiator decomposes by a first-order reaction to form two free radicals

$$I \rightarrow 2R^{\bullet}$$

The rate constant for this reaction (based on AIBN) is k_d. Each free radical starts one polymer chain.

The rate equation for the disappearance of styrene monomer is

$$-r_M = k_p[M][I]^{1/2}$$

The initial AIBN concentration is 0.010 g-mol/l, and the initial styrene concentration is 8.23 g-mol/l. The reactor will be operated isothermally at 60 °C. At this temperature,

$$k_d = 8.0 \times 10^{-6}\,\mathrm{s}^{-1}$$
$$k_p = 7.5 \times 10^{-4}\mathrm{l}^{1/2}/\mathrm{mol}^{1/2}\text{-s}$$

1. How much time is required to achieve a monomer conversion of 60%?
2. What is the concentration of initiator at the time that was calculated in Question 1?
3. Assume that "dead" polymer is formed by the combination of two growing polymer chains. What is the *average* number of monomer molecules contained in each "dead" polymer molecule?

Problem 7-11 (Level 1) Calculate the fractional conversion of aniline (A), and the yield and selectivity of cyclohexyl amine (CHA), cyclohexane (CH), and dicyclohexylamine (DCHA) using the data in Problem 1-6. The yields and selectivities of CHA, CH, and DCHA should be based on aniline.

Problem 7-12 (Level 3) The gas-phase reactions

$$\text{isobutanol(B)} \rightarrow \text{isobutene (IB)} + H_2O$$
$$(CH_3)_2CHCH_2OH \rightarrow (CH_3)_2C{=}CH_2 + H_2O$$

$$2\,\text{methanol} \rightarrow \text{dimethyl ether (DME)} + H_2O$$
$$2CH_3OH \rightarrow CH_3OCH_3 + H_2O$$

$$\text{isobutanol} + \text{methanol} \rightarrow \text{methylisobutyl ether (MIBE)} + H_2O$$
$$(CH_3)_2CHCH_2OH + CH_3OH \rightarrow (CH_3)_2CHCH_2OCH_3 + H_2O$$

$$2\,\text{isobutanol} \rightarrow \text{diisobutyl ether (DIBE)} + H_2O$$
$$2(CH_3)_2CHCH_2OH \rightarrow (CH_3)_2CHCH_2OCH_2CH(CH_3)_2 + H_2O$$

take place in a fluidized bed that can be treated as an ideal CSTR. The reactor contains 100 g Nafion H catalyst and operates at a temperature of 400 K and a total pressure of 1.34×10^3 kPa. Nitrogen gas is used a diluent. The alcohol plus nitrogen feed rate is 125 mol/h, with a 1:4 molar alcohol/ N_2 ratio. The alcohol feed contains 1 mol of isobutanol for every 2 mol of methanol.

The rate equations for the formation of IB, DME, MIBE, and DIBE are

$$r_{DME} = \frac{k_1 K_M^2 p_M^2}{(1 + K_M p_M + K_B p_B)^2}$$

$$r_{DIBE} = \frac{k_2 K_B^2 p_B^2}{(1 + K_M p_M + K_B p_B)^2}$$

$$r_{IB} = \frac{k_3 K_B p_B}{(1 + K_M p_M + K_B p_B)^2}$$

$$r_{MIBE} = \frac{k_4 K_M p_M K_B p_B}{(1 + K_M p_M + K_B p_B)^2}$$

where $k_1 = 0.23\,\mathrm{mol/g_{cat}\text{-}h}$, $k_2 = 0.10\,\mathrm{mol/g_{cat}\text{-}h}$, $k_3 = 1.10\,\mathrm{mol/g_{cat}\text{-}h}$, $k_4 = 2.78\,\mathrm{mol/g_{cat}\text{-}h}$, $K_B = 0.0243\,\mathrm{kPa^{-1}}$, and $K_M = 0.0137\,\mathrm{kPa^{-1}}$.[3]

What is the composition of the stream leaving the reactor?

Problem 7-13 (Level 1) Dimethyl ether (DME, CH_3OCH_3) is made commercially by the reaction of 2 mol of methanol (CH_3OH).

$$2CH_3OH \rightarrow CH_3OCH_3 + H_2O \quad (1)$$

Methanol is made by the reaction of CO and H_2

$$CO + 2H_2 \rightleftarrows CH_3OH \quad (2)$$

The equilibrium for Reaction (2) is relatively unfavorable. The methanol synthesis reactor must be operated at relatively high pressures to obtain a reasonable conversion of CO and H_2.

A process for manufacturing DME has been developed that couples Reactions (1) and (2) in series in a single reactor, i.e.,

$$2CO + 4H_2 \rightleftarrows 2CH_3OH \rightarrow CH_3OCH_3 + H_2O$$

In this process, Reaction (1) drives the equilibrium of Reaction (2) to the right by removing the product, methanol.

Let's generalize this system as follows:

$$A \rightleftarrows B \rightarrow C$$

Assume that all three of these generalized reactions are first-order. At a certain temperature, the rate constant (k_1) for the forward reaction A → B is $100\,\mathrm{min^{-1}}$, and the equilibrium constant (K_1) for the reversible reaction A ⇄ B is 1.0.

1. Suppose that only the reversible reaction A ⇄ B is carried out in an ideal CSTR. Assume that the reaction B → C does not take place. There is no B in the feed to the reactor. What value of the space time τ is required if the actual conversion of A leaving the CSTR must be at least 90% of the equilibrium conversion of A?
2. A catalyst must be developed for the reaction B → C. The rate constant for this reaction is designated k_2. If the reactor is operated at the space time that you calculated in Part (a), and at the same temperature, what value of k_2 is required to achieve a 75% conversion of A? There is no B in the feed to the reactor.

Problem 7-14 (Level 3) A study was conducted to investigate the kinetics of the hydrolysis of corn cob hemicellulose, catalyzed

[3] Nunan, J. G., Klier, K., and Herman, R. G., Methanol and 2-methyl-1-propanol (isobutanol) coupling to ethers and dehydration over nafion H: selectivity, kinetics, and mechanism, *J. Catal.* 139, 406–420 (1993).

by dilute sulfuric acid, at 98 °C.[4] The hemicellulose portion of the plant cell wall is made up primarily of the sugar polymer xylan. During hydrolysis, xylan reacts with water to form xylose, a monomeric sugar. The xylose then decomposes to furfural. Furfural is toxic to many microorganisms, and interferes with the later fermentation of xylose to xylitol, a sugar substitute.

The following model was used to represent the experimental data:

$$\text{xylan} \xrightarrow{k_1} \text{xylose} \xrightarrow{k_2} \text{furfural}$$

Both reactions were found to be first order and irreversible. The overall rate equation for the formation of xylose is

$$r_X = k_1 H - k_2 X$$

Since the structure of xylan is not well defined, its rate of disappearance was expressed in terms of *mass* concentration. The units of r_X are (mass xylose/vol solution-min). The units of k_1 are (mass xylose/mass xylan-min), "H" is the mass concentration of xylan (mass xylan/vol solution), and X is the mass concentration of xylose. The two rate constants (k_1, k_2) were determined experimentally as a function of acid concentration at 98 °C and were correlated by the following relationship:

$$k_i = a_i [\text{Ac}]^{b_i} \text{ min}^{-1}$$

Here, [Ac] is the H_2SO_4 concentration in g/ml. The values of a, i, and b are as follows:

i	a	b
1	5.52	1.43
2	0.0113	1.38

The molecular weight of xylose is 150 and the molecular weight of furfural is 96.

1. An experiment was carried out in an ideal batch reactor operating isothermally at 98 °C. At the beginning of the experiment, 8 g of corn cob (dry basis) was placed in the reactor and 100 ml of H_2SO_4 solution (3.00 g/100 ml) was added to the reactor. The xylan content of the feed material was 42 wt% based on dry material. There was no xylose or furfural in the corn cob. (Assume that the corn cob dissolves completely and that the volume of solution after the corn cob is dissolved is the same as the volume of the dilute H_2SO_4. You also will have to assume that the mass of xylose formed is equal to the mass of xylan hydrolyzed.) How much time is required to reach 90% conversion of xylan? What are the mass concentrations of xylose and furfural at this time?
2. If the concentration of furfural cannot exceed 105 μg/ml when the fractional conversion of xylan is 95%, what is the minimum acid concentration (g/100 ml) that can be used? At this concentration, how long will it take to reach 95% xylan conversion in an ideal batch reactor operating at 98 °C?

[4] Eken-Saracoglu, N., and Mutlu, F., Kinetics of dilute acid catalyzed hydrolysis of corn cob hemicellulose at 98 °C, *J. Qafqaz University*, 1 (1) 92–102 (1997).

Problem 7-15 (Level 3) Conventionally, aromatic compounds are benzylated by reacting them with benzyl chloride. Hydrochloric acid is a by-product of this reaction, and strong liquid acids are often used as a catalyst. A "greener" alternative is to use benzyl alcohol as the benzylating agent, along with a solid acid catalyst.

Naphthalene reacts with benzyl alcohol as shown in Reaction (1):

$$\text{naphthalene(N)} + \text{benzyl alcohol(B)} \rightarrow \text{benzyl naphthalene(M)} + \text{water} \quad (1)$$

Benzyl naphthalene (monobenzyl naphthalene) is the desired product. However, several side reactions occur:

$$\text{benzyl naphthalene} + \text{benzyl alcohol} \rightarrow \text{dibenzylnaphthalene (D)} + \text{water} \quad (2)$$

$$2\,\text{benzyl alcohol} \rightarrow \text{dibenzylether (BOB)} + \text{water} \quad (3)$$

As part of a kinetic study of the reaction of naphthalene with benzyl alcohol, Beltrame and Zuretti[5] developed the following rate equations for these reactions, using a Nafion®/silica composite catalyst.

$$-r_N = \frac{k_{am} C_B C_N}{(1 + K_B C_B)^2}$$

$$r_{BOB} = \frac{k_{ef} C_B^2}{(1 + K_B C_B)^2}$$

$$r_D = \frac{k_{ad} C_B C_M}{(1 + K_B C_B)^2}$$

The units of these rates are mol/g cat-min. At 80 °C,

$$k_{am} = 0.5\,\text{l}^2/\text{mol-min-g}$$

$$k_{ad} = 2.3\,\text{l}^2/\text{mol-min-g}$$

$$k_{ef} = 4\,\text{l}^2/\text{mol-min-g}$$

$$K_B = 110\,\text{l/mol}$$

The feed rate of naphthalene to a CSTR operating at 80 °C is 120 mol/h and the feed rate of benzyl alcohol is 24 mol/h. The feed is compromised of N and B dissolved in cyclohexane. The concentrations in the feed are $C_N^0 = 0.30$ mol/l and $C_B^0 = 0.06$ mol/l. There are 2.0 g/l of solid catalyst (the Nafion®/silica composite) suspended in the reactor, which has a volume of 1000 l.

1. What is the composition of the stream leaving the reactor?
2. What is the selectivity to monobenzyl naphthalene based on benzyl alcohol?

[5] Beltrame, P. and Zuretti, G., The reaction of naphthalene with benzyl alcohol over a Nafion-silica composite: a kinetic study, *Appl. Catal. A: Gen.*, 248, 75–83 (2003).

Problem 7-16 (Level 2) Ethylene glycol ($C_2H_6O_2$) is manufactured by the hydrolysis of ethylene oxide (C_2H_4O)

$$C_2H_4O + H_2O \rightarrow C_2H_6O_2 \qquad (1)$$

However, ethylene glycol, the desired product, can react with ethylene oxide to give diethylene glycol ($C_4H_{10}O_3$).

$$C_2H_6O_2 + C_2H_4O \rightarrow C_4H_{10}O_3 \qquad (2)$$

Let's represent these irreversible reactions as

$$A + B \rightarrow R \qquad (1)$$

$$A + R \rightarrow S \qquad (2)$$

where A = ethylene oxide, B = water, R = ethylene glycol, and S = diethylene glycol.

In aqueous solution at 120 °C, the rate equation for Reaction (1) is

$$-r_B = k_1[A][B]$$

$$k_1 = 1 \times 10^{-6} \text{l/mol-s}$$

The rate equation for Reaction (2) is

$$r_s = k_2[A][R]$$

1. Classify this network of reactions.
2. The concentrations of A and B in the feed to a CSTR are 5.0 and 40.0 mol/l respectively. There is no R or S in the feed. The concentrations of B and R in the effluent from the CSTR are 38.0 and 1.90 mol/l, respectively. What is the value of k_2/k_1?
3. The volumetric flow rate to the CSTR is 1.0 l/s. For the same inlet and outlet concentrations given in Part 2, what is the volume of the CSTR?
4. If $k_2/k_1 = 2.0$, what is the concentration of R in the effluent from an isothermal PFR that has the same feed as the CSTR above, and has a concentration of B in the effluent of 38.0 mol/l.

Problem 7-17 (Level 3$^+$) Dimethyl ether (DME, CH_3OCH_3) is made commercially by the reaction of 2 mol of methanol (CH_3OH).

$$2CH_3OH \rightarrow CH_3OCH_3 + H_2O \qquad (1)$$

Methanol is made by the reaction of CO and H_2.

$$CO + 2H_2 \rightleftarrows CH_3OH \qquad (2)$$

The equilibrium for Reaction (2) is relatively unfavorable. The methanol synthesis reactor must be operated at relatively high pressures to obtain a reasonable conversion of CO and H_2.

A process for manufacturing DME has been developed that couples Reactions (1) and (2) in series in a single reactor, i.e.,

$$2CO + 4H_2 \rightleftarrows 2CH_3OH \rightarrow CH_3OCH_3 + H_2O$$

In this process, Reaction 1 drives the equilibrium of Reaction (2) to the right by removing the product methanol.

Let's generalize this system as follows:

$$A \rightleftarrows B \rightarrow C$$

Assume that all the three reactions are first order. At a certain temperature, the rate constant (k_1) for the forward reaction A → B is 100 min^{-1}, the equilibrium constant (K_1) for the reversible reaction $A \rightleftarrows B$ is 1.0, and the rate constant for the reaction B → C also is 100 min^{-1}.

1. Suppose that the reversible reaction $A \rightleftarrows B$ is carried out in an ideal, isothermal PFR. Assume that the reaction B → C does not take place. There is no B in the feed to the reactor. What value of the space time τ is required if the actual conversion of A leaving the PFR must be at least 90% of the equilibrium conversion of A?
2. Now, consider the situation where the reaction B → C does take place. The PFR is operated at the space time that you calculated in Part (a), it remains isothermal, and there is no B or C in the feed. What is the conversion of A? What are the overall yields of B and C based on A? What are the selectivities to B and C, based on A?

Problem 7-18 (Level 2) Transesterification of vegetable oils with alcohols such as methanol is a key step in producing "bio-diesel" fuel. Lopez et al.[6] have used the reaction of the triglyceride triacetin with methanol as a model reaction to evaluate the performance of various solid transesterification catalysts.

The reactions taking place are

$$\underset{(TA)}{\text{triacetin}} + \underset{(MeOH)}{CH_3OH} \rightarrow \underset{(DA)}{\text{diacetin}} + \underset{(MeAc)}{\text{methyl acetate}} \qquad (1)$$

$$\text{diacetin} + CH_3OH \rightarrow \underset{(MA)}{\text{monoacetin}} + \text{methyl acetate} \qquad (2)$$

$$\text{monoacetin} + CH_3OH \rightarrow \underset{(G)}{\text{glycerol}} + \text{methyl acetate} \qquad (3)$$

Lopez et al. studied these liquid-phase reactions using an ETS-10 (Na,K) catalyst at 60 °C in an ideal batch reactor. Some of their data are shown in the following table.

Time (min)	Concentrations (mol/l)					
	TA	DA	MA	G	MeAc	MeOH
0	2.33	0.00	0.00	0.00	0.00	13.44
5	1.56	0.51	0.06	0.00	1.29	
10	1.29	0.77	0.20	0.00	1.89	
15	1.10	0.90	0.37	0.04	2.29	
20	0.99	0.99	0.50	0.10	2.56	
30	0.80	0.94	0.71	0.17	2.96	
40	0.67	0.86	0.83	0.24	3.26	
50	0.54	0.74	0.89	0.33	3.57	
60	0.43	0.66	0.90	0.43	3.89	

[6] Lopez, D. E., Goodwin, James G., Jr., Bruce, D. A., and Lotero, E., Transesterification of triacetin with methanol on solid acid and base catalysts, *Appl. Catal. A: Gen.*, 295, 97–105 (2005).

1. Based on the high initial ratio of methanol to triacetin, it might be hypothesized that the reaction is psuedo-first-order in triacetin. Plot the data so as to test this hypothesis. How well does the model fit the data?
2. What is the ratio of the methanol concentration after 60 min of reaction to the initial methanol concentration? In view of your answer, how reasonable is the psuedo-first-order assumption, i.e., that the methanol concentration was essentially constant over the first 60 min of reaction?
3. What is the selectivity to glycerol based on triacetin after 60 min of reaction? What is the yield of glycerol based on triacetin after 60 min?
4. The rate equation for triacetin disappearance may be

$$-r_{TA} = k[TA][MeOH]$$

Think of a way to test this rate equation against the above data graphically, using the integral method. Explain *clearly* and *concisely* what you would do.

Problem 7-19 (Level 1) The liquid-phase reactions

$$A \rightleftarrows B \quad (1)$$

$$A \rightarrow C \quad (2)$$

are taking place in an ideal CSTR. Reaction (1) is reversible; Reaction (2) is irreversible. "B" is the desired product. Each reaction is first order. The value of the forward rate constant for Reaction (1), k_{1f}, is 100 min^{-1} and the value of the reverse rate constant k_{1r} is 10 min^{-1}. The value of the rate constant for Reaction (2), k_2, is 10 min^{-1}. The CSTR operates at a space time τ of 0.10 min. There is no B or C in the feed to the reactor.

1. Classify the above system of reactions.
2. What is the *maximum* yield of B that can be obtained, ignoring Reaction (2) and letting the space time be very large?
3. Now, consider the case where both reactions take place. For the values given, calculate the
   (a) fractional conversion of A;
   (b) selectivity, $S(B/A)$;
   (c) yield, $Y(B/A)$.
4. Suppose that the space time τ were increased substantially. What would you expect to happen to the:
   (a) fractional conversion of A;
   (b) selectivity, $S(B/A)$;
   (c) yield, $Y(B/A)$.

Your answer for each question should be either: "increase," "no change," or "decrease." Calculations are not necessary, but you should explain each of your answers physically.

Multiple-Reactor Problems

Problem 7-20 (Level 2) The liquid-phase reactions $A \xrightarrow{k_1} R \xrightarrow{k_2} S$ are taking place at steady state in a system of reactors consisting of an ideal stirred-tank reactor followed by an ideal plug-flow reactor. Both reactions are irreversible and first order, with $k_1 = k_2 = 0.50$ min^{-1}. Component R is the desired product.

The feed to the CSTR contains A at a concentration of 2.0 g mol/l. There is no R or S in the feed to the CSTR.

The CSTR has a volume of 200 l and the PFR also has a volume of 200 l. The molar feed rate of A to the CSTR, F_{A0}, is 200 g mol/min.

1. What are the concentrations of A, R, and S in the stream leaving the plug-flow reactor?
2. Can the production rate of R be improved by using the same reactors, except that the PFR would precede the CSTR? How would this change affect the concentration of A in the effluent from the last reactor?

Problem 7-21 (Level 2) The irreversible, first-order, liquid-phase reactions

$$A \xrightarrow{k_1} R \xrightarrow{k_2} S$$

are currently carried out in two ideal CSTRs in series. The desired product is R. Two streams are fed to the first reactor. One is a solution of A at a volumetric flow rate of 100 gal/min. The concentration of A in this stream is 0.75 lb-mol/gal. The second stream is pure solvent at a flow rate of 200 gal/min. There is no R or S in either stream. Both reactors operate at 200 °C; at this temperature $k_1 = k_2 = 0.25$ min^{-1}. The volume of the first reactor is 150 gal and the volume of the second reactor is 1000 gal.

1. What are the concentrations of A, R, and S in the effluent from the first reactor?
2. What are the concentrations of A, R, and S in the effluent from the second reactor?
3. Would the concentration of R in the effluent from the second reactor increase, decrease or stay the same if the order of the two reactors was switched (large reactor first, small reactor last), with all other conditions remaining the same?
4. The second reactor in the original configuration is to be replaced by an ideal plug-flow reactor operating isothermally at 200 °C. What should the volume of this reactor be in order to maximize the production rate of R, i.e., the concentration of R in the effluent from the plug-flow reactor? How much will the production rate of R increase relative to the original configuration?

Problem 7-22 (Level 3) Conventionally, aromatic compounds are benzylated by reacting them with benzyl chloride. Hydrochloric acid is a by-product of this reaction, and strong liquid acids are often used as a catalyst. A "greener" alternative is to use benzyl alcohol as the benzylating agent, along with a solid acid catalyst.

Naphthalene reacts with benzyl alcohol as shown in Reaction (1)

$$\text{naphthalene (N)} + \text{benzyl alcohol (B)} \rightarrow \text{benzyl naphthalene (M)} + \text{water} \quad (1)$$

Benzyl naphthalene (monobenzyl naphthalene) is the desired product. However, several side reactions occur

$$\text{benzyl naphthalene} + \text{benzyl alcohol} \rightarrow \text{dibenzylnaphthalene (D)} + \text{water} \quad (2)$$

$$2\ \text{benzyl alcohol} \rightarrow \text{dibenzylether (BOB)} + \text{water} \quad (3)$$

As part of a kinetic study of the reaction of naphthalene with benzyl alcohol, Beltrame and Zuretti[5] developed the following rate equations for these reactions, using a Nafion®/silica composite catalyst.

$$-r_N = \frac{k_m C_B C_N}{(1 + K_B C_S)^2}$$

$$r_{BOB} = \frac{(k_{ai} C_B^2)}{(1 + K_B C_S)^2}$$

$$r_D = \frac{(k_{ad} C_B C_M)}{(1 + K_B C_B)^2}$$

The units of these rates are mol/g-cat min. At 80 °C,

$$k_{am} = 0.5\ \text{l}^2/\text{mol-min-g}$$

$$k_{ad} = 2.3\ \text{l}^2/\text{mol-min-g}$$

$$k_{ef} = 4\ \text{l}^2/\text{mol-min-g}$$

$$K_B = 110\ \text{l/mol}$$

Consider 2 CSTRs in series, each with a volume of 200 l and operating at 80 °C. There are 2.0 g/l of solid catalyst (the Nafion®/silica composite) suspended in each reactor. The feed rate of naphthalene to the first reactor is 120 mol/h and the feed rate of benzyl alcohol is 24 mol/h. The feed is compromised of N and B dissolved in cyclohexane. The concentrations in the feed to the first reactor are $C_{N,0} = 0.30$ mol/l and $C_{B,0} = 0.06$ mol/l. The outlet stream from the first reactor flows directly into the second reactor.

1. What is the composition of the stream leaving the first reactor?
2. What is the composition of the stream leaving the second reactor?
3. What is the selectivity to monobenzyl naphthalene based on benzyl alcohol for the overall system of two CSTRs?

APPENDIX 7-A NUMERICAL SOLUTION OF ORDINARY DIFFERENTIAL EQUATIONS

7-A.1 Single, First-Order Ordinary Differential Equation

Suppose we have a differential equation of the form

$$\frac{dy}{dx} = f(x, y) \tag{7A-1}$$

The value of y at $x = x_0$ is y_0. We are interested in calculating the value of y at some value of x, designated x_f, that is greater than x_0.

Begin by dividing the interval between x_0 and x_f into N equal segments, so that

$$\frac{x_f - x_0}{N} = \Delta x = h$$

The value of h is called the step size. The change in y when x is changed by $\Delta x = h$ will be denoted Δy.

Now, let's approximate the derivative on the left-hand side of Eqn. (7A-1) as

$$\frac{dy}{dx} \cong \frac{\Delta y}{\Delta x} = \frac{\Delta y}{h}$$

Let y_n be the value of y when $x = x_n$, and let $\bar{f}$ be the *average* value of $f(x, y)$ over the interval between x_n and x_{n+1}. The value of y_{n+1}, that is, the value of y at $x = x_{n+1} = x_n + h$, can be calculated using

$$\Delta y = y_{n+1} - y_n = \bar{f} \times h$$

$$y_{n+1} = y_n + \bar{f} \times h \tag{7A-2}$$

Calculating an accurate value of $\bar{f}$ is the most challenging part of solving differential equations numerically. Textbooks on numerical analysis discuss this issue in detail. For many problems, the fourth-order Runge–Kutta method can be used. In this method,

$$\bar{f} = \tfrac{1}{6}(f_1 + 2f_2 + 2f_3 + f_4) \tag{7A-3}$$

The parameters f_1, f_2, f_3, and f_4 in Eqn. (7A-3) are various approximations to the function $f(x, y)$. The parameter f_1 is $f(x, y)$ evaluated at x_n and y_n, i.e., at the start of the interval between x_n and x_{n+1}.

$$f_1 = f(x_n, y_n) \tag{7A-4a}$$

Once f_1 has been calculated, a better approximation of the function $f(x, y)$ can be calculated.

$$f_2 = f\left(x_n + \frac{h}{2}, y_n + \frac{hf_1}{2}\right) \tag{7A-4b}$$

Then, the value of f_2 can be used to generate a third approximation

$$f_3 = f\left(x_n + \frac{h}{2}, y_n + \frac{hf_2}{2}\right) \tag{7A-4c}$$

Finally, the process is completed by calculating f_4, which is an approximation to the function at x_{n+1}.

$$f_4 = f(x_n + h,\ y_n + h f_3) \tag{7A-4d}$$

The following spreadsheet shows how the fourth-order Runge–Kutta technique can be used to solve Example 7-2. After entering the values of the known parameters (C_{A0}, C_{B0}, k_1, and k_2) into the spreadsheet, a step size $h = \Delta\tau$ was selected. The value of h was chosen arbitrarily to be 4 min, so that the interval between the reactor inlet ($\tau = 0$) and the reactor outlet ($\tau = 40$) was divided into 10 slices. The corresponding values of τ were entered in the first column of the spreadsheet. The second column was set up to contain values of C_B. The initial value of C_B (0.10 mol/l) was entered into the first row of this column. The other values were calculated as described below.

The formula for $f(x, y)$ is given by the right-hand side of Eqn. (7-12). The value of f_1 at $\tau = 0$ was calculated from this formula and multiplied by h. The result is the first entry in the third column, labeled $h \times f_1$. Once the value of f_1 was known, the values of f_2, f_3, and f_4 were calculated in sequence. These values, multiplied by h, are shown in the fourth, fifth, and sixth columns of the spreadsheet.

Once these calculations were complete, the value of $\bar{f} \times h$ for the interval between $\tau = 0$ and $\tau = 4$ min was calculated by multiplying Eqn. (7A-3) by h. The value of C_B at $\tau = 4$ min then was calculated from Eqn. (7A-2), and the result was entered into the spreadsheet in the box corresponding to the second column, second row (C_B column, $\tau = 4$ min row).

The same procedure then was repeated for each of the remaining values of τ. The results are shown in the following spreadsheet. The calculated value of C_B at $\tau = 40$ min is 0.215 mol/l.

Solution To Example 7-2

Numerical solution to Eqn. 7-11 (using the fourth-order Runge–Kutta method)
$C_{A0} = 1$ mol/l   A → B (first order)
$C_{B0} = 0.1$ mol/l   2B → C (second order)
$k_1 = 0.5$ min^{-1}
$k_2 = 0.1$ l/mol-min Ideal, isothermal PFR
First attempt
Delta tau = h = 4 min

Tau (x) (min)	CB (y) (mol/l)	$h \times f_1$	$h \times f_2$	$h \times f_3$	$h \times f_4$
0	0.1000	1.9960	0.2535	0.7152	0.0049
4	0.7564	0.0418	−0.1421	−0.0883	−0.1419
8	0.6629	−0.1391	−0.1273	−0.1302	−0.1086
12	0.5358	−0.1099	−0.0907	−0.0944	−0.0773
16	0.4429	−0.0778	−0.0650	−0.0671	−0.0564
20	0.3765	−0.0566	−0.0485	−0.0496	−0.0427
24	0.3272	−0.0428	−0.0374	−0.0381	−0.0334
28	0.2894	−0.0335	−0.0297	−0.0301	−0.0269
32	0.2594	−0.0269	−0.0242	−0.0245	−0.0221
36	0.2350	−0.0221	−0.0201	−0.0202	−0.0184
40	**0.2148**				

In general, the accuracy of any numerical method depends on the value of the step size, h. If h is too large, $\Delta y/h$ will not be an accurate approximation of the derivative, dy/dx. To test the accuracy of the above solution, the procedure was repeated for a smaller value of h, 2 min, half of the original value. Now, the interval between $t = 0$ and $t = 40$ has been divided into 20 steps instead of the original 10. The results are shown in the following spreadsheet.

Solution to Example 7-2

Second attempt
Delta tau = h = 2 min

Tau (x) (min)	CB (y) (mol/l)	$h \times f_1$	$h \times f_2$	$h \times f_3$	$h \times f_4$
0	0.1000	0.9980	0.5348	0.5795	0.2755
2	0.6837	0.2744	0.0884	0.1172	0.0071
4	0.7991	0.0076	−0.0469	−0.0383	−0.0660
6	0.7610	−0.0660	−0.0758	−0.0744	−0.0760
8	0.6873	−0.0762	−0.0732	−0.0736	−0.0686
10	0.6142	−0.0687	−0.0632	−0.0638	−0.0581
12	0.5508	−0.0582	−0.0529	−0.0535	−0.0485
14	0.4975	−0.0486	−0.0442	−0.0446	−0.0407
16	0.4530	−0.0407	−0.0372	−0.0375	−0.0344
18	0.4156	−0.0344	−0.0317	−0.0319	−0.0294
20	0.3838	−0.0294	−0.0272	−0.0274	−0.0254
22	0.3564	−0.0254	−0.0236	−0.0237	−0.0221
24	0.3327	−0.0221	−0.0207	−0.0208	−0.0195
26	0.3120	−0.0195	−0.0183	−0.0183	−0.0172
28	0.2936	−0.0172	−0.0162	−0.0163	−0.0154
30	0.2774	−0.0154	−0.0145	−0.0146	−0.0138
32	0.2628	−0.0138	−0.0131	−0.0131	−0.0125
34	0.2497	−0.0125	−0.0119	−0.0119	−0.0113
36	0.2378	−0.0113	−0.0108	−0.0108	−0.0103
38	0.2270	−0.0103	−0.0098	−0.0099	−0.0094
40	**0.2171**				

In going from $h\,(\Delta\tau) = 4$ min to $h\,(\Delta\tau) = 2$ min, the final answer for C_B changed by only two places in the third significant figure. However, the values of C_B at lower values of τ are more sensitive to the value of h. For example, the value of C_B at $\tau = 4$ min is about 5% higher for $h = 2$ min than for $h = 4$ min. Therefore, we will try an even smaller value of h, 1 min. The results are shown in the following spreadsheet.

Solution to Example 7-2

Third attempt
Delta tau = h = 1 min

Tau (x) (min)	CB (y) (mol/l)	$h \times f_1$	$h \times f_2$	$h \times f_3$	$h \times f_4$
0	0.1000	0.4990	0.3772	0.3811	0.2801
1	0.4826	0.2800	0.1974	0.2024	0.1370
2	0.6854	0.1370	0.0864	0.0902	0.0514

Solution to Example 7-2

Third attempt Continued
Delta tau = h = 1 min

Tau (x) (min)	CB (y) (mol/l)	$h \times f_1$	$h \times f_2$	$h \times f_3$	$h \times f_4$
3	0.7756	0.0514	0.0227	0.0250	0.0036
4	0.8007	0.0036	−0.0117	−0.0105	−0.0214
5	0.7903	−0.0214	−0.0288	−0.0282	−0.0332
6	0.7622	−0.0332	−0.0362	−0.0360	−0.0376
7	0.7263	−0.0377	−0.0383	−0.0383	−0.0382
8	0.6882	−0.0382	−0.0376	−0.0377	−0.0368
9	0.6506	−0.0368	−0.0356	−0.0357	−0.0344
10	0.6149	−0.0344	−0.0331	−0.0332	−0.0318
11	0.5818	−0.0318	−0.0304	−0.0305	−0.0292
12	0.5513	−0.0292	−0.0278	−0.0279	−0.0266
13	0.5234	−0.0266	−0.0254	−0.0255	−0.0243
14	0.4980	−0.0243	−0.0232	−0.0233	−0.0223
15	0.4747	−0.0223	−0.0213	−0.0213	−0.0204
16	0.4534	−0.0204	−0.0195	−0.0195	−0.0187
17	0.4338	−0.0187	−0.0179	−0.0180	−0.0172
18	0.4159	−0.0172	−0.0165	−0.0166	−0.0159
19	0.3993	−0.0159	−0.0153	−0.0153	−0.0147
20	0.3840	−0.0147	−0.0142	−0.0142	−0.0137
21	0.3698	−0.0137	−0.0132	−0.0132	−0.0127
22	0.3566	−0.0127	−0.0123	−0.0123	−0.0119
23	0.3444	−0.0119	−0.0115	−0.0115	−0.0111
24	0.3329	−0.0111	−0.0107	−0.0107	−0.0104
25	0.3222	−0.0104	−0.0100	−0.0101	−0.0097
26	0.3121	−0.0097	−0.0094	−0.0094	−0.0092
27	0.3027	−0.0092	−0.0089	−0.0089	−0.0086
28	0.2938	−0.0086	−0.0084	−0.0084	−0.0081
29	0.2854	−0.0081	−0.0079	−0.0079	−0.0077
30	0.2775	−0.0077	−0.0075	−0.0075	−0.0073
31	0.2700	−0.0073	−0.0071	−0.0071	−0.0069
32	0.2629	−0.0069	−0.0067	−0.0067	−0.0066
33	0.2562	−0.0066	−0.0064	−0.0064	−0.0062
34	0.2498	−0.0062	−0.0061	−0.0061	−0.0059
35	0.2437	−0.0059	−0.0058	−0.0058	−0.0057
36	0.2379	−0.0057	−0.0055	−0.0055	−0.0054
37	0.2324	−0.0054	−0.0053	−0.0053	−0.0052
38	0.2271	−0.0052	−0.0050	−0.0050	−0.0049
39	0.2220	−0.0049	−0.0048	−0.0048	−0.0047
40	**0.2172**				

The calculations for $h = 1$ min and $h = 2$ min agree to within 1 place in the third significant figure, at all values of τ. This confirms the accuracy of the calculations at $h = 2$ min.

The effect of the interval or step size, h, on the solution is an issue with *all* numerical methods for solving differential equations. The step size must be reduced until a solution is obtained that is independent of h.

7-A.2 Simultaneous, First-Order, Ordinary Differential Equations

Suppose we have

$$\frac{dy}{dx} = f(x, y, z) \tag{7A-5}$$

$$\frac{dz}{dx} = g(x, y, z) \tag{7A-6}$$

The Runge–Kutta method can be extended as shown below:

$$y_{n+1} = y_n + \bar{f} \times h \tag{7A-2}$$

$$z_{n+1} = z_n + \bar{g} \times h \tag{7A-7}$$

$$\bar{f} = \tfrac{1}{6}(f_1 + 2f_2 + 2f_3 + f_4) \tag{7A-3}$$

$$\bar{g} = \tfrac{1}{6}(g_1 + 2g_2 + 2g_3 + g_4) \tag{7A-8}$$

$$f_1 = f(x_n, y_n, z_n) \tag{7A-9a}$$

$$g_1 = g(x_n, y_n, z_n) \tag{7A-9b}$$

$$f_2 = f\left(x_n + \frac{h}{2}, y_n + \frac{hf_1}{2}, z_n + \frac{hg_1}{2}\right) \tag{7A-9c}$$

$$g_2 = g\left(x_n + \frac{h}{2}, y_n + \frac{hf_1}{2}, z_n + \frac{hg_1}{2}\right) \tag{7A-9d}$$

$$f_3 = f\left(x_n + \frac{h}{2}, y_n + \frac{hf_2}{2}, z_n + \frac{hg_2}{2}\right) \tag{7A-9e}$$

$$g_3 = g\left(x_n + \frac{h}{2}, y_n + \frac{hf_2}{2}, z_n + \frac{hg_2}{2}\right) \tag{7A-9f}$$

$$f_4 = f(x_n + h, y_n + hf_3, z_n + hg_3) \tag{7A-9g}$$

$$g_4 = g(x_n + h, y_n + hf_3, z_n + hg_3) \tag{7A-9h}$$

The following spreadsheet shows how the Runge–Kutta technique for simultaneous differential equations can be used to solve Example 7-4.

First, we note that this problem is more complex than Example 7-2. The value of τ was known in that problem, and the only challenge was to find a value of h that was small enough. In this problem, the value of τ is not known, so no guidance is available in selecting h.

Let's begin by finding an approximate value of τ. Better yet, let's find values of τ that bracket the true value. Equation (7-24) can be rearranged and integrated from the reactor inlet to the reactor outlet to give

$$\int_{10^{-2}}^{10^{-5}} \frac{dp_{CO}}{p_{CO}} = \ln(10^{-3}) = -6.91 = -(k_1RT)\int_0^{\tau} p_{O_2}^{1/2}d\tau$$

The last integral in the above equation may be evaluated by taking an average value of p_{O_2} and substituting the value of k_1RT.

$$\tau = \frac{6.91}{4743\bar{p}_{O_2}^{1/2}} \text{ g-min/l}$$

The value of $\bar{p}_{O_2}^{1/2}$ must lie somewhere between $(10^{-4}\text{ atm})^{1/2}$ and $(0.0118\text{ atm})^{1/2}$. Therefore, the true value of τ must be bracketed by

$$0.014 \leq \tau \leq 0.15 \text{ g-min/l}$$

Let's begin the numerical integration with a step size h of 0.002 g-min/l. With this step size, the above calculation suggests that somewhere between 7 and 75 steps will be required to reach a carbon monoxide partial pressure of 10^{-5} atm. The result of this calculation is shown in the table on the following page, labeled "First attempt." This table shows that between 12 and 13 steps were required. Linear interpolation between the values of p_{CO} at $\tau = 0.024$ and $\tau = 0.026$ g-min/l gives a value of $\tau = 0.0256$ g-min/l for the space time required to reach a CO concentration of 10 ppm.

The effect of the step size h now must be investigated. The table labeled "Second attempt" shows the results of calculations with $h = 0.001$ g-min/l. This calculation shows that the required value of τ is between 0.024 and 0.025 g-min/l. Linear interpolation gives a value of $\tau = 0.0247$ g-min/l. This value is about 4% lower than the "First attempt" value.

Just to be sure, a third calculation was run with $\tau = 0.0005$ g-min/l. The result is shown in the table labeled "Third attempt." This calculation gives a final value of $\tau = 0.0246$ g-min/l, which we will take as the final answer to the problem.

Solution to Example 7-4

Numerical solution to Eqns. (7-23) and (7-24)

p_{H_2} (in) = 0.3000 atm k_1 (RT) = 4743.72075 l/g-min-atm$^{0.5}$

p_{CO} (in) = 0.100 atm k_2 (RT) = 59.6790675 l/g-min-atm$^{0.5}$

p_{O_2} (in) = 0.0118 atm K_{CO} = 1000 atm^{-1}

First attempt:

Delta tau = h = 0.002 g-min/l

Tau (g-min/l)	p_{H_2} (atm)	p_{CO} (atm)	p_{O_2} (atm)	$h \times f_1$	$h \times g_1$	$h \times f_2$	$h \times g_2$	$h \times f_3$	$h \times g_3$	$h \times f_y$	$h \times g_4$
0.000	0.30000	1.0000E−02	1.1800E−02	−1.031E−02	−3.215E−05	−4.414E−03	−1.005E−04	−7.637E−03	−4.784E−05	−1.999E−03	−2.825E−04
0.002	0.29990	3.9318E−03	8.7149E−03	−3.482E−03	−1.374E−04	−1.837E−03	−3.107E−04	−2.585E−03	−2.009E−04	−1.093E−03	−5.560E−04
0.004	0.29961	1.6952E−03	7.4536E−03	−1.389E−03	−4.250E−04	−7.945E−04	−7.468E−04	−1.035E−03	−5.687E−04	−5.106E−04	−1.057E−03
0.006	0.29893	7.6873E−04	6.6477E−03	−5.946E−04	−9.299E−04	−3.540E−04	−1.303E−03	−4.432E−04	−1.109E−03	−2.366E−04	−1.550E−03
0.008	0.29771	3.6442E−04	5.8368E−03	−2.641E−04	−1.458E−03	−1.621E−04	−1.716E−03	−1.970E−04	−1.576E−03	−1.118E−04	−1.825E−03
0.010	0.29606	1.8209E−04	4.9234E−03	−1.212E−04	−1.774E−03	−7.688E−05	−1.869E−03	−9.078E−05	−1.794E−03	−5.466E−05	−1.861E−03
0.012	0.29424	9.6888E−05	3.9674E−03	−5.790E−05	−1.839E−03	−3.810E−05	−1.814E−03	−4.372E−05	−1.784E−03	−2.788E−05	−1.739E−03
0.014	0.29244	5.5321E−05	3.0488E−03	−2.898E−05	−1.731E−03	−1.979E−05	−1.641E−03	−2.212E−05	−1.634E−03	−1.484E−05	−1.532E−03
0.016	0.29081	3.4049E−05	2.2204E−03	−1.522E−05	−1.530E−03	−1.074E−05	−1.407E−03	−1.175E−05	−1.413E−03	−8.214E−06	−1.283E−03
0.018	0.28940	2.2644E−05	1.5101E−03	−8.349E−06	−1.284E−03	−6.038E−06	−1.145E−03	−6.510E−06	−1.159E−03	−4.661E−06	−1.015E−03
0.020	0.28825	1.6294E−05	9.3142E−04	−4.718E−06	−1.017E−03	−3.438E−06	−8.686E−04	−3.693E−06	−8.914E−04	−2.630E−06	−7.358E−04
0.022	0.28737	1.2692E−05	4.9025E−04	−2.666E−06	−7.405E−04	−1.880E−06	−5.844E−04	−2.067E−06	−6.206E−04	−1.348E−06	−4.483E−04
0.024	0.28677	1.0707E−05	1.8937E−04	−1.398E−06	−4.611E−04	−8.154E−07	−2.879E−04	−1.058E−06	−3.628E−04	−2.495E−07	−9.138E−05
0.026	0.28646	9.8082E−06	3.4425E−05								

By linear interpolation, tau = 0.0256 g-min/l.

Solution to Example 7-4

Second attempt
Delta tau = h = 0.001 g-min/l

Tau (g-min/l)	p_{H_2} (atm)	p_{CO} (atm)	p_{O_2} (atm)	$h \times f_1$	$h \times g_1$	$h \times f_2$	$h \times g_2$	$h \times f_3$	$h \times g_3$	$h \times f_4$	$h \times g_4$
0.000	0.30000	1.0000E−02	1.1800E−02	−5.153E−03	−1.607E−05	−3.610E−03	−2.586E−05	−4.057E−03	−2.210E−05	−2.785E−03	−3.669E−05
0.001	0.29998	6.1214E−03	9.8483E−03	−2.882E−03	−3.503E−05	−2.120E−03	−5.298E−05	−2.316E−03	−4.700E−05	−1.681E−03	−7.216E−05
0.002	0.29992	3.8822E−03	8.7031E−03	−1.718E−03	−7.005E−05	−1.303E−03	−1.005E−04	−1.401E−03	−9.138E−05	−1.050E−03	−1.317E−04
0.003	0.29983	2.5196E−03	7.9731E−03	−1.067E−03	−1.290E−04	−8.253E−04	−1.758E−04	−8.784E−04	−1.628E−04	−6.721E−04	−2.213E−04
0.004	0.29966	1.6619E−03	7.4585E−03	−6.808E−04	−2.180E−04	−5.332E−04	−2.821E−04	−5.638E−04	−2.654E−04	−4.372E−04	−3.407E−04
0.005	0.29938	1.1099E−03	7.0447E−03	−4.419E−04	−3.369E−04	−3.490E−04	−4.142E−04	−3.673E−04	−3.946E−04	−2.875E−04	−4.797E−04
0.006	0.29897	7.4953E−04	6.6617E−03	−2.902E−04	−4.758E−04	−2.306E−04	−5.571E−04	−2.419E−04	−5.367E−04	−1.907E−04	−6.206E−04
0.007	0.29843	5.1186E−04	6.2692E−03	−1.923E−04	−6.169E−04	−1.536E−04	−6.914E−04	−1.606E−04	−6.723E−04	−1.275E−04	−7.446E−04
0.008	0.29775	3.5383E−04	5.8495E−03	−1.284E−04	−7.415E−04	−1.031E−04	−8.008E−04	−1.075E−04	−7.847E−04	−8.589E−05	−8.387E−04
0.009	0.29695	2.4791E−04	5.4006E−03	−8.642E−05	−8.363E−04	−6.982E−05	−8.768E−04	−7.261E−05	−8.642E−04	−5.840E−05	−8.984E−04
0.010	0.29608	1.7629E−04	4.9300E−03	−5.872E−05	−8.967E−04	−4.774E−05	−9.186E−04	−4.951E−05	−9.096E−04	−4.012E−05	−9.256E−04
0.011	0.29517	1.2740E−04	4.4490E−03	−4.031E−05	−9.244E−04	−3.300E−05	−9.306E−04	−3.413E−05	−9.245E−04	−2.788E−05	−9.257E−04
0.012	0.29424	9.3662E−05	3.9688E−03	−2.799E−05	−9.249E−04	−2.308E−05	−9.187E−04	−2.380E−05	−9.149E−04	−1.960E−05	−9.047E−04
0.013	0.29333	7.0101E−05	3.4990E−03	−1.967E−05	−9.043E−04	−1.634E−05	−8.888E−04	−1.681E−05	−8.867E−04	−1.396E−05	−8.684E−04
0.014	0.29244	5.3447E−05	3.0470E−03	−1.400E−05	−8.681E−04	−1.171E−05	−8.461E−04	−1.202E−05	−8.452E−04	−1.006E−05	−8.210E−04
0.015	0.29160	4.1529E−05	2.6184E−03	−1.008E−05	−8.209E−04	−8.499E−06	−7.942E−04	−8.695E−06	−7.942E−04	−7.334E−06	−7.660E−04
0.016	0.29080	3.2895E−05	2.2171E−03	−7.348E−06	−7.660E−04	−6.236E−06	−7.361E−04	−6.367E−06	−7.367E−04	−5.406E−06	−7.057E−04
0.017	0.29007	2.6568E−05	1.8459E−03	−5.415E−06	−7.057E−04	−4.623E−06	−6.736E−04	−4.711E−06	−6.748E−04	−4.024E−06	−6.419E−04
0.018	0.28939	2.1884E−05	1.5065E−03	−4.029E−06	−6.419E−04	−3.457E−06	−6.083E−04	−3.518E−06	−6.099E−04	−3.018E−06	−5.756E−04
0.019	0.28878	1.8385E−05	1.2002E−03	−3.021E−06	−5.757E−04	−2.601E−06	−5.410E−04	−2.644E−06	−5.430E−04	−2.274E−06	−5.077E−04
0.020	0.28824	1.5754E−05	9.2798E−04	−2.277E−06	−5.079E−04	−1.962E−06	−4.723E−04	−1.944E−06	−4.749E−04	−1.714E−06	−4.388E−04
0.021	0.28777	1.3770E−05	6.9023E−04	−1.716E−06	−4.390E−04	−1.475E−06	−4.028E−04	−1.501E−06	−4.059E−04	−1.284E−06	−3.692E−04
0.022	0.28736	1.2278E−05	4.8735E−04	−1.286E−06	−3.695E−04	−1.097E−06	−3.327E−04	−1.118E−06	−3.365E−04	−9.448E−07	−2.990E−04
0.023	0.28703	1.1168E−05	3.1955E−04	−9.471E−07	−2.995E−04	−7.932E−07	−2.620E−04	−8.141E−07	−2.670E−04	−6.692E−07	−2.284E−04
0.024	0.28676	1.0363E−05	1.8698E−04	−6.722E−07	−2.292E−04	−5.413E−07	−1.908E−04	−5.647E−07	−1.978E−04	−4.355E−07	−1.572E−04
0.025	0.28657	9.8099E−06	8.9736E−05	−4.408E−07	−1.589E−04	−3.214E−07	−1.185E−04	−3.546E−07	−1.300E−04	−2.224E−07	−8.317E−05

By linear interpolation, tau = 0.0247 g-min/l.

Solution to Example 7-4

Third attempt
Delta tau = h = 0.0005 g-min/l

Tau (g-min/l)	p_{H_2} (atm)	p_{CO} (atm)	p_{O_2} (atm)	$h \times f_1$	$h \times g_1$	$h \times f_2$	$h \times g_2$	$h \times f_3$	$h \times g_3$	$h \times f_4$	$h \times g_4$
0.0000	0.30000	1.0000E−02	1.1800E−02	−2.576E−03	−8.037E−06	−2.182E−03	−1.002E−05	−2.241E−03	−9.671E−06	−1.901E−03	−1.206E−05
0.0005	0.29999	7.7791E−03	1.0685E−02	−1.907E−03	−1.201E−05	−1.635E−03	−1.477E−05	−1.673E−03	−1.431E−05	−1.437E−03	−1.759E−05
0.0010	0.29998	6.1189E−03	9.8472E−03	−1.440E−03	−1.753E−05	−1.247E−03	−2.129E−05	−1.272E−03	−2.071E−05	−1.103E−03	−2.512E−05
0.0015	0.29995	4.8553E−03	9.2048E−03	−1.105E−03	−2.505E−05	−9.640E−04	−3.006E−05	−9.817E−04	−2.934E−05	−8.569E−04	−3.514E−05
0.0020	0.29992	3.8798E−03	8.7021E−03	−8.584E−04	−3.506E−05	−7.536E−04	−4.160E−05	−7.662E−04	−4.070E−05	−6.728E−04	−4.817E−05
0.0025	0.29988	3.1180E−03	8.3006E−03	−6.738E−04	−4.808E−05	−5.944E−04	−5.640E−05	−6.035E−04	−5.529E−05	−5.325E−04	−6.467E−05
0.0030	0.29983	2.5176E−03	7.9724E−03	−5.332E−04	−6.456E−05	−4.722E−04	−7.486E−05	−4.790E−04	−7.354E−05	−4.242E−04	−8.499E−05
0.0035	0.29975	2.0410E−03	7.6969E−03	−4.247E−04	−8.486E−05	−3.774E−04	−9.725E−05	−3.825E−04	−9.569E−05	−3.397E−04	−1.093E−04
0.0040	0.29966	1.6603E−03	7.4582E−03	−3.401E−04	−1.091E−04	−3.030E−04	−1.236E−04	−3.068E−04	−1.218E−04	−2.732E−04	−1.373E−04
0.0045	0.29953	1.3548E−03	7.2440E−03	−2.735E−04	−1.372E−04	−2.441E−04	−1.535E−04	−2.471E−04	−1.515E−04	−2.205E−04	−1.688E−04
0.0050	0.29938	1.1087E−03	7.0447E−03	−2.207E−04	−1.686E−04	−1.974E−04	−1.864E−04	−1.997E−04	−1.843E−04	−1.785E−04	−2.028E−04
0.0055	0.29920	9.0982E−04	6.8525E−03	−1.786E−04	−2.026E−04	−1.600E−04	−2.214E−04	−1.618E−04	−2.191E−04	−1.448E−04	−2.383E−04
0.0060	0.29897	7.4866E−04	6.6618E−03	−1.449E−04	−2.381E−04	−1.300E−04	−2.572E−04	−1.314E−04	−2.549E−04	−1.178E−04	−2.741E−04
0.0065	0.29872	6.1776E−04	6.4683E−03	−1.178E−04	−2.739E−04	−1.058E−04	−2.926E−04	−1.069E−04	−2.903E−04	−9.594E−05	−3.089E−04
0.0070	0.29843	5.1123E−04	6.2693E−03	−9.601E−05	−3.087E−04	−8.629E−05	−3.265E−04	−8.718E−05	−3.243E−04	−7.832E−05	−3.416E−04
0.0075	0.29810	4.2435E−04	6.0632E−03	−7.837E−05	−3.414E−04	−7.052E−05	−3.577E−04	−7.122E−05	−3.556E−04	−6.406E−05	−3.711E−04
0.0080	0.29775	3.5336E−04	5.8495E−03	−6.410E−05	−3.710E−04	−5.774E−05	−3.853E−04	−5.830E−05	−3.834E−04	−5.250E−05	−3.969E−04
0.0085	0.29736	2.9525E−04	5.6283E−03	−5.254E−05	−3.968E−04	−4.738E−05	−4.090E−04	−4.783E−05	−4.072E−04	−4.313E−05	−4.185E−04
0.0090	0.29695	2.4757E−04	5.4005E−03	−4.315E−05	−4.184E−04	−3.897E−05	−4.283E−04	−3.932E−05	−4.268E−04	−3.551E−05	−4.357E−04
0.0095	0.29653	2.0836E−04	5.1672E−03	−3.552E−05	−4.356E−04	−3.212E−05	−4.432E−04	−3.241E−05	−4.419E−04	−2.930E−05	−4.486E−04
0.0100	0.29608	1.7605E−04	4.9298E−03	−2.932E−05	−4.485E−04	−2.655E−05	−4.540E−04	−2.677E−05	−4.528E−04	−2.425E−05	−4.574E−04
0.0105	0.29563	1.4934E−04	4.6898E−03	−2.426E−05	−4.573E−04	−2.200E−05	−4.607E−04	−2.218E−05	−4.598E−04	−2.012E−05	−4.624E−04
0.0110	0.29517	1.2722E−04	4.4487E−03	−2.013E−05	−4.623E−04	−1.828E−05	−4.639E−04	−1.843E−05	−4.632E−04	−1.674E−05	−4.640E−04
0.0115	0.29471	1.0884E−04	4.2078E−03	−1.675E−05	−4.640E−04	−1.524E−05	−4.639E−04	−1.535E−05	−4.633E−04	−1.397E−05	−4.626E−04
0.0120	0.29424	9.3526E−05	3.9684E−03	−1.397E−05	−4.625E−04	−1.273E−05	−4.611E−04	−1.283E−05	−4.606E−04	−1.169E−05	−4.585E−04
0.0125	0.29378	8.0728E−05	3.7317E−03	−1.170E−05	−4.585E−04	−1.068E−05	−4.558E−04	−1.075E−05	−4.554E−04	−9.817E−06	−4.522E−04
0.0130	0.29333	6.9999E−05	3.4985E−03	−9.820E−06	−4.522E−04	−8.979E−06	−4.484E−04	−9.039E−06	−4.482E−04	−8.268E−06	−4.440E−04
0.0135	0.29288	6.0979E−05	3.2699E−03	−8.271E−06	−4.440E−04	−7.575E−06	−4.393E−04	−7.623E−06	−4.391E−04	−6.985E−06	−4.341E−04
0.0140	0.29244	5.3370E−05	3.0465E−03	−6.987E−06	−4.341E−04	−6.410E−06	−4.287E−04	−6.449E−06	−4.286E−04	−5.919E−06	−4.228E−04
0.0145	0.29201	4.6932E−05	2.8290E−03	−5.921E−06	−4.228E−04	−5.441E−06	−4.168E−04	−5.473E−06	−4.167E−04	−5.031E−06	−4.105E−04
0.0150	0.29159	4.1469E−05	2.6179E−03	−5.033E−06	−4.104E−04	−4.632E−06	−4.039E−04	−4.658E−06	−4.039E−04	−4.290E−06	−3.971E−04
0.0155	0.29119	3.6818E−05	2.4137E−03	−4.290E−06	−3.971E−04	−3.956E−06	−3.901E−04	−3.976E−06	−3.901E−04	−3.667E−06	−3.830E−04

Solution to Example 7-4

Third attempt Continued
Delta tau = h = 0.0005 g-min/l

Tau (g-min/l)	p_{H_2} (atm)	p_{CO} (atm)	p_{O_2} (atm)	$h \times f_1$	$h \times g_1$	$h \times f_2$	$h \times g_2$	$h \times f_3$	$h \times g_3$	$h \times f_4$	$h \times g_4$
0.0160	0.29080	3.2848E−05	2.2167E−03	−3.668E−06	−3.830E−04	−3.387E−06	−3.756E−04	−3.404E−06	−3.757E−04	−3.144E−06	−3.682E−04
0.0165	0.29042	2.9449E−05	2.0271E−03	−3.145E−06	−3.682E−04	−2.908E−06	−3.605E−04	−2.922E−06	−3.606E−04	−2.703E−06	−3.528E−04
0.0170	0.29006	2.6531E−05	1.8454E−03	−2.703E−06	−3.528E−04	−2.503E−06	−3.449E−04	−2.514E−06	−3.451E−04	−2.329E−06	−3.371E−04
0.0175	0.28972	2.4020E−05	1.6717E−03	−2.329E−06	−3.371E−04	−2.159E−06	−3.290E−04	−2.169E−06	−3.291E−04	−2.011E−06	−3.209E−04
0.0180	0.28939	2.1854E−05	1.5061E−03	−2.012E−06	−3.209E−04	−1.867E−06	−3.127E−04	−1.875E−06	−3.128E−04	−1.740E−06	−3.045E−04
0.0185	0.28908	1.9981E−05	1.3488E−03	−1.741E−06	−3.045E−04	−1.617E−06	−2.961E−04	−1.623E−06	−2.963E−04	−1.508E−06	−2.878E−04
0.0190	0.28878	1.8360E−05	1.1999E−03	−1.508E−06	−2.878E−04	−1.402E−06	−2.793E−04	−1.408E−06	−2.795E−04	−1.309E−06	−2.709E−04
0.0195	0.28850	1.6954E−05	1.0595E−03	−1.309E−06	−2.709E−04	−1.217E−06	−2.623E−04	−1.222E−06	−2.626E−04	−1.136E−06	−2.539E−04
0.0200	0.28824	1.5733E−05	9.2762E−04	−1.137E−06	−2.539E−04	−1.057E−06	−2.452E−04	−1.061E−06	−2.455E−04	−9.869E−07	−2.367E−04
0.0205	0.28799	1.4673E−05	8.0443E−04	−9.871E−07	−2.367E−04	−9.180E−07	−2.280E−04	−9.216E−07	−2.283E−04	−8.567E−07	−2.195E−04
0.0210	0.28777	1.3753E−05	6.8991E−04	−8.568E−07	−2.195E−04	−7.963E−07	−2.106E−04	−7.995E−07	−2.110E−04	−7.425E−07	−2.021E−04
0.0215	0.28756	1.2954E−05	5.8412E−04	−7.426E−07	−2.021E−04	−6.893E−07	−1.932E−04	−6.922E−07	−1.936E−04	−6.418E−07	−1.847E−04
0.0220	0.28736	1.2263E−05	4.8707E−04	−6.419E−07	−1.847E−04	−5.946E−07	−1.757E−04	−5.973E−07	−1.762E−04	−5.525E−07	−1.672E−04
0.0225	0.28719	1.1666E−05	3.9880E−04	−5.526E−07	−1.672E−04	−5.103E−07	−1.582E−04	−5.129E−07	−1.587E−04	−4.726E−07	−1.497E−04
0.0230	0.28703	1.1155E−05	3.1932E−04	−4.728E−07	−1.497E−04	−4.347E−07	−1.406E−04	−4.372E−07	−1.412E−04	−4.007E−07	−1.321E−04
0.0235	0.28689	1.0718E−05	2.4864E−04	−4.009E−07	−1.321E−04	−3.662E−07	−1.230E−04	−3.688E−07	−1.237E−04	−3.353E−07	−1.145E−04
0.0240	0.28676	1.0351E−05	1.8678E−04	−3.355E−07	−1.146E−04	−3.036E−07	−1.054E−04	−3.063E−07	−1.062E−04	−2.753E−07	−9.689E−05
0.0245	0.28666	1.0045E−05	1.3375E−04	−2.756E−07	−9.697E−05	−2.458E−07	−8.772E−05	−2.488E−07	−8.864E−05	−2.196E−07	−7.925E−05
0.0250	0.28657	9.7981E−06	8.9546E−05	−2.199E−07	−7.936E−05	−1.918E−07	−6.999E−05	−1.953E−07	−7.116E−05	−1.672E−07	−6.156E−05

By linear interpolation, tau = 0.0246 g-min/l.

Chapter 8

Use of the Energy Balance in Reactor Sizing and Analysis

LEARNING OBJECTIVES

After completing this chapter, you should be able to

1. explain why designing an isothermal PFR or isothermal batch reactor is not practical on a commercial scale, if the enthalpy change on reaction is significant;
2. explain why adiabatic operation always should be considered in reactor sizing and analysis, and discuss the factors that might make it impractical to operate adiabatically;
3. size and analyze adiabatic plug-flow and batch reactors;
4. size and analyze continuous stirred-tank reactors that operate either adiabatically or with heating/cooling;
5. analyze the behavior of the combination of an adiabatic reactor and a feed/product heat exchanger;
6. use a graphical technique to predict whether a CSTR, or the combination of an adiabatic reactor and a feed/product heat exchanger, can exhibit multiple steady states;
7. explain the phenomena of "blowout" and feed-temperature hysteresis.

8.1 INTRODUCTION

When we sized and analyzed ideal reactors in Chapters 4 and 7, we made a rather naïve assumption. We assumed that we knew the temperature at which the reactor would operate. Moreover, when we treated ideal PFRs and batch reactors, we compounded our naiveté with a second assumption. With batch reactors, we usually assumed that the temperature did not vary with time. With PFRs, we assumed that the temperature did not vary with position in the direction of flow. We referred to this temperature invariance with time or position as "isothermal" operation.

In Chapters 4 and 7, we did not consider *how* to make a reactor operate at a specified temperature, or whether it was even *possible* to make the reactor operate at that temperature. Furthermore, we did not ask how to make PFRs and batch reactors operate isothermally.

The energy balance ultimately determines whether or not a reactor can be operated at a specified temperature, and whether a batch reactor or a PFR can operate isothermally. Detailed reactor sizing and analysis requires solution of the energy balance *in conjunction with* one or more material balances. In Chapters 4 and 7, we considered only the material balances. This chapter deals with the additional complexity of the energy balance.

8.2 MACROSCOPIC ENERGY BALANCES

8.2.1 Generalized Macroscopic Energy Balance

The energy balance is covered in detail in introductory courses on material and energy balances, and in later courses on thermodynamics. The following is a brief summary of the energy balance for chemically reacting systems.

For a reactor in which one or more reactions are taking place,

$$Q - W_s + \dot{H}_{in} - \dot{H}_{out} = dU_{sys}/dt \tag{8-1}$$

The terms in Eqn. (8-1) have the following meanings:

- Q is the rate of heat transfer *into* the reactor, e.g., in kJ/s. The value of Q is positive if heat is transferred into the reactor, and negative if heat is removed from the reactor.
- W_s is the rate at which shaft work is done *by* the system (the contents of the reactor) on the surroundings. If shaft work is done on the contents of the reactor, e.g., by an agitator, the value of W_s is negative.
- $\dot{H}_{in}$ is the rate at which enthalpy is transported into the reactor.
- $\dot{H}_{out}$ is the rate at which enthalpy is transported out of the reactor.
- dU_{sys}/dt is the rate at which the total internal energy of the system, U_{sys}, changes with time.

In formulating the energy balance in Eqn. (8-1), the kinetic and potential energy of the feed and product Streams, and of the reactor contents, have been neglected. In most cases, this is a valid assumption.

The rate at which enthalpy enters the reactor is given by $\sum_{i=0}^{N} F_{i0}\overline{H}_{i0}$. The summation includes all species in the system, not just the compounds that participate in reactions. The symbol $\overline{H}_{i0}$ denotes the partial molar enthalpy of species "i", at inlet conditions. In previous chapters, we used the relationship $F_{i0} = \upsilon C_{A0}$. This relationship is valid when bulk flow (convection) is the only mechanism for mass transport across the system boundaries, i.e., when diffusion of chemical species across the system boundaries can be neglected. This assumption was made in Chapter 3 during the derivation of the design equations for the ideal CSTR and ideal PFR. We will continue to assume that convection is the only mechanism for enthalpy transport across the system boundaries. The enthalpy flow associated with mass diffusion is neglected.

The rate at which enthalpy leaves the reactor is given by $\sum_{i=1}^{N} F_i\overline{H}_i$, where F_i is the molar flow rate of "i" out of the reactor, and $\overline{H}_i$ is the partial molar enthalpy of "i" in the outlet Stream. Again, diffusion of mass and enthalpy across the system boundaries will be ignored.

8.2.1.1 Single Reactors

The inlet and outlet molar flow rates are related through the extents of reaction. If "R" independent reactions take place, and if the extents of reaction are zero in the Stream that enters the reactor,

$$F_i - F_{i0} = \sum_{k=1}^{R} \nu_{ki}\xi_k \tag{1-10}$$

Therefore,

$$\dot{H}_{in} - \dot{H}_{out} = \sum_{i=1}^{N} F_{i0}\dot{H}_{i0} - \sum_{i=1}^{N}\left(F_{i0} + \sum_{k=1}^{R} \nu_{ki}\xi_k\right)\overline{H}_i$$

We will assume that the feed and product Streams are ideal solutions, so that the partial molar enthalpies, $\overline{H}_i$, can be replaced by pure component enthalpies, H_i. The ideal solution assumption usually is reasonably good for gas-phase reactions, provided that the pressure is not too high. Moreover, this assumption may be necessary for reactions in solution, at least for preliminary calculations, until a sound thermodynamic description of non-idealities is available. Thus,

$$\dot{H}_{\text{in}} - \dot{H}_{\text{out}} = \sum_{i=1}^{N} F_{i0}(H_{i0} - H_i) - \sum_{k=1}^{R} \xi_k \sum_{i=1}^{N} \nu_{ki} H_i \tag{8-2}$$

The term $\sum_{i=1}^{N} \nu_{ki} H_i$ is just the enthalpy change on reaction, i.e., the heat of reaction, for Reaction "k", evaluated at exit conditions. Since the temperature of the effluent Stream is T, we write

$$\sum_{i=1}^{N} \nu_{ki} H_i = \Delta H_{\text{R},k}(T) \tag{8-3}$$

where $\Delta H_{\text{R},k}(T)$ is the heat of reaction for Reaction "k", evaluated at the temperature T.

If the pressure difference between the feed and product Streams is not substantial, and if there are no phase changes

$$H_{i0} - H_i = \int_{T}^{T_0} c_{p,i} \mathrm{d}T = \overline{c}_{p,i}(T_0 - T) \tag{8-4}$$

In this equation, T_0 is the temperature of the inlet Stream, $c_{p,i}$ is the constant-pressure molar heat capacity of species "i", and $\overline{c}_{p,i}$ is the average constant-pressure molar heat capacity over the temperature range from T_0 to T. Combining Eqns. (8-1) through (8-4) gives

Energy balance—whole reactor—multiple reactions

$$\boxed{Q - W_{\text{s}} - \left(\sum_{i=1}^{N} F_{i0}\overline{c}_{p,i}\right)(T - T_0) - \sum_{k=1}^{R} \xi_k \Delta H_{\text{R},k}(T) = \mathrm{d}U_{\text{sys}}/\mathrm{d}t} \tag{8-5}$$

The enthalpy difference, $\dot{H}_{\text{in}} - \dot{H}_{\text{out}}$, appears in two pieces in Eqn. (8-5). The term $\left(\sum_{i=1}^{N} F_{i0}\overline{c}_{p,i}\right)(T - T_0)$ is the rate at which sensible heat must be added (or removed) to bring the feed from the inlet temperature, T_0, to the outlet temperature, T. The term $\sum_{k=1}^{R} \xi_k \Delta H_{\text{R},k}(T)$ is the rate of enthalpy change associated with the conversion of reactants into products. This term often is referred to as the rate of *heat generation* (or consumption). This terminology is not rigorous in a thermodynamic sense. However, it can lead to a better understanding of reactor behavior, and we will use it in the discussion of reactor energy balances.

In most cases, the shaft work, W_{s} in Eqn. (8-5), can be neglected compared to the other terms in the energy balance. There are a few exceptions, e.g., when a very viscous liquid, for example, a concentrated polymer solution, is vigorously stirred with an agitator. However, these exceptions are relatively rare.

In Eqn. (8-5), the sensible heat term $\left(\sum_{i=1}^{N} F_{i0}\overline{c}_{p,i}\right)(T - T_0)$ is written on a *molar* basis. However, in some problems, it is more convenient to write this term on a *mass* basis. For example, if the total mass feed rate is $\dot{W}$, and if the average heat capacity of the feed Stream on a *mass* basis is $\overline{c}_{p,\text{m}}$, then

$$\left(\sum_{i=1}^{N} F_{i0}\overline{c}_{p,i}\right)(T - T_0) = \dot{W} \int_{T_0}^{T} c_{p,\text{m}} \mathrm{d}T = \dot{W}\overline{c}_{p,\text{m}}(T - T_0)$$

In writing Eqn. (8-5), it is important to use the *same* stoichiometric coefficients, ν_{ki}, that are used to compute the enthalpies of reaction from Eqn. (8-3) and to relate the molar flows and the extents of reaction through Eqn. (1-10). Once a set of coefficients has been assumed, e.g., to calculate the ξ_k's from the F_i's, the *same* coefficients must be used to calculate the ΔH_R's.

If only one reaction is taking place, Eqn. (8-5) can be written in terms of the fractional conversion. For a single reaction, where A is a reactant,

$$\xi = -(F_{A0} - F_A)/\nu_A = -F_{A0}x_A/\nu_A$$

Now, choose $\nu_A = -1$, so that

$$\xi = F_{A0}x_A$$

Substituting this relationship into Eqn. (8-5) gives

Energy balance—whole reactor—single reaction

$$\boxed{Q - W_s - \left(\sum_{i=1}^{N} F_{i0}\bar{c}_{p,i}\right)(T - T_0) - F_{A0}x_A\Delta H_R(T) = dU_{sys}/dt} \tag{8-6}$$

When we chose $\nu_A = -1$, we fixed the basis that must be used to calculate ΔH_R, i.e., $\nu_A = -1$ must be used to calculate ΔH_R from Eqn. (8-3). This means that the basis for ΔH_R is 1 mole of A, i.e., the units of ΔH_R must be energy/mole *of A*.

8.2.1.2 Reactors in Series

In Eqns. (8-5) and (8-6), the conversion, x_A and/or the extent of reaction, ξ_k, in the Stream entering the reactor was taken to be zero. These equations can be applied to a single reactor, to the first reactor in a series of reactors, to the first portion of an ideal PFR, or to the first time segment of a batch reactor. However, we may need to carry out an energy balance on a reactor, other than the first, in a series of reactors. We also may need to carry out an energy balance on a section of a PFR that does not include the inlet, or on a batch reactor for a time segment that does not start with $t = 0$. As we learned in Chapter 4, the bookkeeping for such problems is simpler if we base the conversion and/or the extent of reaction on the feed to the *first* reactor. As a consequence, the conversion and/or extent of reaction will be nonzero in the feed that enters a reactor, or a segment of a reactor, other than the first.

If the conversion of A in the feed that enters a continuous reactor, or a segment of a continuous reactor, is x_A, and if only one reaction is taking place, then Eqn. (8-6) becomes

Energy balance—single reaction—reactors in series

$$\boxed{\begin{aligned} &Q - W_s - \left(\sum_{i=1}^{N} F_{i0}\bar{c}_{p,i}\right)(T_{out} - T_{in}) - F_{A0}(x_{A,out} - x_{A,in})\Delta H_R(T) \\ &= dU_{sys}/dt \end{aligned}} \tag{8-7}$$

If more than one reaction takes place, and if the extents of reaction in the feed to a reactor or reactor segment are denoted ξ_k^0, then Eqn. (8-5) becomes

Energy balance—multiple reactions—reactors in series

$$\boxed{\begin{aligned} &Q - W_s - \left(\sum_{i=1}^{N} F_{i0}\bar{c}_{p,i}\right)(T_{out} - T_{in}) - \sum_{k=1}^{R} (\xi_k - \xi_k^0)\Delta H_{R,k}(T) \\ &= dU_{sys}/dt \end{aligned}} \tag{8-8}$$

In Eqns. (8-7) and (8-8), Q is the rate at which heat is transferred into the section of reactor under consideration, i.e., for which the inlet temperature is T_{in} and the outlet temperature is T_{out}.

Finally, to use the energy balances derived above, the values of the various heats of reaction, $\Delta H_{R,k}(T)$, must be known. As noted in Chapter 1, enthalpies of formation $(\Delta H^0_{f,i})$ have been tabulated for a large number of organic and inorganic compounds, at a standard or reference set of conditions. The reference temperature (T_{ref}) is almost always 298 K. The heat of reaction at the reference conditions is calculated from

$$\Delta H^0_R = \sum_i \nu_i \Delta H^0_{f,i} \tag{1-3}$$

where the superscript "0" denotes the reference conditions.

Except in unusual cases, e.g., reactions at supercritical conditions, a temperature correction is the only thing that is necessary to calculate the value of $\Delta H_{R,k}(T)$ from ΔH^0_R. The relationship is

$$\Delta H_{R,k}(T) = \Delta H^0_R + \int_{T_{ref}}^{T} \nu_i c_{p,i} \mathrm{d}T \tag{8-9}$$

Equation (8-9) is based on the assumption that no phase change takes place between T_{ref} and T.

8.2.2 Macroscopic Energy Balance for Flow Reactors (PFRs and CSTRs)

For an ideal CSTR or PFR operating *at steady state*, $\mathrm{d}U_{sys}/\mathrm{d}t = 0$. The energy balances derived above (Eqns. (8-5)–(8-8)) can be applied directly by setting the right-hand side to zero. For a PFR, the shaft work will be zero, and for a CSTR the shaft work usually can be neglected.

8.2.3 Macroscopic Energy Balance for Batch Reactors

For a batch reactor, there is no flow across the system boundaries and no transport of enthalpy across the system boundaries. From Eqn. (8-1),

$$Q - W_s = \mathrm{d}U_{sys}/\mathrm{d}t$$

Neglecting the shaft work,

$$Q = \mathrm{d}U_{sys}/\mathrm{d}t$$

The total internal energy of the system, i.e., the reactor contents, is related to the total enthalpy of the system, H_{sys}, by

$$U_{sys} = H_{sys} - PV$$

where P is the total pressure and V is the volume of the reactor contents. In most cases, changes in PV are small, and the approximation $\mathrm{d}U_{sys}/\mathrm{d}t = \mathrm{d}H_{sys}/\mathrm{d}t$ is reasonable. Thus,

$$Q = \mathrm{d}H_{sys}/\mathrm{d}t$$

The total system enthalpy, H_{sys}, can be written as

$$H_{sys} = \sum_{i=1}^{N} N_i \overline{H}_i$$

where N_i is the number of moles of compound "i" present at any time and $\overline{H}_i$ is the partial molar enthalpy of component "i". For an ideal solution, the partial molar enthalpy is equal to the pure component enthalpy, H_i. With this assumption, the energy balance for the ideal batch reactor becomes

$$Q = \mathrm{d}\left(\sum_{i=1}^{N} N_i H_i\right)/\mathrm{d}t = \sum_{i=1}^{N} N_i(\mathrm{d}H_i/\mathrm{d}t) + \sum_{i=1}^{N} H_i(\mathrm{d}N_i/\mathrm{d}t) \tag{8-10}$$

If the change in pressure is small, the enthalpy will depend only on temperature so that

$$\sum_{i=1}^{N} N_i \frac{\mathrm{d}H_i}{\mathrm{d}t} = \left(\sum_{i=1}^{N} N_i c_{p,i}\right)\frac{\mathrm{d}T}{\mathrm{d}t} \tag{8-11}$$

If "R" reactions are occurring,

$$\frac{\mathrm{d}N_i}{\mathrm{d}t} = \sum_{k=1}^{R} \nu_{ki}\frac{\mathrm{d}\xi_k}{\mathrm{d}t}$$

so that

$$\sum_{i=1}^{N} H_i \frac{\mathrm{d}N_i}{\mathrm{d}t} = \sum_{i=1}^{N} \mathrm{H}_i \sum_{k=1}^{R} \nu_{ki}\frac{\mathrm{d}\xi_k}{\mathrm{d}t} = \sum_{k=1}^{R} \Delta H_{\mathrm{R}}(T)\frac{\mathrm{d}\xi_k}{\mathrm{d}t} \tag{8-12}$$

Combining Eqns. (8-10)–(8-12) gives

Energy balance—batch reactor—multiple reactions

$$\boxed{Q = \left(\sum_{i=1}^{N} N_i c_{p,i}\right)\frac{\mathrm{d}T}{\mathrm{d}t} + \sum_{k=1}^{R} \Delta H_{\mathrm{R},k}(T)\frac{\mathrm{d}\xi_k}{\mathrm{d}t}} \tag{8-13}$$

For a single reaction,

$$\frac{\mathrm{d}\xi}{\mathrm{d}t} = \frac{1}{\nu_{\mathrm{A}}}\frac{\mathrm{d}N_{\mathrm{A}}}{\mathrm{d}t}$$

Setting $\nu_{\mathrm{A}} = -1$,

$$\frac{\mathrm{d}\xi}{\mathrm{d}t} = -\frac{\mathrm{d}N_{\mathrm{A}}}{\mathrm{d}t} = N_{\mathrm{A}0}\frac{\mathrm{d}x_{\mathrm{A}}}{\mathrm{d}t}$$

Therefore, for a single reaction, Eqn. (8-13) becomes

Energy balance—batch reactor—single reaction

$$\boxed{Q = \left(\sum_{i=1}^{N} N_i c_{p,i}\right)\frac{\mathrm{d}T}{\mathrm{d}t} + \Delta H_{\mathrm{R}}(T) N_{\mathrm{A}0}\frac{\mathrm{d}x_{\mathrm{A}}}{\mathrm{d}t}} \tag{8-14}$$

Since this equation is based on $\nu_{\mathrm{A}} = -1$, the heat of reaction, ΔH_{R}, must be based on the same stoichiometric coefficients.

The design equation for an ideal batch reactor is

$$\frac{N_{\mathrm{A}0}}{V}\frac{\mathrm{d}x_{\mathrm{A}}}{\mathrm{d}t} = -r_{\mathrm{A}} \tag{3-6}$$

so that an equivalent form of Eqn. (8-14) is

$$\boxed{Q = \left(\sum_{i=1}^{N} N_i c_{p,i}\right)\frac{\mathrm{d}T}{\mathrm{d}t} + (-r_{\mathrm{A}})[\Delta H_{\mathrm{R}}(T)]V} \tag{8-15}$$

8.3 ISOTHERMAL REACTORS

To illustrate the use of the macroscopic energy balance, and to confront the challenge of designing an isothermal batch or plug-flow reactor, consider the following problem.

EXAMPLE 8-1
Isothermal Reactor?

The irreversible, liquid-phase reaction $A \rightarrow R$ is to be carried out in an ideal PFR. The inlet concentration of A, C_{A0}, is 2500 mol/m^3, the volumetric flow rate is 1.0 m^3/h, and the feed temperature is 150 °C. The fractional conversion of A in the reactor effluent must be at least 0.95. The reactor will be a tube with an inside diameter of 0.025 m. The reactor will operate isothermally at 150 °C.

The reaction is second-order in A. The rate constant at 150 °C is 1.40×10^{-4} m^3/mol A-s. At 150 °C, $\Delta H_R = -165$ kJ/mol. You may assume that the physical properties of the liquid flowing through the reactor are the same as those of water at room temperature.

A. Calculate the length of pipe required for 95% conversion.
B. Calculate the rate (kJ/min) at which heat must be removed from the whole reactor, and the fraction of the total heat that must be removed in each quarter of the reactor.
C. Calculate an approximate value of the overall heat-transfer coefficient, assuming that the controlling resistance is heat transfer to the wall of the pipe from the fluid flowing through the pipe.
D. A separate jacket will be installed on each quarter of the reactor in order to remove the heat generated in that portion of the reactor. Consider the first (inlet) quarter. A cooling fluid with an average mass heat capacity ($\bar{c}_{p,m}$) of 4.20 J/g-K is available at a temperature of 60 °C. The cooling fluid will flow through the jacket *cocurrent* to the fluid flowing through the reactor. What must the flow rate of the cooling fluid be in order to remove heat at the required rate?
E. Will this heat-transfer system maintain the first quarter of the reactor *exactly* isothermal?
F. Discuss the feasibility of operating a PFR isothermally. Suppose the reactor were an ideal batch reactor instead of a PFR. Would isothermal operation be easier to achieve?

Part A: Calculate the length of pipe required for 95% conversion.

APPROACH

The design equation for an ideal, isothermal PFR will be used to calculate the required reactor volume. The required tube length then can be calculated, since the inner diameter of the tube is given.

SOLUTION

The cross-sectional area of the tube is $A_c = \pi D^2/4 = 4.91 \times 10^{-4}\text{m}^2$. Let L be the length of the tube. Using the design equation for an ideal PFR gives

$$\tau = \frac{V}{v} = \frac{A_c L}{v} = C_{A0} \int_0^{x_{A,\text{out}}} \frac{dx}{kC_{A0}^2(1-x_A)^2}$$

$$L = \frac{v}{A_c k C_{A0}} \left[\frac{1}{1-x_A}\right]_0^{0.95}$$

$$L = \frac{1(\text{m}^3/\text{h}) \times (1/3600)(\text{h/s}) \times (20-1)}{1.40 \times 10^{-4}(\text{m}^3/\text{mol-s}) \times 2500(\text{mol/m}^3) \times 4.91 \times 10^{-4}(\text{m}^2)} = 30.7\,\text{m}$$

Part B: Calculate the rate (kJ/min) at which heat must be removed from the whole reactor, and the fraction of the total heat that must be removed in each quarter of the reactor.

APPROACH

The overall energy balance for a single reaction taking place in one reactor, Eqn. (8-6), will be used to calculate the required rate of heat removal for the reactor as a whole. The energy balance for one reactor in a series of reactors, Eqn. (8-7), then will be used to calculate the fraction of the heat removed in each quarter of the reactor.

SOLUTION For an isothermal reactor, Eqn. (8-6), with $dU_{sys}/dt = 0$ and $W_s = 0$ becomes

$$Q - F_{A0}x_A\Delta H_R(T) = 0$$

For isothermal operation, the rate of heat transfer is exactly equal to the rate at which heat is "generated" (or consumed) by the reaction.

The molar feed rate of A (F_{A0}) $= \upsilon C_{A0} = 41.7$ (mol/min). For the whole reactor, $x_A = 0.95$, so that

$$Q = 41.7\,(\text{mol/min}) \times 0.95 \times (-165)(\text{kJ/mol}) = -6530\,(\text{kJ/min})$$

The value of Q is negative because heat is removed from the reactor.

Equation (8-7), with $dU_{sys}/dt = 0$ and $W_s = 0$, can be applied to the nth quarter of the isothermal reactor.

$$Q(n) = F_{A0}[x_{A,out} - x_{A,in}]\Delta H_R$$

The symbol $Q(n)$ denotes the rate of heat removal in the nth quarter, and $x_{A,out}$ and $x_{A,in}$ are the fractional conversions leaving and entering the nth quarter, respectively.

The fraction of the overall heat that is removed in the nth quarter of the reactor is given by

$$\frac{Q(n)}{Q} = \frac{F_{A0}(x_{A,out} - x_{A,in})\Delta H_R}{F_{A0}x_{A,tot}\Delta H_R} = \frac{x_{A,out} - x_{A,in}}{x_{A,tot}} \tag{8-16}$$

The problem now has been reduced to calculating the conversion for 1/4, 1/2, and 3/4 of the total reactor length.

The design equation can be integrated to give

$$x_A = \frac{kC_{A0}\tau}{1 + kC_{A0}\tau}$$

This equation can be solved for x_A when $\tau = \tau_{tot}/4$, $\tau_{tot}/2$, and $3\tau_{tot}/4$. The value of $\tau_{tot} = A_cL/\upsilon = 54.3$ s. These conversions then can be used in Eqn. (8-16) to calculate the fraction of the overall heat that must be removed in each quarter of the reactor. The results are

Quarter of reactor	Fractional conversion of A leaving quarter	Fraction of total heat removed in quarter
First	0.826	0.869
Second	0.905	0.083
Third	0.935	0.032
Fourth	0.950	0.016

Most of the heat is generated and must be removed in the first quarter of the reactor because most of the reaction takes place there.

Part C: Calculate an approximate value of the overall heat-transfer coefficient, assuming that the controlling resistance is heat transfer to the wall of the pipe from the fluid flowing through the pipe.

APPROACH First, the Reynolds number will be calculated to determine whether flow in the tube is laminar or turbulent. Then an appropriate correlation will be used to calculate a value of the heat-transfer coefficient.

SOLUTION The Reynolds number (Re) for the reactor is given by

$$\mathrm{Re} = D_{\mathrm{in}} v\rho/\mu = D_{\mathrm{in}}(\upsilon/A_c)/\nu$$

where v is the average velocity of the fluid, D_{in} is the inside diameter of the tube, and ν is the kinematic viscosity. For water at room temperature, the kinematic viscosity is about $1 \times 10^{-6}\mathrm{m}^2/\mathrm{s}$. Therefore,

$$\mathrm{Re} = 0.025(\mathrm{m}) \times 1(\mathrm{m}^3/\mathrm{h})/[4.91 \times 10^{-4}(\mathrm{m}^2) \times 10^{-6}(\mathrm{m}^2/\mathrm{s}) \times 3600(\mathrm{s/h})] = 14{,}100$$

Flow is turbulent in the pipe, consistent with the assumption of an ideal PFR. For turbulent flow in pipes, the inside heat-transfer coefficient, h_{in}, can be approximated from

$$\mathrm{Nu_h} = 0.023\mathrm{Re}^{0.80}\mathrm{Pr}^{0.30}$$

where $\mathrm{Nu_h}$ is the Nusselt Number for heat transfer ($h_{\mathrm{in}}D_{\mathrm{in}}/k_{\mathrm{th}}$) and Pr is the Prandtl number. For water at room temperature, the thermal conductivity, k_{th}, is about 0.65 J/s-m-K and the Prandtl number is about 7.0. Therefore,

$$\mathrm{Nu_h} = 0.023(14100)^{0.80}(7.0)^{0.30} = 86$$

$$h_{\mathrm{in}} = 86 \times 0.65(\mathrm{J/s\text{-}m\text{-}K}) \times 60(\mathrm{s/min})/0.025(\mathrm{m}) \times 1000(\mathrm{J/kJ}) = 134\,\mathrm{kJ/m^2\text{-}min\text{-}K}$$

Since the resistance between the flowing fluid and the tube wall is the controlling resistance, the overall heat-transfer coefficient, U, is almost the same as the individual coefficient, h_{in}.

Part D: A separate jacket will be installed on each quarter of the reactor in order to remove the heat generated in that portion of the reactor. Consider the first (inlet) quarter. A cooling fluid with an average mass heat capacity ($\bar{c}_{p,\mathrm{m}}$) of 4.20 J/g-K is available at a temperature of 60 °C. The cooling fluid will flow through the jacket *cocurrent* to the fluid flowing through the reactor. What must the flow rate of the cooling fluid be in order to remove heat at the required rate?

APPROACH The rate of heat transfer in the first quarter of the PFR can be calculated from the results of Part B. The required temperature difference between the fluid in the reactor and the cooling fluid can be calculated from the rate law for heat transfer, $Q = UA_{\mathrm{h}}\Delta T$. Since the temperature difference will vary along the length of the reactor, ΔT_{lm} (the log-mean temperature difference) must be used in this equation. The value of U was calculated in Part C, and the dimensions of the reactor are known, permitting A_{h} to be calculated. Since ΔT_{lm} and the coolant temperature, T_{in}, are known, the outlet temperature of the cooling fluid can be calculated. Finally, the required coolant flow rate can be calculated from an energy balance on the coolant.

SOLUTION The rate law for heat transfer is

$$Q = UA_{\mathrm{h}}\Delta T_{\mathrm{lm}}$$

where ΔT_{lm} is the log-mean temperature difference,

$$\Delta T_{\mathrm{lm}} = (\Delta T_2 - \Delta T_1)/\ln(\Delta T_2/\Delta T_1)$$

In this equation, ΔT_2 is the temperature difference between the reactor Stream and the coolant Stream at the inlet end of the reactor and ΔT_1 is the temperature difference at the outlet end. From Part B, the rate of heat transfer in the first quarter of the reactor is $Q = 0.869 \times (6530) = 5675$ kJ/min. The area for heat transfer in the first quarter of the reactor is $A_{\mathrm{h}} = \pi \times 30.7(\mathrm{m}) \times 0.025(\mathrm{m})/4 = 0.602\,\mathrm{m}^2$, so that

$$\Delta T_{\mathrm{lm}} = Q/UA_{\mathrm{h}} = 5675\,(\mathrm{kJ/min})/134\,(\mathrm{kJ/min\text{-}m^2\text{-}K}) \times 0.602\,(\mathrm{m}^2) = 70.4\,\mathrm{K}$$

Let T_{in} be the inlet temperature of the cooling fluid and T_{out} be the outlet temperature, both in °C. Since the reactor is supposed to be isothermal, the temperature of the fluid flowing through the reactor is taken to be 150 °C along the whole length, so that

$$\Delta T_{\mathrm{lm}} = 70.4 = [(150 - T_{\mathrm{in}}) - (150 - T_{\mathrm{out}})]/\ln[(150 - T_{\mathrm{in}})/(150 - T_{\mathrm{out}})]$$

Now $T_{\mathrm{in}} = 60$ °C,

$$70.4 = [(150 - 60) - (150 - T_{\mathrm{out}})]/\ln[(150 - 60)/(150 - T_{\mathrm{out}})]$$

This equation can be solved to give $T_{out} = 95$ °C.

The required coolant flow rate ($\dot{M}$) can be calculated from

$$Q = \dot{M}\bar{c}_{p,m}(T_{out} - T_{in})$$
$$\dot{M} = Q/\bar{c}_{p,m}(T_{out} - T_{in}) = 5675 \times 10^3 (\text{J/min})/4.2(\text{J/g-K}) \times (95 - 60)(\text{K})$$
$$\dot{M} = 38{,}600 \text{ g/min}$$

Part E: Will this heat-transfer system maintain the first quarter of the reactor *exactly* isothermal?

APPROACH The energy balance will be examined at several points along the length of the first quarter of the reactor to determine whether this balance is *exactly* satisfied.

SOLUTION Consider a differential element of volume at some point along the length of the reactor, where $dV = A_c dL$ and $A_c = \pi D_{in}^2/4$. The rate at which heat is "generated" in this element by the exothermic chemical reaction is

$$\text{rate of heat "generation"} = (-r_A)(-\Delta H_R)A_c dL = kC_{A0}^2(1 - x_A)^2(-\Delta H_R)\pi D_{in}^2 dL/4$$

The rate of heat removal is

$$\text{Rate of heat removal} = U\pi D_{in}(T_R - T_c)dL$$

In this expression, T_R is the temperature of the fluid flowing through the reactor and T_c is the temperature of the coolant, both at the *same position* along the length of the reactor. Equating the rates of generation and removal, and solving for $T_R - T_c$,

$$T_R - T_c = (1 - x_A)^2[kC_{A0}^2(-\Delta H_R)D/4U] \tag{8-17}$$

This is a very interesting relationship! The term in square brackets on the right-hand side does not vary with position along the length of the tube. Therefore, Eqn. (8-17) tells us that $T_R - T_c$ must be directly proportional to $(1 - x_A)^2$ at *every* point along the length of the reactor, in order for the reactor to be isothermal! In more general terms, the temperature difference must *exactly* track the kinetics of the reaction.

Let's check to see whether Eqn. (8-17) is obeyed at the inlet and outlet of the first quarter of the reactor. For this problem, the term in the square brackets has a value of 404 K. At the reactor inlet

$$T_R - T_c = (150 - 60)\text{ K} = 90\text{ K} \neq 404 \times (1 - 0)^2 = 404\text{ K}$$

This inequality shows that the rate of heat generation at the inlet to the reactor is more than four times greater than what the heat-transfer system can handle at the reactor inlet. Therefore, the temperature of the fluid flowing through the reactor will start to rise above 150 °C at the inlet.

At the reactor outlet,

$$T_R - T_c = (150 - 95)\text{K} = 55\text{ K} \neq 404 \times (1 - 0.826)^2 = 12\text{ K}$$

According to this calculation, the rate of heat removal right at the outlet from the first quarter of the reactor would be more than four times greater than necessary.

Part F: Discuss the feasibility of operating a PFR isothermally. Suppose the reactor were an ideal batch reactor instead of a PFR. Would isothermal operation be easier to achieve?

SOLUTION These calculations show that isothermal operation of a PFR is an idealization. Isothermality is very difficult to achieve in practice, especially for reactions that have a significant enthalpy change. In the present problem, the number of jacketed reactor segments could be increased. However, a very large number of segments would be required to closely approach isothermal operation, and each segment would have to have a different coolant flow rate and/or inlet temperature.

The problem of isothermal operation is only slightly less formidable for a batch reactor. Here, the coolant flow rate and/or temperature would have to be adjusted continuously with time in order to "match" the heat removal rate with the rate of heat generation.

Isothermal operation can be approached closely in small-scale, experimental reactors, where practical and economic issues are not of great concern. In designing an experimental PFR to operate isothermally, several "tricks" can be employed that are not usually feasible on a larger scale. The object of these "tricks" is to increase the value of *UA* to the point that the required value of ΔT is very small. In that case, the temperature of the reactor contents approaches the temperature of the heat-transfer medium. The temperature of the heat-transfer medium is kept nearly constant, e.g., by using a large flow rate.

The devices that are available to increase *UA* include selecting the smallest feasible tube diameter, in order to create a high surface area for heat transfer per unit volume of reactor. Also, the reactor can be filled with an inert packing that occupies volume inside the tube, increasing the length of tube that is required to achieve a given conversion. This also helps to increase the inside heat-transfer coefficient. Finally, in a catalytic reactor, the ratio of inert packing to catalyst can be changed along the length of the reactor so that A is approximately proportional to Q.

Despite the difficulty of achieving isothermal operation, this is an important limiting case of reactor analysis. It provides an easy-to-calculate bound. For example, suppose an irreversible, exothermic reaction is carried out adiabatically, with an initial or inlet temperature of T_0. Since the temperature will increase as the reaction proceeds, so will the rate constant. A calculation based on isothermal operation at T_0 will provide a lower bound on the conversion that can be achieved in a given volume, and it will provide an upper bound on the volume required to achieve a specified fractional conversion.

8.4 ADIABATIC REACTORS

The simplest and cheapest mode of reactor design and operation is the adiabatic mode. If heat is added to or withdrawn from the reactor as the reaction proceeds, heat-exchange surface must be designed *into* the reactor. This adds substantially to the capital cost. Moreover, operation of a reactor to which heat is being added or withdrawn can be complex, and can require a relatively sophisticated control system. The construction of an adiabatic reactor is relatively simple, and the capital cost is relatively low. In fact, the advantages of adiabatic reactors are so substantial that adiabatic operation must be treated as the default mode. An adiabatic design always should be considered, and a more complex design should be used only when there are compelling reasons.

8.4.1 Exothermic Reactions

If an exothermic reaction is carried out adiabatically, the temperature will rise with time in a batch reactor, or in the direction of flow in a steady-state, plug-flow reactor. In a CSTR, the temperature of the reactor may be substantially higher than the temperature of the feed. In general, the temperature increases as the conversion increases. If the heat of reaction is high, very high temperatures may result.

High temperatures can cause significant problems, which may be severe enough to preclude adiabatic operation. Some of the more common disadvantages of adiabatic operation for exothermic reactions include

- potential for unsafe conditions due to excessive temperature, e.g.,
  - → rapid pressure increases due to vaporization of liquid reactants, products, or solvents;
  - → runaway side reactions, e.g., explosions or polymerizations;
  - → damage to reactor materials leading to vessel rupture.
- loss of reaction selectivity at high temperatures;

- damage to a temperature-sensitive catalyst;
- unfavorable equilibrium at high temperatures.

The last point is worthy of some discussion, since there are several important examples of this phenomenon. For an exothermic reaction, the equilibrium constant decreases with increasing temperature. Therefore, the equilibrium conversion of the limiting reactant, x_{eq}, decreases as the temperature increases. If the temperature in an adiabatic reactor increases too much, the equilibrium conversion may be reached and the reaction will stop, perhaps well short of the desired conversion.

This situation is approached in two important commercial processes, the synthesis of methanol and the oxidation of sulfur dioxide (SO_2) to sulfur trioxide (SO_3). The latter reaction is a key step in the manufacture of sulfuric acid. The reaction is

$$SO_2 + 1/2O_2 \rightleftarrows SO_3$$

The reaction is carried out over a "vanadium pentoxide" catalyst at essentially atmospheric pressure. The temperature of the feed must be about 400 °C, because the reaction is quite slow below that temperature. There are no side reactions that might complicate adiabatic operation, and the catalyst can tolerate very high temperatures. However, if the reactor is adiabatic and the feed enters at 400 °C, the reaction will come to equilibrium at a temperature of about 600 °C and an SO_2 conversion of about 75%. A conversion well in excess of 99% is required commercially.

A solution to this problem is to use a series of adiabatic reactors with interstage cooling as shown in the following figure.

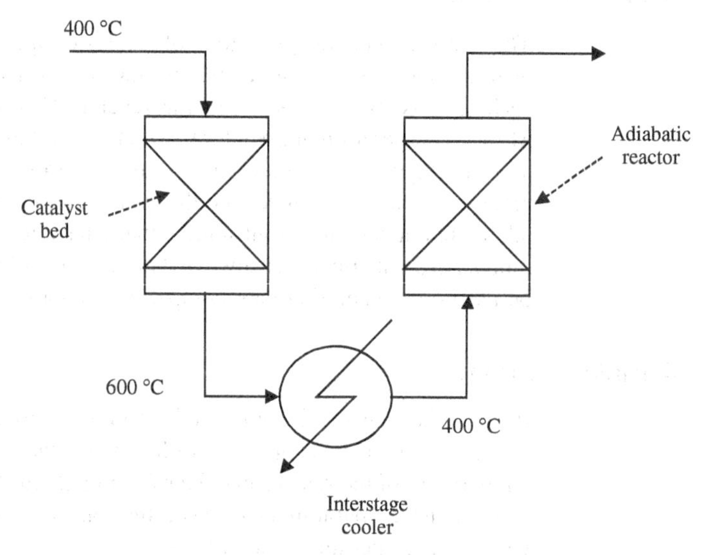

This figure shows two adiabatic reactors, with an interstage cooler between them. In a sulfuric acid plant, there usually are four or more reactors in series, with interstage cooling between them.

8.4.2 Endothermic Reactions

If an endothermic reaction is carried out adiabatically, the temperature will decrease as the reaction proceeds. The temperature will decrease with time in a batch reactor, and with axial position in a plug-flow reactor.

Usually, the decrease in temperature does not lead to catalyst damage or loss of selectivity. However, the equilibrium conversion decreases as the temperature decreases, as does the reaction rate. Because of the decrease of rate with temperature, the reaction can "self-extinguish" as the kinetics become slower and slower with decreasing temperature.

An important commercial example of an endothermic reaction that is carried out adiabatically is the catalytic reforming of petroleum naphtha to produce high-octane gasoline. In this process, the naphtha is mixed with hydrogen and passed over a heterogeneous catalyst that contains platinum, and perhaps other metals such as rhenium or tin, on a ceramic support such as alumina. The temperature is in the region of 800–900 °F.

Naphtha is a complex mixture of compounds, mainly paraffins, naphthenes, and aromatics, generally containing 6–10 carbon atoms. Several different types of reaction take place in the reforming process. Two of the most important classes of reaction are dehydrogenation and dehydrocyclization, examples of which are given by the following reactions.

Dehydrogenation

$$\text{Methyl cyclohexane } (C_6H_{11}CH_3) \longrightarrow \text{Toluene } (C_6H_5CH_3) + 3H_2$$

Dehydrocyclization

$$\underset{n\text{-Heptane}}{C_7H_{16}} \longrightarrow \text{Toluene } (C_6H_5CH_3) + 4H_2$$

Both dehydrogenation and dehydrocyclization are endothermic. If naphtha reforming is carried out adiabatically, the temperature will decrease as the reaction proceeds. Almost all commercial naphtha reforming processes employ a series of adiabatic reactors with interstage heating, as shown below.

Usually, there are three to five reactors in series, with an interstage heater between each pair of reactors.

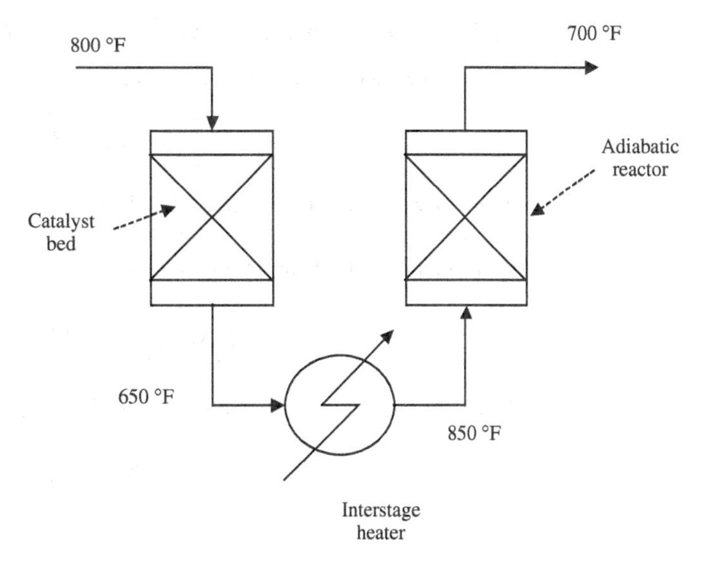

8.4.3 Adiabatic Temperature Change

When a reactor is operated adiabatically, and when only one reaction takes place, there is a simple relationship between the temperature and the fractional conversion.

For an adiabatic flow reactor operating at steady state with no shaft work, Eqn. (8-6) simplifies to

$$-\left(\sum_{i=1}^{N} F_{i0}\overline{c}_{p,i}\right)(T - T_0) - F_{A0}x_A \Delta H_R(T) = 0$$

Rearranging,

$$T = T_0 + \left(\frac{F_{A0}(-\Delta H_R(T))}{\sum\limits_{i=1}^{N} F_{i0}\overline{c}_{p,i}}\right) x_A \tag{8-18a}$$

Equation (8-18a) shows that the temperature, T, is proportional to the fractional conversion, x_A, for an adiabatic flow reactor at steady state. If the fractional conversion is known, the corresponding temperature can be calculated. Of course, the thermochemical data required to calculate ΔH_R and all of the $\overline{c}_{p,i}$, as functions of T, must be available.

Let T_{ad} be the temperature that corresponds to complete conversion, i.e., $x_A = 1$. When the group $\left(F_{A0}(-\Delta H_R(T))/\sum_{i=1}^{N} F_{i0}\overline{c}_{p,i}\right)$ is evaluated at T_{ad}, it is known as the *adiabatic temperature change*, denoted ΔT_{ad}. Physically, the adiabatic temperature change is the amount that the temperature will increase or decrease as the reaction goes to completion under adiabatic conditions. For an exothermic reaction, this quantity is called the *adiabatic temperature rise*.

$$\Delta T_{ad} = \left(\frac{F_{A0}(-\Delta H_R(T_{ad}))}{\sum\limits_{i=1}^{N} F_{i0}\overline{c}_{p,i}(T_0 \rightarrow T_{ad})}\right) \tag{8-19a}$$

The symbol $\overline{c}_{p,i}(T_0 \rightarrow T_{ad})$ indicates that the average heat capacities, i.e., the $\overline{c}_{p,i}$, are averages *over the temperature range from T_0 to T_{ad}*. If ΔH_R and the $\overline{c}_{p,i}$ are strong functions of temperature, a trial-and-error solution of Eqn. (8-19a) is required to obtain a value of T_{ad}.

Fortunately, for many systems, $\left(F_{A0}(-\Delta H_R(T))/\sum_{i=1}^{N} F_{i0}\overline{c}_{p,i}\right)$ is not a strong function of temperature, and can be assumed to be constant. If so, Eqn. (8-18a) can be written as

$$\boxed{T = T_0 + (\overline{\Delta T}_{ad})x_A} \tag{8-20}$$

In this case, the reaction temperature, T, is a linear function of the fractional conversion, x_A.

For an ideal, adiabatic ($Q = 0$) batch reactor, with one reaction taking place, time can be eliminated from Eqn. (8-14) to give

$$\left(\sum_{i=1}^{N} N_i c_{p,i}\right) \mathrm{d}T + \Delta H_R(T) N_{A0} \mathrm{d}x_A = 0$$

When only one reaction is taking place, $N_i = N_{i0} + N_{A0}\nu_i x_A$, so that

$$\left(\sum_{i=1}^{N} N_{i0}c_{p,i} + N_{A0}x_A \sum_{i=1}^{N} \nu_i c_{p,i}\right) \mathrm{d}T + \Delta H_R(T) N_{A0} \mathrm{d}x_A = 0$$

From thermodynamics,

$$\sum_{i=1}^{N} \nu_i c_{p,i} = \left(\frac{\partial \Delta H_R(T)}{\partial T}\right)_P \cong \frac{d\Delta H_R(T)}{dT}$$

so that

$$\left(\sum_{i=1}^{N} N_{i0} c_{p,i}\right) dT + N_{A0} x_A \frac{d\Delta H_R(T)}{dT} dT + \Delta H_R(T) N_{A0} dx_A = 0$$

$$\left(\sum_{i=1}^{N} N_{i0} c_{p,i}\right) dT + N_{A0} d[\Delta H_R(T) x_A] = 0$$

Integrating from $T_0, x_A = 0$, to T, x_A,

$$\left(\sum_{i=1}^{N} N_{i0} \overline{c}_{p,i}\right)(T - T_0) + N_{A0} \Delta H_R(T) x_A = 0$$

Rearranging,

$$T = T_0 + \left(\frac{N_{A0}[-\Delta H_R(T)]}{\sum_{i=1}^{N} N_{i0} \overline{c}_{p,i}}\right) x_A \tag{8-18b}$$

Equations (8-18a) and (8-18b) are identical since

$$\frac{N_{i0}}{N_{A0}} = \frac{y_{i0}}{y_{A0}} = \frac{F_{i0}}{F_{A0}}$$

where y_i is the mole fraction of component "i". Both of these equations could be written as

$$T = T_0 + \left(\frac{y_{A0}[-\Delta H_R(T)]}{\sum_{i=1}^{N} y_{i0} \overline{c}_{p,i}}\right) x_A \tag{8-18c}$$

Equation (8-18a) is based on molar flow rates, and it applies to a flow reactor. Equation (8-18b) applies to a batch reactor. Equation (8-18c) may be used in place of either equation, depending on convenience.

Once again, if $\left(N_{A0}[-\Delta H_R(T)] / \sum_{i=1}^{N} N_{i0} \overline{c}_{p,i}\right)$ is constant, independent of temperature, then Eqns. (8-18b) and (8-18c) can be written as

$$\boxed{T = T_0 + (\overline{\Delta T}_{ad}) x_A} \tag{8-20}$$

where $\overline{\Delta T}_{ad}$ is calculated from Eqns. (8-21a), (8-21b), or (8-21c), as appropriate and convenient.

$$\Delta T_{ad} = F_{A0}(-\overline{\Delta H}_R) / \sum_{i=1}^{N} F_{i0} \overline{c}_{p,i} \tag{8-21a}$$

$$\Delta T_{ad} = N_{A0}(-\overline{\Delta H}_R) / \sum_{i=1}^{N} N_{i0} \overline{c}_{p,i} \tag{8-21b}$$

$$\Delta T_{ad} = y_{A0}(-\overline{\Delta H}_R) / \sum_{i=1}^{N} y_{i0} \overline{c}_{p,i} \tag{8-21c}$$

The overbars on ΔH_R and $c_{p,i}$ indicate that these quantities are averages over the range from T_0 to T_{ad}.

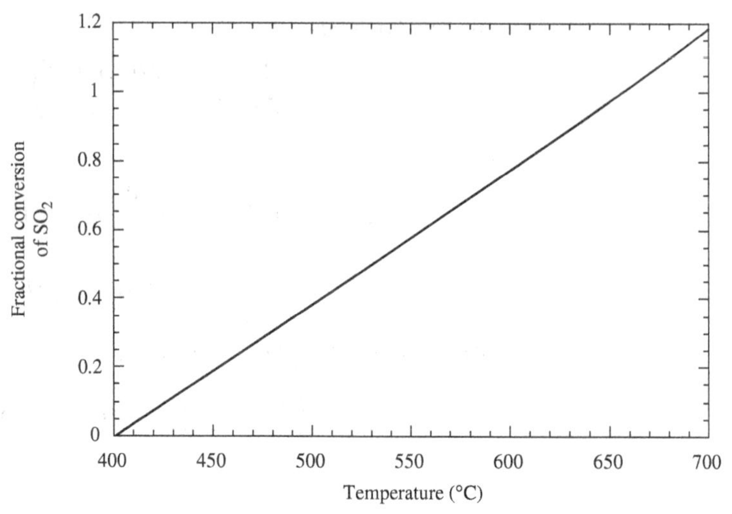

Figure 8-1 Graphical representation of the energy balance for an adiabatic SO_2 oxidation reactor: inlet temperature—400 °C; inlet mole fractions—$SO_2 = 0.090$; $O_2 = 0.13$; $N_2 = 0.78$; atmospheric pressure.

Equations (8-18a)–(8-18c) are nothing more than energy balances for *adiabatic* reactors, and Eqn. (8-20) is a simplified version of these balances, valid when ΔT_{ad} is constant. These equations show the relationship between the temperature, T, the conversion, x_A, and the initial temperature, T_0, for an adiabatic reactor. These energy balances can be represented on a plot of x_A versus T. If Eqn. (8-20) is obeyed, the result is a straight line.

Figure 8-1 shows the energy balance for an adiabatic SO_2 oxidation reactor, operating at conditions that are typical of the first reactor in a sulfuric acid plant. The reaction is

$$SO_2 + 1/2O_2 \rightleftarrows SO_3$$

In preparing Figure 8-1, the heat of reaction and the average heat capacities of N_2, O_2, and SO_2 were calculated rigorously, as functions of temperature. For all practical purposes, the temperature is linear in the fractional conversion of SO_2.

The operating point of an adiabatic CSTR at steady state *must* lie somewhere on the line that represents the adiabatic energy balance. For an adiabatic PFR at steady state, or for an adiabatic batch reactor, the energy balance line describes the *path* of the reaction, including the exit condition for a PFR and the final condition for a batch reactor. For any type of adiabatic reactor, if a given point (x, T) does not lie on the line, the energy balance is not satisfied.

8.4.4 Graphical Analysis of Equilibrium-Limited Adiabatic Reactors

Suppose we are asked to analyze the behavior of a single reaction, taking place in an adiabatic reactor. One of the questions we might want to answer is: "What is the *maximum* conversion that can be achieved?" Of course, the maximum conversion is obtained when the reaction reaches chemical equilibrium.

The dashed line in Figure 8-2 shows the *equilibrium* fractional conversion of SO_2 as a function of temperature, for conditions that are typical of the first SO_2 oxidation reactor in a sulfuric acid plant. The solid line shows the energy balance for adiabatic operation. This is the same line shown in Figure 8-1. The two lines intersect at one point: $x_{SO_2} = 0.74$; $T = 593\,°C$. At this condition, *both* the energy balance and the equilibrium relationship are satisfied simultaneously. The intersection of the two lines is the *only* place where this requirement is met.

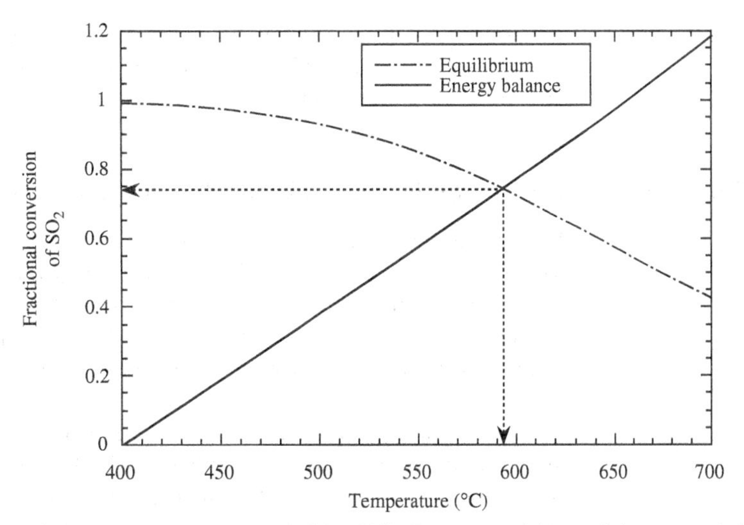

Figure 8-2 Graphical representation of the equilibrium composition and the energy balance for an adiabatic SO_2 oxidation reactor: inlet temperature—400 °C; inlet mole fractions—$SO_2 = 0.090$; $O_2 = 0.13$; $N_2 = 0.78$; atmospheric pressure.

If there is a large amount of catalyst in the reactor, and if the catalyst is very active, the reaction might be close to equilibrium at the reactor effluent conditions. The assumption that equilibrium is achieved provides an upper limit on conversion, and either an upper or lower limit on temperature, depending on whether the reaction is exothermic or endothermic. In this example, the outlet composition will correspond to a fractional SO_2 conversion of 0.74, and the outlet temperature will be 593 °C, *if* equilibrium is achieved at the outlet conditions of the reactor.

The analysis that is illustrated in Figure 8-2 can be performed for *any* type of reactor. The only requirements are that the reactor be adiabatic and that the reaction be at chemical equilibrium at the exit condition of a continuous reactor, or at the final condition of a batch reactor.

The dashed line in Figure 8-2 was calculated using five relationships from thermodynamics:

$$\Delta G_R(T_{ref}) = \sum_i \nu_i \Delta G_{f,i}(T_{ref})$$

$$\ln K_{eq}(T_{ref}) = -\Delta G_R(T_{ref})/RT_{ref}$$

$$\left(\frac{\partial \ln K_{eq}}{\partial T}\right)_P = \frac{\Delta H_R(T)}{RT^2}$$

$$\Delta H_R(T_{ref}) = \sum_i \nu_i \Delta H_{f,i}(T_{ref})$$

$$\left(\frac{\partial \Delta H_R(T)}{\partial T}\right)_P = \sum_i \nu_i c_{p,i}(T)$$

These five equations, which were introduced in Chapter 2, permit the equilibrium constant, K_{eq}, to be calculated at any temperature. The equilibrium conversion, x_{eq}, then can be calculated from the equilibrium expression:

$$K_{eq} = \sum_i a_i^{\nu_i}$$

where a_i is the activity of species "i". The activity, a_i, can be related to the mole fraction of species "i". For this example, the ideal gas laws are valid, so that $a_i = p_i = Py_i$. Here, p_i is the partial pressure of species "i", y_i is its mole fraction, and P is the total pressure. A stoichiometric table is used to express the y_i's as a function of x_{SO_2}, which then is calculated from the equilibrium expression.

8.4.5 Kinetically Limited Adiabatic Reactors (Batch and Plug Flow)

An infinite reactor volume, or an infinite weight of catalyst, is required to bring the effluent from a reactor *exactly* into chemical equilibrium. The composition and temperature of the effluent from any real reactor will be determined by the kinetics of the reaction. Nevertheless, the type of equilibrium analysis illustrated in the preceding section can be valuable as a means of understanding an important limiting case of reactor behavior.

In order to size or analyze a nonisothermal reactor, the material balance(s) and the energy balance must be solved simultaneously. This solution is relatively straightforward if the reactor is adiabatic, so that's where our discussion will begin. For an adiabatic reactor, with a single reaction taking place, the temperature is directly related to the fractional conversion, as shown in Section 8.4.3. Let's consider a situation where the energy balance reduces to a linear relationship between temperature and fractional conversion, i.e., where Eqn. (8-20) is valid.

$$\boxed{T = T_0 + (\overline{\Delta T}_{\mathrm{ad}})x_{\mathrm{A}}} \tag{8-20}$$

The analyses of an ideal batch reactor and an ideal PFR are essentially identical, so let's illustrate with the PFR. The design equation for an ideal PFR in differential form is

$$\frac{\mathrm{d}V}{F_{\mathrm{A0}}} = \frac{\mathrm{d}x_A}{-r_{\mathrm{A}}(x_{\mathrm{A}}, T)} \tag{3-27}$$

The dependence of $-r_{\mathrm{A}}$ on x_{A} and T is shown explicitly in this equation to remind us that, in general, the reaction rate depends on both temperature and the various species concentrations. The reaction rate depends on temperature because the constants that appear in the rate equation, e.g., the rate constant, the equilibrium constant, and the adsorption/binding constants, generally depend on temperature.

For adiabatic operation, there is a direct relationship between T and x_{A}. The relationship may not always be linear, as given by Eqn. (8-20). However, if either T or x_{A} is known, the other can be calculated. This allows either x_{A} or T to be eliminated from the design equation. For example, if Eqn. (8-20) is valid, the rate constant

$$k = A\mathrm{e}^{-E/RT}$$

can be written as a function of x_{A} by substituting Eqn. (8-20) for T to give

$$k = A\mathrm{e}^{-E/R[T_0 + (\overline{\Delta T}_{\mathrm{ad}})x_{\mathrm{A}}]}$$

If all of the temperature-dependent constants in the rate equation are transformed in this manner, the design equation can be written as

$$\mathrm{d}V = \frac{F_{\mathrm{A0}}\mathrm{d}x_{\mathrm{A}}}{-r_{\mathrm{A}}(x_{\mathrm{A}})}$$

The dependence of $-r_{\mathrm{A}}$ on T has been eliminated by expressing T as a function of x_{A}.

If the feed composition and molar flow rate are fixed, the volume of reactor, V, that is required to achieve a specified final conversion, $x_{A,f}$, can be calculated via numerical integration,

$$V = F_{A0} \int_0^{x_{A,f}} \frac{dx_A}{-r_A(x_A)}$$

Alternatively, it might be more convenient to solve the differential equation

$$\frac{dx_A}{d\left(\frac{VC_{A0}}{F_{A0}}\right)} = \frac{dx_A}{d\tau} = [-r_A(x_A)]/C_{A0} \tag{8-22}$$

numerically, using the techniques described in Chapter 7. Consider the following example.

EXAMPLE 8-2
Reversible Reaction in an Adiabatic Batch Reactor

The reversible, liquid-phase reaction

$$A + B \rightleftarrows R + S$$

is being carried out in an *adiabatic*, ideal *batch* reactor. The reaction obeys the rate equation

$$-r_A = k\left(C_A C_B - \frac{C_R C_S}{K_{eq}}\right)$$

The value of the rate constant, k, is 0.050 l/mol-h at 100 °C and the activation energy is 80 kJ/mol. The value of the equilibrium constant, K_{eq}, is 500 at 100 °C and the heat of reaction, ΔH_R, is −60 kJ/mol, independent of temperature.

Initially, the reactor is at a temperature of 100 °C and the initial concentrations are $C_{A0} = C_{B0} = 4.0$ mol/l; $C_{R0} = C_{S0} = 0$. The average heat capacity of the contents of the reactor is 3.4 J/g-K and the density is 800 g/l.

A. What is the fractional conversion of A after 1.5 h?

B. What is the temperature of the reactor after 1.5 h?

Part A: What is the fractional conversion of A after 1.5 h?

APPROACH

The design equation for an ideal batch reactor will be solved numerically, using the relationship $T = T_0 + (\overline{\Delta T}_{ad})x_A$ to express T as a function of x_A. A numerical value of $\overline{\Delta T}_{ad}$ can be calculated from the given values of C_{A0}, $-\Delta H_R$, ρ, and $\bar{c}_{p,m}$.

SOLUTION

Since the reaction takes place in the liquid phase, the density of the system is essentially constant. The rate equation can be written in terms of fractional conversion as

$$-r_A = k[C_{A0}^2(1-x_A)^2 - \{C_{A0}^2 x_A^2/K_{eq}^C\}] = kC_{A0}^2[(1-x_A)^2 - \{x_A^2/K_{eq}^C\}]$$

The design equation for a constant-density, ideal batch reactor is

$$C_{A0}\frac{dx_A}{dt} = -r_A \tag{3-9}$$

Substituting the expression for $-r_A$ and canceling one C_{A0}:

$$\frac{dx_A}{dt} = kC_{A0}[(1-x_A)^2 - \{x_A^2/K_{eq}^C\}] \tag{8-23}$$

The constants k and K_{eq} depend on temperature. The rate constant, k, is given by the Arrhenius relationship, which may be written as

$$k(T) = k(T_0)e^{-(E/R)[(1/T)-(1/T_0)]} \tag{8-24}$$

In this equation, T_0 is a reference temperature. Since the rate and equilibrium constants are known at 100 °C (373 K), and since the initial temperature of the reactor is 100 °C, it is convenient to take $T_0 = 100°C$ for this problem. When the heat of reaction, ΔH_R, is constant, the variation of the equilibrium constant, K_{eq}^{C}, with temperature is given by the van't Hoff relationship, which may be written as

$$K_{eq}^{C}(T) = K_{eq}^{C}(T_0)e^{-(\Delta H_R/R)[(1/T)-(1/T_0)]} \tag{8-25}$$

For an adiabatic reactor with constant density, heat capacity, and ΔH_R, x_A and T are related by

$$T = T_0 + (\overline{\Delta T}_{ad})x_A \tag{8-20}$$

Substituting Eqn. (8-20) into Eqns. (8-24) and (8-25), and then substituting the resulting equations into Eqn. (8-23) gives

$$\frac{dx_A}{dt} = C_{A0}k(T_0)e^{-(E/R)[(1/(T_0+\Delta T_{ad}x_A)-(1/T_0)]}[(1-x_A)^2 - \{x_A^2/K_{eq}^{C}(T_0)e^{-(\Delta H_R/R)[(1/(T_0+\Delta T_{ad}x_A))-(1/T_0)]}\}] \tag{8-26}$$

To calculate the value of $\overline{\Delta T}_{ad}$, choose 1 l of solution as a basis. Then,

$$\overline{\Delta T}_{ad} = C_{A0}(-\Delta H_R)/\rho\bar{c}_{p,m} = 4(\text{mol/l})(+60{,}000)(\text{J/mol})/800(\text{g/l})(3.4)(\text{J/g-K}) = 88.2\ \text{K}$$

Equation (8-26) now can be solved numerically. It could be solved directly as a differential equation, using any available program, including the techniques described in Chapter 7. Use of the fourth-order Runge–Kutta technique in EXCEL to solve Eqn. (8-26) is illustrated in Appendix 8-A. The numerical solution gives $x_{A,f} = 0.70$ when $t = 1.5$ h. Direct solution of the differential equation is straightforward because the final value of the independent variable, time, is known.

Part B: What is the temperature of the reactor after 1.5 h?

APPROACH The value of T when $x_{A,f} = 0.70$ (the conversion when $t = 1.5$ h) will be calculated from Eqn. (8-20).

SOLUTION

$$T = T_0 + (\overline{\Delta T}_{ad})x_A$$
$$T = 373\ \text{K} + (88.2) \times (0.70)\ \text{K} = 435\ \text{K}$$

Equation (8-26) also could have been solved by numerical integration. If the right-hand side of this equation is denoted $f(x_A)$, then the equation can be put into the form

$$\int_0^{x_{A,f}} \frac{dx_A}{f(x_A)} = \int_0^{1.5\ \text{h}} dt = 1.5\ \text{h}$$

If a value of $x_{A,f}$ is assumed, the left-hand side of this equation can be integrated numerically. The correct value of $x_{A,f}$ is the one that makes the value of the left-hand side equal to 1.5 h. Obviously, this technique would require trial and error and/or interpolation.

This problem might have been posed differently by specifying the final conversion, $x_{A,f}$, and asking what time is required to achieve that conversion. For this variation, a standard numerical integration would be more convenient than a numerical solution of Eqn. (8-26).

8.5 CONTINUOUS STIRRED-TANK REACTORS (GENERAL TREATMENT)

The use of the energy balance to analyze or size an ideal CSTR is not as complex mathematically as for an ideal batch reactor or an ideal PFR. Therefore, in this section, we will not limit ourselves to an adiabatic CSTR. The adiabatic case will be covered as part of a more general treatment.

Consider the ideal CSTR shown in Figure 8-3.

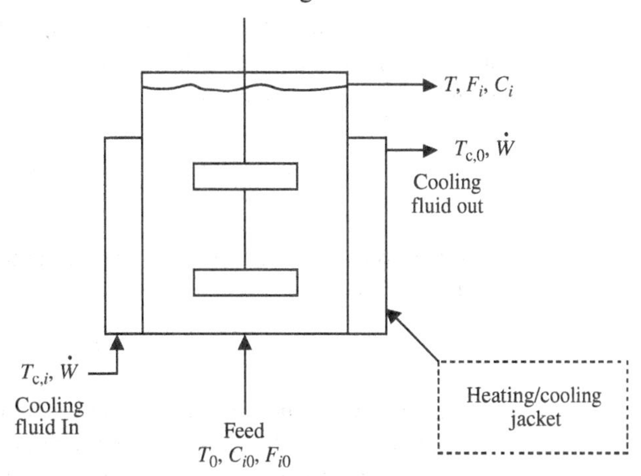

Figure 8-3 Schematic diagram of an ideal CSTR with heat transfer through a heating/cooling jacket.

Heat is added to or removed from the reactor through a jacket, which is in intimate contact with the walls of the reactor. A heat-transfer fluid flows through the jacket. Depending on the reactor temperature, and on whether the reactor is being heated or cooled, the heat-transfer fluid may be cooling water, chilled brine, chilled glycol solution, hot oil, or some other fluid. It is also possible to use a fluid that undergoes a phase change, e.g., boiling water or condensing steam. The fluid is a source of (or a sink for) the heat that is transferred through the reactor wall.

The *mass* flow rate of the heat-transfer fluid is $\dot{W}$. The temperature of the fluid at a point in the jacket is T_c. The fluid enters the jacket at $T_{c,i}$ and leaves at $T_{c,0}$. The feed enters the reactor at a temperature, T_0. The molar flow rates of the various components of the feed are designated F_{i0} and the corresponding concentrations are designated C_{i0}. The reactor operates at a temperature, T, which is the temperature of the Stream leaving the reactor. The molar flow rates in the effluent Stream are designated F_i and the corresponding concentrations are C_i. In the following analysis, we will assume that the reactor is operating at steady state.

A jacket is not the only device that can be used to transfer heat in a stirred-tank reactor, or in a batch reactor. It is also common to use a coil of tubing that is inserted into the reactor through the top head. In fact, it is not uncommon to use both a jacket and a coil. A third possibility is to circulate the reactor contents through an external heat exchanger, although this alternative is less common. The analysis of all three of these heat-transfer techniques is similar.

Another option for heat removal from a CSTR or batch reactor is to vaporize some of the contents of the reactor, condense some or all of the vapor in an external condenser, and return the liquid condensate to the reactor. This technique is feasible when the reactor can be operated at a temperature where the rate of vaporization is large enough to allow a significant rate of heat removal. Analyzing vaporization/condensation heat removal is more complex than analyzing heat transfer through a jacket or an internal coil. The following development is based on the latter means of heat transfer.

8.5.1 Simultaneous Solution of the Design Equation and the Energy Balance

If a single reaction takes place in a CSTR, the energy balance on the whole reactor is given by

$$Q - W_s - \left(\sum_{i=1}^{N} F_{i0}\bar{c}_{p,i}\right)(T - T_0) - F_{A0}x_A\Delta H_R(T) = \mathrm{d}U_{sys}/\mathrm{d}t \tag{8-6}$$

At steady state, $\mathrm{d}U_{sys}/\mathrm{d}t = 0$. For many situations, the shaft work W_s is negligible compared to the other terms in the energy balance. We will make this assumption and simplify the energy balance to

$$Q - \left(\sum_{i=1}^{N} F_{i0}\bar{c}_{p,i}\right)(T - T_0) - F_{A0}x_A\Delta H_R(T) = 0 \tag{8-27}$$

If heat is exchanged through a jacket, a coil, or an external exchanger, the rate law for heat transfer is

$$q = U\Delta T = U(T_c - T) \tag{8-28}$$

In this equation, q is the heat *flux* into the reactor (e.g., J/m^2-h), U is the overall heat-transfer coefficient (e.g., J/m^2-h-°C), T is the temperature of the reactor contents (e.g., °C), and T_c is the temperature of the heat-transfer fluid at some point in the exchanger. If $T < T_c$, heat is transferred into the reactor, and $q > 0$. The total rate of heat transfer, Q, is obtained by integrating the flux, q, over the whole area of the exchanger.

There are two limiting cases of the behavior of the heat-transfer fluid. First, in a coil or in a one-pass exchanger, there is little or no mixing of the heat-transfer fluid in the direction of flow. In other words, the heat-transfer fluid flows through the heat-exchange device essentially in plug flow. If all of the heat transferred goes into increasing the sensible heat of the heat-transfer fluid, then integration of Eqn. (8-28) gives

$$Q = UA_h\Delta T_{lm} \tag{8-29}$$

where A_h is the total area of the heat-exchange device and the log-mean temperature difference is given by

$$\Delta T_{lm} = \frac{[(T_{c,i} - T) - (T_{c,0} - T)]}{\ln\left(\frac{(T_{c,i}-T)}{(T_{c,0}-T)}\right)}$$

This situation also might exist in a jacket, if it were designed to minimize mixing of the heat-transfer fluid in the direction of flow.

In the second limiting case, the temperature of the fluid is the same at every point in the exchanger. For this case, Eqn. (8-28) integrates to

$$Q = UA_h(\overline{T}_c - T)$$

where $\overline{T}_c$ is the constant temperature of the heat-transfer fluid. The temperature of the heat-transfer fluid can be constant, for example, (a) in boiling/condensing heat transfer, if the pressure is constant, or (b) when all of the heat transferred goes into increasing the sensible heat of the heat-transfer fluid, but the heat-transfer fluid is completely mixed, e.g., in the jacket. In either case, $\overline{T}_c$ is equal to the outlet temperature, $T_{c,0}$.

Consider another situation where all of the heat transferred goes into increasing the sensible heat of the heat-transfer fluid. If the flow rate of the fluid is high, the difference between the inlet and the outlet temperatures will be small. For this situation, the total rate of heat transfer is given by

$$Q = UA_{\mathrm{h}}(T_{\mathrm{c},i} - T) \tag{8-30}$$

Equation (8-30) leads to a simple, graphical interpretation of the behavior of a CSTR. Substituting Eqn. (8-30) into Eqn. (8-27) and rearranging gives

$$[-\Delta H_{\mathrm{R}}(T)]F_{\mathrm{A0}}x_{\mathrm{A}} = UA_{\mathrm{h}}(T - T_{\mathrm{c}}) + \left(\sum_{i=1}^{N} F_{i0}\bar{c}_{p,i}\right)(T - T_0) \tag{8-31}$$

The subscript "i" has been dropped from $T_{\mathrm{c},i}$ for simplicity, so that T_{c} is the constant temperature of the heat-transfer fluid.

To understand the physical meaning of the terms in Eqn. (8-31), consider an exothermic reaction. The term $[-\Delta H_{\mathrm{R}}(T)]F_{\mathrm{A0}}x_{\mathrm{A}}$ is the part of the overall enthalpy change that results from the change in composition between the outlet and inlet Streams, i.e., from the reaction. The term $UA_{\mathrm{h}}(T - T_{\mathrm{c}})$ is the rate at which heat is transferred out of the CSTR. Finally, the term $\left(\sum_{i=1}^{N} F_{i0}\bar{c}_{p,i}\right)(T - T_0)$ is the increase in sensible heat *of the feed* as it goes from T_0, the reactor inlet temperature, to T. Thermodynamically, the sum of the first and the third terms is the overall enthalpy change when the feed Stream at temperature T_0 reacts to give the product Stream at temperature T.

Conceptually, the term $(-\Delta H_{\mathrm{R}}(T))F_{\mathrm{A0}}x_{\mathrm{A}}$ can be regarded as a rate of "heat generation." For an exothermic reaction, this term will be >0. Let's label this term $G(T)$. The "T" in parenthesis is a reminder that x_{A} will be a strong function of temperature, and that, in general, ΔH_{R} also depends on temperature.

At steady state, this generation term must be exactly balanced by a "removal" term, which is the sum of the two terms on the right-hand side of Eqn. (8-31). Let's label this sum $R(T)$, the total rate of heat "removal." The term $UA_{\mathrm{h}}(T - T_{\mathrm{c},0})$ is the rate at which heat is removed by heat transfer. The term $\left(\sum_{i=1}^{N} F_{i0}\bar{c}_{p,i}\right)(T - T_0)$ is the rate at which heat is "absorbed" by heating the feed Stream from the inlet temperature, T_0, to the outlet temperature, T.

Thus, Eqn. (8-31), the overall energy balance for an ideal CSTR, becomes

Simplified energy balance—ideal CSTR

$$\boxed{G(T) = R(T)} \tag{8-32}$$

where

Definition of $G(T)$

$$G(T) = [-\Delta H_{\mathrm{R}}(T)]F_{\mathrm{A0}}x_{\mathrm{A}} \tag{8-33}$$

and

$$R(T) = UA_{\mathrm{h}}(T - T_{\mathrm{c}}) + \left(\sum_{i=1}^{N} F_{i0}\bar{c}_{p,i}\right)(T - T_0) \tag{8-34}$$

Definition of $R(T)$

$$R(T) = \left[UA_{\mathrm{h}} + \left(\sum_{i=1}^{N} F_{i0}\bar{c}_{p,i}\right)\right]T - \left[UA_{\mathrm{h}}T_{\mathrm{c}} + \left(\sum_{i=1}^{N} F_{i0}\bar{c}_{p,i}\right)T_0\right] \tag{8-34a}$$

The term $G(T)$ depends directly on x_{A}. The term $R(T)$ does not depend on x_{A}, but it varies directly with T, as shown by Eqn. (8-34a). The values of x_{A} and T *must* be such that

$G(T) = R(T)$, since the energy balance is not satisfied unless these two terms are equal. If $G(T) \neq R(T)$, the CSTR is *not* at steady state.

Suppose that we were analyzing the behavior of an exothermic, irreversible reaction in an ideal CSTR. The volume, V, and the inlet molar flow rate of A, F_{A0}, are fixed. Assume that ΔH_R is not a function of temperature. For this situation, a plot of $G(T)$ versus reactor temperature, T, would have the characteristics shown in the following graph.

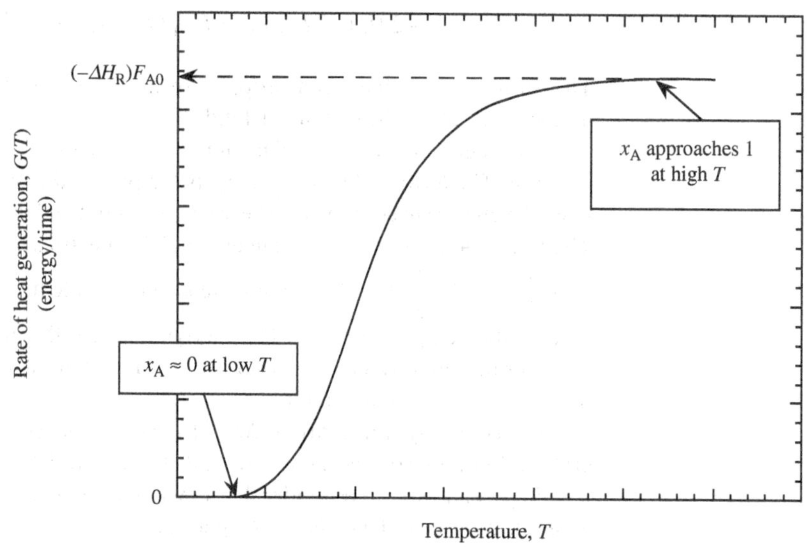

Figure 8-4 Rate of heat "generation," $G(T)$, versus reactor temperature, T, for an irreversible, exothermic reaction in an ideal CSTR.

The shape of the curve in Figure 8-4 is easy to understand. When the temperature of the reactor is low, the rate constant will be small and the reaction will be slow. From the design equation, the fractional conversion, x_A, will be close to zero. As the temperature is increased, the rate constant increases exponentially. The reaction rate increases rapidly, giving rise to a sharp increase in x_A. As the conversion approaches 1, the rate of change of x_A slows. At very high temperatures, x_A is close to 1, if the reaction is irreversible.

Suppose that the rate equation for a given reaction is

$$-r_A = kC_A/(1 + K_A C_A)$$

Further suppose that the constants k and K_A are known as functions of temperature. If this reaction is carried out in an ideal CSTR with a volume of V and an inlet molar flow rate of A, F_{A0}, then the $G(T)$ versus T curve can be constructed in a straightforward manner. Set up a table as shown below.

Temperature (K)	Rate constant, k	Adsorption constant, K_A	Fractional conversion, x_A	$G(T) = -x_A \times F_{A0} \times \Delta H_R$
T_1	$k(T_1)$	$K_A(T_1)$	$x_A(T_1)$	$G(T_1)$
T_2	$k(T_2)$	$K_A(T_2)$	$x_A(T_2)$	$G(T_2)$
T_3	—	—	—	—

Choose a temperature, say T_1. Calculate the value of the rate constant, k, at T_1 using the Arrhenius equation. Calculate the value of the constant K_A at T_1 using an equation that expresses K_A as a function of temperature, e.g., the van't Hoff equation. Then use the design equation for an ideal CSTR to calculate the outlet conversion, x_A, for T_1. Finally, calculate $G(T_1)$ from $G(T) = -x_A \times F_{A0} \times \Delta H_R$. Now, pick another value of T, say T_2, and repeat the process. Keep choosing new T's and repeating the calculations until the curve of $G(T)$ versus T has sufficient detail.

The following figure summarizes the algorithm for generating the $G(T)$ curve.

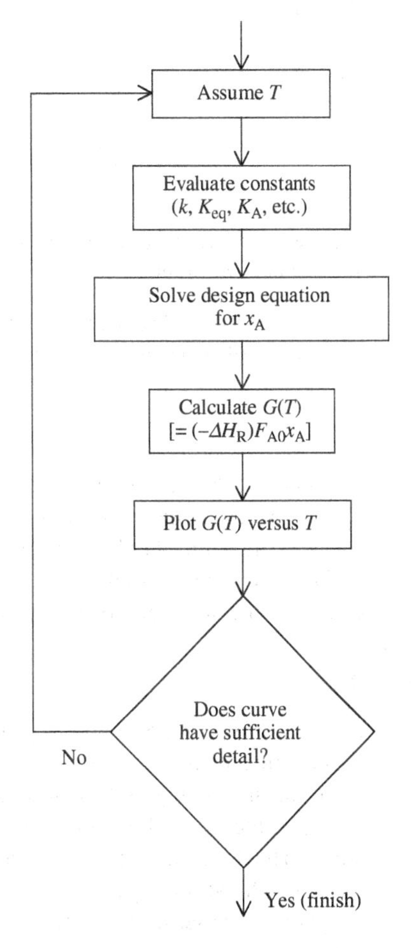

It is important to recognize that the design equation is "contained" in $G(T)$ since the design equation must be used to calculate x_A.

Now, let's consider $R(T)$. If U and $\left(\sum_{i=1}^{N} F_{i0}\bar{c}_{p,i}\right)$ are independent of temperature, then $R(T)$ is linear in T. For this situation, Eqn (8-34a) shows that a plot of $R(T)$ versus T will be a straight line with a slope of $\left[UA_h + \left(\sum_{i=1}^{N} F_{i0}\bar{c}_{p,i}\right)\right]$ and an intercept of $\left[UA_hT_c + \left(\sum_{i=1}^{N} F_{i0}\bar{c}_{p,i}\right)T_0\right]$. Let's plot $R(T)$ on the same graph as $G(T)$, as shown in Figure 8-5.

The intersection of the $G(T)$ curve with the $R(T)$ line is the "operating point" of the reactor. The x-coordinate of this point is the temperature in the CSTR, T_{op}. The y-coordinate is the rate

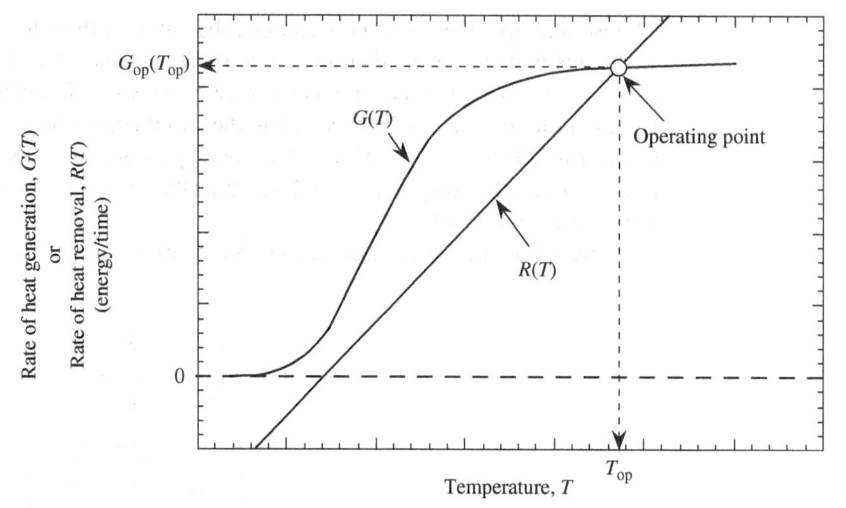

Figure 8-5 Rate of heat "generation," $G(T)$, and rate of heat "removal," $R(T)$, versus reactor temperature, T, for an irreversible, exothermic reaction in an ideal CSTR.

of heat generation, $G_{op}(T_{op})$, *and* the rate of heat removal, $R(T)$, for the reactor. The fractional conversion of A, x_A, in the reactor effluent at the "operating point" can be calculated from

$$G_{op}(T_{op}) = (-\Delta H_R)F_{A0}x_{A,op}$$
$$x_{A,op} = G_{op}(T_{op})/(-\Delta H_R)F_{A0}$$

Figure 8-5 is a *graphical* solution to the design equation and the energy balance for the CSTR. We could have solved these two simultaneous algebraic equations by other means, for example, with GOALSEEK, to find $x_{A,op}$ and T_{op}. The advantage of the graphical method is that we obtain a visual picture of the situation, and we can easily evaluate the impact of changing the slope and/or intercept of the $R(T)$ line.

8.5.2 Multiple Steady States

Let's consider some changes to the situation shown in Figure 8-5. Suppose that the temperature of the feed, T_0, or the temperature of the cooling fluid, T_c, is reduced. The effect of this change will be to raise the intercept of the $R(T)$ line on the y-axis, without changing the slope of the line. This can be seen by examining Eqn. (8-34a). The $R(T)$ curve shifts to the left, parallel to the original curve. If the decrease in T_0 and/or T_c is sufficient, the situation shown in Figure 8-6 might result.

For this case, there are three possible combinations of x_A and T that *simultaneously* satisfy both the design equation and the energy balance. These are shown as the three circled "operating points" in Figure 8-6. This situation is known as "multiple steady states" or "multiplicity."

Let's examine this result more carefully. In order to fix the position of both $G(T)$ and $R(T)$, we specified all of the following variables: the CSTR volume, V; each of the inlet molar flow rates, F_{i0}, the rate equation, all of the constants in the rate equation as functions of temperature, the inlet temperature, T_0, the coolant temperature, T_c, and the product of the exchanger area and the overall heat-transfer coefficient, UA. Despite fixing all of these parameters, the graphical analysis shows that, at steady state, the CSTR can operate at any one of three combinations of x_A and T. Quite remarkable!

Figure 8-6 illustrates why it is desirable to carry out a complete graphical analysis when sizing or analyzing a CSTR. The algebraic equations that describe this problem, i.e., the

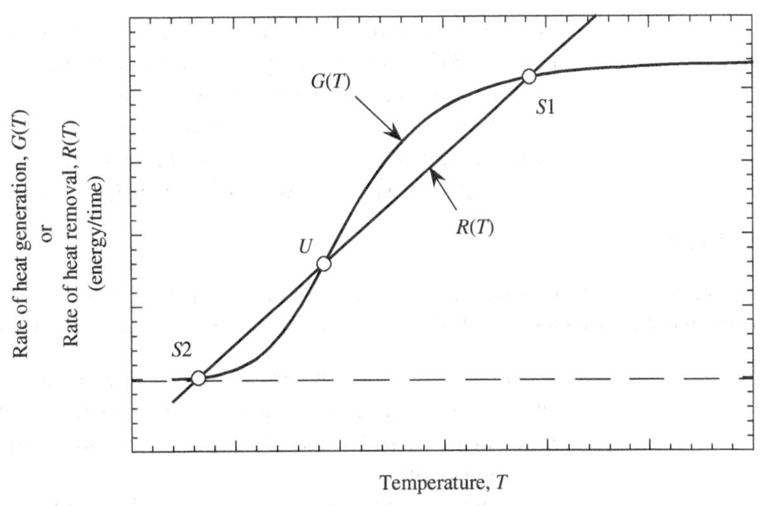

Figure 8-6 Rate of heat "generation," $G(T)$, and rate of heat "removal," $R(T)$, versus reactor temperature, T, for an irreversible, exothermic reaction in an ideal CSTR. Three steady-state operating points are possible.

energy balance and the design equation for A, could have been solved simultaneously by numerical techniques to find *one* of the three possible solutions, depending on the initial estimate of T or x_A. However, unless we deliberately looked for the second and third solutions, they might never have been discovered. The graphical analysis automatically identified all three steady-state solutions.

8.5.3 Reactor Stability

At which of the three points in Figure 8-6 will the CSTR *actually* operate?

First, consider the middle point, labeled "U". Figure 8-7 shows a blowup of the region around this point. Suppose that a CSTR that is operating at point "U" experiences a small,

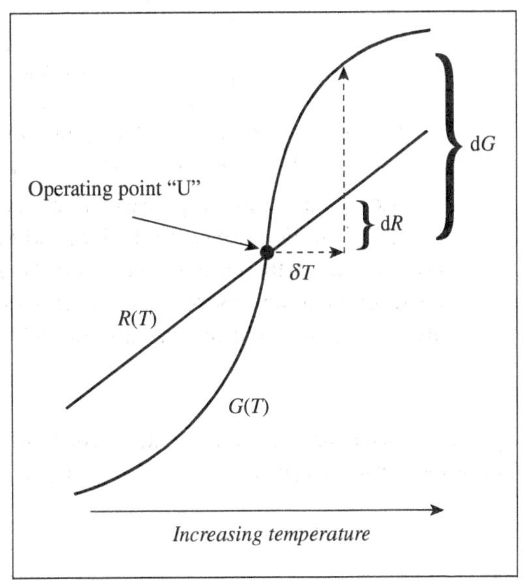

Figure 8-7 Expanded view of the region around point "U" in Figure 8-6 showing the response to a small, positive fluctuation of the reactor temperature.

positive temperature fluctuation, δT. The temperature perturbation causes the rate of heat removal, $R(T)$, to increase by an amount δR. It also causes the rate of heat generation, $G(T)$, to increase by an amount δG. However, $\delta G > \delta R$.

The reactor is no longer at steady state, and the rate of heat generation exceeds the rate of heat removal. As a result, the reactor temperature must increase. The increase continues until the point "S1" in Figure 8-6 is reached, at which point a new steady state is established.

EXERCISE 8-1

Suppose that the temperature fluctuation, δT, was in the opposite direction, such that the reactor temperature dropped. Analyze the stability of the operating point "U" to this change. What operating point is reached eventually?

Because of this behavior, we say that the point "U" is "intrinsically unstable." A very small fluctuation in temperature (or composition) will cause the reactor to wander away from an intrinsically unstable operating point until it reaches some other steady-state operating point.

Now consider an operating point such as "S1" in Figure 8-6. A blowup of this point is shown in Figure 8-8.

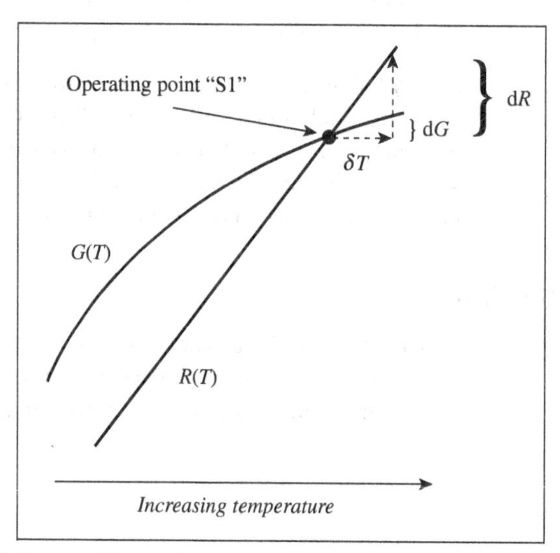

Figure 8-8 Expanded view of the region around point "S1" in Figure 8-6 showing the response to a small, positive fluctuation of the reactor temperature.

Let's analyze the response to a small, positive change in temperature, δT, as we did with point "U" in Figure 8-7. Both $G(T)$ and $R(T)$ increase. However, the increase in $G(T)$, δG, now is *smaller* than the increase in $R(T)$, δR. Therefore, $R(T)$ is greater than $G(T)$, and the unsteady-state energy balance requires that the temperature decrease toward the temperature of the original operating point, S1.

EXERCISE 8-2

Suppose that the temperature fluctuation, δT, was in the opposite direction, such that the reactor temperature dropped. Analyze the stability of the operating point "S1" to this change.

EXERCISE 8-3

Analyze the behavior of the operating point S2 in Figure 8-6.

Both S1 and S2 are "intrinsically stable" operating points.

The question of whether the reactor operates at S1 or S2 depends on how it is started up. In order to determine the effect of startup conditions, the *unsteady-state* energy and material balances must be solved simultaneously. This task is beyond the scope of this chapter.

The type of stability analysis carried out above is not mathematically rigorous. However, it is a very useful way to understand the concept of stable and unstable operating points. In mathematical terms, the above analysis has shown that a steady-state operating point will be unstable if

$$\left[\frac{\partial G(T)}{\partial T}\right]_{\text{operating point}} > \left[\frac{\partial R(T)}{\partial T}\right]_{\text{operating point}}$$

and will be stable if

$$\left[\frac{\partial G(T)}{\partial T}\right]_{\text{operating point}} < \left[\frac{\partial R(T)}{\partial T}\right]_{\text{operating point}}$$

A more rigorous analysis[1] would show that the first inequality is a sufficient criterion for *instability*. If this inequality is satisfied, the operating point will be intrinsically unstable. The rigorous analysis also shows that the second criterion is a sufficient criterion for *stability*, provided that the CSTR is adiabatic. If the reactor is not adiabatic, the second condition is necessary, but not sufficient.

8.5.4 Blowout and Hysteresis

The existence of multiple steady states for a CSTR gives rise to several interesting and important phenomena, two of which are known as "blowout" and feed-temperature hysteresis.

8.5.4.1 Blowout

Consider a CSTR that is operating at steady state at point 1 in Figure 8-9a. The reactor temperature is about 453 K and the conversion of A is about 84%. The *G*(*T*) curve in this figure was calculated for a second-order, irreversible reaction ($A \rightarrow B$) taking place in the liquid phase in a CSTR with a volume of 0.40 m^3. The inlet concentration of A is 16 kmol/m^3, and the volumetric feed rate to the reactor is 1.3 m^3/ks. The heat of reaction is –21 kJ/mol A, independent of temperature. The rate constant is $3.20 \times 10^9 \exp(-12{,}185/T)$ (m^3/mol A-ks), where T is in K.

The CSTR is operating adiabatically with a feed temperature of 312 K. The heat capacity of the feed Stream is 2.0 J/cm^3-K, independent of temperature. The *R*(*T*) line in Figure 8-9a was constructed from these values.

EXERCISE 8-4

Verify that the *G*(*T*) curve in Figure 8-9a is correct by calculating the value of *G*(*T*) for a reactor temperature of 450 K.

[1] See, for example, Froment, G. F. and Bischoff, K. B., *Chemical Reactor Analysis and Design, 2nd edition*, John Wiley & Sons, Inc., New York (1990), pp. 376–381.

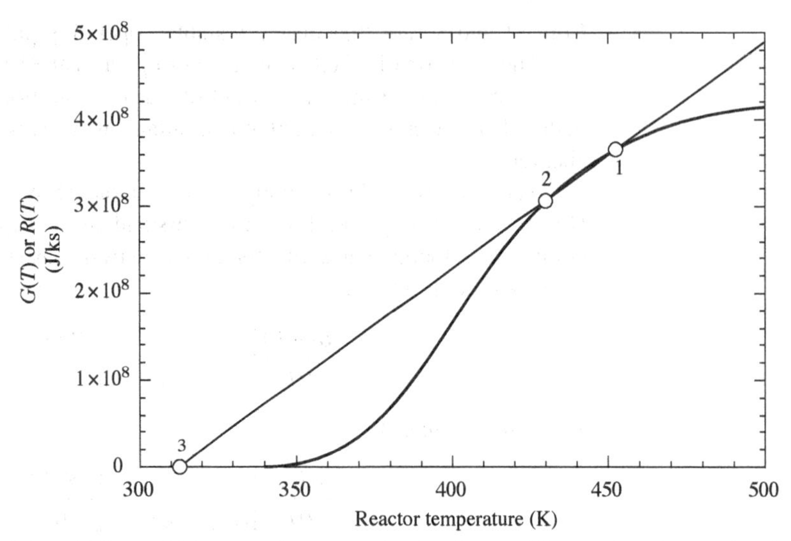

Figure 8-9a $G(T)$ and $R(T)$ curves for a second-order, irreversible, exothermic reaction: base case.

EXERCISE 8-5

Verify that the $R(T)$ curve in Figure 8-9a is correct by calculating the value of $R(T)$ for a reactor temperature of 450 K.

There are three possible steady-state operating points, labeled 1, 2, and 3. Point 1, at which the CSTR is operating, is intrinsically stable. Point 3 also is intrinsically stable, and point 2, which is located close to point 1, is intrinsically unstable. The $G(T)$ and $R(T)$ curves barely intersect in the high-temperature (high-conversion) region. Clearly, it is risky to operate at point 1. A slight shift in the $G(T)$ or $R(T)$ curves might eliminate the intersection in the high conversion region, thereby eliminating the high-conversion operating point and leaving only a point at essentially zero conversion.

EXAMPLE 10-3 ***"Blowout" of a CSTR***

Suppose that the volumetric flow rate to the reactor described above is increased by 20%, with no change in the feed concentrations. A new steady state then is allowed to establish. At what temperature and fractional conversion does the CSTR operate, at this new steady state?

APPROACH

The average residence time, τ, has decreased as a result of the increase in the volumetric feed rate. Therefore, the conversion of A, x_A, will decrease, at a fixed temperature, leading to a new $G(T)$ curve. The location of this curve will be calculated and $G_{new}(T)$ will be plotted versus T.

Even though the feed temperature (T_0) remains constant, the increase in feed rate will cause the position of the $R(T)$ line to change because $\sum_{i=1}^{N} F_{i0}\bar{c}_{p,i}$ changes. For this example, where the reactor is adiabatic, the slope of $R(T)$ increases by 20% and the intercept decreases by 20%. This is because both the slope and intercept of the $R(T)$ line depend on the feed rate, as shown by Eqn. (8-34a). The $R_{new}(T)$ curve will be calculated and plotted on the same graph that was used to plot $G_{new}(T)$.

The new operating point(s) will lie at the intersection(s) of G_{new} with R_{new}.

SOLUTION

The calculations of $G_{new}(T)$ and $R_{new}(T)$ are shown in Appendix 8-B. The $G_{new}(T)$ and $R_{new}(T)$ lines are shown in Figure 8-9b, along with the original $G(T)$ and $R(T)$ curves. The $G_{new}(T)$ and $R_{new}(T)$ curves intersect at only one point, where the temperature, T, is about 312 K and x_A is about 0.

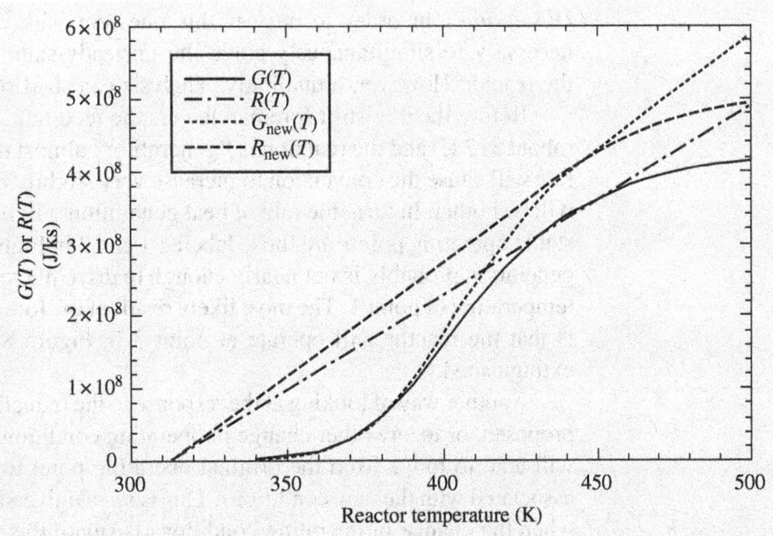

Figure 8-9b $G(T)$ and $R(T)$ curves for a second-order, irreversible, exothermic reaction. Base case [$G(T)$ and $R(T)$] and for a 20% increase in the feed rate to the reactor [$G_{new}(T)$ and $R_{new}(T)$].

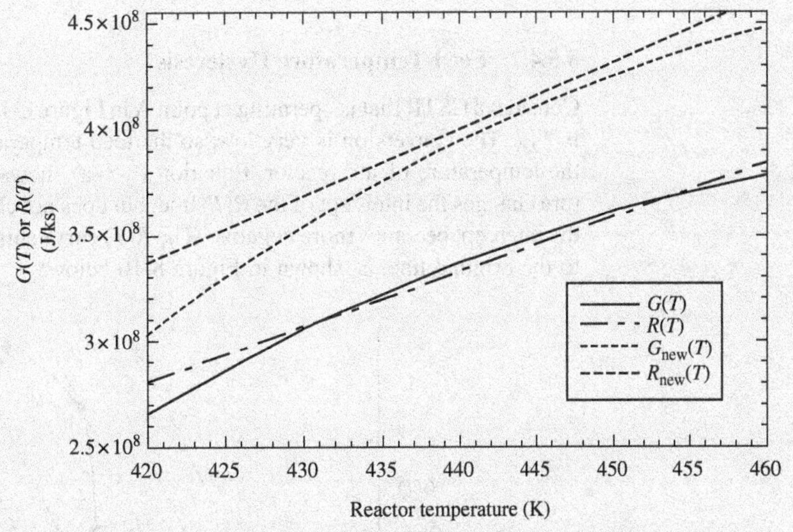

Figure 8-9c Expanded view of the high-conversion region in Figure 8-9b.

Figure 8-9c is a blow up of the $G(T)$ and $R(T)$ lines in the high-conversion region, in the vicinity of the original operating point. It is clear that G_{new} and R_{new} do not intersect in this region.

The 20% increase in feed rate has effectively extinguished, or "blown out," the reaction. The conversion of A has dropped from about 84% to almost 0, and the reactor temperature has dropped from about 453 K to about 312 K.

Extension The day-shift foreman arrives in the morning to find the reactor operating at steady state at the new condition, i.e., at a feed rate of 1.56 m^3/ks, a temperature of 312 K (the feed temperature), and a conversion of essentially zero. The foreman decides to reduce the feed rate to its original value, in the hope of returning the reactor to its original operating point ($x_A = 0.84$, $T = 453\text{ K}$). What do you think will happen?

Discussion In order to answer this question with complete confidence, it would be necessary to simultaneously solve the unsteady-state energy and material balances for the reactor. However, a qualitative analysis can shed some light on the situation.

Before the day-shift foreman lowers the feed rate, the reactor is at a low temperature (about 312 K) and the reaction is "generating" almost no heat ($x_A \cong 0$). Reducing the feed rate will cause the conversion to increase very slightly because the average residence time will be higher. In turn, the rate of heat generation will increase slightly. However, the only stable operating points are those labeled 1 and 3 in Figure 8-9a. The increased rate of heat generation probably is not nearly enough to drive the reactor temperature up to 453 K, the temperature of point 1. The most likely result of the foreman's proposed change in feed rate is that the reactor will operate at point 3 in Figure 8-9a, and the reaction will remain extinguished.

Another way of looking at the response to the reduction in feed rate that the foreman has proposed, or to any other change in operating conditions for that matter, is that the system will tend to move from the original operating point to the *closest* stable operating point associated with the new conditions. This is *not* a universally valid generalization. However, when the change in operating conditions is small, this generalization often holds.

It is likely that more drastic steps will be required to restore the original operation. One option would be to temporarily preheat the feed Stream. Again, the exact strategy for restoring operation should be developed by solving the unsteady-state energy and material balances.

8.5.4.2 Feed-Temperature Hysteresis

Consider a CSTR that is operating at point A in Figure 8-10. The temperature of the feed (T_0) is $T_{0,A}$. The conversion is very low, so the feed temperature is increased in order to raise the temperature of the reactor. Equation (8-34a) shows that changing the feed temperature changes the intercept of the $R(T)$ line, but does not change its slope. As T_0 is increased, the intercept becomes more negative. The $R(T)$ curve shifts to the right and remains parallel to the original line, as shown in Figure 8-10 below.

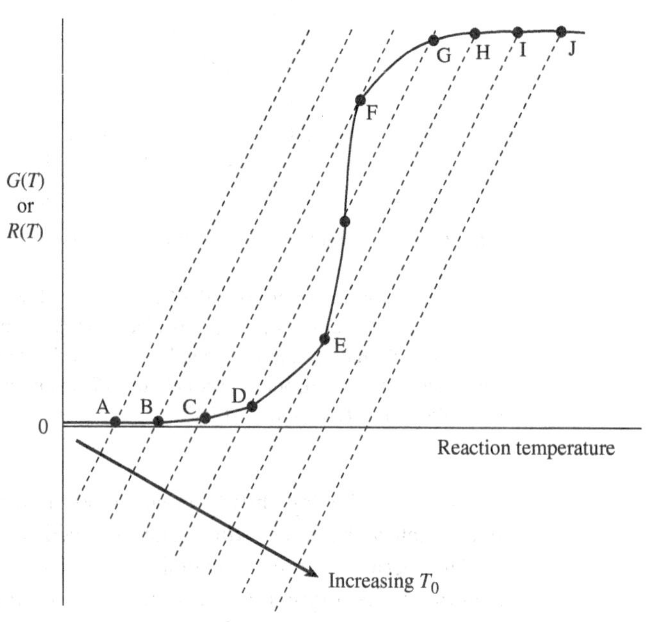

Figure 8-10 Operating points for a CSTR with different feed temperatures.

Assume that the change in T_0 is small, so that the new steady-state operating point is point B in Figure 8-10. The feed temperature at this point is $T_{0,B}$. Figure 8-10 shows that the conversion x_A still is very low. Therefore, the feed temperature is increased by an additional small amount, so that the $R(T)$ line now intersects the $G(T)$ curve at point C and is tangent to $G(T)$ at point F. The feed temperature for this point is $T_{0,CF}$. If the change in the feed temperature is small, we would expect the reactor to operate at point C, rather than at point F.

Now suppose that the feed temperature is increased again, by a very small amount to $T_{0,DG}$, such that the possible *stable* operating points are D and G. For a small change in the feed temperature, the expected operating point would be D. The feed temperature is increased again, to a $T_{0,EH}$ where the stable operating points are E, at which the $R(T)$ line is just tangent to the $G(T)$ curve, and H. If the increase in feed temperature were small enough, the reactor would operate at point E, still a relatively low conversion.

At this point, the situation changes! Any further increase in the feed temperature, however small, causes the conversion to increase substantially, e.g., to point I and then to point J. The corresponding feed temperatures are $T_{0,I}$ and $T_{0,J}$.

The behavior of the CSTR so far is summarized in Figure 8-11, a plot of the fractional conversion of A versus the temperature of the feed to the CSTR. The above discussion started with point A and has moved A → B → C → D → E → H → I → J. The conversion was low until point E ($T_{0,EH}$) was reached. A further increase in feed temperature, to $T_{0,I}$, caused the conversion to increase very substantially, to essentially 100%.

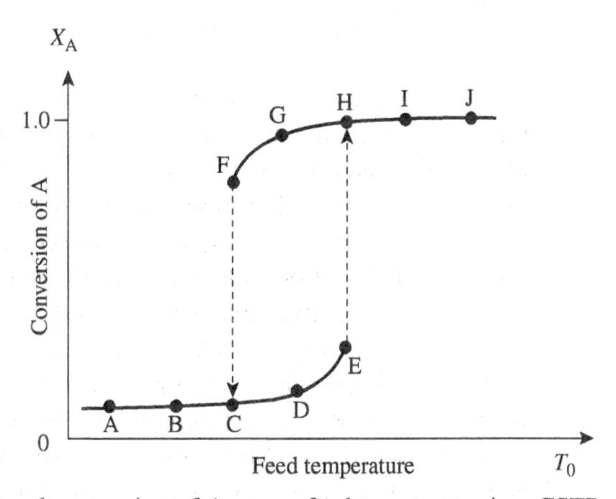

Figure 8-11 Fractional conversion of A versus feed temperature in a CSTR.

Now consider what will happen as the temperature is decreased, starting at point J ($T_{0,J}$). When the feed temperature is $T_{0,I}$, the steady-state conversion is still high. If the difference between $T_{0,I}$ and $T_{0,EH}$ is small, a reduction to $T_{0,EH}$ should cause the reactor to operate at point H in Figure 8-11, rather than at point E. Similarly, when the feed temperature is reduced to $T_{0,DG}$, the reactor should operate at point G, and when the temperature is reduced to $T_{0,CF}$, the reactor should operate at F. However, a further reduction below $T_{0,CF}$ will cause the conversion to drop substantially.

Figure 8-11 shows *hysteresis*. Over a certain region of feed temperature, the conversion is not the same when the feed temperature is being increased as it is when the feed temperature is being decreased, even though the actual feed temperature is the same in both cases.

The sudden increase in conversion when the feed temperature is being increased from a low initial value is known as "light off," and the feed temperature at which this rapid jump is observed, $T_{0,EH}$ in this example, is called the "light-off temperature." The sudden decrease in conversion that is observed when the feed temperature is being decreased from a high value is called "extinction," and the feed temperature at which this occurs, $T_{0,CF}$ in this example, is called the "extinction temperature." When hysteresis occurs, the extinction and light-off temperatures will be different.

8.6 NONISOTHERMAL, NONADIABATIC BATCH, AND PLUG-FLOW REACTORS

8.6.1 General Remarks

Despite the simplicity of adiabatic reactors, there are many situations where adiabatic operation is not feasible or practical. Careful control of temperature may be required to avoid deactivating a catalyst by overheating it, or to control the selectivity of a reaction. This may require that heat be supplied or removed through the wall of the reactor as the reaction proceeds.

For a well-agitated batch reactor, heat can be removed or supplied through a jacket on the reactor, or through coils immersed in the reactor, without violating the assumption that the temperature is the same everywhere in the reactor at any time. Although temperature gradients will exist close to the heat-transfer surfaces, the volume of the reactor in which these gradients exist is small compared to the total volume of the reactor. As long as the temperature in a nonisothermal, nonadiabatic batch reactor is spatially uniform, the reactor is still *ideal*, and the design equation and the energy balance are the same as formulated earlier.

The situation is not as straightforward for a continuous tubular reactor. When heat is removed or supplied through the walls of the tube, temperature gradients are established in the radial direction. These gradients can cause the assumption of *ideal* plug flow to be violated. The ideal plug-flow model is a *one-dimensional* model, in that all variations of temperature, concentration, and reaction rate are confined to a single dimension, the direction of flow. For almost all nonadiabatic, tubular reactors, a *two-dimensional* model is required. The two-dimensional model permits temperature, concentration, and the reaction rate to vary in *both* the direction of flow (i.e., the axial dimension in a tubular reactor) and in one direction normal to flow (i.e., the radial dimension in a tubular reactor). In short, the PFR design equations that were formulated in Chapter 3 are not valid when radial gradients exist. Most tubular reactors with heat transfer through the walls cannot be considered to be plug-flow reactors. Both the material balance(s) and the energy balance must be (re)formulated to include both radial and axial gradients.

The use of two-dimensional models to size and analyze tubular reactors is beyond the scope of the present text. Froment and Bischoff[2] discuss such models in more detail.

8.6.2 Nonadiabatic Batch Reactors

Sizing or analysis of a nonadiabatic batch reactor, with a single homogeneous reaction taking place, requires the simultaneous solution of a mass balance and the macroscopic energy balance.

Mass balance (design equation)

$$\frac{-1}{V}\frac{dN_A}{dt} = -r_A \tag{3-5}$$

[2] Froment, G. F. and Bischoff, K. B., *Chemical Reactor Analysis and Design*, 2nd edition, John Wiley & Sons, 1990, Chapter 11.

Macroscopic energy balance

$$Q = \left(\sum_{i=1}^{N} N_i c_{p,i}\right)\frac{dT}{dt} + (-r_A)[\Delta H_R(T)]V \tag{8-15}$$

Substituting $Q = UA_h \Delta T$ and rearranging,

$$\frac{dT}{dt} = \frac{-(-r_A)[\Delta H_R(T)]V}{\left(\sum_{i=1}^{N} N_i c_{p,i}\right)} + \frac{UA_h \Delta T}{\left(\sum_{i=1}^{N} N_i c_{p,i}\right)} \tag{8-35}$$

The last step is to express ΔT in terms of the temperature of the contents of the reactor at any time. If the flow rate of heat-transfer fluid through the jacket or coils is relatively high, its temperature will not change significantly between the inlet and the outlet. Then

$$\Delta T = T_c(t) - T(t) \tag{8-36}$$

Here, $T_c(t)$ is the temperature of the heat-transfer fluid, and $T(t)$ is the temperature of the reactor contents. In general, both the reactor temperature and the heat-transfer fluid temperature will depend on time. The variation of T_c with time will be specified as part of the design. This permits Eqn. (8-36) to be substituted into Eqn. (8-35).

For this case, Eqns. (3-5) and (8-35) are a pair of simultaneous, first-order, ordinary differential equations. They are subject to the initial conditions: $T = T_0,\ N_A = N_{A0},\ t = 0$. These equations can be solved for T and x_A as a function of time, using the numerical techniques discussed in Chapter 7, Section 7.4.3.2 and Appendix 7-A.2.

If the flow rate of heat-transfer fluid is relatively low, then its temperature will change in passing through the jacket or coil. If the heat-transfer fluid is in plug flow,

$$\Delta T = \Delta T_{lm}$$

where ΔT_{lm} is the log-mean temperature difference, as discussed in Section 8.3 of this chapter. In general, ΔT_{lm} will depend on time. In this case, three ordinary differential equations must be solved simultaneously. The third equation, in addition to Eqns. (3-5) and (8-35), is

$$\frac{dT_c}{dA_h} = \frac{U(T_c - T)}{\dot{W} c_{p,m}}$$

Here, A_h is the area of the heat-exchange surface, $\dot{W}$ is the mass flow rate of the heat-transfer fluid, and $c_{p,m}$ is the mass heat capacity of the heat-transfer fluid.

8.7 FEED/PRODUCT (F/P) HEAT EXCHANGERS

8.7.1 Qualitative Considerations

The performance of a reactor is very sensitive to the temperature of the inlet Stream. This is especially true for plug-flow reactors, and for reactors that operate adiabatically.

Consider a reactor in which a single, exothermic reaction takes place. Figure 8-12 shows how the conversion of reactant A changes when the temperature of the feed Stream is varied, at constant space time and feed composition.

The conversion of reactant A in the reactor effluent (x_A) will be close to zero when the feed temperature is very low. As the feed temperature is increased, the outlet conversion increases rapidly. For a fixed space time and feed composition, the rate at which x_A increases with feed temperature will depend on the activation energy. The higher the activation energy, the steeper the slope of the x_A versus feed temperature curve. If the reaction is

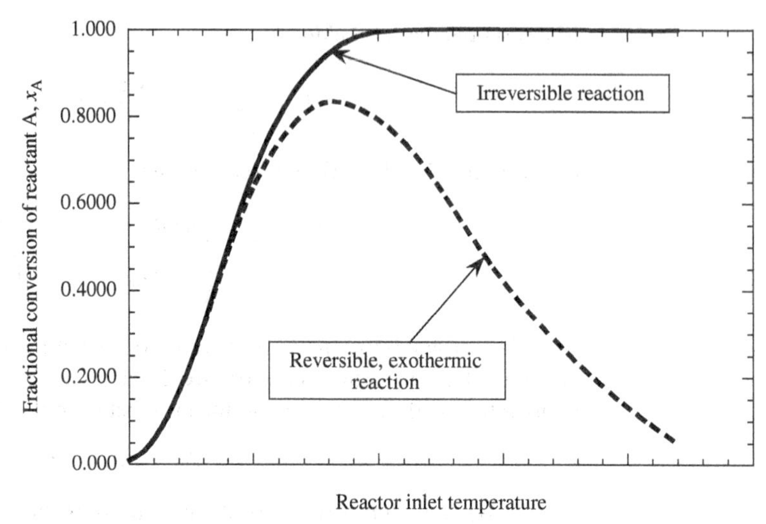

Figure 8-12 Effect of feed temperature on fractional conversion of reactant A, at constant space time and constant feed composition.

essentially irreversible over the temperature range of interest, the outlet conversion approaches 1 as the feed temperature continues to increase.

If the reaction is exothermic and *reversible*, the conversion versus feed temperature curve at low temperatures is similar to that of an irreversible reaction. However, as the temperature increases, the reverse reaction becomes increasingly important. Eventually, the outlet conversion goes through a maximum as the equilibrium becomes increasingly unfavorable. At very high temperatures, the reaction is limited by chemical equilibrium and the outlet conversion declines with feed temperature.

EXERCISE 8-6

Sketch the x_A versus feed temperature curve for a reversible, *endothermic* reaction.

For any reaction, the conversion is very small if the feed temperature is too low. For example, the feed to the reactor may come from storage tanks at essentially ambient temperature. Unless the reaction is very fast at room temperature, the feed will have to be preheated to some extent in order to obtain a reasonable reaction rate.

When the reaction is exothermic, a common approach to preheating the feed Stream is to use a feed/product (F/P) heat exchanger. This arrangement is shown schematically in Figure 8-13. The feed is heated by exchanging it against the hotter Stream leaving the reactor. Flow of the two Streams is countercurrent.

The system consisting of the reactor and the F/P heat exchanger is said to be *autothermal* if the desired outlet conversion can be achieved without addition of heat from an "outside" source. In other words, the system is autothermal if the Stream leaving the reactor is hot enough to heat the feed Stream to the temperature that is necessary to achieve the desired conversion, at steady state.

8.7.2 Quantitative Analysis

Let's consider an exothermic, *reversible* reaction, such as sulfur dioxide oxidation or methanol synthesis, occurring in an adiabatic, ideal PFR. A feed/product heat exchanger is

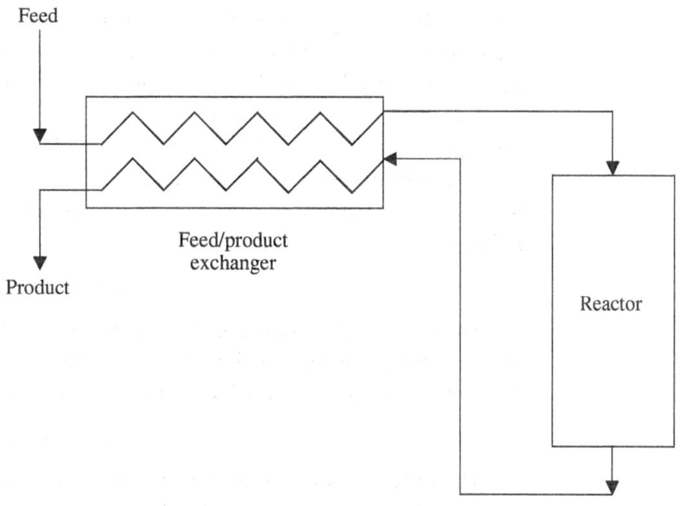

Figure 8-13 Schematic diagram of reactor and feed/product heat exchanger.

used to preheat the feed and to cool the product. Figure 8-14 is a sketch of the two items of equipment, and the profiles of temperature and fractional conversion of reactant A in each.

The "cold" (feed) Stream that enters the F/P heat exchanger, and eventually passes into the reactor, is labeled Stream 1. Stream 2 is the product Stream, i.e., the "hot" Stream. Let z' designate the axial dimension of the F/P exchanger. Stream 1 enters the F/P exchanger at

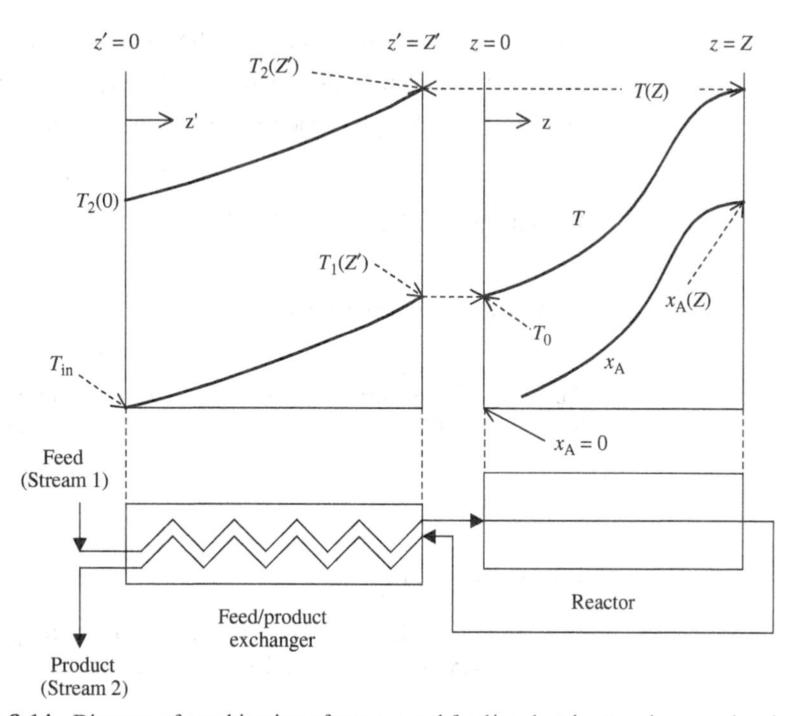

Figure 8-14 Diagram of combination of reactor and feed/product heat exchanger, showing temperature profiles in both pieces of equipment, and the profile of fractional conversion of reactant A in reactor.

$z' = 0$ with a temperature of T_{in}, which is assumed to be known. Stream 1 leaves the exchanger at $z' = Z'$ with a temperature $T_1(Z')$. This temperature will depend on how much heat is transferred in the F/P exchanger, and is not known at this point in the analysis.

Stream 2, which comes from the reactor, enters the F/P exchanger at $z' = Z'$ with a temperature of $T_2(Z')$. It leaves at $z' = 0$ with a temperature of $T_2(0)$.

The axial dimension of the reactor is designated z. The preheated feed enters the reactor at $z = 0$ with a temperature T_0. Stream 1 does not lose any heat between the F/P exchanger and the reactor. Therefore,

$$T_1(Z') = T_0 \tag{8-37}$$

The product Stream leaves the reactor at $z = Z$, with a temperature $T(Z)$. By the same logic, the temperature at which Stream 2 enters the F/P heat exchanger must be the same as the temperature at which it leaves the reactor. Therefore,

$$T(Z) = T_2(Z') \tag{8-38}$$

The only temperature that is known at this point is T_{in}. The reactor inlet temperature, T_0, is not known. Therefore, the design equation cannot be integrated to determine the composition and temperature of the reactor effluent.

In order to analyze the behavior of the reactor/exchanger combination, three "tools" are required: an energy balance on the reactor, an energy balance on the F/P exchanger, and the design equation for the reactor. These equations will have to be solved simultaneously. We will solve them graphically because of the valuable, visual insight that such a solution can provide.

8.7.2.1 Energy Balance—Reactor

The energy balance for an adiabatic flow reactor at steady state was developed in Section 8.4.3 of this chapter. If $F_{A0}(-\Delta H_R(T))/\sum_{i=1}^{N} F_{i0}\bar{c}_{p,i}$ is constant, then the temperature, T, at any point along the length of the reactor and the conversion, x_A, at that point are related by

$$T = T_0 + (\overline{\Delta T}_{ad})x_A \tag{8-20}$$

Rearranging and letting $\lambda = 1/\overline{\Delta T}_{ad}$,

$$x_A = \lambda[T - T_0]$$

In the present nomenclature,

$$x_A(z) = \lambda[T(z) - T_0] \tag{8-39}$$

Applying this equation to the whole reactor gives a relationship between the reactor inlet temperature, T_0, the reactor outlet temperature, $T(Z)$, and the overall conversion, $x_A(Z)$,

$$x_A(Z) = \lambda[T(Z) - T_0] \tag{8-40}$$

8.7.2.2 Design Equation

The solution of the design equation for an adiabatic plug-flow reactor was discussed in Section 8.4.5 of this chapter. The design equation for an ideal PFR can be written as

$$\frac{dx_A}{d\tau'} = [-r_A(x_A, T)]/C_{A0} \tag{8-41}$$

When the reactor is adiabatic, T and x_A are related through Eqn. (8-39).

If the space time, τ, for the whole PFR is fixed, then the outlet conversion, $x_A(Z)$, and the temperature of the stream leaving the reactor, $T(Z)$, depend only on the feed temperature, T_0. The values of $x_A(Z)$ and $T(Z)$ can be determined by integrating Eqn. (8-41), in conjunction with Eqn. (8-39), from $\tau' = 0$ to $\tau' = \tau$, using the initial conditions $x_A = 0$, $T = T_0$ at $\tau' = 0$. An example of this type of calculation was given in Section 8.4.5 of this chapter, for the case of an ideal, adiabatic batch reactor.

A graph of $x_A(Z)$ versus $T(Z)$ can be generated by assuming different values of T_0 and repeating the integration of Eqn. (8-41). An illustration of the result of such calculations, for a reversible reaction, is given by the curve in Figure 8-15. This curve "contains" both the design equation and the reactor energy balance.

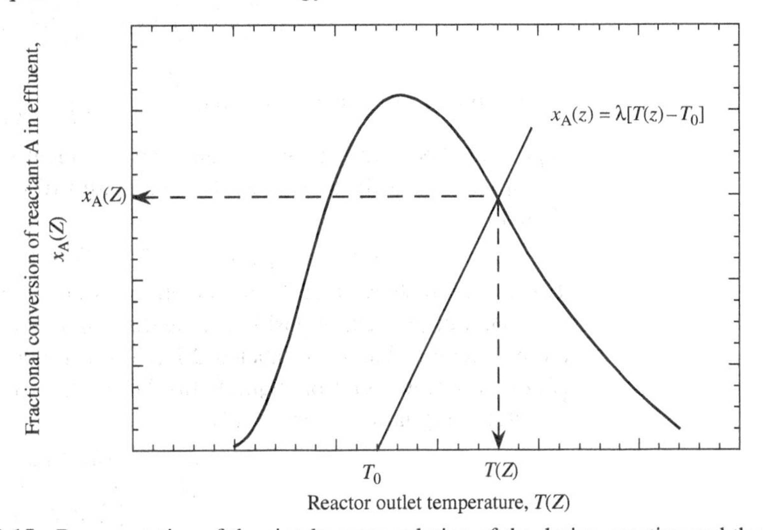

Figure 8-15 Representation of the simultaneous solution of the design equation and the energy balance for an adiabatic PFR. Each point on the curve corresponds to a different value of the inlet temperature, T_0. The straight line is the overall energy balance for the reactor and can be used to locate the T_0 corresponding to a given point on the curve.

The straight line in Figure 8-15 is the overall reactor energy balance, Eqn. (8-40). This equation serves to locate the value of T_0 that corresponds to a given outlet condition, $(x_A(Z), T(Z))$.

8.7.2.3 Energy Balance—F/P Heat Exchanger

Let's carry out an energy balance on a differential slice of the F/P exchanger, normal to the direction of flow. This control volume is shown in the following figure.

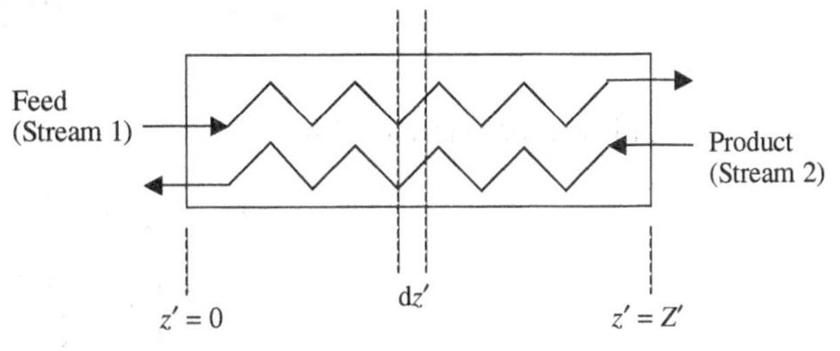

For Stream 1, the energy balance is

$$\left(\sum_{i=1}^{N} F_i \bar{c}_{p,i}\right)_1 dT_1 = U a_h (T_2(z') - T_1(z')) dz' \tag{8-42}$$

Here, U is the overall heat-transfer coefficient between the hot Stream and the cold Stream, a_h is the heat-exchange area per unit length of exchanger (e.g., m^2/m), dz' is the exchanger length over which the energy balance is taken, $T_2(z')$ is the temperature of the hot (product) Stream at z', and $T_1(z')$ is the temperature of the cold (feed) Stream at z'. Simply stated, the rate at which the sensible heat of Stream 1 increases is equal to the rate at which heat is transferred between Streams.

The same balance for Stream 2 is

$$\left(\sum_{i=1}^{N} F_i \bar{c}_{p,i}\right)_2 \mathrm{d}T_2 = Ua_\mathrm{h}(T_2(z') - T_1(z'))\mathrm{d}z' \tag{8-43}$$

Eliminating $Ua_\mathrm{h}(T_2(z') - T_1(z'))\mathrm{d}z'$ between Eqns. (8-42) and (8-43) gives

$$\left(\sum_{i=1}^{N} F_i \bar{c}_{p,i}\right)_1 \mathrm{d}T_1 = \left(\sum_{i=1}^{N} F_i \bar{c}_{p,i}\right)_2 \mathrm{d}T_2$$

At this point, we will assume that $\left(\sum_{i=1}^{N} F_i \bar{c}_{p,i}\right)_1 = \left(\sum_{i=1}^{N} F_i \bar{c}_{p,i}\right)_2$, i.e., that the total heat capacities of both Streams are the same. This frequently is a good, first-pass assumption, and it simplifies the analysis considerably. Integrating $\mathrm{d}T_1 = \mathrm{d}T_2$ between an arbitrary position, z', and Z' gives

$$T_2(z') - T_1(z') = T_2(Z') - T_1(Z') = \text{constant} = \Delta T_\mathrm{ex}$$

This equation shows that the temperature difference between Stream 1 and Stream 2 is constant, independent of position along the length of the exchanger. This constant temperature difference has been labeled ΔT_ex. From Figure 8-14, $[T_2(Z') - T_1(Z')] = \Delta T_\mathrm{ex} = [T(Z) - T_0]$. However, from Eqn. (8-40), $T(Z) - T_0 = x_\mathrm{A}(Z)/\lambda$. Therefore, $\Delta T_\mathrm{ex} = x_\mathrm{A}(Z)/\lambda$.

Returning now to Eqn. (8-42),

$$\frac{\mathrm{d}T_1}{\mathrm{d}z'} = \frac{Ua_\mathrm{h}\Delta T_\mathrm{ex}}{\left(\sum_i F_1 \bar{c}_{p,i}\right)}$$

Integrating from $z' = 0$, $T_1 = T_\mathrm{in}$ to $z' = Z'$, $T_1 = T_1(Z')$, and recognizing that $a_\mathrm{h}Z' = A_\mathrm{h}$, the total heat-transfer area of the F/P exchanger,

$$T_1(Z') - T_\mathrm{in} = \frac{UA_\mathrm{h}\Delta T_\mathrm{ex}}{\left(\sum_i F_i \bar{c}_{p,i}\right)}$$

Adding $T(Z) - T_0 = x_\mathrm{A}(Z)/\lambda$ (see Eqn. (8-40)) to both sides of the above and recognizing that $T_1(Z') = T_0$ and that $\Delta T_\mathrm{ex} = x_\mathrm{A}(Z)/\lambda$ gives

$$T(Z) - T_\mathrm{in} = \frac{x_\mathrm{A}(Z)}{\lambda}\left(1 + \frac{UA_\mathrm{h}}{\sum_{i=1}^{N} F_i \bar{c}_{p,i}}\right)$$

This can be rearranged to

$$x_\mathrm{A}(Z) = \frac{\lambda}{\left(1 + \dfrac{UA_\mathrm{h}}{\sum_{i=1}^{N} F_i \bar{c}_{p,i}}\right)}(T(Z) - T_\mathrm{in}) \tag{8-44}$$

Equation (8-44) is a straight line on a graph of $x_\mathrm{A}(Z)$ versus $T(Z)$, such as Figure 8-15. The line has an intercept of T_in on the temperature axis, and a slope of $\lambda/\left[1 + UA_\mathrm{h}/\sum_{i=1}^{N} F_i \bar{c}_{p,i}\right]$. Since $UA_\mathrm{h}/\sum_{i=1}^{N} F_i \bar{c}_{p,i} > 0$, the slope of this line is less than λ, i.e., the slope is less than that of Eqn. (8-40).

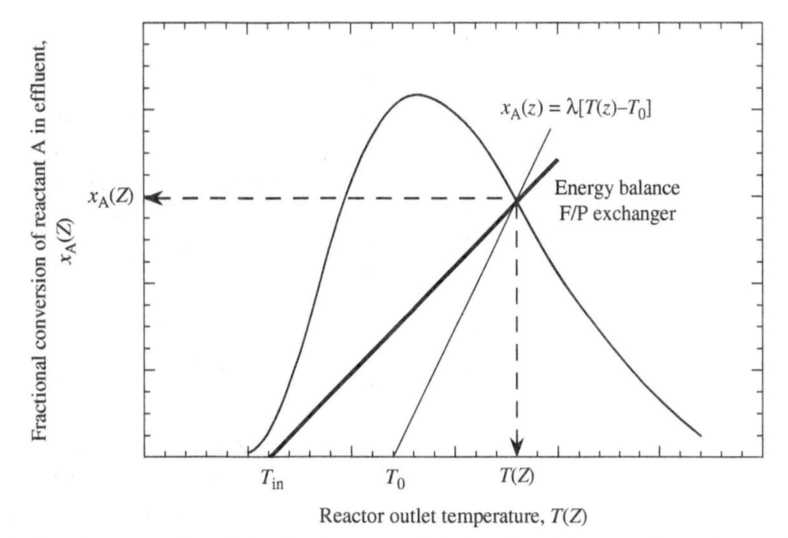

Figure 8-16 Representation of the simultaneous solution of the design equation and energy balance for an adiabatic PFR, plus the energy balance for a feed/product heat exchanger.

8.7.2.4 Overall Solution

Figure 8-16 is an extension of Figure 8-15, with Eqn. (8-44) added.

The intersection of the curve and the straight line is the *only* point at which the reactor energy balance, the design equation, and the F/P exchanger energy balance are satisfied *simultaneously*. The point of intersection gives the values of $x_A(Z)$ and $T(Z)$ that result from a specified value of T_{in}.

8.7.2.5 Adjusting the Outlet Conversion

Figure 8-16, as constructed, shows a conversion of A that is well below the maximum possible value. The outlet temperature from the reactor, $T(Z)$, is too high, so that the reactor operates at a temperature where the equilibrium is relatively unfavorable.

The design of the F/P exchanger can be changed to reduce the reactor outlet temperature, $T(Z)$, and thereby increase the conversion, x_A. If T_{in} is fixed, the slope of the F/P energy balance can be increased, as shown in Figure 8-17.

The slope of the F/P energy balance can be increased by reducing the overall area, A_h, of the F/P exchanger. This can be seen from Eqn. (8-43). Reducing the exchanger area reduces the quantity of heat transferred into the feed Stream entering the reactor. As a consequence, T_0, the inlet temperature to the reactor, is reduced.

In an existing facility, with the F/P exchanger already in place, it still is possible to change the slope of the F/P energy balance. The quantity of heat transferred into the feed Stream can be decreased by allowing only a portion of the feed Stream to pass through the F/P exchanger. The remainder of the feed is bypassed around the exchanger. This configuration is shown in Figure 8-18.[3]

[3] The usefulness of partial feed bypass is discussed in more detail in Froment, G. F. and Bischoff, K. B., *Chemical Reactor Analysis and Design,* 2nd edition, John Wiley & Sons (1990), Chapter 11, Section 5.5, pp. 423–437.

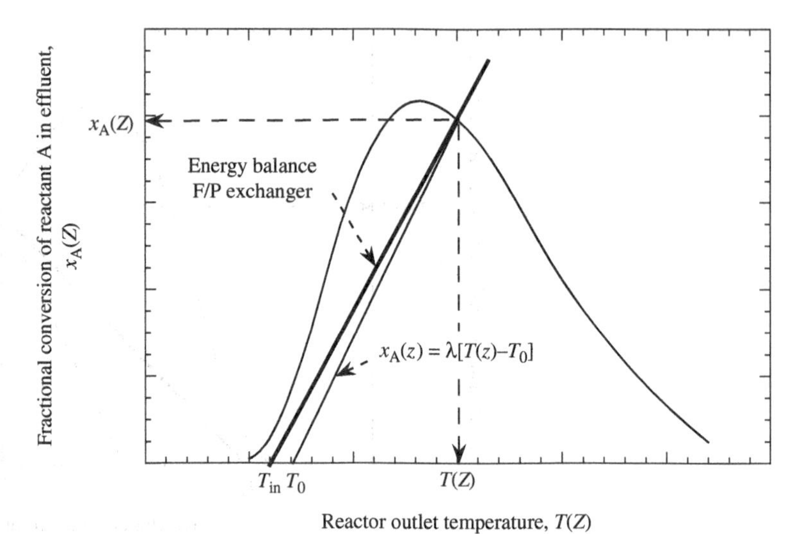

Figure 8-17 Solution of the design equation and energy balance for an adiabatic PFR, plus the energy balance for a feed/product heat exchanger. The area of the F/P exchanger is lower than the one in Figure 8-16.

8.7.2.6 Multiple Steady States

Let's return to Figure 8-16 and consider another possibility. The position of the heavy line representing the energy balance on the F/P exchanger is determined by the values of T_{in}, λ, UA_h, and $\sum_{i=1}^{N} F_i\bar{c}_{p,i}$. Depending on these values, the energy balance line can fall into a different position relative to the curve for the adiabatic reactor, as shown in Figure 8-19.

This figure shows three points of intersection between the energy balance for the F/P exchanger and the design equation for the adiabatic PFR, i.e., there are three possible steady states for the reactor–F/P exchanger combination. The *system* exhibits multiplicity, as discussed in Section 8.5.2 of this chapter. Point 1 is a low-conversion, low-temperature solution. It is the operating point that would be reached if a feed Stream at T_{in} were passed through a "cold" system, that was at a temperature in the region of T_{in}. Practically, this point tells us that the system cannot be started up simply by feeding a cold Stream into a cold F/P exchanger and reactor.

Points 2 and 3 have higher conversions and temperatures, and merit further analysis. A physical interpretation of Figure 8-19 can be developed by first recognizing that the

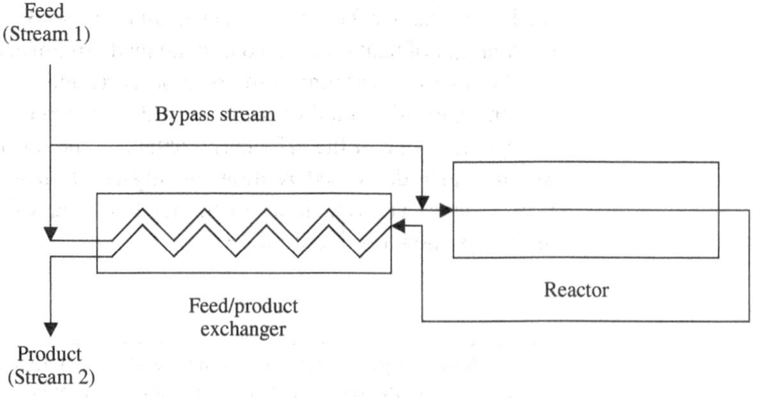

Figure 8-18 Feed/product heat exchanger with partial bypass of the feed stream.

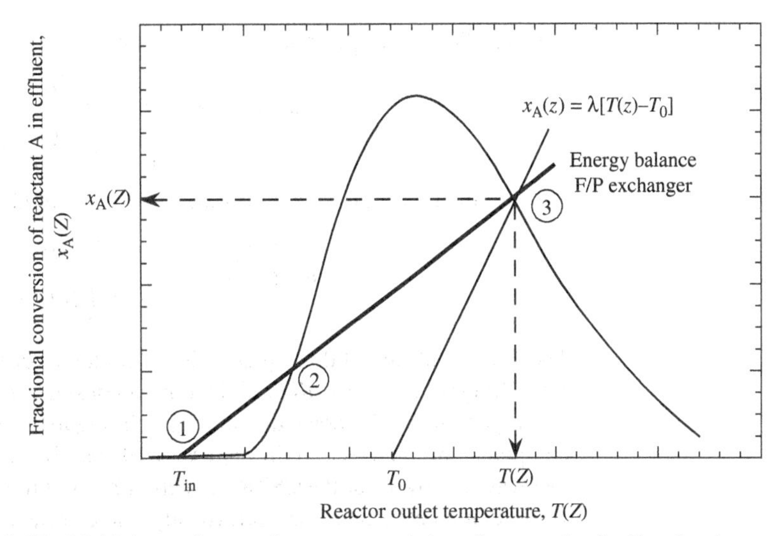

Figure 8-19 Multiple steady states in system consisting of reactor plus feed/product heat exchanger.

combination of the adiabatic reactor and the F/P heat exchanger is adiabatic. Therefore, the enthalpy change associated with the difference in composition between the Streams leaving and entering the reactor must equal the sensible heat difference between the fluid leaving the F/P exchanger at $T_2(0)$ and the fluid entering at T_{in}.

$$F_{A0}(-\Delta H_R)x_A(Z) = \left(\sum_{i=1}^{N} F_i\bar{c}_{p,i}\right)(T_2(0) - T_{in}) \tag{8-45}$$

Following the logic of Section 8.5.1 of this chapter, let's label the right-hand side of this equation $R(T)$,

$$R(T) = \left(\sum_{i=1}^{N} F_i\bar{c}_{p,i}\right)(T_2(0) - T_{in}) \tag{8-46}$$

and let's label the left-hand side $G(T)$.

$$G(T) = F_{A0}(-\Delta H_R)x_A(Z) \tag{8-47}$$

Physically, $R(T)$ is the difference in sensible heat between the feed Stream entering the exchanger at T_{in} and the product Stream leaving the exchanger at $T_2(0)$. It is one part of the overall enthalpy change for the reactor–F/P exchanger system. Similarly, $G(T)$ is the portion of the enthalpy change that results from the difference in composition between the outlet and the inlet Streams, i.e., from the reaction. As with the single CSTR that was treated earlier in this chapter, it is convenient to refer to $G(T)$ as the rate of "heat generation" and to refer to $R(T)$ as the rate of "heat removal." Combining Eqns. (8-45)–(8–47),

$$G(T) = R(T)$$

Now, the heat transferred in the F/P exchanger must be equal to the change in the sensible heat of the hot Stream.

$$UA_h\Delta T_{ex} = UA_h(T_2(0) - T_{in}) = \left(\sum_{i=1}^{N} F_i\bar{c}_{p,i}\right)(T(Z') - T_2(0))$$

$$T_2(Z') - T_2(0) = \frac{UA_h}{\sum_{i=1}^{N} F_i\bar{c}_{p,i}}(T_2(0) - T_{in})$$

Adding $[T_2(0) - T_{in}]$ to both sides of this equation and rearranging gives

$$T_2(0) - T_{in} = \frac{(T_2(Z') - T_{in})}{\left[1 + \left(UA_h \Big/ \sum_{i=1}^{N} F_i \bar{c}_{p,i}\right)\right]} \tag{8-34}$$

Substituting this expression into Eqn. (8-46) and dividing both sides by $F_{A0}(-\Delta H_R)$ gives

$$\frac{R(T)}{F_{A0}(-\Delta H_R)} = \frac{\lambda(T_2(Z') - T_{in})}{\left[1 + \left(UA_h \Big/ \sum_{i=1}^{N} F_i \bar{c}_{p,i}\right)\right]}$$

The right-hand side of this equation is the equation for the straight line 1–3 in Figure 8-19. This straight line is just $R(T)$ divided by a constant $(F_{A0}(-\Delta H_R))$.

Equation (8-47) shows that the curve in Figure 8-19 is $G(T)/F_{A0}(-\Delta H_R)$. Therefore, the points of intersection in Figure 8-19 can be analyzed using the same approach that was employed for the CSTR in Section 8.5.3 of this chapter. This approximate analysis shows that points 1 and 3 are intrinsically stable because a slight temperature excursion will cause a bigger change in R than in G. This causes the temperature to change in a direction that is *opposite* to the original perturbation. Point 2 is intrinsically unstable since the change in G, for a given temperature fluctuation, is greater than the change in R. This causes the temperature to continue to move in the direction of the original change.

8.8 CONCLUDING REMARKS

In this chapter, we have seen that multiple steady states can occur when an exothermic reaction takes place in an ideal CSTR, and when a feed/product heat exchanger is used in conjunction with an adiabatic PFR. We might ask whether this multiplicity is coincidental, or whether there is a link between these two examples.

This question was addressed in a now-classic paper by van Heerden.[4] The answer is that multiple steady states and associated phenomena such as hysteresis can arise in situations where energy is transferred from a late stage of reaction to an earlier stage. In the case of an ideal CSTR with an exothermic reaction taking place, fluid mixing is the mechanism of energy feedback. In the case of the F/P exchanger/PFR combination, heat is transferred from the product Stream (late stage of reaction) to the feed Stream (early stage of reaction) via the F/P exchanger.

There are other configurations that can lead to energy transfer from a late reaction stage to an early one. Examples of such process and reactor configurations are

1. flow reactors with *some* mixing in the direction of flow (mixing does not have to be complete in the sense of the CSTR);
2. flames, where heat can be transferred "backward" by conduction and/or radiation;
3. direct heat transfer between the reactor and the inlet Stream, as illustrated in Figure 8-20a;
4. "backward" conduction of heat, as might occur when an exothermic reaction takes place in a monolithic catalyst support, as illustrated in Figure 8-20b (monolithic catalyst supports are discussed in Chapter 9).

Although processes and/or equipment configurations such as those discussed above *can* lead to multiple steady states, the mere existence of energy transfer from a late stage of

[4] van Heerden, C., The character of the stationary state of exothermic processes, *Chem. Eng. Sci.*, 8, 133–145 (1958).

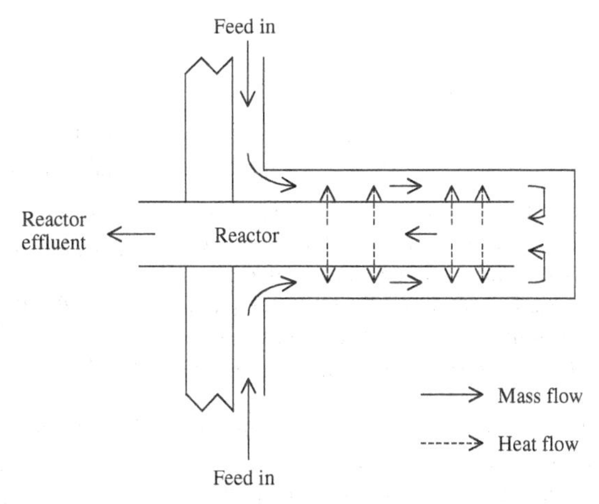

Figure 8-20a Direct heat transfer between a reactor and the fluid entering the reactor.

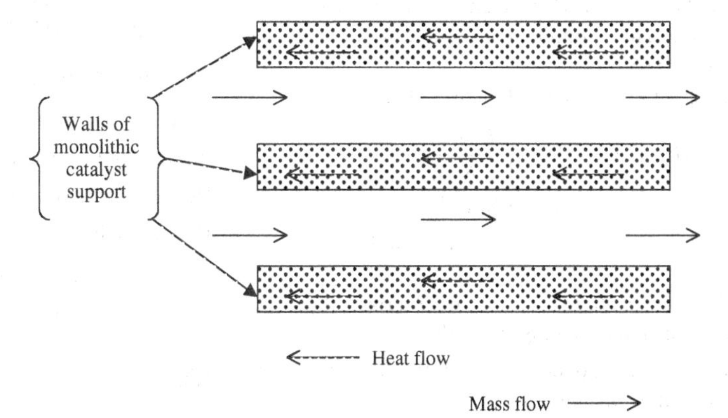

Figure 8-20b "Backward" conduction of heat along the walls of a monolithic catalyst support, in which an exothermic reaction is taking place. The reaction occurs on the surface of the walls of the support.

reaction to an earlier one does not *guarantee* multiplicity. As with the CSTR and the F/P heat exchanger/adiabatic PFR combination, the parameters of the problem will determine whether or not multiple steady states occur. Nevertheless, a flow of energy from a late to an early stage of reaction should serve as a red flag to make us aware of the need to search for multiplicity by carrying out a deeper and more thorough analysis of the problem.

SUMMARY OF IMPORTANT CONCEPTS

- Isothermal operation of a batch reactor or a plug-flow reactor is an important limiting case for reactor sizing and analysis. However, it is difficult and costly to even approach isothermal operation on a commercial scale.
- Adiabatic operation of a reactor usually is the simplest and most cost-effective mode of operation, and should always be considered. However, there are practical constraints that may preclude adiabatic operation.
- In order to size and analyze nonisothermal reactors, the design equation(s) and the energy balance must be solved simultaneously.
- There is a simple relationship between fractional conversion and temperature for adiabatic reactors with a single reaction taking place ($T = T_0 + (\overline{\Delta T_{ad}})x_A$). This relationship simplifies the simultaneous solution of the energy balance and the design equation for adiabatic reactors.
- For any CSTR, a graphical technique can be used to solve the energy balance and the design equation(s) simultaneously.
- CSTRs and combinations of an adiabatic reactor and a feed/product heat exchanger can exhibit multiple steady states and associated phenomena, such as feed-temperature hysteresis and "blowout."

PROBLEMS

Problem 8-1 (Level 1) The reaction

$$\underset{(MW = 95)}{A} + \underset{(MW = 134)}{B} \rightarrow \underset{(MW = 229)}{R}$$

is essentially irreversible at all conditions of interest. A stoichiometric mixture of A and B (with no diluent) contains 0.035 lb·mol/gal of each component and has a density of 8.00 lb/gal. The heat capacities of A and B are both equal to 68 cal/g-mol-°C and are essentially independent of temperature. The heat of reaction, ΔH_R, is -65 kcal/g mol, and also is independent of temperature.

You have recently been put in charge of a facility that manufactures R. The facility operates 7890 h/year. The reaction is carried out in a single, ideal CSTR with a volume of 1000 gal. The reactor operates at 550 °F. The feed is a stoichiometric mixture of A and B at 100 °F. Heat is removed through a cooling coil in the reactor. The fractional conversion of A, x_A, in the reactor effluent is 0.95.

The rate equation for the reaction at 550 °F is

$$-r_A(\text{lb}\cdot\text{mol/h gal}) = \frac{kC_A}{1 + KC_A}$$

$$k = 1.82\,\text{h}^{-1}$$

$$K = 85.0\,\text{gal/lb}\cdot\text{mol}$$

1. What is the annual production rate of R (100% basis) in pounds per year in the existing reactor?
2. What is the required rate of heat removal (BTU/h) through the cooling coil in the existing reactor?
3. The capacity of the plant is to be increased to 10,000,000 lbs/year, with the same fractional conversion of A. Propose a reactor system that will do the job, and specify the volume of any additional reactors that you add. The system should include the existing CSTR. Minimizing the additional reactor volume is a sufficient design criterion. Assume that the new reactor system will operate isothermally at 550 °F.
4. What is the required rate of heat removal through the cooling coil in the existing CSTR, as it operates in the expanded plant? The feed to the CSTR will remain at 100 °F.
5. Do the results of Questions 2 and 4 suggest a potential problem? What problem?

Problem 8-2 (Level 2) The Marketing Department of Cauldron Chemical Company's Catalyst Division has persuaded a potential customer to build a small pilot plant to demonstrate experimental catalyst EXP-37A for the gas-phase isomerization:

$$A \rightleftarrows R$$

Cauldron's customer is operating a fluidized-bed reactor. For a preliminary analysis, assume that this reactor behaves as an ideal CSTR. There is no provision for removing heat from the reactor. The operating conditions of the pilot plant are

Reactor temperature, T—300 °C

Total pressure, P—1 atm absolute

Total feed rate—50,000 l/h (at 1 atm, 300 °C)

Feed composition—A = 40 mol%, H_2 = 60 mol%

Catalyst weight—10 kg

Some problems have developed in the pilot plant, which the customer is blaming on the catalyst and which the Marketing Department is blaming on "sloppy operations" by the customer. You have been asked to "check out" the customer's design. Specifically,

- What fractional conversion of A would you expect for the specified conditions?
- What feed temperature is required to operate at steady state at the specified conditions?

The following information is available:

Ideal gas law is valid

Equilibrium constant (based on pressure), $K_P = 4.19$ at 300 °C

Heat of reaction, $\Delta H_R = +3.500$ kcal/mol (independent of T)

Heat capacities

$$C_{p,H_2} = 7.00\,\text{cal/mol-°C (independent of } T)$$
$$C_{p,A} = C_{p,R} = 17.5\,\text{cal/mol-°C (independent of } T)$$

Kinetics

$$-r_A = k_f\left(C_A - \frac{C_R}{K_P}\right)$$

$$k_f = 14.9\,\text{l/g}\cdot\text{cat-h at } 300\,°\text{C}$$

Problem 8-3 (Level 1) Wolftenic acid is made by the reaction of lobonic acid and formaldehyde in aqueous solution, using a homogeneous catalyst:

lobonic acid + formaldehyde $\rightleftarrows$ wolftenic acid + water

The reaction is reversible at reaction conditions. The forward reaction is first order in lobonic acid and first order in formaldehyde. The reverse reaction is first order in wolftenic acid and first order in water. There are no side reactions. Some data on the rate and equilibrium constants for this reaction are given in the following table.

Temperature (°C)	Forward rate constant (l/mol-h)	Equilibrium constant, K_c
60	0.00900	60
90	0.110	20

Wolftenic acid is being produced in an ideal CSTR with a volume of 10,000 l. The composition of the feed to the reactor is

Lobonic acid	10 mol/l
Formaldehyde	5 mol/l
Wolftenic acid	0 mol/l
Water	5 mol/l
Solvent	25 mol/l

The molar feed rate of formaldehyde to the CSTR (F_{F0}) is 2000 mol/h. The temperature of the feed to the reactor is 32 °C. As a rough first approximation, you may assume that the molar heat capacity is the same for all species at 15 cal/mol-°C. All solutions are ideal.

The reactor contains a cooling coil to remove heat. The heat-transfer area of the coil is 2.0 m^2. The overall heat-transfer coefficient between the coil and the reactor contents is 200,000 cal/m^2-h-°C. The flow rate of coolant through the coil is very high; the coolant enters and leaves at 32 °C.

Following is a graph of the *product* of the enthalpy change of the reaction, the fractional conversion of formaldehyde (x_F) and the molar feed rate of formaldehyde ($-\Delta H_R \cdot x_F \cdot F_{F0}$) as a function of temperature.

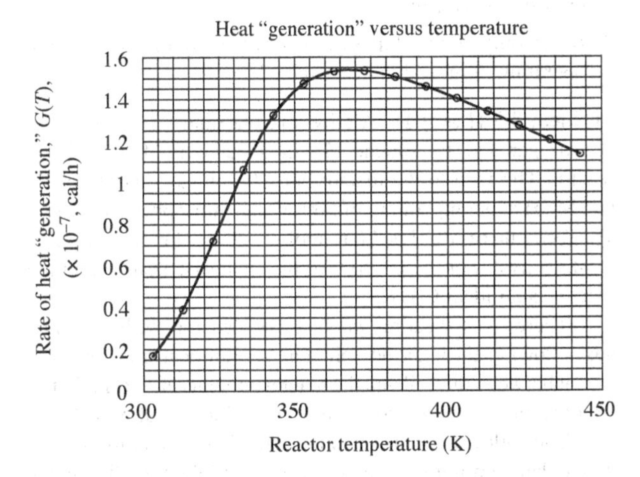

1. Verify the above curve by calculating the value of $G(T)$ $(= -\Delta H_R \cdot x_F \cdot F_{F0})$ at 100 °C.
2. What is the approximate fractional conversion of formaldehyde in the effluent from the CSTR?
3. What is the approximate temperature of the effluent from the reactor?
4. What is the best action that could be taken to increase the fractional conversion to the maximum value? (You may *not* change F_{F0}, V, or the feed composition.)
5. Is the reaction exothermic or endothermic?

Problem 8-4 (Level 3) A pilot plant is being operated to test a new catalyst for the partial oxidation of naphthalene to phthalic anhydride. The chemistry of this process can be approximated as two first-order reactions in series:

$$\text{naphthalene} \xrightarrow{k_1} \text{phthalic anhydride} \xrightarrow{k_2} CO_2$$

This reaction is carried out in an atmospheric-pressure, fluidized-bed reactor that can be treated as an ideal CSTR for preliminary analysis. The feed to the reactor is a naphthalene/air mixture at 150 °C with a naphthalene concentration of 1.0 mol%. The reactor contains 100 kg of catalyst. The molar flow rate of naphthalene to the reactor is 0.12 g·mol/s.

Data

$$\Delta H_R\ (\text{Rxn.1}) = -1881 \times 10^6\ \text{J/kmol(naphthalene)}$$
$$\Delta H_R\ (\text{Rxn.2}) = -3282 \times 10^6\ \text{J/kmol(phthalic anhydride)}$$
$$c_{P,m}\ (\text{feed}) = 1040\ \text{J/kg-°C}$$
$$k_1 = 1.61 \times 10^{33} \exp(-E_1/RT)\text{l/s-kg}_{cat}$$
$$E_1 = 3.50 \times 10^5\ \text{J/mol}$$
$$k_2 = 5.14 \times 10^{13} \exp(-E_2/RT)\text{l/s-kg}_{cat}$$
$$E_2 = 1.65 \times 10^5\ \text{J/mol}$$

In the expressions for k_1 and k_2, T is the absolute temperature. The enthalpies of reaction and the heat capacity may be assumed to be constant, independent of temperature.

1. Comment on the feasibility of operating the reactor adiabatically.
2. What is the conversion of naphthalene and the selectivity to phthalic anhydride if the reactor operates at 600 K?
3. Suppose that an unlimited flow rate of a coolant is available at a temperature of 310 K, and that a heat exchanger is installed inside the fluidized bed. What value of *UA* would be required to operate the reactor at 600 K?
4. Comment on the feasibility of this design.
5. Specify what you feel is the best cooling system design, if the reactor must operate at 600 K. (UA = ?; coolant temperature = ?)

Problem 8-5 (Level 1) Hydrodealkylation is a reaction that can be used to convert toluene to benzene, which usually is more valuable than toluene. The reaction is

$$\underset{\text{(toluene)}}{C_7H_8} + H_2 \rightarrow \underset{\text{(benzene)}}{C_6H_6} + CH_4$$

Zimmerman and York[5] have studied this reaction at temperatures between 700 and 950 °C, in the absence of any catalyst. They found that the rate of toluene disappearance was well correlated by

[5] Zimmerman, C. C. and York, R., I&EC Process Design Dev.. *3* (1), 254–258 (1962).

$$-r_T = k_T[H_2]^{1/2}[C_7H_8]$$
$$k_T = 3.5 \times 10^{10} \exp(-E/RT)\left(\frac{1}{\text{mol}}\right)^{1/2} \text{s}^{-1}$$
$$E = 50{,}900 \text{ cal/mol}$$

The reaction is essentially irreversible at the conditions of the study, and the ideal gas laws are valid. You may assume that the heat of reaction is −12.9 kcal/mol, independent of temperature. You also may assume that the heat capacity of a mixture of C_7H_8, H_2, C_6H_6, and CH_4 is 36 cal/K-mol of mixture, independent of temperature and the exact mixture composition.[6]

The feed to the reactor consists of 1 mol of H_2 per mol of toluene and the toluene feed rate is 1000 mol/h. A reactor is to be designed that operates at atmospheric pressure and an inlet temperature of 850 °C. Assume that the reactor is an ideal, plug-flow reactor.

1. A reactor volume of 133 l is required to achieve a toluene conversion of 0.50, if the reactor is operated isothermally at 850 °C, with exactly identical feed conditions. If that reactor is insulated very heavily, so that it operates adiabatically instead of isothermally, will the fractional conversion of toluene go up, stay the same or go down? Assume that all other conditions remain the same, and that the feed enters the adiabatic reactor at 850 °C. Explain your answer. No calculations are necessary.
2. If the reaction is carried out adiabatically, what will the gas temperature be when the fractional conversion of toluene is 0.30?
3. What volume of reactor is required to reach a toluene conversion of 0.50 if the reactor is operated adiabatically?

Problem 8-6 (Level 2) Methyl cyclohexane (MCH) is being dehydrogenated to toluene (T) in a catalytic, fluidized-bed reactor. The feed to the reactor is a 2/1 (molar) mixture of H_2 and MCH at a temperature of 500 °C and atmospheric pressure. The feed rate of MCH is 2.0 lb·mol/s. The reactor contains 20.0 lb of catalyst and operates adiabatically. For the purpose of this analysis, assume that the reactor is an ideal CSTR.

The following table contains some kinetic and thermodynamic data pertaining to the reaction.

Kinetic and thermodynamic data for methyl cyclohexane dehydrogenation

$$\text{Methyl cyclohexane } (CH_3) \rightleftharpoons \text{Toluene } (CH_3) + 3H_2$$

Heat of reaction ≅ +54.3 kcal/mol (you may assume that ΔH_R is constant even though $\sum \nu_i c_{p,i} \neq 0$)

Equilibrium constant (based on pressure)

$$\ln K_{eq} \cong 53.47 - (27{,}330/T) \text{ atm}^3, \text{ where } T \text{ is in K}$$

Heat capacities (cal/mol-K)

H_2	6.89
MCH	32.3
Toluene	24.3

Rate equation

$$-r_{MCH} = kC_{MCH}\left(1 - \frac{p_T p_{H_2}^3}{p_{MCH}K_{eq}}\right)$$

p_i denotes partial pressure of species "i", and C_{MCH} is the concentration of MCH.

$$k = 3.35 \times 10^{24}\left(\frac{\text{ft}^3}{\text{s-lb}\cdot\text{cat}}\right)e^{-E/RT}$$

$E = 70.0$ kcal/mol

The ideal gas law is valid.

1. At what temperature and fractional conversion of MCH does the reactor operate?
2. Based only on the slopes of the $G(T)$ and $R(T)$ curves, is the operating point stable or unstable?

Problem 8-7 (Level 1) Methyl cyclohexane (MCH) is being dehydrogenated to toluene (T) in a catalytic, fluidized-bed reactor. The feed to the reactor is a 2/1 (molar) mixture of H_2 and MCH at a temperature of 500 °C and atmospheric pressure. The feed rate of MCH is 2.0 lb·mol/s. The reactor contains 20.0 lb of catalyst. For the purpose of this analysis, assume that the reactor is an ideal CSTR.

There is a heating coil in the reactor with hot flue gases flowing through the coil. The overall heat-transfer coefficient between the coil and the reactor contents is 25 BTU/h-ft^2-°F. Flue gas enters the coil at 600 °C and leaves at 550 °C.

The following table contains some kinetic and thermodynamic data pertaining to the reaction. The following figure is a plot of the rate at which heat is generated in the reactor, $G(T)$, as a function of temperature, for the conditions stated. $\{(G(T) = X_A(-\Delta H_R)F_{A0}\}$

If the fractional conversion of MCH is 0.50, how much heat-transfer area (ft^2) is required?

Kinetic and thermodynamic data for methyl cyclohexane dehydrogenation

$$\text{Methyl cyclohexane } (CH_3) \rightleftharpoons \text{Toluene } (CH_3) + 3H_2$$

[6] Provided that the mixture results from an initial composition of 1 mol of toluene and 1 mol of H_2.

Heat of reaction $\cong +54.3$ kcal/mol
Equilibrium constant (based on pressure)

$$\ln K_{eq} \cong 53.47 - (27{,}330/T)\ \text{atm}^3, \text{ where } T \text{ is in K}$$

Heat capacities (cal/mol-K)

H_2	6.89
MCH	32.3
Toluene	24.3

Rate equation

$$-r_{MCH} = kC_{MCH}\left(1 - \frac{p_T p_{H_2}^3}{p_{MCH} K_{eq}}\right)$$

Here, p_i denotes partial pressure of species "i", and C_{MCH} is the concentration of MCH.

$$k = 3.35 \times 10^{24}\left(\frac{\text{ft}^3}{\text{s-lb}\cdot\text{cat}}\right)e^{-E/RT}$$

$E = 70.0$ kcal/mol

The ideal gas law is valid.

Heat generation $G(T)$ versus temperature for methyl cyclohexane dehydrogenation

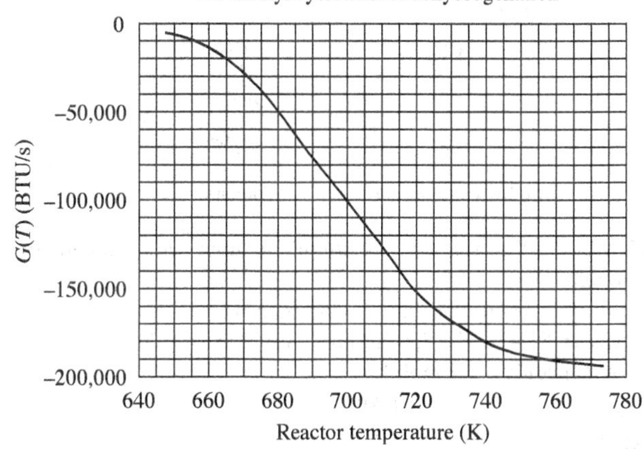

Problem 8-8 (Level 1) The homogeneous reaction of A with B takes place in an organic solution. An ideal CSTR with a volume of 10,000 l is being used. The molar feed rate of A to the CSTR (F_{A0}) is 2000 mol/h and the mole fraction of A in the feed is 0.20. The temperature of the feed to the reactor is 50 °C. As a first approximation, you may assume that the molar heat capacity for all species is 60 J/mol-°C. The enthalpy change on reaction (ΔH_R) is approximately constant and is equal to -37 kJ/mol A.

The reactor contains a cooling coil to remove heat. The heat-transfer area of the coil is 1.0 m^2. The overall heat-transfer coefficient between the coil and the reactor contents is 800 kJ/m^2-h-°C. The flow rate of coolant through the coil is very high; the coolant enters and leaves the coil at 30 °C.

A graph of the *product* of the enthalpy change of the reaction, the fractional conversion of A (x_A), and the molar feed rate of A ($-\Delta H_R \cdot x_A \cdot F_{A0}$) as a function of temperature follows.

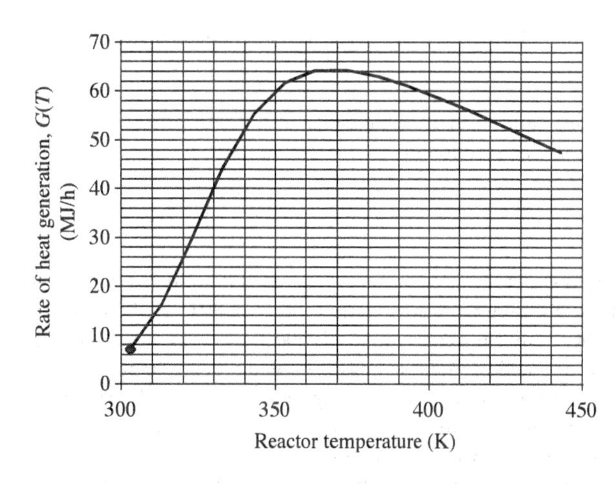

1. What might cause the attached curve go through a maximum with temperature?
2. What is the approximate fractional conversion of A in the effluent from the reactor?
3. What is the approximate temperature of the effluent from the reactor?
4. Is the operating point intrinsically stable?

Problem 8-9 (Level 1) The liquid phase reaction

$$A \longrightarrow R$$

is taking place in an ideal batch reactor. The reaction is homogeneous and irreversible, and the rate equation for the disappearance of A is

$$-r_A = kC_A^2$$

At 150 °C, the value of k is 7.60 l/mol-h. The enthalpy change on reaction, ΔH_R, is -37.5 kcal/mol A at 150 °C. The initial concentration of A is 0.65 mol/l.

1. If the reactor is operated isothermally at 150 °C, how much time is required to achieve a fractional conversion of A, $x_A = 0.95$?
2. If the reactor is operated isothermally at 150 °C, what *percentage* of the total heat will be transferred in the first quarter of the total time, i.e., of the time you calculated in Part 1? The second quarter? The third quarter? The last quarter?
3. The reactor is to be kept isothermal at 150 °C over the whole course of the reaction. Cooling water is available at 30 °C at a flow rate of 6000 kg/h. If the reactor volume is 1000 l and the overall heat-transfer coefficient is 500 kcal/m^2-h-°C, how much heat-transfer area must be installed in the reactor?
4. If the reactor is operated *adiabatically* and the initial temperature is 150 °C, will the time required to achieve $x_A = 0.95$ be greater, the same, or less than what you calculated in Part 1? Explain your reasoning.

Problem 8-10 (Level 2) The reversible, liquid-phase reaction

$$A + B \rightleftarrows R + S$$

is being carried out in an adiabatic, ideal, continuous stirred-tank reactor. The reaction obeys the rate equation

$$-r_A = k\left(C_A C_B - \frac{C_R C_S}{K_{eq}}\right)$$

The value of the rate constant, k, is 0.050 l/mol-h at 100 °C and the rate constant has an activation energy of 80 kJ/mol. The value of the equilibrium constant, K_{eq}, is 500 at 100 °C. The value of the heat of reaction, ΔH_R, is −60 kJ/mol and is constant.

The reactor operates at a space time (τ) of 2 h. The inlet concentrations are $C_{A0} = C_{B0} = 4.0$ mol/l; $C_{R0} = C_{S0} = 0$. The molar feed rate of A (F_{A0}) is 2000 mol/h. The average heat capacity of the feed to the reactor is 3.4 J/g-K and the mass feed rate is 211,000 g/h. The feed temperature is 50 °C.

1. What are the possible operating points (x_A, T) at steady state?
2. Are any of the possible operating points unstable? Which ones? Explain your reasoning.

Problem 8-11 (Level 1) The reversible, liquid-phase reaction

$$A + B \rightleftarrows R + S$$

is being carried out in an ideal, continuous stirred tank-reactor. The reaction follows the rate equation

$$-r_A = k\left(C_A C_B - \frac{C_R C_S}{K_{eq}}\right)$$

The value of the rate constant, k, is 0.050 l/mol-h at 100 °C, and the rate constant has an activation energy of 80 kJ/mol. The value of the equilibrium constant, K_{eq}, is 500 at 100 °C. The value of the heat of reaction, ΔH_R, is –60 kJ/mol and is constant.

The reactor operates at a space time, τ, of 2.0 h. The inlet concentrations are $C_{A0} = C_{B0} = 4.0$ mol/l; $C_{R0} = C_{S0} = 0$. The molar feed rate of A (F_{A0}) is 2000 mol/h. The average heat capacity of the feed to the reactor is 3.4 J/g-K and the mass feed rate is 211,000 g/h. The feed temperature is 400 K.

A cooling coil inside the reactor is used to remove heat. A heat-transfer fluid enters the cooling coil with an inlet temperature of 400 K. The flow rate of the cooling fluid is so high that the temperature of the cooling fluid that leaves the coil is essentially 400 K.

The following graph shows the value of $F_{A0}x_A(-\Delta H_R)$ as a function of reactor temperature for this problem.

1. Verify that the attached graph is correct by calculating the value of $F_{A0}x_A(-\Delta H_R)$ at 450 K, using the information given above.
2. It is desired to operate the reactor at 450 K. What value of the product of the coil area, A, and the overall heat-transfer coefficient, U, is required?

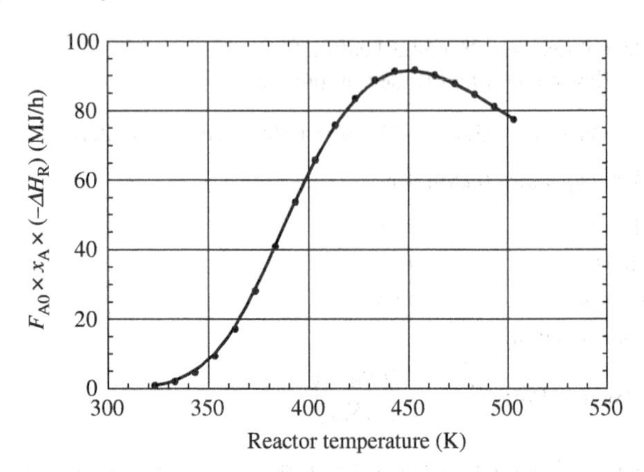

Problem 8-12 (Level 2) Methyl cyclohexane (MCH) is being dehydrogenated to toluene (T) in an ideal, plug-flow reactor that is packed with catalyst particles. The feed to the reactor is a 20/1 (molar) mixture of H_2 and MCH at a temperature of 500 °C and atmospheric pressure. The feed rate of MCH is 1000 g mol/s.

The following table contains some kinetic and thermodynamic data pertaining to the reaction.

Kinetic and thermodynamic data for methyl cyclohexane dehydrogenation

CH_3 CH_3

$\rightleftarrows$ + $3H_2$

Methyl cyclohexane Toluene

Heat of reaction $\cong +54.3$ kcal/mol (assume independent of temperature)

Equilibrium constant (based on pressure):

$\ln K_{eq} \cong 53.47 - (27{,}330/T)$ atm^3, where T is in K

Heat capacities (cal/mol, K)

H_2	6.89
MCH	32.3
Toluene	24.3

Rate equation

$$-r_{MCH} = kC_{MCH}\left(1 - \frac{p_T p_{H_2}^3}{p_{MCH} K_{eq}}\right)$$

In this equation, p_i is the partial pressure of species "i", and C_{MCH} is the concentration of MCH.

$$k = 2.09 \times 10^{23} e^{-E/RT}\left(\frac{1}{\text{s-g cat}}\right); \quad E = 70.0 \text{ kcal/mol}$$

The ideal gas law is valid.

1. What is the adiabatic temperature change of the reaction?
2. The PFR operates isothermally at 500 °C. The outlet conversion of MCH is 50%. What is the *highest* value of the term $(p_T p_{H_2}^3 / p_{MCH} K_{eq})$ that occurs anywhere along the length of the reactor?

3. Based on your answer to Question 2, simplify the rate equation. Then calculate the weight of catalyst that is required to obtain an outlet MCH conversion of 50% if the reactor is operated isothermally at 500 °C? You may assume that transport effects are not important, and you may neglect pressure drop through the reactor.
4. What weight of catalyst is required to obtain an outlet MCH conversion of 50% if the reactor is operated adiabatically with an inlet temperature of 500 °C? You may assume that transport effects are not important, and you may neglect pressure drop through the reactor.

Problem 8-13 (Level 1)

1. The irreversible, gas-phase reaction

$$A \rightarrow 2B + C$$

is being carried out in an adiabatic, plug-flow reactor. Pressure drop in the reactor can be neglected. The feed to the reactor consists of A and H_2O in a 1/1 molar ratio. No other species are present in the feed. Water is an inert diluent; it is not a reactant or a product. The ideal gas laws are valid.

Write an expression for the concentration of A (C_A) at any point in the reactor as a function of the fractional conversion of A (x_A). This expression may include the inlet temperature (T_0) and the adiabatic temperature change (ΔT_{ad}). However, it may *not* include any other temperature.

2. The irreversible, liquid-phase reactions

$$A \rightarrow B + C$$

$$D \rightarrow E + F$$

are taking place in an adiabatic, plug-flow reactor. The feed to the reactor consists of an equimolar mixture A and D. The inlet temperature is T_0. The enthalpy change on reaction for Reaction (1) is $\Delta H_{R,1}$ and the enthalpy change on reaction for Reaction (2) is $\Delta H_{R,2}$. Let the extent of Reaction (1) be ξ_1 and let the extent of Reaction (2) be ξ_2.

Write an expression for the concentration of A (C_A) in terms of ξ_1, ξ_2, $\Delta H_{R,1}$, $\Delta H_{R,2}$, and T_0.

APPENDIX 8-A NUMERICAL SOLUTION TO EQUATION (8-26)

$k(373\text{ K}) = 0.05$ l/mol-h $C_{A0} = 4$ mol/l

$K_{eq}(373\text{ K}) = 500$ $-\Delta H_R = 60{,}000$ J/mol

$\rho \times C_p = 2720$ J/mol-K

$E_a = 80{,}000$ J/mol $\Delta T_{ad} = 88.24$

$R = 8.314$ J/mol-K

First attempt

$\Delta t = h = 0.1$ h

Time (h)	x	T (K)	$h \times f_1$	$h \times f_2$	$h \times f_3$	$h \times f_4$
0	0	373	0.02	0.02083245	0.02086762	0.02176434
0.1	0.02086075	374.840654	0.02176374	0.02272531	0.02276845	0.02380828
0.2	0.04362067	376.848883	0.02380749	0.02492609	0.02497948	0.02619396
0.3	0.06858944	379.052009	0.0261929	0.02750348	0.02757012	0.02899869
0.4	0.0961459	381.483462	0.02899726	0.03054334	0.03062711	0.03231853
0.5	0.12675535	384.184296	0.03231655	0.03415123	0.03425703	0.03626972
0.6	0.16098915	387.204925	0.03626696	0.03845212	0.03858567	0.04098481
0.7	0.19954371	390.606798	0.0409809	0.04358109	0.04374793	0.04659369
0.8	0.24324915	394.46316	0.04658814	0.04965102	0.04985342	0.05316803
0.9	0.29304333	398.856764	0.05316018	0.05666596	0.05689552	0.06058412
1	0.34985454	403.869518	0.06057342	0.06432098	0.06454518	0.06823338
1.1	0.41427773	409.553917	0.06822025	0.07161419	0.07176665	0.07454917
1.2	0.48586624	415.870551	0.07453771	0.07632988	0.07635666	0.07665787
1.3	0.56196102	422.584796	0.07666185	0.07504805	0.07510493	0.07126043
1.4	0.63666573	429.176388	0.07131507	0.06514148	0.06578588	0.05763765
1.5	0.7018003	434.923556	0.05782436	0.04839633	0.05009839	0.03989195

The sensitivity of the solution to step size must now be explored. Choose a step size that is half of the original, i.e., choose $h = 0.05$ h, and repeat the calculation.

Second attempt
$\Delta t = h = 0.05$ h

Time (h)	x	T (K)	$h \times f_1$	$h \times f_2$	$h \times f_3$	$h \times f_4$
0	0	373	0.01	0.01020662	0.01021092	0.01042513
0.05	0.01021003	373.900885	0.0104251	0.01064694	0.0106517	0.01088191
0.1	0.02086075	374.840654	0.01088187	0.01112048	0.01112575	0.01137359
0.15	0.0319854	375.822241	0.01137355	0.01163064	0.0116365	0.0119038
0.2	0.04362067	376.848883	0.01190375	0.01218127	0.01218779	0.0124766
0.25	0.05580708	377.924154	0.01247653	0.01277664	0.01278391	0.01309653
0.3	0.06858944	379.05201	0.01309645	0.01342156	0.0134297	0.01376868
0.35	0.08201739	380.236828	0.01376859	0.01412141	0.01413052	0.01449873
0.4	0.09614592	381.483463	0.01449863	0.01488216	0.01489239	0.015293
0.45	0.11103604	382.797298	0.01529287	0.01571044	0.01572193	0.01615842
0.5	0.12675538	384.184298	0.01615828	0.01661351	0.01662643	0.0171026
0.55	0.14337884	385.651074	0.01710243	0.01759925	0.01761379	0.01813369
0.6	0.1609892	387.20493	0.01813348	0.01867602	0.01869237	0.01926021
0.65	0.17967762	388.853907	0.01925997	0.01985244	0.01987079	0.02049075
0.7	0.19954381	390.606807	0.02049046	0.02113692	0.02115743	0.02183334
0.75	0.22069589	392.473167	0.02183299	0.0225369	0.02255971	0.02329449
0.8	0.24324934	394.463177	0.02329408	0.02405766	0.02408277	0.02487764
0.85	0.26732477	396.587479	0.02487715	0.0257003	0.02572757	0.02658069
0.9	0.2930437	398.856797	0.02658012	0.02745889	0.02748788	0.02839246
0.95	0.32052138	401.281298	0.0283918	0.02931605	0.02934594	0.03028752
1	0.34985526	403.869582	0.03028675	0.03123702	0.03126643	0.03221921
1.05	0.38110741	406.627124	0.03221836	0.03316172	0.03318868	0.03411111
1.1	0.41427912	409.55404	0.0341102	0.03499568	0.03501773	0.03584804
1.15	0.44927663	412.642056	0.03584713	0.03660163	0.03661637	0.03726972
1.2	0.48586877	415.870774	0.03726894	0.03779617	0.03780254	0.038173
1.25	0.523642	419.203706	0.03817257	0.0383588	0.03835921	0.03833062
1.3	0.5619652	422.585165	0.03833089	0.03806173	0.03806452	0.03753332
1.35	0.59998465	425.939822	0.03753492	0.03672444	0.03674516	0.03565317
1.4	0.63667254	429.176989	0.03565707	0.03428332	0.03434235	0.03270865
1.45	0.67094205	432.200769	0.03271627	0.0308472	0.03096273	0.02889724
1.5	0.70181428	434.924789	0.02891018	0.02670186	0.02688079	0.02456646

The value of x_A at $t = 1.5$ h changed in the fifth significant figure when the step size was changed from 0.10 to 0.050 h. The numerical solution does not depend on step size.

APPENDIX 8-B CALCULATION OF *G*(*T*) AND *R*(*T*) FOR "BLOWOUT" EXAMPLE

Calculations for Figure 8-9

A→B; second order in A, liquid phase
$[A]_{in} = 16$ kmol/m^3
$V = 0.40$ m^3

ΔH_R @ 300 K = –21 kJ/mol A
Liquid heat capacity = 2.0 J/cm^3-K
$k = 3.20 \times 10^9 \exp(-12{,}185/T)$(m^3/mol A-ks)

Base case:
Inlet volumetric flow rate = 1.3 m^3/ks
Tau (base) = 0.3077 ks
$[A]_{in}$ × Tau (base) = 4923 mol-ks/m^3

+20% Feed rate case:
Inlet volumetric flow rate = 1.56 m^3/ks
Tau(+20%) = 0.2564 ks
$[A]_{in}$ × Tau(+20%) = 4103 mol-ks/m^3

Solution to Design Equation:
$\alpha(T) = k(T) \times [A]_{in} \times$ Tau
$x_A = (((2 \times \alpha(T)+1) - \text{sqrt}(4 \times \alpha(T)+1))/(2 \times \alpha(T)))$

	Base Case					+20% Feed Rate			
Reactor temperature, T(K)	Rate constant, $k(T)$ (m^3/mol M ks)	$\alpha(T)$	Conversion (x_A)	Heat "generation", $G(T)$(J/ks)	Heat removal, $R(T)$ (J/ks)	$\alpha(T)$	Conversion (x_A)	Heat "generation", $G(T)$(J/ks)	Heat removal, $R(T)$ (J/ks)
312	3.50E–08	1.72E–04	0.0017	7.52E+04	0.00E+00	1.44E–04	0.00014	7.53E+04	0.00E+00
315	5.08E–08	2.50E–04	0.00025	1.09E+05	7.80E+06	2.08E–04	0.00021	1.09E+05	9.36E+06
320	9.29E–08	4.57E–04	0.00046	2.00E+05	2.08E+07	3.81E–04	0.00038	2.00E+05	2.50E+07
340	8.73E–07	4.29E–03	0.00426	1.86E+06	7.28E+07	3.58E–03	0.00355	1.86E+06	8.74E+07
360	6.39E–06	3.14E–02	0.02960	1.29E+07	1.25E+08	2.62E–02	0.02493	1.31E+07	1.50E+08
370	1.60E–05	7.85E–02	0.06815	2.98E+07	1.51E+08	6.54E–02	0.05807	3.04E+07	1.81E+08
380	3.79E–05	1.87E–01	0.13854	6.05E+07	1.77E+08	1.56E–01	0.12044	6.32E+07	2.12E+08
390	8.63E–05	4.25E–01	0.24327	1.06E+08	2.03E+08	3.54E–01	0.21713	1.14E+08	2.43E+08
400	1.89E–04	9.28E–01	0.36918	1.61E+08	2.29E+08	7.74E–01	0.33852	1.78E+08	2.75E+08
410	3.96E–04	1.95E+00	0.49580	2.17E+08	2.55E+08	1.63E+00	0.46519	2.44E+08	3.06E+08
420	8.04E–04	3.96E+00	0.60803	2.66E+08	2.81E+08	3.30E+00	0.58057	3.04E+08	3.37E+08
430	1.58E–03	7.77E+00	0.69988	3.06E+08	3.07E+08	6.48E+00	0.67681	3.55E+08	3.68E+08
440	3.01E–03	1.48E+01	0.77163	3.37E+08	3.33E+08	1.23E+01	0.75296	3.95E+08	3.99E+08
450	5.56E–03	2.74E+01	0.82627	3.61E+08	3.59E+08	2.28E+01	0.81147	4.26E+08	4.31E+08
460	1.00E–02	4.93E+01	0.86739	3.79E+08	3.85E+08	4.11E+01	0.85576	4.49E+08	4.62E+08
470	1.76E–02	8.67E+01	0.89819	3.93E+08	4.11E+08	7.23E+01	0.88908	4.66E+08	4.93E+08
480	3.02E–02	1.49E+02	0.92129	4.03E+08	4.37E+08	1.24E+02	0.91415	4.79E+08	5.24E+08
500	8.34E–02	4.11E+02	0.95185	4.16E+08	4.89E+08	3.42E+02	0.94740	4.97E+08	5.87E+08
520	2.13E–01	1.05E+03	0.96959	4.24E+08	5.41E+08	8.74E+02	0.96674	5.07E+08	6.49E+08
540	5.07E–01	2.50E+03	0.98019	4.28E+08	5.93E+08	2.08E+03	0.97832	5.13E+08	7.11E+08
560	1.14E+00	5.59E+03	0.98671	4.31E+08	6.45E+08	4.66E+03	0.98546	5.17E+08	7.74E+08
580	2.41E+00	1.18E+04	0.99085	4.33E+08	6.97E+08	9.87E+03	0.98998	5.19E+08	8.36E+08
600	4.85E+00	2.38E+04	0.99354	4.34E+08	7.49E+08	1.99E+04	0.99293	5.12E+08	8.99E+08

Adapted from Hill, C. G., Jr., *An Introduction to Chemical Engineering Kinetics and Reactor Design*, John Wiley & Sons (1977), Problem 10-14.

Chapter 9

Heterogeneous Catalysis Revisited

LEARNING OBJECTIVES

After completing this chapter, you should be able to

1. estimate quantitatively the effect of concentration gradients within a porous catalyst particle on the rate of reaction;
2. estimate quantitatively the value of the diffusion coefficient of a compound within the porous structure of a catalyst particle;
3. calculate the weight of catalyst required to achieve a specified conversion for an isothermal reactor, when concentration gradients are present within the catalyst particles;
4. explain why true activation energies and reaction orders are not observed when catalyst performance is studied under conditions where transport resistances are significant;
5. explain how internal and external transport limitations can affect the observed selectivity of a catalyst;
6. determine whether the performance of a catalyst is affected by an external transport resistance;
7. estimate the reaction rate for a reaction that is controlled by external mass transport.

9.1 INTRODUCTION

The interactions among heat transfer, mass transfer, and chemical reaction in heterogeneous catalysts were discussed in Chapter 4, from a *qualitative* point of view. In all of our calculations up to this point, we ignored any temperature and concentration differences between the bulk fluid stream and the interior of the catalyst particle, where reaction actually takes place. To solve reactor sizing and analysis problems involving solid catalysts, we assumed that the concentrations and the temperature throughout the catalyst particle were identical to those in the bulk fluid.

In this chapter, we revisit the subject of reaction/transport interactions in heterogeneous catalysts, this time from a quantitative standpoint. The topic must be examined from two perspectives. First, a researcher that is studying the kinetics of a heterogeneous catalytic reaction (or reactions) must ensure that his or her experiments are free of transport effects. In other words, the experiments must be conducted under conditions where intrinsic chemical kinetics determines the reaction rate(s). The researcher may have to make calculations to estimate the magnitude of heat and mass transport influence. He or she may also have to carry out "diagnostic" experiments in order to define a region of operation where transport does not affect the reaction rate and selectivity.

Second, the optimum commercial reactor may operate at conditions where transport influences the behavior of the reaction(s) to some extent. This is especially true with single reactions, where selectivity is not an issue. Therefore, the engineer who is designing a reactor must be able to take transport effects into account in calculating the required amount of catalyst, and in calculating the expected performance of the catalytic reactor.

In Chapter 4, two different transport regimes were identified: transport inside the particles and transport between the bulk fluid and the surface of the catalyst particles. Transport inside the catalyst particles is known as internal or intraparticle transport, or as pore diffusion. Transport between the bulk fluid stream and the external surface of the catalyst particles is known as external or interparticle transport. The mechanisms of transport are different for these two regimes, and the rates of transport are influenced by different variables. Internal transport will be treated first, followed by external transport. These discussions will be preceded by a brief overview of the physical nature of heterogeneous catalysts.

9.2 THE STRUCTURE OF HETEROGENEOUS CATALYSTS

9.2.1 Overview

Many heterogeneous catalysts consist of a ceramic or metallic material that contains an interconnected network of irregular pores, as illustrated in Figure 9-1. The pore structure can vary considerably from one type of catalyst to another.

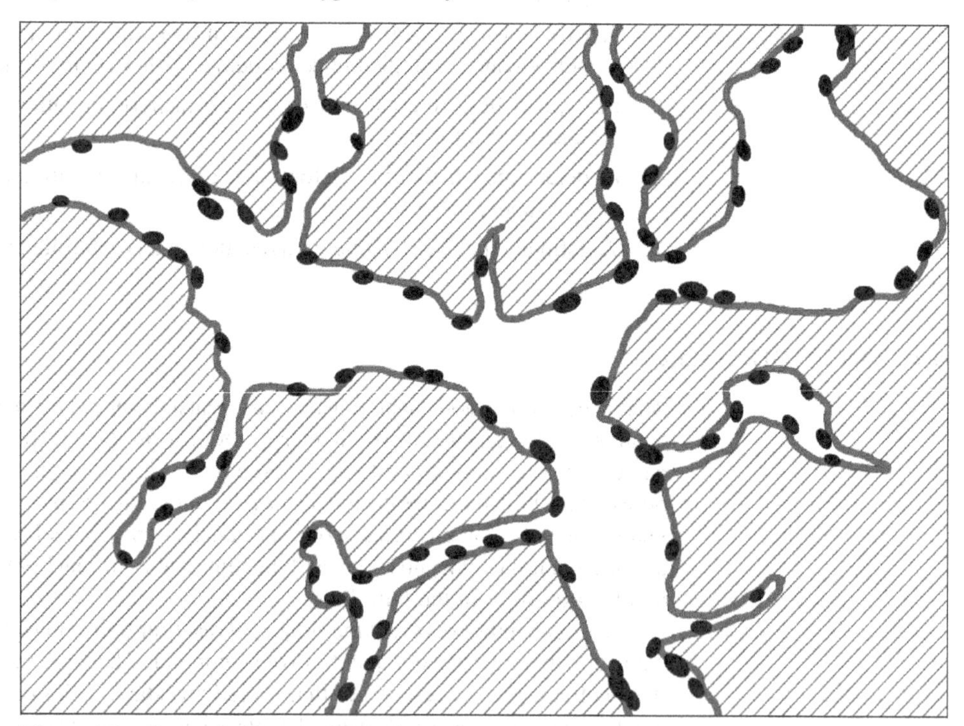

Figure 9-1 The pores in a catalyst. The shaded area is solid material. The unshaded area is the pores, which contain a gas or liquid. The solid, black areas represent "islands" or "clusters" of an "active component" attached to the walls of the pores.

In some cases, the ceramic or metallic material that comprises most of the catalyst actually catalyzes the reaction. For example, γ-alumina ($\gamma\text{-}Al_2O_3$) catalyzes the formation of dimethyl ether from methanol.

$$2CH_3OH \rightarrow CH_3OCH_3 + H_2O$$

The reaction takes place on acidic "sites" on the walls of the pores in the alumina. The exact nature of these sites is relatively unimportant at this stage of discussion. However, the sites generally will be distributed more or less evenly along the pore walls.

In other cases, the material that makes up most of the catalyst does not catalyze the reaction. For example, the hydrogenation of benzene to cyclohexane

$$\text{C}_6\text{H}_6 + 3\text{H}_2 \longrightarrow \text{C}_6\text{H}_{12}$$

would be very slow at 200 °C on a catalyst comprised only of alumina. However, if Pt or Ni is added in the form of small "islands" or "clusters" of metal that are attached to the walls of the pores, the hydrogenation of benzene can proceed at a rate that is satisfactory for the commercial manufacture of cyclohexane. The presence of small metal islands (nanoparticles) on the walls of the catalyst pores is illustrated in Figure 9-1.

In this example, the metal (e.g., Pt or Ni) is referred to as the "active component." The alumina is referred to as a "support." Its function is to provide the surface to which the "active component" is anchored, as well as the mechanical strength that is required to use the catalyst in a reactor. Desirably, the metal clusters should be very small in order to make effective use of an expensive metal such as Pt. It is common to find metal nanoparticles with dimensions as small as a few nanometers in heterogeneous catalysts.

Catalysts can be formulated with active components other than metals. Metal salts are the active component in some catalysts. Moreover, an enzyme can be attached to the pore walls of an inorganic support to form a "supported enzyme catalyst." The advantage of this structure is that the enzyme can be used in a continuous process, e.g., in a fixed-bed reactor, or can be recovered and recycled if the catalyst is used in a batch process. However, the process of attaching the enzyme to the inorganic support can cause a partial or complete loss of enzyme activity.

In most cases, the catalyst in a reactor is in the form of particles. Figure 9-2 shows some different kinds of heterogeneous catalyst particles.

The catalyst particles in a fluidized-bed reactor, or in a slurry reactor, are relatively small. They are roughly spherical in shape and have diameters in the range of 1–100 μm. Agitation keeps the small particles suspended in the fluid in the reactor. Agitation can be provided either mechanically, e.g., by a stirrer, or by turbulence that is created by the fluid stream entering the reactor.

The catalyst particles that are used in a fixed-bed reactor are larger, in the region of 1–10 mm in diameter. Fixed-bed catalysts come in a variety of shapes, as shown in Figure 9-3. The rationale for choosing one shape over another is difficult to understand completely at this early stage of discussion. One important point to recognize is that the different shapes will pack differently in the reactor. For example, cylindrical rings will have a higher void fraction, i.e., a higher interstitial volume ε_i, in the reactor than solid cylinders with the same length and outer diameter. This higher interstitial volume will lead to a lower pressure drop per unit length of reactor, for a given superficial fluid velocity. Rings will also have a higher external surface area per unit volume of reactor than solid cylinders of the same length and outer diameter. However, the rings will have less actual volume of catalyst per unit volume of reactor.

Some shapes, for example, the "trilobe" extrudates shown in the left-hand portion of Figure 9-3, are designed to increase the amount of external catalyst area that is wetted by a flowing liquid. The roughly v-shaped indentation that runs parallel to the length of the extrudate provides a region where capillarity holds the liquid in contact with the catalyst particle. Trilobes are sometimes used in so-called "trickle-bed" reactors, in which a gas and

Figure 9-2 Catalyst particles come in a wide range of sizes, for use in different types of reactor (*Source:* BASF Catalysts, LLC).

a liquid flow concurrently down through a fixed bed of catalyst. In this type of reactor, good and controlled contacting between the liquid and the solid is quite important.

It is common to form a catalyst from small, primary particles, as depicted in Figure 9-4. The primary particles are porous. They usually contain most of the surface area, and most of the reaction occurs in these particles. The larger pores that connect the primary particles are known as "feeder" pores.

In this chapter, we will assume that the active component is uniformly distributed throughout the catalyst particle, unless otherwise stated. However, it is not unusual for catalysts to be designed with an uneven distribution. Catalysts with most of the active

Figure 9-3 Fixed-bed catalysts are manufactured in many different shapes to accommodate different reactor designs and reaction characteristics (*Source:* BASF Catalysts, LLC).

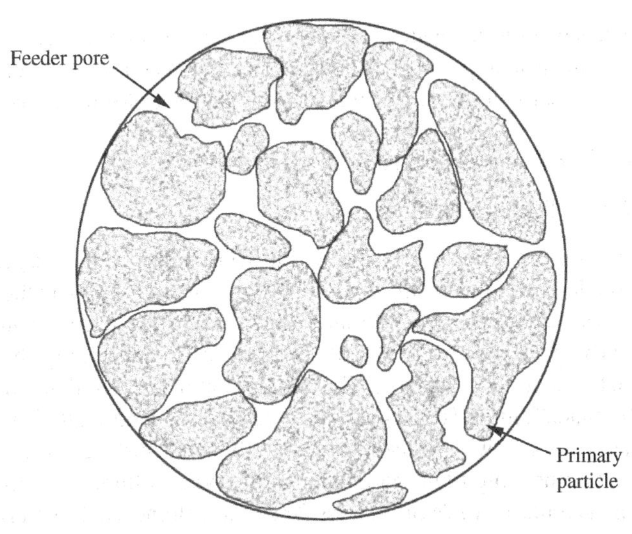

Figure 9-4 A spherical catalyst particle consisting of smaller, "primary" particles.

component deposited near the geometrical (external) surface of the particle are known as "eggshell" catalysts, and those with the active component concentrated in the interior are known as "egg yolk" catalysts.

Another form of catalyst is the so-called "monolithic" or "honeycomb" catalyst, as shown in Figure 9-5.

Figure 9-5 Ceramic monolith catalyst supports. Channels that are square in shape run straight through the ceramic blocks. A catalytic layer is deposited on the walls of the channels. (Photo: Advanced Catalyst Systems, Inc.)

Monoliths can be made completely from a catalytic material, or they can consist of an inert material with a layer of catalytic material applied to the walls of the channels.

Monolithic catalysts are used as automobile exhaust catalysts, and in several other pollution-control applications. For example, a monolithic catalyst is used for the selective reduction of NO_x by ammonia. The chemistry of this process is

$$6NO_2 + 8NH_3 \rightarrow 7N_2 + 12H_2O$$
$$6NO + 4NH_3 \rightarrow 5N_2 + 6H_2O$$

Selective catalytic reduction is used in some power-generation plants to reduce the concentration of NO_x discharged to the atmosphere, ideally to 10 ppm or less.

Compared with particulate supports such as those shown in Figure 9-4, monolithic supports contain much more external surface area per unit volume of reactor, i.e., the area of the channel walls is much greater per unit of reactor volume than the external area of the particles. This can be an important advantage for certain reactions. Conversely, the amount of catalytic material per unit volume of reactor is much lower for the monolith than for the particles.

9.2.2 Characterization of Catalyst Structure

9.2.2.1 Basic Definitions

The walls of the pores in a typical industrial catalyst contain a great deal of surface area. For example, there may be from 10 to 1000 m^2 of surface area in the interior of a single gram of catalyst particles. The internal surface area of a porous catalyst can be measured by adsorbing N_2 onto the walls of the pores and determining the amount of N_2 required to exactly cover the surface. The surface area measured in this way is called the BET (Brunauer/Emmett/Teller) surface area and conventionally is reported as m^2/g of catalyst. The symbol S_p will be used to denote the BET surface area.

Another important parameter is the pore volume per gram of catalyst, denoted V_p. Typical units are m^3/g of catalyst. The pore volume can be determined in several ways. One is to measure the amount of a wetting liquid that is imbibed into the pores of the catalyst by capillarity, i.e., the volume of liquid that is taken up by a known weight of catalyst.

Three different densities are used in the characterization of heterogeneous catalysts. The first is the particle density ρ_p. *Particle* density is defined as the weight of catalyst per unit *geometric* volume of particle. If a catalyst particle was perfectly spherical, with a radius R, its *geometric* volume would be $\frac{4\pi R^3}{3}$. The second is the *skeletal* density ρ_s, which is the density of the *solid material* of which the catalyst particle is comprised. The particle density can be measured by mercury displacement, i.e., by determining how much Hg is displaced by a known weight of catalyst. Mercury is a nonwetting liquid and will not enter the catalyst pores except under high pressure. The skeletal density can be measured by helium displacement. With common materials, for example, alumina, the skeletal density can also be looked up in common references. The third density is the bulk density ρ_B, which is defined as the weight of catalyst per unit *geometrical volume of reactor.*

The pore volume, the skeletal density, and the particle density are related:

$$V_p\left(\frac{\text{volume pores}}{\text{weight catalyst}}\right)+\frac{1}{\rho_s}\left(\frac{\text{volume solids}}{\text{weight catalyst}}\right)=\frac{1}{\rho_p}\left(\frac{\text{volume catalyst}}{\text{weight catalyst}}\right)$$

$$V_p=\frac{1}{\rho_p}-\frac{1}{\rho_s}$$

The *porosity* of the catalyst ε is defined as the volume of the pores divided by the geometric volume of the catalyst particle:

$$\varepsilon \equiv \text{volume pores/geometric volume of particle}$$

For example, if a spherical catalyst particle with a radius of 1 mm contained 1.89 mm^3 of pores, its porosity would be 0.45, or 45%. The porosity is related to the pore volume and the particle and skeletal densities:

$$\varepsilon = V_p\rho_p = 1-(\rho_p/\rho_s)$$

The bulk density is related to the particle density and the *interstitial volume* of the catalyst bed. The interstitial volume ε_i is defined as

$$\varepsilon_i=\frac{\text{volume of ``free space'' between catalyst particles in a vessel}}{\text{geometrical volume of vessel}}$$

This definition leads to

$$\rho_B = \rho_p(1 - \varepsilon_i)$$

The volume of "free space" between catalyst particles does *not* include the volume of the pores.

9.2.2.2 Model of Catalyst Structure

Assume that each gram of catalyst contains "n_p" straight, round pores of radius $\bar{r}$ and length L_p. Since V_p is the volume of pores per gram of catalyst,

$$V_p = n_p(\pi \bar{r}^2 L_p)$$

The surface area of the pores, per gram of catalyst, is given by

$$A_p = n_p(2\pi \bar{r} L_p)$$

so that

$$\bar{r} = 2V_p/A_p \tag{9-1}$$

Values of the BET surface area (A_p) and the pore volume per gram (V_p) are available (or easily measured) for most heterogeneous catalysts. Therefore, a value of $\bar{r}$ can be estimated quite readily.

The *distribution* of pore sizes can be measured by mercury porosimetry. In this technique, Hg, a nonwetting liquid, is forced into the pores of the catalyst by applying pressure. The pressure required depends on the radius of the pore, with the smallest pores requiring the highest pressure. The cumulative amount of Hg that has entered the particle is measured as a function of the applied pressure. The data can be processed to obtain the *distribution function* for pore radii $f(r)$, which is frequently referred to as the pore-size distribution. This distribution function is defined as

$$f(r)\mathrm{d}r = \text{fraction of total pore volume in pores with radii between } r \text{ and } r + \mathrm{d}r$$

By the definition of a distribution function,

$$\bar{r} = \int_0^\infty rf(r)\mathrm{d}r$$

If the parallel-pore model is valid,

$$\int_0^\infty rf(r)\mathrm{d}r = 2V_p/A_p$$

9.3 INTERNAL TRANSPORT

9.3.1 General Approach—Single Reaction

Concentration and temperature gradients inside a catalyst particle can influence the rate of reaction, i.e., the apparent catalyst activity. They can also influence the product distribution, i.e., the apparent catalyst selectivity. First, let's deal with the reaction rate.

Consider a spherical catalyst particle of radius R, in which a single, exothermic reaction is taking place at steady state. The concentration of reactant A at the external surface of the particle is $C_{A,s}$ and the temperature at the external surface is T_s. If the reaction is fast, the concentration profile of reactant A inside the particle, and the temperature profile inside the particle, might look like those shown in Figure 9-6.

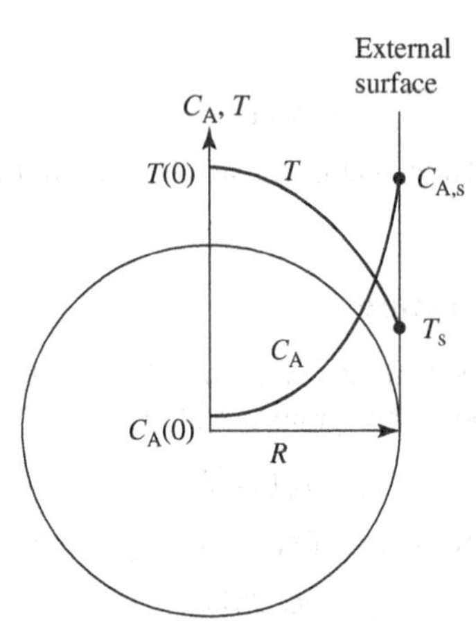

Figure 9-6 Profiles of the concentration of reactant A (C_A) and the temperature (T) inside a porous spherical catalyst particle in which an exothermic reaction is taking place at steady state. $T(0)$ is the temperature at the center of the particle ($r = 0$) and $C_A(0)$ is the concentration at the center of the particle.

The most common approach to quantifying the effect of internal concentration and temperature gradients on the reaction rate is to apply a correction factor.

$$\left\{ \begin{array}{c} \text{actual reaction} \\ \text{rate} \end{array} \right\} = \eta \times \left\{ \begin{array}{c} \text{rate with} \\ \text{no internal gradients} \end{array} \right\} \tag{9-2}$$

In this equation, the "actual reaction rate" is the rate that *is* observed, i.e., measured, in a catalyst particle in which gradients are present. The "rate with no gradients" is the rate that *would be* observed if the concentrations and the temperature throughout the particle were equal to their respective values at the external surface. Both of these rates are intensive variables, i.e., rate per unit weight or per unit volume of catalyst. For example, if the reaction was irreversible and first order in A, the "rate with no internal gradients" would be $k(T_s)C_{A,s} = A_{\exp}(-E/RT_s)C_{A,s}$.

The parameter η is the "correction factor" that accounts for the effect of internal transport on the reaction rate. This parameter is known as the "internal effectiveness factor," or simply the "effectiveness factor." According to Eqn. (9-2), the actual reaction rate can be obtained by multiplying the "rate with no internal gradients" by η. The problem of accounting for internal temperature and concentration gradients then boils down to predicting the value of η.

The effectiveness factor can be related to the system parameters by solving the differential equations that describe mass and energy transport inside the catalyst particle. To illustrate, consider a control volume that consists of a spherical shell of thickness dr within a spherical catalyst particle, as shown below.

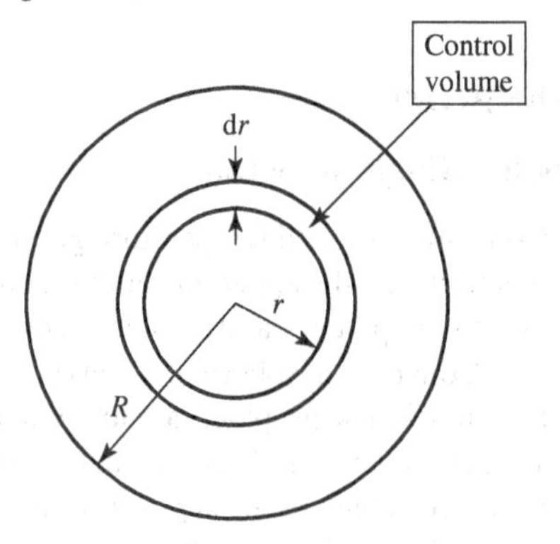

We will assume that the molar flux of A in the radial direction $(\vec{N}_{A,r})$, across a control surface at $r = r$, is purely diffusive and is given by

$$\vec{N}_{A,r} = -D_{A,eff} \left.\frac{\partial C_A}{\partial r}\right|_r \tag{9-3}$$

The parameter $D_{A,eff}$ is the "effective" diffusion coefficient of A inside the catalyst particle and is based on the *geometric* area of the control surface. In this case, the geometric area is $4\pi r^2$. The value of $D_{A,eff}$ depends on the radii of the pores in the catalyst particle, the porosity of the particle, and on other features of the particle structure. Procedures to estimate a value of $D_{A,eff}$ are discussed in detail in Section 9.3.4.

The steady-state material balance on A for this control volume is

$$\frac{\partial}{\partial r}\left(r^2 D_{A,eff} \frac{\partial C_A}{\partial r}\right) = r^2(-R_{A,v}) \tag{9-4}$$

The term on the left-hand side of this equation is the *net* flux of A into the control volume. The term on the right-hand side is the rate of disappearance of A due to the chemical reaction.

The parameter $-R_{A,v}$ is the rate of disappearance of A per unit of *geometric* catalyst volume, i.e., moles A/time-unit geometric volume of particle. Up to this point, we have always expressed the rate of a catalytic reaction $(-r_A)$ on a weight basis. The relationship between these two rates is $-R_{A,v} = \rho_p(-r_A)$, where ρ_p is the particle density.

The boundary conditions on Eqn. (9-4) are

$$\frac{\partial C_A}{\partial r} = 0; \quad r = 0 \tag{9-4a}$$

$$C_A = C_{A,s}; \quad r = R \tag{9-4b}$$

The first boundary condition (9-4a) results from the fact that the net diffusive flux at the exact center of the catalyst particle must be zero. The second boundary condition is based on the presumption that the concentrations at the external surface of the particle are known. If the resistance to mass transfer at the external surface is negligible, then the surface concentration $C_{A,s}$ will be equal to the concentration of A in the bulk fluid $C_{A,B}$. However, if the resistance to mass transfer at the external surface is important, $C_{A,s}$ will be less than $C_{A,B}$, and the exact value of $C_{A,s}$ will be determined in part by the external transport resistance.

If conduction is the only significant mechanism of heat transfer inside the catalyst particle, the steady-state energy balance for the same control volume is

$$\frac{\partial}{\partial r}\left(r^2 k_{eff} \frac{\partial T}{\partial r}\right) + r^2(-\Delta H_R)(-R_{A,v}) = 0 \tag{9-5}$$

The first term in Eqn. (9-5) is the net rate of heat conduction into the control volume. The second term is the rate at which heat is "consumed" by the reaction. In Eqn. (9-5), k_{eff} is the "effective" thermal conductivity of the catalyst particle, based on *geometric* area, and ΔH_R is the enthalpy change on reaction.

The boundary conditions for Eqn. (9-5) are

$$\frac{\partial T}{\partial r} = 0; \quad r = 0 \tag{9-5a}$$

$$T = T_s; \quad r = R \tag{9-5b}$$

If the rate equation on a volumetric basis $(-R_{A,v})$ is known, along with $D_{A,eff}$ and k_{eff}, Eqns. (9-4) and (9-5) can be solved simultaneously, at least in principle, to yield the concentration and temperature profiles inside the catalyst particle. Once these profiles are known, values of the effectiveness factor can be calculated.

9.3.2 An Illustration: First-Order, Irreversible Reaction in an Isothermal, Spherical Catalyst Particle

To illustrate how the effectiveness factor is obtained and to identify some of the variables that determine its value, let's take a simple example. Suppose that an irreversible, first-order reaction is taking place in a spherical catalyst particle. Moreover, suppose that the temperature difference $(T(0) - T_s)$ is very small, so that the catalyst particle is essentially isothermal, with a temperature of T_s throughout. If the particle is isothermal and its temperature is T_s, there is no need to solve the energy balance, Eqn. (9-5). The effectiveness factor is determined by the concentration gradients only.

In this case, the effectiveness factor can be obtained simply by solving Eqn. (9-4), subject to the boundary conditions of Eqns. (9-4a) and (9-4b). For a first-order reaction, $-R_{A,v} = k_v C_A$. Here, k_v is the rate constant based on *geometric* volume of catalyst. This rate constant is related to the one that we have used previously, k (moles/time-weight of catalyst), by $k_v = k\rho_p$.

If the effective diffusivity is constant, i.e., independent of concentration and position, Eqn. (9-4) becomes

$$\frac{d^2C_A}{dr^2} + \frac{2}{r}\frac{dC_A}{dr} = \frac{k_v C_A}{D_{A,eff}} \tag{9-4c}$$

Assume that the rate constant k_v is independent of position, for example, the catalyst is neither an "eggshell" nor an "egg yolk" design. Then the solution to this ordinary differential equation, subject to the boundary conditions of Eqns. (9-4a) and (9-4b), is

$$C_A(r) = \frac{C_{A,s} R \sinh(\phi_{s,1} r/R)}{r \sinh \phi_{s,1}} \tag{9-6}$$

The details of the solution of Eqn. (9-4c) are given in Appendix 9-A.

In this equation, $\phi_{s,1}$ is a dimensionless group known as the Thiele modulus. (Yep, it's the same Thiele!) For a spherical catalyst particle and for a first-order, irreversible reaction, the Thiele modulus is defined as

$$\phi_{s,1} \equiv R\sqrt{\frac{k_v}{D_{A,eff}}} \tag{9-7}$$

The Thiele modulus is subscripted to indicate that the definition of Eqn. (9-7) applies only to an irreversible, *first-order* reaction in a *spherical* catalyst particle.

EXERCISE 9-1

Using Eqn. (9-6), plot $C_A/C_{A,s}$ versus r/R for $\phi_{s,1} = 0.01$, 1, and 10. Based on these graphs, which of these three values of $\phi_{s,1}$ would you expect to have the lowest value of η?

To complete the analysis, Eqn. (9-6) must be used to derive an expression for the effectiveness factor. From Eqn. (9-2)

Definition of effectiveness factor

$$\eta = \frac{\text{actual reaction rate}}{\text{rate with no internal gradients}} \tag{9-8}$$

For this example, the "rate with no internal gradients" is just $4\pi R^3 k_v C_{A,s}/3$.

There are two possible approaches to calculating the "actual reaction rate." First, we could integrate the rate equation over the whole volume of the catalyst particle, using Eqn. (9-6) to express the dependence of C_A on radial position.

$$\text{actual reaction rate} = 4\pi \int_0^R r^2 k_v C_A(r) dr$$

The second approach generally is a bit easier mathematically. *At steady state*, the reaction rate in the whole particle must be equal to the rate at which A diffuses into the particle through the external surface at $r = R$. Therefore,

$$\text{actual reaction rate} = 4\pi R^2 D_{A,\text{eff}} \left(\frac{dC_A}{dr} \right) \bigg|_{r=R}$$

Either approach gives

$$\eta = \frac{3}{\phi_{s,1}} \left[\frac{1}{\tanh\phi_{s,1}} - \frac{1}{\phi_{s,1}} \right] \tag{9-9}$$

For a first-order irreversible reaction taking place in an isothermal, spherical catalyst particle, the effectiveness factor depends on a *single* dimensionless variable, the Thiele modulus, $\phi_{s,1}$. Using Eqn. (9-9), it can be shown that $\eta \to 1$ as $\phi_{s,1} \to 0$. It can also be shown that $\eta \to 3/\phi_{s,1}$ as $\phi_{s,1} \to \infty$.

EXERCISE 9-2

Prove that $\eta \to 1$ as $\phi_{s,1} \to 0$. Prove that $\eta \to 3/\phi_{s,1}$ as $\phi_{s,1} \to \infty$.

9.3.3 Extension to Other Reaction Orders and Particle Geometries

Expressions for the effectiveness factor, similar to Eqn. (9-9), can be derived for other particle shapes, other reaction orders, and for reversible as well as irreversible reactions. Fortunately, if the Thiele modulus is redefined somewhat, all of these solutions can be approximated by a single curve of η versus ϕ.

First, the *characteristic dimension* l_c of a catalyst particle is

Definition of characteristic dimension

$$l_c = \frac{\text{geometric volume of particle}}{\text{geometric surface area of particle}} = \frac{V_G}{A_G} \tag{9-10}$$

The geometric surface area (A_G) is the external area of the catalyst particle, i.e., the area that is in contact with the fluid. For example, the characteristic dimension of a sphere is $l_c = (4/3)\pi R^3/4\pi R^2 = R/3$.

EXAMPLE 9-1 ***Calculation of Characteristic Dimension***

Calculate the numerical value of l_c for a ring that is 1.0 cm in length (L), with an outer diameter (D_o) of 1.0 cm and an inner diameter (D_i) of 0.50 cm.

APPROACH

The given dimensions of the catalyst particle will be used to calculate the geometric volume V_G and the geometric area A_G. Then l_c will be calculated from Eqn. (9-10).

SOLUTION

The geometric volume (V_G) of the particle is

$$V_G = \pi(D_o/2)^2 L - \pi(D_i/2)^2 L = \pi(1.0)[(1.0/2)^2 - (0.50/2)^2] = 0.59 \text{ cm}^3.$$

The geometric area (A_G) of the particle is

$$A_G = 2\pi(D_o/2)L + 2\pi(D_i/2)L + 2\pi[(D_o/2)^2 - (D_i/2)^2]$$

$$A_G = 2\pi[(1.0)(1.0/2) + (1.0)(0.50/2) + \{(1.0/2)^2 - (0.50/2)^2\}] = 5.9 \text{ cm}^2.$$

Note that the outer area, the inner area, and the area of the ends of the ring all were taken into account in the calculation of A_G. From Eqn. (10-9),

$$l_c = V_G/A_G = 0.59 \text{ cm}^3/5.9 \text{ cm}^2 = 0.10 \text{ cm}$$

Now, let's define a Thiele modulus that is not tied to a particular geometry or reaction order and applies to both irreversible and reversible reactions. Consider the reversible reaction

$$A \rightleftarrows B$$

which is nth order in both directions. The equilibrium constant based on concentration is K_{eq}^C and the forward rate constant based on geometrical catalyst volume is k_v. Let α_s be the ratio of the concentration of B at the external surface to that of A,

$$\alpha_s = C_{B,s}/C_{A,s} \tag{9-11}$$

and let β be the ratio of the effective diffusion coefficient of B to that of A,

$$\beta = D_{B,eff}/D_{A,eff} \tag{9-12}$$

The new Thiele modulus is defined by

Generalized Thiele modulus

$$\phi = l_c \left(\frac{(n+1)k_v C_{A,s}^{n-1}}{2D_{A,eff}} \right)^{1/2} \Psi \tag{9-13}$$

$$\Psi = \left[\frac{(K_{eq}^C - \alpha_s^n)(1 + \beta\sqrt[n]{K_{eq}^C})^{(n+1)/2}}{K_{eq}^C\{(1+\beta\sqrt[n]{K_{eq}^C})^{n+1}(1 + (\beta\alpha_s^{n+1}/K_{eq}^C)) - (1+\beta\alpha_s)^{n+1}(1+\beta\sqrt[n]{K_{eq}^C})\}^{1/2}} \right] \tag{9-13a}$$

This version of the Thiele modulus looks complicated, primarily because three new parameters (α_s, β, and K_{eq}^C) are required to account for reversibility. If the reaction is essentially irreversible ($K_{eq}^C \rightarrow \infty$), then $\Psi = 1$ and Eqn. (9-13) reduces to

Generalized Thiele modulus-irreversible reaction

$$\phi = l_c \left(\frac{(n+1)k_v C_{A,s}^{n-1}}{2D_{A,eff}} \right)^{1/2} \tag{9-14}$$

When Eqns. (9-13) and (9-13a) are used to define the Thiele modulus, the effectiveness factor for an isothermal catalyst particle is *almost* independent of particle geometry and

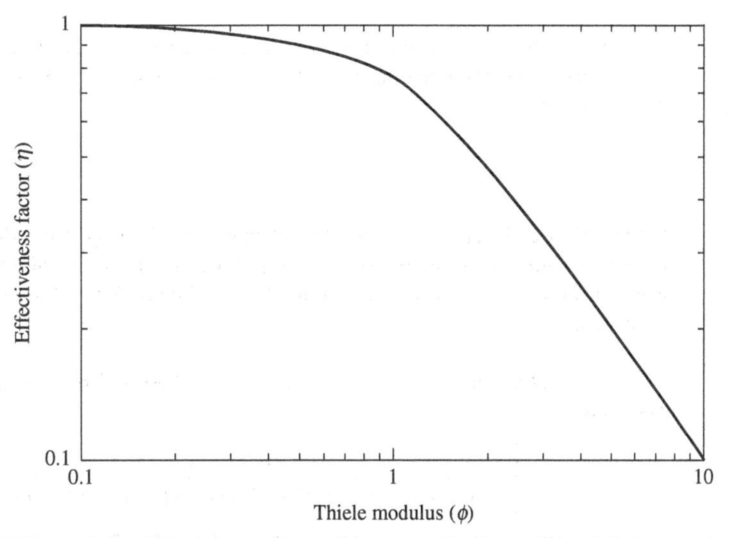

Figure 9-7 Effectiveness factor (η) versus Thiele modulus (ϕ) for an nth-order reaction in an isothermal catalyst particle. The modulus is defined by Eqn. (9-13), which applies to both reversible and irreversible reactions. For $\phi < 0.10$, $\eta \cong 1$. For $\phi > 10$, $\eta \cong 1/\phi$.

reaction order. The relationship between η and ϕ is shown in Figure 9-7, which covers a range of ϕ from 0.1 to 10. When ϕ is less than 0.1, the effectiveness factor is essentially unity. When ϕ is greater than 10, η is approximately equal to $1/\phi$. Figure 9-7 shows that $\eta \rightarrow 1$ as ϕ becomes very small and that $\eta \rightarrow 1/\phi$ as ϕ becomes very large.

The equation for the line in Figure 9-7 is

Relationship between effectiveness factor and generalized Thiele modulus

$$\eta = \tanh(\phi)/\phi \tag{9-15}$$

The *exact* relationship between η and ϕ depends to a small extent on particle geometry and reaction order. However, the differences between Eqn. (9-15) and the exact solutions are not significant, given the normal uncertainties in k_v and $D_{A,eff}$.

EXAMPLE 9-2 ***Calculating the Effectiveness Factor for an Irreversible Reaction***

A second-order irreversible reaction ($A \rightarrow B$) is taking place in an isothermal catalyst particle in the shape of a ring that is 1.0 cm in length (L), with an outer diameter (D_o) of 1.0 cm and an inner diameter (D_i) of 0.50 cm. The value of the rate constant k_v is 46 cm^3/mol-s, the value of the effective diffusivity of reactant A is 5×10^{-4} cm^2/s and the concentration of A at the external surface of the particles is 1.1×10^{-4} mol/cm^3. Estimate the value of the effectiveness factor.

APPROACH

First, the value of ϕ will be calculated from Eqn. (9-14). The value of the characteristic dimension l_c was calculated to be 0.10 cm in the preceding example. All of the other parameters in Eqn. (9-14) are known. Then Figure 9-7 will be used to obtain the value of η.

SOLUTION

Using the given values of l_c, k_v, $D_{A,eff}$, and $C_{A,s}$ in Eqn. (9-14) leads to

$$\phi = 0.10(\text{cm})\sqrt{\frac{3 \times 46(\text{cm}^3/\text{mol-s}) \times 1.1 \times 10^{-4}(\text{mol/cm}^3)}{2 \times 5 \times 10^{-4}(\text{cm}^2/\text{s})}} = 0.40$$

From Figure 9-7 (or Eqn. (9-15)), the value of the effectiveness factor is about 0.95.

EXAMPLE 9-3
Calculating the Effectiveness Factor for a Reversible Reaction

Suppose that the reaction in the previous example was *reversible*, with an equilibrium constant of 1.0 and a surface concentration of B, $C_{B,s}$, of 5.5×10^{-5} mol/cm^3. Estimate the effectiveness factor for the case where $D_{A,eff} = D_{B,eff}$.

APPROACH

In Example 9-2, Eqn. (9-14) was used to calculate the value of ϕ. When the reaction is reversible, the value of ϕ is given by Eqn. (9-13), which is just the right-hand side of Eqn. (9-14) multiplied by the value of the parameter Ψ given by Eqn. (9-13a). Figure 9-7 or Eqn. (9-15) then can be used to estimate η.

SOLUTION

For the present example, $n = 2$, $\alpha_s = C_{B,s}/C_{A,s} = 0.50$, and $\beta = D_{B,eff}/D_{A,eff} = 1$. Therefore,

$$\Psi = \left[\frac{[1-(0.50)^2](1+1)^{3/2}}{1\{(1+1)^3(1+(1\times 0.50)^3/1)) - (1+1\times 0.50)^3(1+1)\}^{1/2}}\right] = 1.4$$

The value of the Thiele modulus is $\phi = 0.40 \times 1.4 = 0.56$. Figure 9-7 or Eqn. (9-15) gives $\eta = 0.91$.

Figure 9-7 can be used to estimate the effectiveness factor for a wide range of situations, as long as the catalyst particle is essentially isothermal. For example, even though the definition of ϕ is based on an nth-order reaction, the behavior of many Langmuir–Hinshelwood and Michaelis–Menten rate equations can be bracketed by two values of n, say 0 and 1 or 0 and 2. This type of rate equation can be treated by using an intermediate order, say $1/2$, in Eqns. (9-13) and (9-13a), or Eqn. (9-14).

There are some situations for which Figure 9-7, in conjunction with the definition of ϕ given by Eqn. (9-13), should *not* be used. These cases are relatively rare but are worth noting. First, Figure 9-7 plus Eqns. (9-13) and (9-13a) do not apply if the effective order of the reaction is less than zero. As noted in Chapter 5, certain Langmuir–Hinshelwood rate equations can behave as though the reaction order is negative with respect to one of the reactants, at least over some range of concentration. Second, Figure 9-7 plus Eqns. (9-13) and (9-13a) do not apply if the effective order of the reaction exceeds about 3. Some Langmuir–Hinshelwood and Michaelis–Menten rate equations contain a product inhibition term. Such a term can give rise to very high apparent reaction orders. Rate equations containing significant product inhibition should not be analyzed using Figure 9-7. Third, Figure 9-7 should not be used if the effective diffusion coefficient is not approximately constant. The effective diffusivity is treated in the next section.

The question of when the catalyst particle can be treated as isothermal is discussed in more detail in Section 9.3.7.

9.3.4 The Effective Diffusion Coefficient

9.3.4.1 Overview

Estimation of the effective diffusion coefficient $D_{A,eff}$ begins with the equation

Starting point—estimation of effective diffusivity

$$D_{A,eff} = D_{A,p}(r)\varepsilon/\tau_p \quad (9\text{-}16)$$

Here, $D_{A,p}(r)$ is the diffusion coefficient of A in an assembly of straight, round pores that has the same distribution of pore sizes as the catalyst for which $D_{A,eff}$ is to be calculated. The diffusivity $D_{A,p}(r)$ is based on the cross-sectional area of the *pores* and not on the geometric area of the catalyst particle. The parameter ε is the porosity of the catalyst particle, as defined earlier in this chapter. The presence of ε in Eqn. (9-16) corrects the diffusion coefficient from an "area of pores" basis to a geometric area basis.

The parameter τ_p is referred to as the "tortuosity" of the catalyst particle. Originally, τ_p was intended to correct for the fact that the pores in a catalyst are not straight and are not all parallel to the direction of diffusion. As a result, the diffusing molecules must follow a tortuous path and must travel a longer distance than a straight line in the direction of the net diffusive flux. In reality, τ_p corrects for many other nonidealities, such as the variation in cross section along the length of a pore.

For many commercial catalysts, the value of τ_p lies in the region

Range of tortuosity values

$$2 \leq \tau_p \leq 10$$

Ideally, the value of τ_p should be obtained from diffusion measurements on the catalyst in question. In the absence of experimental data, an approximate value of τ_p can be used. Of course, this limits the accuracy to which both $D_{A,eff}$ and η can be predicted. The uncertainty in τ_p also justifies the use of certain approximations in the calculation of $D_{A,p}(r)$, as will be discussed below. Satterfield[1] recommends using a tortuosity factor in the range of 2–6. A value of $\tau_p = 4$ may provide a reasonable starting point for many problems.

9.3.4.2 Mechanisms of Diffusion[2]

The prediction of $D_{A,p}(r)$ in Eqn. (9-16) requires some understanding of how a molecule diffuses in a porous material. The nature of diffusion will depend on the size of the pores. Consider a straight round pore of radius r as shown below. The pore contains molecules of a fluid that are diffusing along the axial coordinate (z-dimension) of the pore.

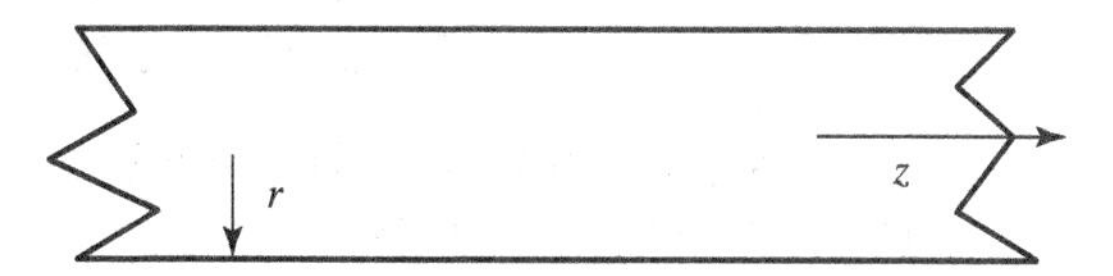

There are three possible modes of diffusion in such a pore.

Configurational (Restricted) Diffusion With some catalysts, the diameter of the diffusing molecule can be close to the diameter of the pore, as depicted below.

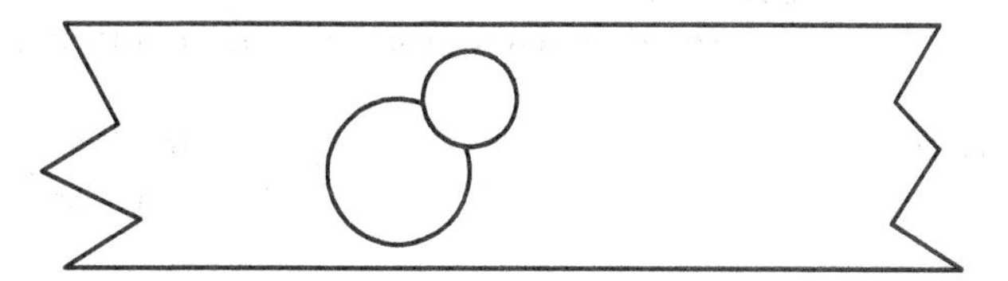

[1] Satterfield, C. N., *Mass Transfer in Heterogeneous Catalysis,* MIT Press, Cambridge, MA, (1970), p. 157.

[2] For a more detailed discussion of diffusion in porous media, see Krishna, R., A unified approach to the modeling of intraparticle diffusion in adsorption processes, *Gas Separat. Purificat.* 7 (2), 91–104 (1993).

Zeolite or so-called "molecular sieve" catalysts are an important class of materials for which this type of diffusion is prevalent. Zeolite catalysts are widely used in refineries for the catalytic cracking of petroleum naphthas.

Diffusion coefficients in the configurational regime can be very low (10^{-14} to 10^{-6} cm^2/s), and they depend on the pore size, the size of the diffusing molecule, and the intermolecular forces between the diffusing molecule and the walls of the pore. At present, there is no theory available to predict diffusion coefficients in the configurational regime. Experimental data are necessary to begin the process of estimating the effective diffusion coefficient.

Knudsen Diffusion (Gases) Suppose that a *gas* is diffusing in a straight, round pore with a radius r that is much larger than molecular dimensions. The molecules in the pore will be in random thermal motion and will collide with other gas molecules and with the walls of the pore. The *mean free path* is defined as the average distance that a molecule travels before it collides with another molecule. The mean free path can be predicted approximately from kinetic theory, for a pure gas.

$$\lambda_m = 1/\sqrt{2}\pi d^2 C_A N_a \tag{9-17}$$

In Eqn. (9-17), λ_m is the mean free path, d is the equivalent diameter of the molecule, N_a is Avogadro's number, and C_A is the molar concentration of the pure gas. The product $C_A N_a$ is the number density (molecules/volume) of the gas. For most gases at atmospheric conditions, λ_m is between 10 and 100 nm ($100\ \text{Å} \leq \lambda \leq 1000\ \text{Å}$).

If $\lambda_m \gg r$, collisions of molecules with the walls of the pore will be much more frequent than collisions with other molecules. Diffusion will take place through molecule–wall interactions rather than through molecule–molecule interactions, as depicted below.

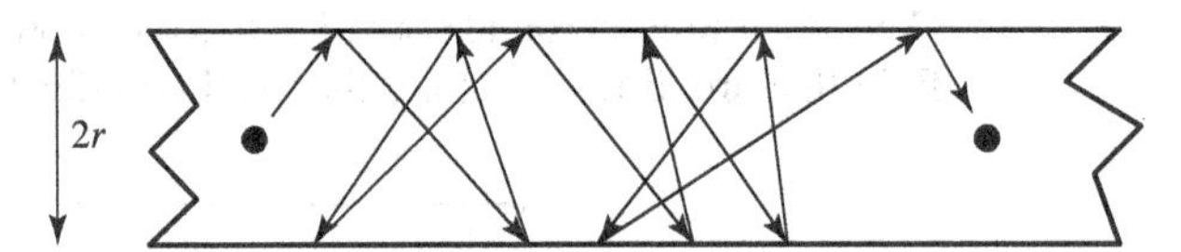

This type of diffusion is known as Knudsen diffusion. In the Knudsen diffusion regime, the equation for the flux is

$$\vec{N}_{A,z} = -D_{A,k}\frac{dC_A}{dz}$$

In this equation, $\vec{N}_{A,z}$ is the flux of A in the z-direction, i.e., along the length of the pore. In the Knudsen regime, the flux of any component is purely diffusive, exactly as assumed in Eqn. (9-3).

The Knudsen diffusion coefficient for species A, $D_{A,k}$ is given by

Knudsen diffusion coefficient

$$D_{A,k} = \frac{4r}{3}\sqrt{\frac{2RT}{\pi M_A}} \tag{9-18}$$

Here, r is the radius of the pore, R is the gas constant, T is the absolute temperature, and M_A is the molecular weight of species A. The Knudsen diffusion coefficient increases linearly with the pore radius. It also increases as the square root of the absolute temperature. It does *not* depend on the gas composition, on the total pressure, or on the fluxes of the other compounds.

Bulk (Molecular) Diffusion Bulk or molecular diffusion is the predominant mode of diffusion in large pores. For a gas, molecular diffusion occurs when $r \gg \lambda_m$. For liquids, molecular diffusion occurs when r is much greater than the radius of the diffusing molecule. In the molecular diffusion regime, collisions between molecules are much more frequent than collisions between molecules and the walls of the pore. Diffusion takes place through molecule–molecule interactions.

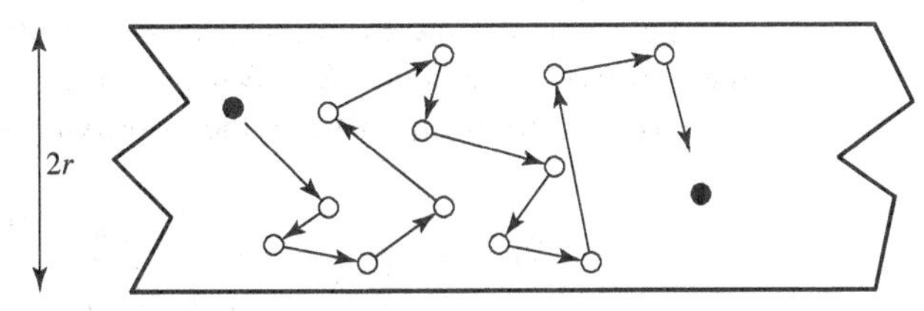

The equation for the flux of component A is more complex for bulk diffusion than for Knudsen diffusion. In bulk diffusion, the flux of A in the z-direction is given by

Flux equation—bulk diffusion

$$\vec{N}_{A,z} = -D_{A,m}\frac{dC_A}{dz} + y_A \sum_{i=1}^{N} \vec{N}_{i,z} \tag{9-19}$$

In this equation, $\vec{N}_{i,z}$ is the flux of component "i" in the z-direction, $D_{A,m}$ is the diffusivity of A in the mixture, y_A is the mole fraction of A, and "N" is the number of compounds in the mixture. The first term on the right-hand side of Eqn. (9-19) is the diffusive flux of A and the second term is the flux of A due to bulk flow. The total flux $\vec{N}_{i,z}$ is a vector that is directed either out of the pore, if "i" is a product, or into the pore, if "i" is a reactant. If "i" is a compound that does not participate in the reaction, $\vec{N}_{i,z} = 0$ at steady state.

For a single reaction taking place in a catalyst particle *at steady state*, the molar fluxes are related through stoichiometry, i.e.,

$$\vec{N}_{i,z}/\vec{N}_{A,z} = \nu_i/\nu_A; \quad \vec{N}_{i,z} = (\vec{N}_{A,z}/\nu_A)\nu_i$$

Substituting this relationship into Eqn. (9-19) and rearranging gives

$$\vec{N}_{A,z} = -\left(\frac{D_{A,m}}{1-(y_A\Delta\nu/\nu_A)}\right)\frac{dC_A}{dz} \tag{9-20}$$

Here $\Delta\nu$ is the change in the number of moles on reaction, i.e., $\Delta\nu = \sum_{i=1}^{N}\nu_i$.

In order to use Eqn. (9-20), a value of $D_{A,m}$ must be available. A variation of the Stefan–Maxwell equations can be used to predict the mixture diffusivity.[3] For a single reaction taking place at steady state:

Diffusion coefficient—molecular regime

$$D_{A,m} = \frac{1-(y_A\Delta\nu/\nu_A)}{\sum_{i=1}^{N}\frac{1}{D_{Ai}}[y_i - (y_A\nu_i/\nu_A)]} \tag{9-21}$$

If a gas or liquid mixture is ideal, then D_{Ai} is the binary molecular diffusivity of A in species "i". Binary molecular diffusivities are almost independent of concentration for ideal gas and liquid systems. Values of D_{ij} for many common binary pairs are tabulated in handbooks, and methods for predicting binary diffusivities as a function of temperature and pressure are

[3] See, for example, Bird, R. B., Stewart, W. E., and Lightfoot, E. N., *Transport Phenomena*, Wiley, New York (1960) p. 571.

well established. If the mixture is not ideal, the form of Eqn. (9-21) is valid, but the D_{ij}'s are not binary molecular diffusion coefficients. Experimental data for the D_{ij}'s are necessary to use Eqn. (9-21) for nonideal mixtures.

Equation (9-21) can be difficult to use when the numerator is small, i.e., when $y_A \Delta\nu/\nu_A$ is close to 1. In this case, the denominator will also be small, and very accurate values of the D_{Ai} are required to obtain an accurate value of $D_{A,m}$.

Let's look at a few illustrations. First, consider an ideal, binary mixture of A and B, with the isomerization reaction A → B taking place. For this case, $\nu_B = -\nu_A = 1$, $\Delta\nu = 0$, and $y_A + y_B = 1$. from Eqn. (9-21), the diffusivity of A in the mixture is

$$D_{A,m} = \frac{1 - y_A \times (0/-1)}{\dfrac{1}{D_{AA}}\left[y_A - y_A\left(\dfrac{-1}{-1}\right)\right] + \dfrac{1}{D_{AB}}\left[y_B - y_A\left(\dfrac{1}{-1}\right)\right]} = D_{AB}$$

From Eqn. (9-20),

$$N_{A,z} = -D_{A,m}\frac{dC_A}{dz} = -D_{AB}\frac{dC_A}{dz}$$

For this case of binary, equimolar counterdiffusion, the diffusivity of A in the mixture ($D_{A,m}$) is just the binary molecular diffusivity of A in B (D_{AB}). Moreover, the flux $\vec{N}_{A,z}$ is purely diffusive. The bulk flow term in Eqn. (9-19) is zero because $\sum_{i=1}^{N} \vec{N}_{i,z} = (\vec{N}_{A,z}/\nu_A)\Delta\nu$ and $\Delta\nu = 0$ for this example.

Now consider the same situation, except that an inert I is also present. Once again, $\nu_B = -\nu_A = 1$ and $\Delta\nu = 0$. Also $\nu_I = 0$. However, $y_A + y_B \neq 1$ because the inert is present. From Eqn. (9-21), the diffusivity of A in the mixture is

$$D_{A,m} = \frac{1 - y_A \times (0/-1)}{\dfrac{0}{D_{AA}} + \dfrac{y_A + y_B}{D_{AB}} + \dfrac{y_I}{D_{AI}}} = \frac{1}{\dfrac{y_A + y_B}{D_{AB}} + \dfrac{y_I}{D_{AI}}}$$

The presence of the inert gas influences the value of $D_{A,m}$. If D_{AI} is greater than D_{AB}, then the value of $D_{A,m}$ for the ternary system is greater than the value of $D_{A,m}$ for the binary system. The opposite is true if D_{AI} is less than D_{AB}. Moreover, the value of $D_{A,m}$ depends on the composition of the system unless $D_{AI} = D_{AB}$.

From Eqn. (9-20),

$$\vec{N}_{A,z} = -D_{A,m}\frac{dC_A}{dz}$$

However, the value of $D_{A,m}$ will vary along the length of the pore, if y_A, y_B, and y_I vary.

For a final illustration, consider the reaction A → νB taking place in an ideal, binary system. From stoichiometry, $\nu_A = -1$, $\nu_B = \nu$, and $\Delta\nu = (\nu - 1)$. Since the system contains only A and B, $y_A + y_B = 1$.

In this case, the bulk flow term in Eqn. (9-19) is not zero because $\sum_{i=1}^{N} \vec{N}_{i,z} = (\vec{N}_{A,z}/\nu_A)\Delta\nu = -(\vec{N}_{A,z})(\nu - 1)$. If $\nu > 1$, the net flux $\sum_{i=1}^{N} \vec{N}_{i,z}$ and the flux of A, N_A, are in opposite directions, i.e., N_A is directed into the pore and the net flux is directed out of the pore. If $\nu > 1$, the number of moles of products leaving the pore at steady state is greater than the number of moles of reactant A that enter the pore. If $\nu < 1$, the opposite is true; the net flux is in the same direction as N_A, into the pore.

The value of $D_{A,m}$ from Eqn. (9-21) is

$$D_{A,m} = D_{AB}$$

and Eqn. (9-20) becomes

$$\vec{N}_{A,z} = -\left\{\frac{D_{AB}}{1 + y_A(\nu - 1)}\right\}\frac{dC_A}{dz}$$

For this example, the flux $(\vec{N}_{A,z})$ depends on the composition of the system and on the stoichiometry of the reaction. If the change in moles on reaction, $\Delta\nu = (\nu - 1)$, is large and positive, the flux of A will be much lower than for a situation where $\Delta\nu = 0(\nu = 1)$. The value of $D_{A,m}$ will vary along the length of the pore if y_A varies.

The Transition Region For certain pore sizes, diffusion will not be purely molecular or purely Knudsen. Both types of diffusion will contribute to the overall flux. The theory for this "transition region" is complex. A workable approach is to assume that the two types of diffusion occur in parallel. This leads to

Diffusion coefficient—transition regime

$$\vec{N}_{A,z} = D_{A,t}\frac{dC_A}{dz}$$
$$\frac{1}{D_{A,t}} = \frac{1}{D_{A,k}} + \sum_{i=1}^{N}\frac{1}{D_{Ai}}[y_i - (y_A\nu_i/\nu_A)] \tag{9-22}$$

The diffusion coefficient in the transition regime, $D_{A,t}$, can depend on composition, as a result of the second term on the right-hand side of Eqn. (9-22), i.e., the resistance to molecular diffusion. The diffusion coefficient in the transition regime also depends on the pore radius, since $D_{A,k}$ depends on pore radius.

Concentration Dependence One important conclusion from the preceding discussion is that the diffusivity in either the molecular or transition regime probably will depend on the mixture composition. However, the method that we developed for calculating the effectiveness factor is based on the assumption of a constant effective diffusivity, independent of position and/or concentration. We might consider (very briefly!) resolving the basic differential equations that led to the effectiveness factor, for a concentration-dependent diffusion coefficient. However, this would create a lot of additional complexity. Moreover, for many nonideal mixtures, the necessary relationships between the diffusivities and concentration are not readily available. Therefore, for a first-pass, an acceptable procedure is to use an average diffusion coefficient, calculated over the relevant range of concentration. Such a calculation is illustrated below.

Illustration: Composition dependence of diffusion coefficient
The irreversible cracking reaction $C_3H_8 \rightarrow C_2H_4 + CH_4$ is taking place on the walls of a pore of radius *r*, in the presence of steam, an inert gas. The system is shown in the following figure.

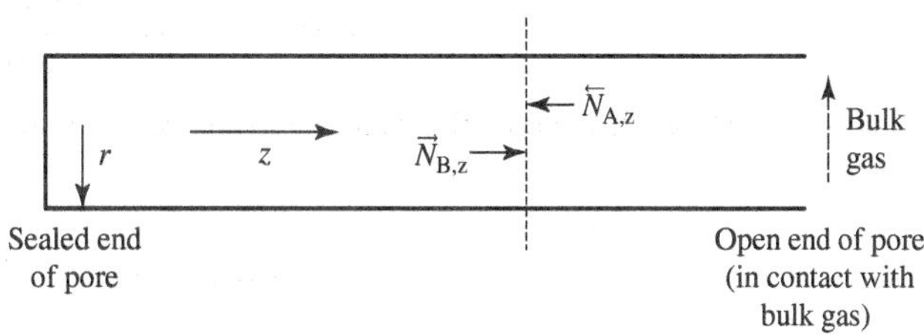

Let A denote propane (C_3H_8), B denote ethylene (C_2H_4), C denote methane (CH_4), and I denote steam. The mole fractions of these components in the bulk gas stream, i.e., at the surface of the catalyst particle, are given in the following table, along with the values of the binary diffusivities. The values of r, T, and M_A are such that $D_{A,k} = 0.22\ cm^2/s$. An "average" value of $D_{A,t}$ is desired.

Species	Mole fraction in bulk gas	Molecular diffusivity, D_{Ai} (cm^2/s)
A (C_3H_8)	0.30	—
B (C_2H_4)	0.05	0.37
C (CH_4)	0.05	0.86
I (H_2O)	0.60	0.75

For this problem, $\nu_A = -1$, $\nu_B = 1$, $\nu_C = 1$, and $\nu_I = 0$. These stoichiometric coefficients, along with the concentrations and binary diffusivities in the above table, plus the given value of $D_{A,k}$, can be used to predict a value of $D_{A,t}$ at the open end of the pore (i.e., at the surface of the catalyst particle) using Eqn. (9-22). The result is $D_{A,t}(\text{surface}) = 0.15\ cm^2/s$.

Next, we need to calculate a value of $D_{A,t}$ at the sealed end of the pore, which corresponds to the center of the catalyst particle. To make this calculation, values of y_i at the sealed end are required. However, these values depend on the effectiveness factor. If η is close to 1, the mole fractions at the mouth of the pore and at the sealed end of the pore will be similar. If $\eta \ll 1$, the mole fraction of reactant A will approach 0 at the sealed end of the pore, if the reaction is irreversible.

Unfortunately, the value of the effectiveness factor is not known. In fact, the reason for calculating the value of $D_{A,t}$ is so that η can be estimated. Nevertheless, two limiting estimates of the average value of $D_{A,t}$ can be made. If $\eta \cong 1$, the concentration will not vary significantly along the length of the pore, and neither will the diffusivity. For this case, $D_{A,t}(\text{average}) = D_{A,t}(\text{surface}) = D_{A,t}(\text{sealed end}) = 0.15\ cm^2/s$.

The second limit is when $\eta \ll 1$, so that $y_A \cong 0$ at the sealed end of the pore. This case provides the largest possible variation of composition along the length of the pore. Values of y_B, y_C, and y_I at the sealed end of the pore are needed to calculate $D_{A,t}(\text{sealed})$. To obtain these values, consider an arbitrary plane that intersects the pore normal to the z-direction, as shown in the preceding figure. At steady state, by stoichiometry, the flux of A crossing the plane must be equal to the flux of B crossing the plane, but the two fluxes must be in opposite directions (reactant A is moving into the pore and product B is moving out). Mathematically,

$$-D_{A,t}(dC_A/dz) = D_{B,t}(dC_B/dz)$$

Let's integrate this expression from the sealed end of the pore to the surface, neglecting any variation of $D_{A,t}$ and $D_{B,t}$ with position. Since we are only trying to *estimate* the composition at the sealed end of the pore, this assumption is reasonable. After integrating and dividing by the total concentration, the result is

$$y_B(0) = y_B(s) + \left(\frac{D_{A,t}}{D_{B,t}}\right)[y_A(s) - y_A(0)]$$

Here (0) denotes the sealed end of the pore and (s) denotes the open end of the pore. For $y_A(0) = 0$,

$$y_B(0) = y_B(s) + \left(\frac{D_{A,t}}{D_{B,t}}\right) y_A(s)$$

We now assume that $D_{A,t}/D_{B,t} = (M_B/M_A)^{1/2} = (28/44)^{1/2} = 0.80$. This will be true if diffusion is predominantly in the Knudsen regime, and will be approximately true if molecular diffusion predominates.

Substituting $y_B(s) = 0.05$, $y_A(s) = 0.30$, and $D_{A,t}/D_{B,t} = 0.80$ into the above equation gives $y_B(0) = 0.29$. A similar calculation gives $y_C(0) = 0.23$. We take the value of $y_I(0)$ to be $y_I(0) = 1 - y_A(0) + y_B(0) + y_C(0) = 0.48$. This procedure is not strictly valid for a gas-phase reaction if the total pressure changes along the length of the pore. However, this approach is usually reasonable.

Substitution of these values of $y_i(0)$, along with the data in the above table, the stoichiometric coefficients, and the value of $D_{A,k}$ into Eqn. (9-22) results in $D_{A,t}(\text{sealed}) = 0.16$ cm^2/s. Taking $D_{A,t}(\text{average}) = [D_{A,t}(\text{surface}) + D_{A,t}(\text{sealed})]/2$ gives $D_{A,t}(\text{average}) = 0.16$ cm^2/s. For this example, there is essentially no difference between the diffusion coefficient at the surface and in the center of the particle. This is primarily because about 70% of the resistance to diffusion comes from the Knudsen regime, where the diffusion coefficient is independent of fluid composition.

EXERCISE 9-3

Repeat the calculations of $D_{A,t}$(surface) and $D_{A,t}$(sealed) assuming that r, T, and M_A are such that $D_{A,k} = 2.0$ cm^2/s.

Answers: $D_{A,t}(\text{surface}) = 0.38$ cm^2/s, $D_{A,t}(\text{sealed}) = 0.46$ cm^2/s, and $D_{A,t}(\text{average}) = 0.42$ cm^2/s. In this case, only about 20% of the resistance to diffusion comes from the Knudsen regime, and the diffusion coefficients reflect a stronger dependence on concentration. However, the diffusivity values are sufficiently close that taking the arithmetic average of the two, for use in calculating the Thiele modulus, is reasonable.

9.3.4.3 The Effect of Pore Size

Up to this point, attention has been focused on calculating diffusion coefficients in straight, round pores. The final challenge is to use this background to calculate $D_{A,p}(r)$ in Eqn. (9-16). Recall that $D_{A,p}(r)$ is the diffusion coefficient of A in an *assembly* of straight, round pores that has the same distribution of pore sizes as the catalyst. To perform this calculation, something must be known about the pore-size distribution $f(r)$.

Narrow Pore-Size Distribution In many cases, the exact pore-size distribution is not available. Nevertheless, an approximate calculation of the effectiveness factor may be necessary. Such a calculation can be performed if we assume that $D_{A,p}(r)$ is equal to $D_A(\bar{r})$, where $\bar{r}$ is the average pore radius of the catalyst. In words, we assume that the average diffusion coefficient is equal to the diffusion coefficient in a pore with radius $\bar{r}$, the average pore radius of the catalyst. A value of $\bar{r}$ can be calculated as described in Section 9.2.2.2.

Once the value of $\bar{r}$ has been determined, $D_A(\bar{r})$ can be calculated from Eqn. (9-22), taking the concentration dependence of the diffusion coefficient into account, if necessary. The effective diffusivity $D_{A,\text{eff}}$ is then obtained by multiplying $D_A(\bar{r})$ by ε/τ_p, i.e.,

$$D_{A,\text{eff}} = \varepsilon D_A(\bar{r})/\tau_p$$

This approach to calculating the effective diffusivity works fairly well if the catalyst has a narrow distribution of pore radii. However, a more rigorous approach is required if the pore-size distribution is broad, for example, bimodal.

Broad Pore-Size Distribution Several models have been developed to estimate the effective diffusivity in catalysts with broad pore-size distributions.[4] One of the simplest is the parallel pore model of Johnson and Stewart.[5] In their approach, the value of $D_{A,p}(r)$ is calculated by averaging over the whole range of pore sizes:

$$D_{A,p}(r) = \int_{r=0}^{\infty} D_{A,t}(r) f(r) dr$$

For values of r where diffusion is in the Knudsen regime, $D_{A,t}(r)$ does not depend on composition. However, if r is such that bulk diffusion is important, a concentration-average value of $D_{A,t}(r)$ should be calculated for each r before the integration is performed.

9.3.5 Use of the Effectiveness Factor in Reactor Sizing and Analysis

The effectiveness factor can be used to account for internal transport effects in the sizing and analysis of heterogeneous catalytic reactors. At this point, it is no longer necessary to assume that internal concentration gradients are negligible.

The differential form of the design equation for an ideal plug flow reactor in which a heterogeneous catalytic reaction takes place is

$$\frac{dW}{F_{A0}} = \frac{dx_A}{-r_A} \tag{3-27a}$$

The reaction rate $-r_A$ now can be written as $\eta \times (-r_A^{surf})$, where $-r_A^{surf}$ is shorthand for the rate that would exist if there were no gradients inside the catalyst particle.

$$\frac{dW}{F_{A0}} = \frac{dx_A}{\eta \times (-r_A^{surf})} \tag{9-23}$$

If the external transport resistances are negligible, the conditions at the surface of the catalyst particle will be the same as those in the bulk fluid stream. In other words, $T_s = T_B$, $C_{A,s} = C_{A,B}$, etc. Equation (9-23) then can be written

$$\frac{dW}{F_{A0}} = \frac{dx_A}{\eta \times (-r_A^{bulk})} \tag{9-24}$$

Here, $-r_A^{bulk}$ is the rate that would exist if the temperature and the concentrations throughout the particle were the same as those in the bulk fluid. For example, if the reaction were first-order in A and irreversible, the above equation would become

$$\frac{dW}{F_{A0}} = \frac{dx_A}{\eta k(T_B) C_{A,B}}$$

To solve the design equation, we must know how η and $-r_A^{surf}$ depend on temperature and composition. For an nth-order reaction, the effectiveness factor is a function of a single dimensionless variable, the Thiele modulus, as given by Eqns. (9-13) and (9-13a). The Thiele modulus depends on temperature because the defining equations contain k_v, K_{eq}, $D_{A,eff}$, and β, all of which are temperature dependent. The Thiele modulus may also depend

[4] Froment, G. B. and Bischoff, K. B., *Chemical Reactor Analysis and Design*, 2nd edition, John Wiley & Sons, (1990), pp. 145–157.

[5] Johnson, M. F. L. and Stewart, W. E., *J. Catal.* 4, 248 (1965).

on the concentrations in the bulk fluid stream, since Eqns. (9-13) and (9-13a) contain the parameters $C_{A,s}$, $D_{A,eff}$, α_s, and β. However, the effectiveness factor *for an irreversible, first-order reaction* may not be very sensitive to concentration because, for this case, the Thiele modulus does not contain $C_{A,s}$, α_s, or β. For an irreversible, first-order reaction, the only sensitivity to concentration arises from the effective diffusion coefficient.

For a plug-flow reactor, the concentration of A (and all other species) in the bulk fluid and at the surface of the catalyst particle will depend on where the catalyst particle is located in the reactor. If the external transport resistances are negligible, the surface concentrations are the same as the bulk concentrations at every point. Moreover, the bulk concentrations can be written as functions of x_A, as we did in Chapter 4. For example, for a reaction that occurs at constant density,

$$C_{A,s} = C_{A,B} = C_{A0}(1 - x_A)$$

For more complex rate equations, and for situations where the fluid density is not constant, the stoichiometric table approach, as developed in Chapter 4, can be used to relate the various bulk concentrations to x_A.

The following example illustrates how the design equation can be solved for an isothermal reactor when the external transport resistances are negligible and the internal transport resistance is significant.

EXAMPLE 9-4
Calculation of Required Catalyst Weight when $\eta < 1$

The irreversible gas-phase reaction $A \rightarrow R$ is being carried out at steady state in a fixed-bed catalytic reactor that operates as an ideal, plug-flow reactor (PFR). The reaction is second-order in A, i.e., $-r_A = kC_A^2$, and the value of the rate constant is 2.5×10^{-4} m^6/mol-kg·cat-s. The reactor is isothermal.

The concentration of A in the feed to the reactor is 12 mol/m^3 and the volumetric flow rate is 0.50 m^3/s. The total pressure is 1 atm, and pressure drop through the reactor can be neglected. Assume that the external transport resistances are negligible and that the catalyst particles are isothermal.

The catalyst particles are spherical, with a radius (R) of 0.3 cm. The particle density (ρ_p) is 3000 kg/m^3 and the effective diffusivity of A in the catalyst particle is 10^{-2} cm^2/s.

A. Estimate the values of the effectiveness factors at the bed inlet and the bed outlet.

B. What weight of catalyst is required to achieve a conversion of 90%?

Part A: Estimate the values of the effectiveness factors at the bed inlet and the bed outlet.

APPROACH

The value of the generalized Thiele modulus ϕ will be calculated at the bed inlet and the bed outlet using Eqn. (9-14). The characteristic dimension will be calculated from Eqn. (9-10), and the rate constant will be converted from a weight-of-catalyst basis to a volume-of-catalyst basis using the given particle density. Equation (9-15) or Figure 9-7 can then be used to estimate the effectiveness factor at each position.

SOLUTION

The characteristic dimension of the catalyst particle is

$$l_c = V_G/A_G = (4/3)\pi R^3/4\pi R^2 = R/3 = 0.1 \text{ cm} = 0.001 \text{ m}$$

The rate constant based on catalyst *volume* is

$$k_v(\text{m}^3/\text{mol-s}) = k\,(\text{m}^6/\text{mol-kg·cat-s})\ \rho_p\ (\text{kg/m}^3)$$

$$k_v = 2.5 \times 10^{-4} \times 3000 = 0.75\ (\text{m}^3/\text{mol-s})$$

For an irreversible, second-order reaction, Eqn. (9-14) becomes

$$\phi = l_c \left(\frac{3k_V C_{A,s}}{2D_{A,eff}}\right)^{1/2}$$

If the external transport resistance is insignificant, the concentration of A at the surface of the catalyst particle at every point in the catalyst bed is the same as the concentration in bulk gas stream, i.e., $C_{A,s} = C_{A0}(1 - x_A)$. At the bed inlet, where $x_A = 0$, $C_{A,s} = C_{A0} = 12\ \text{mol/m}^3$.

$\phi = 0.001(\text{m})[3 \times 0.75(\text{m}^3/\text{mol-s}) \times 12\,(\text{mol/m}^3)/2 \times 10^{-2}\,(\text{cm}^2/\text{s}) \times 10^{-4}\,(\text{m}^2/\text{cm}^2)]^{1/2}$
$\phi = 3.67$

From Eqn. (9-15),

$$\eta = \tanh\phi/\phi = \tanh(3.67)/3.67 = 0.27$$

At the bed outlet, $x_A = 0.90$, so that $C_{A,s} = 1.2\ \text{mol/m}^3$

$\phi = 0.001(\text{m})[3 \times 0.75\,(\text{m}^3/\text{mol-s}) \times 1.2\,(\text{mol/m}^3)/2 \times 10^{-2}(\text{cm}^2/\text{s}) \times 10^{-4}\,(\text{m}^2/\text{cm}^2)]^{1/2}$
$\phi = 1.16$

Again, from Eqn. (9-15),

$$\eta = \tanh\phi/\phi = \tanh(1.16)/1.16 = 0.71$$

For this second-order reaction, the effectiveness factor changes substantially from the inlet of the catalyst bed to the outlet, because the Thiele modulus depends on the concentration of A at the external surface of the catalyst particle $C_{A,s}$.

Part B: What weight of catalyst is required to achieve a conversion of 90%?

APPROACH The design equation, as given by Eqn. (9-24), will be solved for the catalyst weight W. Since η is a function of C_A, η will be calculated for various values of C_A between 12 and 1.2 mol/m^3, and the design equation will be integrated numerically.

SOLUTION For this problem, Eqn. (9-24) becomes

$$\frac{dW}{F_{A0}} = \frac{dx_A}{\eta(x_A)kC_{A0}^2(1 - x_A)^2}$$

The parenthesis (x_A) after η is a reminder that the effectiveness factor depends on conversion, as shown by the calculations in Part A. Integrating the above equation from $W = 0$, $x_A = 0$ to $W = W$, $x_A = 0.90$,

$$W = (F_{A0}/kC_{A0}^2)\int_0^{0.90} \frac{dx_A}{\eta(x_A)(1 - x_A)^2} \qquad (9\text{-}25)$$

The value of the Thiele modulus at any point along the length of the catalyst bed is given by

$$\phi(x_A) = l_c\left(\frac{3k_V C_{A0}(1 - x_A)}{2D_{A,eff}}\right)^{1/2}$$

The value of $\eta(x_A)$ was calculated from $\phi(x_A)$ via Eqn. (9-15). Finally, the value of the integral in Eqn. (9-25) was obtained by numerical integration and was multiplied by F_{A0}/kC_{A0}^2. The values of k and C_{A0} are given. The value of F_{A0} was calculated from $F_{A0} = \upsilon C_{A0}$, since υ is given. The final result is $W = 2850$ kg.

9.3.6 Diagnosing Internal Transport Limitations in Experimental Studies

9.3.6.1 Disguised Kinetics

If a kinetic study is carried out under conditions where the resistance to internal transport is high, the experimental data will be very misleading!

When ϕ is greater than about 10, $\eta \cong 1/\phi$. Substituting this relationship into Eqn. (9-2) gives

$$\left\{ \begin{matrix} \text{actual reaction} \\ \text{rate} \end{matrix} \right\} = \frac{1}{\phi} \times \left\{ \begin{matrix} \text{rate with} \\ \text{no internal gradients} \end{matrix} \right\}$$

The "actual reaction rate" is the rate that is measured experimentally in the laboratory.

Let's consider an irreversible, nth-order reaction. The Thiele modulus is given by

$$\phi = l_c \left(\frac{(n+1) k_v C_{A,s}^{n-1}}{2 D_{A,eff}} \right)^{1/2} \tag{9-14}$$

On a volumetric basis, the "rate with no internal gradients" is $k_v C_{A,s}^n$. Combining these relationships gives

$$\left\{ \begin{matrix} \text{actual reaction} \\ \text{rate} \end{matrix} \right\} = \frac{(k_v D_{A,eff})^{1/2} C_{A,s}^{(n+1)/2}}{l_c [(n+1)/2]^{1/2}}$$

Now, suppose that the external transport resistance is negligible, so that $C_{A,s} = C_{A,B}$. The above equation becomes

$$\left\{ \begin{matrix} \text{actual reaction} \\ \text{rate} \end{matrix} \right\} = \frac{(k_v D_{A,eff})^{1/2} C_{A,B}^{(n+1)/2}}{l_c [(n+1)/2]^{1/2}} \tag{9-26}$$

Equation (9-26) shows the behavior that will be measured experimentally if the catalyst operates in the regime of strong pore diffusion influence and if external transport resistances are negligible.

Let's use Eqn. (9-26) to examine three aspects of catalyst performance: (1) the effect of concentration; (2) the effect of temperature; and (3) the effect of changing the size of the catalyst particle.

Effect of Concentration Equation (9-26) shows that the observed rate is proportional to $C_{A,B}$ *raised to the* $(n+1)/2$ *power*, in the region of strong pore diffusion influence. In other words, the *observed* order with respect to A will be $(n+1)/2$, while the *true* or *intrinsic* order is n. For example, if the true order is 2, a 3/2 order will be observed experimentally when ϕ is large. This phenomenon is referred to as *disguised* or *falsified* kinetics (*disguised* or *falsified* reaction order). In the regime of strong pore diffusion influence, the experimental data will show the correct (true or intrinsic) concentration dependence only when the reaction is first order ($n = 1$). Otherwise, the order will be *falsified.*

Effect of Temperature The rate constant k_v and the effective diffusion coefficient $D_{A,eff}$ are the only parameters in Eqn. (9-26) that are sensitive to temperature. Let E_{kin} be the true (intrinsic) activation energy for k_v. In general, the effective diffusivity will not follow an Arrhenius-type relationship over a wide range of temperature. However, the temperature dependence of $D_{A,eff}$ can be approximated with an exponential in $1/T$, at least over a small range of temperature. Let E_{diff} be the activation energy for $D_{A,eff}$ over the temperature range in question.

Equation (9-26) shows that the *apparent* activation energy of the reaction, i.e., the activation energy that would be measured experimentally, is

$$E_{app} = (E_{kin} + E_{diff})/2$$

The value of $D_{A,eff}$ generally is not very sensitive to temperature. If diffusion is primarily in the molecular or Knudsen regime, E_{diff} is only about 5–20 kJ/mol. On the other hand, the activation energies for chemical reactions are much higher, of the order of 50–300 kJ/mol. If $E_{kin} > E_{diff}$,

$$E_{app} \cong E_{kin}/2$$

provided that the Thiele modulus is high ($\phi > 10$) and that the external transport resistance is negligible.

If the rate constant for a heterogeneous catalytic reaction was measured over a wide range of temperature, using catalyst particles of a fixed size, an Arrhenius plot might have the characteristics shown in Figure 9-8.

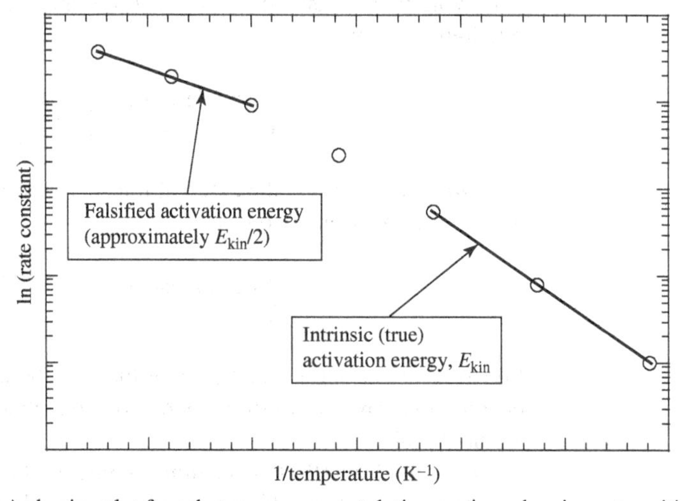

Figure 9-8 Arrhenius plot for a heterogeneous catalytic reaction, showing a transition from the intrinsic kinetic regime at low temperatures to a regime of strong pore diffusion influence at higher temperatures.

At very low temperatures, k_v has a low value and the value of ϕ is low. Pore diffusion is not a significant resistance and the effectiveness factor is essentially 1. Intrinsic (true) kinetics are observed. The measured activation energy is E_{kin}.

As the reaction temperature is increased, the value of k_v increases more rapidly than the value of $D_{A,eff}$, since the activation energy of k_v is much greater than that of $D_{A,eff}$. Therefore, ϕ increases. The effect of pore diffusion becomes more important and the effectiveness factor drops below 1. These trends continue until the value of ϕ is large, and the measured reaction rate is given by Eqn. (9-26). At this point, the measured activation energy is approximately $E_{kin}/2$, i.e., the apparent activation energy is only about half of the true activation energy.

Effect of Particle Size When $\eta = 1$, the rate per unit weight of catalyst (or the rate per unit volume of catalyst) does not depend on the catalyst particle size. However, when the effectiveness factor is low, Eqn. (9-26) shows that the "actual reaction rate" is inversely proportional to l_c, the characteristic dimension of the particle. Consequently, the actual reaction rate (moles/volume-time or moles/weight-time) can be increased by using smaller particles. In the region of strong pore diffusion influence ($\phi >$ ca. 10), the actual reaction rate can be doubled by reducing l_c by a factor of 2.

9.3.6.2 The Weisz Modulus

Suppose that an experiment has been carried out to measure the reaction rate for a heterogeneous catalyst at some specific set of experimental conditions. We might like to perform a calculation to determine whether internal transport had a significant impact on the rate that was measured. However, the catalyst being tested is a new, experimental catalyst, and the value of k_v is not known. In fact, the objective of the experiment might have been to measure k_v. In any event, the Thiele modulus cannot be calculated *a priori*, and Figure 9-7 is not directly useful.

The problem of evaluating the influence of pore diffusion on an experimental result can be simplified through some transformations of the previous equations. Suppose that a reaction rate has been measured in some kind of experimental reactor, preferably an ideal CSTR or a differential PFR. From the experimental data, a rate of reaction per unit of geometrical catalyst volume, designated $-R_{A,v}$, can be calculated. The "v" in the subscript indicates that this is a *volumetric* rate of reaction. The measured rate $(-R_{A,v})$ is not necessarily the same as the intrinsic rate, expressed on a volumetric basis $(-r_{A,v})$. The measured rate may reflect internal transport effects, whereas the intrinsic rate does not.

The measured rate can be expressed in terms of the effectiveness factor.

$$\frac{\text{measured rate}}{\text{geometric catalyst volume}} = -R_{A,v} = \eta \times \{\text{rate with no internal gradients}\}$$

For a reversible reaction, A $\rightleftarrows$ B, that is nth order in both directions,

$$-R_{A,v} = \eta k_v \left[C_{A,s}^n - \frac{C_{B,s}^n}{K_{eq}^C} \right] = \eta k_v C_{A,s}^n \left[1 - \frac{\alpha_s^n}{K_{eq}^C} \right]$$

where α_s is defined by Eqn. (9-11). Multiplying both sides by $(n+1)l_c^2\Psi^2/2D_{A,eff}C_{A,s}$, where Ψ is defined by Eqn. (9-13a),

$$\frac{(n+1)l_c^2(-R_{A,v})\Psi^2}{2D_{A,eff}C_{A,s}} = \eta l_c^2 \left(\frac{(n+1)k_v C_{A,s}^{n-1}}{2D_{A,eff}} \right) \left[\frac{K_{eq}^C - \alpha^n}{K_{eq}^C} \right] \Psi^2 \tag{9-27}$$

From Eqn. (9-13),

$$\phi^2 = l_c^2 \left(\frac{(n+1)k_v C_{A,s}^{n-1}}{2D_{A,eff}} \right) \Psi^2 \tag{9-28}$$

Combining Eqns. (9-27) and (9-28) gives

$$\frac{(n+1)l_c^2(-R_{A,v})}{2D_{A,eff}C_{A,s}} \Psi^2 = \eta \phi^2 \left(\frac{K_{eq}^C - \alpha^n}{K_{eq}^C} \right) \tag{9-29}$$

Let the generalized Weisz modulus Φ be defined by

Definition of Weisz modulus

$$\Phi \equiv \frac{(n+1)l_c^2(-R_{A,v})}{2D_{A,eff}C_{A,s}} \left(\frac{K_{eq}^C}{K_{eq}^C - \alpha^n} \right) \Psi^2 \tag{9-30}$$

Then Eqn. (9-29) becomes

$$\Phi = \eta \phi^2$$

A graph of η versus Φ can now be constructed, as follows: a value of ϕ is assumed, η is calculated from Eqn. (9-15) and Φ is calculated from $\Phi = \eta\phi^2$. The process is repeated until the η versus Φ relationship has been defined over a range of Φ. A graph of η versus Φ is shown as Figure 9-9. For values of Φ greater than 10, $\eta \cong 1/\Phi$.

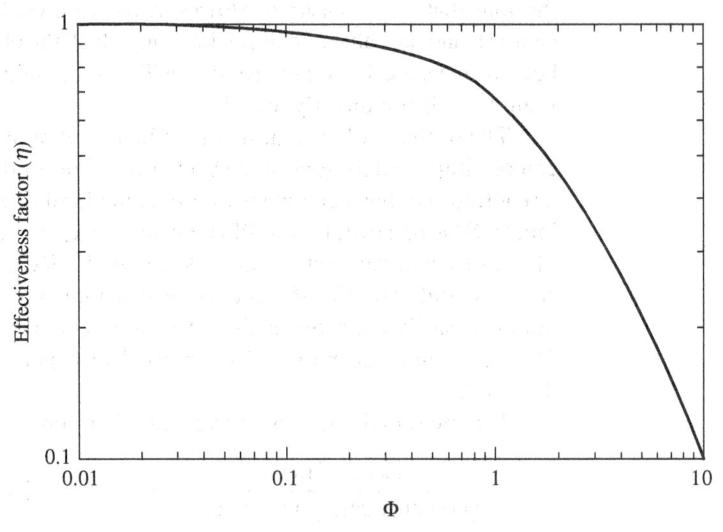

Figure 9-9 The effectiveness factor η as a function of the generalized Weisz modulus Φ. When $\Phi > 10$, $\eta = 1/\Phi$.

Since $-R_{A,v}$ is the *measured* rate of reaction, the value of Φ can be calculated directly from experimental data, as long as the other parameters in Eqn. (9-30) are known or can be estimated. It is not necessary to know the value of the intrinsic rate constant k_v in order to calculate Φ.

The use of the Weisz modulus to estimate the effectiveness factor directly from experimental data is illustrated in the following example.

EXAMPLE 9-5
Use of the Weisz Modulus

Satterfield et al.[6] studied the liquid-phase hydrogenation of α-methyl styrene to cumene over a 1 wt.% Pd/alumina catalyst. The catalyst was spherical, with a diameter of 0.825 cm and a porosity ε of 0.50. The tortuosity τ_p was measured and found to be approximately 8.

At 50 °C and 1 atm of H_2 pressure, the measured rate of disappearance of α-methyl styrene was 3.4×10^{-7} g·mol/s-cc (particle). The solubility of H_2 in α-methyl styrene at these conditions is 3.5×10^{-6} mol/cc and the diffusivity of H_2 in α-methyl styrene is 1.7×10^{-4} cm^2/s. The reaction is irreversible, first order in H_2, and zero order in α-methyl styrene.

If the external transport resistances are negligible, what is the approximate value of the effectiveness factor?

APPROACH

Since the intrinsic rate constant is not known, the effectiveness factor will be estimated via the Weisz modulus. A value of Φ will be calculated via Eqn. (9-30), and Figure 9-9 will be used to estimate η.

SOLUTION

The reaction is essentially irreversible so that $[K_{eq}^C/(K_{eq}^C - \alpha_s^n)]\Psi^2 = 1$. Since the external transport resistance is negligible, the concentration of H_2 at the external surface of the catalyst particle is equal to the solubility of H_2 in α-methyl styrene, i.e., $C_{H_2,s} = 3.5 \times 10^{-6}$ mol/cc. In the hydrogenation of

[6] Satterfield, C. N., Pelossof, A. A., and Sherwood, T. K., *AIChE J.*, 15 (2), 226–234 (1969).

α-methyl styrene to cumene, 1 mol of H_2 reacts with 1 mol of α-methyl styrene. Therefore, the value of $(-R_{H_2,v})$ is 3.4×10^{-7} mol/s-cc (particle). The value of the effective diffusivity of H_2 is

$$D_{H_2,\text{eff}} = \frac{\varepsilon D_{H_2/\alpha MS}}{\tau} = \frac{0.50 \times 1.7 \times 10^{-4}}{8} = 1.1 \times 10^{-5}\ \text{cm}^2/\text{s}$$

Finally, the characteristic dimension l_c is $(0.825/2)/3 = 0.14$ cm.

The value of Φ is

$$\Phi = (1+1) \times (0.14)^2 \times 3.4 \times 10^{-7}/2 \times 1.1 \times 10^{-5} \times 3.5 \times 10^{-6} = 170$$

This value is in the asymptotic region of Figure 9-9, where $\eta \cong 1/\Phi$. The value of η is approximately $1/170 = 0.0059$.

9.3.6.3 Diagnostic Experiments

Suppose that you were studying the behavior of a catalyst in the laboratory and suspected that internal transport was influencing the observed reaction rate. What *experiments* could you run to test whether pore diffusion was an important resistance?

Figure 9-7 contains the answer. If $\eta \cong 1$, the reaction rate per unit volume of catalyst (or per unit weight of catalyst) is not sensitive to ϕ. However, if $\eta < 1$, the rate will depend on the value of ϕ. Therefore, we need to vary ϕ and see if the measured rate changes.

Unfortunately, life is not quite that simple. The Thiele modulus contains some variables that influence the rate, even if $\eta = 1$. Recall the definition of ϕ that applies to essentially irreversible reactions.

$$\phi = l_c \left(\frac{(n+1) k_v C_{A,s}^{(n-1)}}{2 D_{A,\text{eff}}} \right)^{1/2} \tag{9-14}$$

If either k_v or $C_{A,s}$ is changed, the measured rate will change, even if $\eta = 1$, since the "rate with no internal gradients" depends on k_v and $C_{A,s}$. However, if the value of ϕ were changed by varying either $D_{A,\text{eff}}$ or l_c, the "rate with no internal gradients" would not be affected. Thus, if the measured reaction rate was sensitive to l_c or $D_{A,\text{eff}}$, it would indicate that the effectiveness factor was less than 1.

In a practical sense, it is difficult to vary $D_{A,\text{eff}}$ without varying the rate constant k_v. The pore structure of the catalyst particle could be changed during catalyst synthesis. However, in the process of creating more and/or larger pores, the BET surface area might be changed in a way that changed the surface area of the active component of the catalyst, e.g., the metal that was deposited on the walls of the pores. A second consideration is that the value of ϕ depends on the square root of $D_{A,\text{eff}}$, so that a relatively large change in $D_{A,\text{eff}}$ is required to make a relatively modest change in ϕ.

The classical diagnostic experiment to test for the presence of a significant pore diffusion resistance, in the absence of an external resistance, is to vary the size of the catalyst particle l_c. The following example illustrates this approach.

EXAMPLE 9-6
Estimating Effectiveness Factors from Experimental Data

Weisz and Swegler[7] studied the dehydrogenation of cyclohexane to cyclohexene and benzene using a chromia–alumina catalyst in the form of approximately spherical particles. The temperature was 479 °C, the total pressure was atmospheric, and the mole fraction of cyclohexane was about 0.31.

The rate of cyclohexane disappearance was measured with three different sizes of catalyst particle, with the results given in the following table. Note that these rates are per unit weight of catalyst.

[7] Weisz, P. B. and Swegler, E. W., *J. Phys. Chem.*, 59, 823–826 (1955).

Radius of catalyst particle (cm)	Rate of disappearance of cyclohexane (mol/min-g)
0.050	8.8×10^{-5}
0.184	5.7×10^{-5}
0.310	4.2×10^{-5}

Estimate a value of the effectiveness factor for each particle size, assuming that the external transport resistance was negligible in these experiments.

APPROACH

Since the radius of the catalyst particle was the only variable in these experiments, the decrease in reaction rate with increasing particle size can be attributed to the presence of a significant internal transport resistance. We can conclude that the effectiveness factor was significantly less than 1, at least for the two catalysts with the largest radii.

Values of the effectiveness factor and Thiele modulus for each particle size can be estimated from the above data, provided that *all three* experiments are not in the asymptotic region, where $\eta = 1/\phi$. We will first test to ensure that all three data points are not in the asymptotic regime by checking to see whether the measured rates are all proportional to $1/\phi$. We will then determine η and ϕ for each particle size by an iterative procedure, beginning with the assumption that $\eta = 1$ for the smallest particle. The values of η for the two larger particle sizes then will be calculated from the experimental data, and values of ϕ for these two particle sizes will be calculated from the values of η, using either Figure 9-7 or Eqn. (9-15). The value of ϕ for the smallest particle will then be calculated, and used to check the assumption that $\eta = 1$ for the smallest particle. If necessary, a new value of η for the smallest particle will be assumed, and the above process will be repeated.

SOLUTION

Let R denote the radius of a catalyst particle. Then

$$\text{rate}\,(R) = \eta(R) \times \text{rate with no gradients}$$

For the experiment where $R = 0.310$ cm,

$$4.2 \times 10^{-5} = \eta(0.310) \times \text{rate with no gradients}$$

For the experiment where $R = 0.050$ cm,

$$8.8 \times 10^{-5} = \eta(0.050) \times \text{rate with no gradients}$$

Since the experiments were conducted under conditions where the "rate with no gradients" was the same for all three particle sizes,

$$\frac{\eta(0.310)}{\eta(0.050)} = \frac{4.2 \times 10^{-5}}{8.8 \times 10^{-5}} = 0.48$$

If both experiments had been in the asymptotic region, where $\eta \cong 1/\phi$,

$$\frac{\eta(0.310)}{\eta(0.050)} = \frac{l_c(0.050)}{l_c(0.310)} = \frac{0.050}{0.310} = 0.16$$

The actual ratio of effectiveness factors is much higher than the ratio predicted by assuming that both catalysts operate in the asymptotic regime. Therefore, at least the smaller of these two catalysts must have operated outside the asymptotic region.

To estimate the actual effectiveness factors for the above experiments, assume that $\eta = 1$ for the smallest particle ($R = 0.050$ cm). Then,

$$\eta(R) = \eta(0.050) \times \frac{\text{rate}\,(R)}{\text{rate}\,(0.050)} = (1.0)\frac{\text{rate}\,(R)}{\text{rate}\,(0.050)}$$

The effectiveness factors for the two largest particles can be calculated from this relationship, using the experimental data. Then the Thiele moduli for these two particle sizes can be calculated from Eqn. (9-15) or Figure 9-7. The results are

$$\eta(0.184) = 0.65; \quad \phi = 1.35$$
$$\eta(0.310) = 0.48; \quad \phi = 2.02$$

The Thiele modulus for the smallest particle now can be estimated, since the characteristic dimension l_c is the only parameter in the Thiele modulus that changes as the particle size is changed.

$$\phi(0.050) = \phi(0.184) \times l_c(0.050)/l_c(0.184) = 0.37$$
$$\phi(0.050) = \phi(0.310) \times l_c(0.050)/l_c(0.310) = 0.33$$

Finally, using these values of $\phi(0.050)$, the value of $\eta(0.050)$ can be calculated from Eqn. (9-15) and compared with the original assumption of $\eta(0.050) = 1$. For $\phi = 0.37$, $\eta = 0.96$, and for $\phi = 0.33$, $\eta = 0.97$.

Although this is probably close enough for practical purposes, a second iteration was performed, assuming that $\eta = 0.96$ for the smallest particle. The result is

Particle radius (cm)	Thiele modulus (ϕ)	Effectiveness factor (η)
0.050	0.36	0.96
0.184	1.43	0.62
0.310	2.12	0.46

The calculation has converged to the values in the above table.

9.3.7 Internal Temperature Gradients

As noted earlier, temperature gradients can exist inside porous catalyst particles. The magnitude of these gradients is related to the magnitude of the concentration gradients inside the particle, since both are proportional to the rate of the reaction taking place.

Solving Eqns. (9-4) and (9-5) for $(-r_{A,v})$ and equating the results gives

$$\frac{1}{r^2}\frac{\partial}{\partial r}\left(r^2 D_{A,\text{eff}}\frac{\partial C_A}{\partial r}\right) = -r_{A,v} = \frac{-1}{r^2(-\Delta H_R)}\frac{\partial}{\partial r}\left(r^2 k_{\text{eff}}\frac{\partial T}{\partial r}\right)$$

Integrating from $r = 0$ to $r = r$, using the boundary conditions given by Eqns. (9-4a) and (9-5a), and simplifying

$$-k_{\text{eff}}\frac{\partial T}{\partial r} = (-\Delta H_R)D_{A,\text{eff}}\frac{\partial C_A}{\partial r}$$

Assuming that k_{eff}, $D_{A,\text{eff}}$, and ΔH_R are constant, this expression can be integrated from $r = 0$ to $r = R$, using the boundary conditions of Eqns. (9-4b) and (9-5b).

$$T(0) - T_s = \left[\frac{(-\Delta H_R)D_{A,\text{eff}}}{k_{\text{eff}}}\right](C_{A,s} - C_A(0))$$

In this expression, $T(0)$ is the temperature at the center of the catalyst particle, $r = 0$, and T_s is the temperature at the external surface, $r = R$. Similarly, $C_A(0)$ is the concentration of A at $r = 0$, and $C_{A,s}$ is the concentration at $r = R$. Although this equation was derived for a spherical particle, it is valid for any symmetric particle.

If the reaction is exothermic, the temperature at the center of the catalyst particle is higher than at the surface, i.e., $T(0) > T_s$. If the reaction is endothermic, the opposite is true, i.e., $T(0) < T_s$. In either case, $T(0) - T_s$ is proportional to $C_{A,s} - C_A(0)$. As the effectiveness factor decreases, $C_A(0)$ decreases and $(T(0) - T_s)$ increases.

The largest absolute value of $T(0) - T_s$ occurs when $C_A(0) \cong 0$, i.e., when the effectiveness factor is very low. This value is given by

$$(T(0) - T_s)_{\max} = \left[\frac{(-\Delta H_R) D_{A,\text{eff}} C_{A,s}}{k_{\text{eff}}}\right]$$

The subscript "max" denotes the largest absolute value of the temperature difference.

If the external mass-transfer resistance is negligible $(C_{A,s} \cong C_{A,B})$

Maximum ΔT in catalyst particle

$$(T(0) - T_s)_{\max} = \left[\frac{(-\Delta H_R) D_{A,\text{eff}} C_{A,B}}{k_{\text{eff}}}\right] \tag{9-31}$$

Equation (9-31) can be used to estimate the maximum (or minimum) value of $T(0)$, if values of $D_{A,\text{eff}}$ and k_{eff} are available. Estimation of $D_{A,\text{eff}}$ is discussed in Section 9.3.4. The theory for estimating k_{eff} is not as well developed. In the absence of data on the catalyst of interest, a value of $k_{\text{eff}} \cong 5 \times 10^{-4}$ cal/s-cm-K can provide a reasonable starting point for many catalysts that involve a ceramic support.

Let ρ be the density of the fluid in the pores of the catalyst and $c_{p,m}$ be the mass heat capacity of that fluid, so that $(\rho c_{p,m})_f$ is the volumetric heat capacity of the fluid. Equation (9-31) then can be written as

$$(T(0) - T_s)_{\max} = \left[\frac{(-\Delta H_R) C_{A,B}}{(\rho c_{p,m})_f}\right] \Bigg/ \left[\frac{k_{\text{eff}}}{D_{A,\text{eff}} (\rho c_{p,m})_f}\right]$$

The numerator $(-\Delta H_R) C_{A,B}/(\rho c_{p,m})_f$ is the adiabatic temperature change of the fluid (ΔT_{ad}) as discussed in Section 8.4.3.

Now, multiply the denominator of the above expression by $(k_{th,f}/k_{th,f})$ and by $(D_{A,m}/D_{A,m})$, where $k_{th,f}$ is the thermal conductivity *of the fluid in the pores of the catalyst* and $D_{A,m}$ is the *molecular* diffusivity of A in the fluid mixture, as discussed in the section "Bulk (Molecular) Diffusion" in Section 9.3.4.2.

$$(T(0) - T_s)_{\max} = \frac{\Delta T_{ad}}{\left(\dfrac{k_{\text{eff}}}{k_{th,f}}\right)\left(\dfrac{D_{A,m}}{D_{A,\text{eff}}}\right)\left(\dfrac{k_{th,f}}{D_{A,m}(\rho c_{p,m})_f}\right)} \tag{9-32}$$

In this equation, $k_{th,f}/D_{Am}(\rho c_{p,m})_f$ is the Lewis number[8] of the fluid in the pores. The Lewis number (Le) is defined as

$$\text{Le} = \text{Sc}/\text{Pr} = (K_{th,f}/\rho c_{p,m})_f / D_{A,m}$$

Here Pr is the Prandtl number and Sc is the Schmidt number of the fluid in the pores. Physically, Le is the ratio of the thermal diffusivity $(k_{th,f}/(\rho c_{p,m})_f)$ to the molecular diffusivity $(D_{A,m})$. For pure gases, Le is of the order of unity. For pure liquids, Le is of the order of 100–1000.

The effective thermal conductivity of the catalyst particle (k_{eff}) generally is much larger than the thermal conductivity of the fluid $(k_{th,f})$. A significant fraction of heat conduction is through the solid phase of the catalyst particle, rather than through the fluid in the pores.

[8] See Bird, R. B., Stewart, W. E., and Lightfoot, E. N., *Transport Phenomena,* 2nd edition, John Wiley & Sons, Inc. (2002), p. 516.

As noted above, $k_{eff} \cong 5 \times 10^{-4}$ cal/s-cm-K for many heterogeneous catalysts. The thermal conductivity of many pure, light gases is approximately 10% of this value (ca. 5×10^{-5} cal/s-cm-K) at standard conditions. The thermal conductivity of many organic liquids is approximately 50% of the typical value of k_{eff}, ca. 3×10^{-4} cal/s-cm-K. Thus, as a very rough approximation,

$$k_{eff}/k_{th,f} \cong 10 \text{ (gases)}; \quad k_{eff}/k_{th,f} \cong 2 \text{ (liquids)}$$

The effective diffusivity is given by

$$D_{A,eff} = D_{A,p}\varepsilon/\tau_p \tag{9-16}$$

If diffusion is in the molecular regime throughout the catalyst, $D_{A,p} = D_{A,m}$. According to Eqn. (9-16), the ratio $(D_{A,m}/D_{A,eff})$ will be τ_p/ε, which is of the order of 10. For liquids, $D_{A,m}/D_{A,eff} \cong 10$ is a reasonable estimate, provided that diffusion is not configurational.

For gases, if the pores are large and diffusion is completely in the molecular regime, then $D_{A,p} = D_{A,m}$ and $D_{A,m}/D_{A,eff} \cong 10$. However, if the pores are small and diffusion is in the transition or Knudsen regime, then $D_{A,m}/D_{A,eff}$ can be substantially greater than 10. Values of 100 and even larger are possible.

Inserting this information into Eqn. (9-32) yields

$$\begin{aligned}(T(0) - T_s)_{max}(\text{liquids}) &\cong \Delta T_{ad}/[(2) \times (10) \times (100 \text{ to } 1000)] \cong \Delta T_{ad}/(2000 \text{ to } 20{,}000)\\ (T(0) - T_s)_{max}(\text{gases}) &\cong \Delta T_{ad}/[(1) \times (10) \times (10 \text{ to } 100)] \cong \Delta T_{ad}/(100 \text{ to } 1000)\end{aligned} \tag{9-33}$$

If the fluid in the catalyst pores is a liquid, the difference between the temperature at the center of the catalyst particle ($T(0)$) and the temperature at the external surface (T_s) will never be more than a few degrees. For practical purposes, the catalyst particle is isothermal for liquid-phase reactions, as we previously assumed.

For gas-phase reactions, $(T(0) - T_s)_{max}$ can be significant when ΔT_{ad} is large and $D_{A,eff}$ is high. If the reaction is endothermic, the temperature throughout the interior of the particle will be less than T_s. That will cause the actual effectiveness factor to be *lower* than for a catalyst particle that is isothermal at T_s. For this situation, the assumption of isothermality leads to an overestimate of η, i.e., η (actual) $< \eta$ (isothermal). When the reaction is endothermic, the general behavior of the η versus ϕ relationship is similar to that for the isothermal case. The effectiveness factor is 1 at very low values of ϕ and declines monotonically as the Thiele modulus increases.

For an exothermic gas-phase reaction, the temperature inside the catalyst particle is higher than T_s, as discussed in Chapter 4 and shown in Figure 9-6. This causes the actual effectiveness factor to be *higher* than for an isothermal particle. The assumption of isothermality leads to an underestimate of the effectiveness factor for this case.

Furthermore, there are several unique aspects to the behavior of the effectiveness factor with highly exothermic, gas-phase reactions. First, η can be greater than unity. Under some circumstances, the increase in reaction rate due to the higher temperature inside the catalyst particle can more than offset the decrease in reaction rate caused by the lower concentration of reactants. When this occurs, the "actual reaction rate" is *larger* than the "rate with no gradients," leading to $\eta > 1$.

Second, a catalyst particle can exhibit multiple steady states when the particle is not isothermal, i.e., η may not be uniquely determined by the value of the Thiele modulus and the other parameters that are important. A discussion of these effects is beyond the scope of this chapter. There are several excellent sources of further information.[9]

In summary, the assumption of an isothermal catalyst particle is valid for many situations, e.g., most liquid-phase reactions and those gas-phase reactions where ΔT_{ad} is not too large and $D_{A,eff}$ is relatively low ($D_{A,m}/D_{A,eff} \gg 10$). If the reaction is endothermic, the isothermal model can be used to estimate a *maximum* value of the effectiveness factor, even when ΔT_{ad} is large.

The isothermal model loses considerable utility for an exothermic, gas-phase reaction with a large ΔT_{ad} and high $D_{A,eff}$ (low $D_{A,m}/D_{A,eff}$). In this case, the isothermal model underpredicts the effectiveness factor and may fail to capture several important features of catalyst behavior. The following example provides a more quantitative procedure for testing whether the assumption of particle isothermality is appropriate.

EXAMPLE 9-7 ***Estimation of Temperature at Center of Catalyst Particle***

The gas-phase hydrogenolysis of thiophene (C_4H_4S) to normal butane (n-C_4H_{10}) and hydrogen sulfide (H_2S) takes place at 250 °C and 1 atm total pressure over a cobalt molybdate catalyst.

$$C_4H_4S + 4H_2 \rightarrow n\text{-}C_4H_{10} + H_2S$$

The mole fractions of these compounds in the bulk fluid are shown in the following table, along with some thermodynamic data. The effective diffusivity is approximately 0.020 cm^2/s. As a first approximation, assume that the ideal gas laws apply.

Species	mole fraction (y_i)	$\Delta H_{f,i}$ (298 K) (kcal/mol)	$c_{p,i}$ (298 K)[10] (cal/mol-K)
C_4H_4S	0.068	27.66	17.42
H_2	0.788	0.00	6.89
C_4H_{10}	0.072	−30.15	23.29
H_2S	0.072	−4.82	8.17

Estimate the maximum possible difference between the temperature at the center of the catalyst particle and the temperature at its external surface.

APPROACH

First, Eqn. (9-31) will be used directly to estimate $(T(0) - T_s)_{max}$. The approximate effective thermal conductivity, $k_{eff} \cong 5 \times 10^{-4}$ cal/s-cm-K, will be used for this calculation. Then, as a check, Eqn. (9-33) will be applied. Since this calculation is approximate, variation of the heat capacities ($c_{p,i}$) with temperature will be neglected.

Both approaches require that the value of ΔT_{ad} be known. To calculate this parameter, the value of ΔH_R will be calculated at 523 K using thermodynamic relationships, the value of C_A will be calculated using the ideal gas law, and the volumetric heat capacity $\rho c_{p,m}$ will be calculated using the data in the table above.

SOLUTION

$$\Delta H_R \text{ (298 K)} = \sum_i \nu_i \Delta H_{f,i} = -62.63 \text{ kcal/mol}$$

$$\left(\frac{\partial H_R}{\partial T}\right)_P = \sum_i \nu_i c_{p,i} = -13.52 \text{ cal/mol-K}$$

[9] Satterfield, C. N., *Mass Transfer in Heterogeneous Catalysis*, MIT Press (1970), Chapter 4, p. 164; Aris, R., *The Mathematical Theory of Diffusion and Reaction in Permeable Catalysts. Volume I—The Theory of the Steady State*, Clarendon Press—*Oxford, (1975)*, Chapter 4, p. 240.

[10] Dean, J. A. (ed.), *Lange's Handbook of Chemistry,* 13th edition, McGraw-Hill, Inc. (1985) Section 9.

$$\Delta H_R\ (523\ K) = \Delta H_R\ (298\ K) + [\sum_i \nu_i c_{p,i}](523 - 298) = -62{,}630 + (-13.52)(225)$$

$$\Delta H_R\ (523\ K) = -65{,}672\ \text{cal/mol}$$

Neglecting any external transport resistance and letting thiophene be "A",

$$C_{A,B} = (P/RT)y_A = 1.58 \times 10^{-3}\ \text{mol/l}$$

The volumetric heat capacity $\rho c_{p,m}$ is given by

$$\rho c_{p,m} = \sum_i c_{p,i} C_{i,B} = \frac{P}{RT}\sum_i y_i c_{p,i} = 0.207\ \text{cal/l-K}$$

Finally,

$$\Delta T_{ad} = (-\Delta H_R)C_{A,B}/(\rho c_{p,m})_f = 502\ K$$

From Eqn. (9-31),

$$(T(0) - T_s)_{max} = (65,630)(\text{cal/mol})(0.02)(\text{cm}^2/\text{s})(1.58 \times 10^{-6})(\text{mol/cm}^3)/$$
$$5 \times 10^{-4}(\text{cal/s-cm-K})$$

$$(T(0) - T_s)_{max} \cong 4\ K$$

Now, from Eqn. (9-33),

$$(T(0) - T_s) \cong \Delta T_{ad}/(100\ \text{to}\ 1000)$$

$$(T(0) - T_s) \cong 502\ K/(100\ \text{to}\ 1000) \cong 5\ \text{to}\ 0.5\ K$$

The approximate analysis based on Eqn. (9-33) is in reasonable agreement with the analysis based on Eqn. (9-31).

For this example, the temperature at the center of the catalyst particle ($T(0)$) is at most about 5 K higher than the temperature at the surface (T_s). For this small temperature difference, we might hope that the isothermal model would provide a reasonable approximation to actual catalyst behavior. However, ΔT_{max} is not trivially small and, if the activation energy were high enough, the effect of the temperature gradient inside the particle might be appreciable.

For a first-order reaction, it can be shown that the isothermal effectiveness factor will be within 5% of the actual (nonisothermal) effectiveness factor if[11]

$$\left|\frac{E(-\Delta H_R)D_{A,eff}C_{A,s}}{RT_s^2 k_{eff}}\right| < 0.30$$

This criterion should be valid, and even more conservative, for a higher order reaction. However, it cannot be applied directly to reactions with effective orders less than 1.

Suppose that $n = 1$ and $E = 30$ kcal/mol for this example. Then,

$$\left|\frac{E(-\Delta H_R)D_{A,eff}C_{A,s}}{RT_s^2 k_{eff}}\right|$$
$$= \left|\frac{3.0 \times 10^4(\text{cal/mol}) \times -6.6 \times 10^4(\text{cal/mol}) \times 0.020(\text{cm}^2/\text{s}) \times 1.6 \times 10^{-6}(\text{mol/cm}^3)}{1.987(\text{cal/mol-K}) \times 523^2(\text{K})^2 \times 5 \times 10^{-4}(\text{cal/s-cm-K})}\right|$$
$$= 0.23$$

For this example, the "isothermal particle" assumption should be accurate to within 5%.

[11] Peterson, E. E., *Chemical Reaction Analysis*, Prentice-Hall, Inc., Englewood Cliffs, NJ, (1965), p. 79.

9.3.8 Reaction Selectivity

In Chapter 7, we saw how the selectivity of a multiple-reaction system could be affected by temperature and by the species concentrations. Therefore, it comes as no surprise that internal transport resistances can also affect the selectivity. As the Thiele modulus changes, so do the concentration and temperature profiles inside the catalyst particles. These changes, in turn, affect the *relative* rates of the reactions taking place.

Perhaps the easiest way to understand the influence of pore diffusion on selectivity is to examine the behavior of three of the reaction networks that were discussed in Chapter 7. The following treatment is confined to isothermal particles. Even so, the mathematics required for a complete quantitative analysis can be quite formidable. Therefore, the following discussion is primarily qualitative.

9.3.8.1 Parallel Reactions

Consider the irreversible, parallel reactions

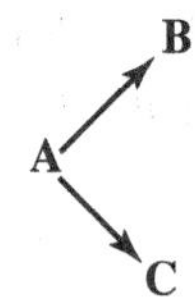

Suppose that the rate of formation of B is given by $r_B = k_B C_A^{\beta}$ and the rate of formation of C is given by $r_C = k_C C_A^{\chi}$. If the Thiele modulus is large, the concentration profile of A inside the particle might look something like the one shown in the following figure.

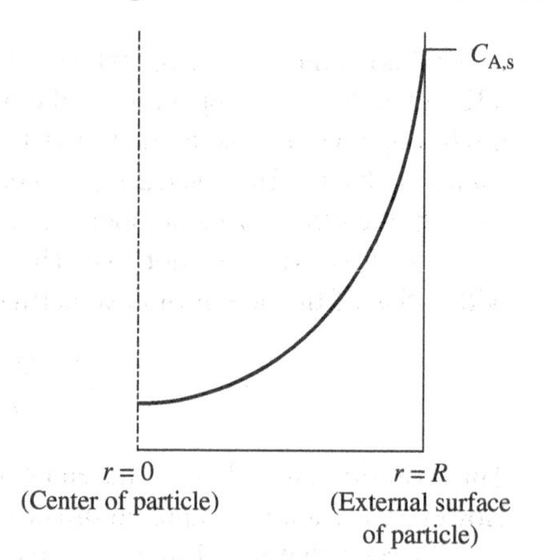

What effect does this concentration gradient have on the reaction selectivity? Of course, the answer is in the rate equations. For the present example

$$\frac{r_B}{r_C} = \left(\frac{k_B}{k_C}\right) C_A^{(\beta-\chi)}$$

If $\beta = \chi$, pore diffusion will not influence the selectivity in an isothermal catalyst particle, since r_B/r_C does not depend on concentration.

On the other hand, if $\beta > \chi$, r_B/r_C will decline as C_A declines. Since C_A is less than $C_{A,s}$ throughout the catalyst particle, an internal transport limitation will cause r_B/r_C to be less than its intrinsic value. The intrinsic value of r_B/r_C is its value when all of the concentrations that influence r_B and r_C are equal to their respective surface concentrations. If B is the desired product and ϕ is reasonably large, the loss of selectivity associated with a significant pore diffusion limitation might not be tolerable. A smaller catalyst particle might have to be considered. Another option would be an "eggshell" catalyst, where the active component is deposited in a thin layer near the surface of the catalyst particle.

If $B < \chi$, r_B/r_C will increase as C_A declines. An internal transport limitation will cause r_B/r_C to be greater than its value at $C_{A,s}$. If B is the desired product, a larger particle might be considered. Alternatively, an "egg yolk" catalyst, where the active component is deposited in the interior of the particle and is surrounded by an inert layer, might give improved results.

To summarize, for parallel reactions, the selectivity to the product that is produced in the higher order reaction will decrease as the Thiele modulus increases, i.e., as the effectiveness factor decreases. If the desired product is produced in the higher order reaction, using a catalyst particle that is too large will increase the amount of catalyst required and simultaneously reduce the selectivity to the desired product. However, if the desired product is produced in the lower order reaction, it may be beneficial to operate with an effectiveness factor that is less than 1. This is because the selectivity to the desired product will be improved if $\eta < 1$. However, there will be an economic tradeoff between improved selectivity and a larger catalyst requirement.

EXAMPLE 9-8
Effect of Internal Concentration Gradients on the Selectivity of Parallel Reactions

Consider the parallel reactions

$$n\text{-}C_6H_{14} \underset{}{\overset{\text{(fast)}}{\rightleftharpoons}} i\text{-}C_6H_{14}$$

$$n\text{-}C_6H_{14} \xrightarrow[H_2]{\text{(slow)}} \text{light paraffins } (CH_4,\ C_2H_6,\ C_3H_8, \text{ etc.})$$

The fast reaction is the reversible skeletal isomerization of normal hexane to isohexane. Isomerization reactions are an important building block in the production of high-octane gasoline, as the octane numbers of branched paraffins are substantially higher than those of normal paraffins with the same number of carbon atoms. The slow reaction is the hydrogenolysis (hydrocracking) of normal hexane to lower paraffins such as methane. This reaction is undesirable, as these light paraffins have a much lower economic value.

Suppose that the forward isomerization reaction is first order in n-C_6H_{14} and that the reverse isomerization reaction is first order in i-C_6H_{14}. Suppose further that the hydrogenolysis reaction is first order in n-C_6H_{14}. If the Thiele modulus is large, will the actual selectivity to i-C_6H_{14} be greater than, less than, or the same as the intrinsic selectivity?

APPROACH

The only difference between this example and the previous discussion is the reversibility of the fast isomerization reaction. The *net* rate of formation of i-C_6H_{16} is the difference between the rates of the forward and reverse isomerization reactions. This net rate will depend on the concentrations of *both* n-C_6H_{16} and i-C_6H_{16}. We will examine the concentration profiles of both isomers and determine qualitatively the effect of a pore diffusion limitation on reaction selectivity. For simplicity, n-C_6H_{14} will be designated as "N" and i-C_6H_{14} as "I".

SOLUTION

For a situation where the Thiele modulus is reasonably high, the concentration profiles might look something like those shown below:

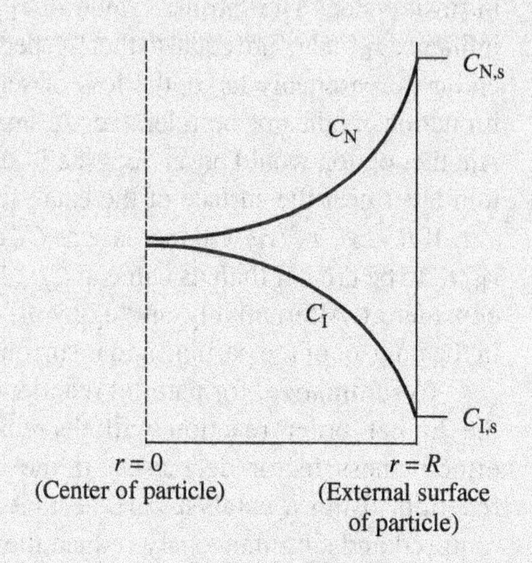

Since C_N is less than $C_{N,s}$ everywhere in the catalyst particle, the rates of both the hydrogenolysis reaction and the *forward* isomerization reaction will be lower than if $C_N = C_{N,s}$. Since both of these reactions are first order in normal hexane, both rates decline by the same percentage amount, no matter what the exact shape of the concentration profiles might be.

The rate of the *reverse* isomerization reaction is not affected by C_N, but it is affected by C_I. When the Thiele modulus is high, the concentration of i-C_6H_{14} is higher than $C_{I,s}$ everywhere inside the catalyst particle. Therefore, the rate of the *reverse* isomerization reaction is *faster* when the Thiele modulus is high ($\eta < 1$) than when the Thiele modulus is low.

In the presence of an internal transport limitation, the *net* rate of formation of i-C_6H_{16} is reduced because C_N inside the catalyst particle is less than $C_{N,s}$, and it is further reduced because C_I inside the particle is greater than $C_{I,s}$. Therefore, the *net* forward rate of the isomerization reaction is reduced more than the rate of the hydrogenolysis reaction, which is reduced in proportion to C_N. The actual selectivity to i-C_6H_{14} should be less than the intrinsic selectivity when the Thiele modulus is high.

9.3.8.2 Independent Reactions

Consider the independent, irreversible reactions

$$\mathrm{A} \rightarrow \mathrm{P}_1 + \mathrm{P}_2 + \cdots$$
$$\mathrm{B} \rightarrow \mathrm{P}_{\mathrm{I}} + \mathrm{P}_{\mathrm{II}} + \cdots$$

Suppose that the rate equation for the first reaction is $-r_A = k_A C_A^{\alpha}$ and the rate equation for the second reaction is $-r_B = k_B C_B^{\beta}$. For a single, isothermal catalyst particle,

$$\frac{\text{rate of disappearance of A (actual)}}{\text{rate of disappearance of B (actual)}} \equiv \Lambda = \frac{\eta_A k_A C_{A,s}^{\alpha}}{\eta_B k_B C_{B,s}^{\beta}}$$

In this equation, η_A is the effectiveness factor for the first reaction and η_B is the effectiveness factor for the second. Let Λ° denote the *intrinsic* ratio of the two reaction rates, i.e., the ratio when internal transport limitations are insignificant and both effectiveness factors are equal to 1. For the present example,

$$\Lambda^\circ = k_A C_{A,s}^{\alpha} / k_B C_{B,s}^{\beta}$$

Therefore,

$$\Lambda = \Lambda^\circ \eta_A / \eta_B \tag{9-34}$$

Suppose that the first reaction, the disappearance of A, is fast relative to the second. For this case, $\Lambda^\circ \gg 1$. Since $\phi = l_c\sqrt{(n+1)k_v C_{A,s}^{(n-1)}/2D_{A,eff}}$ for an irreversible, nth-order reaction, the fact that the first reaction is fast relative to the second implies that $\phi_A \gg \phi_B$, unless $D_{B,eff} \ll D_{A,eff}$. However, it is not likely that $D_{B,eff} \ll D_{A,eff}$ unless the molecular weight of B is much larger than that of A and/or diffusion is in the configurational regime. Therefore, we will ignore this possibility.

Consider a situation where the characteristic dimension of the catalyst particle, l_c, is very small, such that both ϕ_A and ϕ_B are substantially less than 1. In this case, both η_A and η_B will be essentially equal to 1 and $\Lambda \cong \Lambda^\circ$ according to Eqn. (9-34).

Now let the particle size be increased. Since $\phi_A \gg \phi_B$, the effectiveness factor for the fast reaction, η_A, will begin to drop below 1, whereas η_B remains at essentially 1. As the particle size continues to be increased, η_A continues to drop and η_B eventually begins to fall below 1.

Finally, when the characteristic dimension of the catalyst particle becomes so large that *both* effectiveness factors are in the asymptotic region,

$$\eta_A \cong \frac{1}{\phi_A} = \frac{1}{l_c}\sqrt{\frac{2D_{A,eff}}{(\alpha+1)k_A C_{A,s}^{\alpha-1}}}$$

$$\eta_B \cong \frac{1}{\phi_B} = \frac{1}{l_c}\sqrt{\frac{2D_{B,eff}}{(\beta+1)k_B C_{B,s}^{\beta-1}}}$$

$$\Lambda \cong \Lambda^\circ \frac{\sqrt{D_{A,eff}/(\alpha+1)k_A C_{A,s}^{(\alpha-1)}}}{\sqrt{D_{B,eff}/(\beta+1)k_B C_{B,s}^{(\beta-1)}}} = \sqrt{\Lambda^\circ}\sqrt{\frac{D_{A,eff}C_{A,s}/(\alpha+1)}{D_{B,eff}C_{B,s}/(\beta+1)}} \tag{9-35}$$

Once the characteristic dimension has become so large that Eqn. (9-35) is valid, Λ no longer depends on l_c. Therefore, Eqn. (9-35) gives the *lowest possible* value of Λ for a given set of independent reactions.

For independent reactions, an internal transport resistance slows the faster reaction more than it slows the slower reaction. Therefore, the selectivity to the product formed in the faster reaction is decreased. If the fast reaction is desired and the slow reaction is undesired (as we might hope), $\Lambda < \Lambda^\circ$ is an undesirable result. For this situation, the internal transport resistance should be kept low enough that $\eta_A \cong 1$.

EXAMPLE 9-9 ***Selective Hydrogenation of Acetylene***

Small amounts of acetylene (C_2H_2) must be removed from ethylene (C_2H_4) before the latter can be used to make polyethylene. Acetylene removal usually is accomplished by selective catalytic hydrogenation:

$$C_2H_2 + H_2 \rightarrow C_2H_4 \quad \text{(fast)}$$
$$C_2H_4 + H_2 \rightarrow C_2H_6 \quad \text{(slow)}$$

The concentration of acetylene in the stream is small compared with the concentration of ethylene, and we will ignore any effect of the common reactant, H_2, on the kinetics of either reaction. Therefore, this pair of reactions can be treated as an independent sequence, rather than as a mixed sequence.

Consider a situation where the feed to the hydrogenation reactor contains 50 mol% C_2H_4 and 1000 ppm C_2H_2. The effective diffusion coefficients of C_2H_2 and C_2H_4 are approximately equal. If "A" is C_2H_2 and "B" is C_2H_4, the value of Λ° is of the order of 100. The hydrogenation of C_2H_2 is zero order in C_2H_2 and the hydrogenation of C_2H_4 is first order in C_2H_4.

If the particle size is so large that both effectiveness factors are in the asymptotic region, what is the value of Λ?

APPROACH The value of Λ will be calculated directly from Eqn. (9-35), using the information provided.

SOLUTION From Eqn. (9-35),

$$\Lambda = \sqrt{100} \times \sqrt{1000 \times 2/500{,}000} = 0.63$$

For this example, a substantial loss of selectivity results from using a catalyst particle that is so large that both effectiveness factors are in the asymptotic region. The value of Λ in the asymptotic regime is more than two orders of magnitude less than its intrinsic value (Λ°).

9.3.8.3 Series Reactions

Finally, let's return to the series of two reactions that we considered in Chapter 7.

$$\mathrm{A} \xrightarrow{k_1} \mathrm{R} \xrightarrow{k_2} \mathrm{S}$$

Suppose that both reactions are irreversible and first order, with the rate constants shown above.

If there are no gradients in the catalyst particle, the ratio of the rate of formation of R to the rate of disappearance of A for the whole particle is given by

$$\frac{r_{\mathrm{R}}}{-r_{\mathrm{A}}} = s\left(\frac{\mathrm{R}}{\mathrm{A}}\right) = \frac{k_1 C_{\mathrm{A,s}} - k_2 C_{\mathrm{R,s}}}{k_1 C_{\mathrm{A,s}}} = 1 - \left(\frac{k_2}{k_1}\right)\left(\frac{C_{\mathrm{R,s}}}{C_{\mathrm{A,s}}}\right)$$

This ratio of reaction rates is the selectivity to R based on A, as defined in Chapter 7. The symbol for the instantaneous or point selectivity is used here because the selectivity for the particle as a whole corresponds to the selectivity at a point in the reactor.

When an internal transport resistance is present, the concentrations of A, R, and S vary with position inside the catalyst particle. Therefore, the selectivity will also vary with position. Nevertheless, the symbol $s(\mathrm{R/A})$ will be used to represent the selectivity of a *whole particle*. However, when an internal transport resistance is present, the equations that describe reaction and transport inside the catalyst particle must be solved to obtain $s(\mathrm{R/A})$.

If $C_{\mathrm{R,s}} = 0$, and if there are no internal gradients, the rate of the second reaction will be zero throughout the catalyst particle. The above equation then becomes

$$s(\mathrm{R/A}) = 1; \quad C_{\mathrm{R,s}} = 0$$

Now consider a situation where the resistance to internal transport is significant and internal gradients do exist. The concentration profile of A inside the catalyst particle will be the same as for a single, first-order reaction and will be similar to those shown in previous figures. However, the profile for R can have a complex shape. For the situation where $C_{R,s} = 0$, the profiles of A and R might look as shown below.

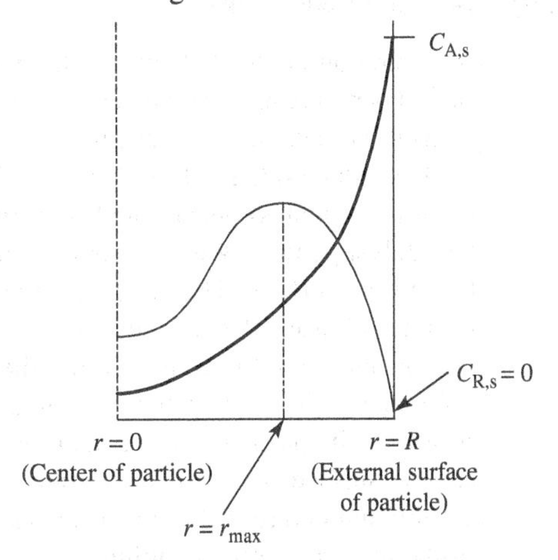

Close to the exterior surface, the profile of C_R looks "normal," similar to that for a product, as shown in earlier figures. However, somewhere in the interior of the particle, at $r = r_{max}$, C_R has a maximum value. In the region between $r = 0$ and $r = r_{max}$, all of the "R" that is formed from "A" is converted to "S". None of it diffuses back to the external surface and into the bulk stream. Between $r = r_{max}$ and $r = R$, some of the "R" that is formed from "A" reaches the exterior surface, although some also reacts to form "S."

EXERCISE 9-4

Sketch the concentration profile of S for this situation.

Qualitatively, an internal transport resistance causes the intermediate R to become "trapped" in the pores of the catalyst, increasing the probability that it will react to form "S." The above figure shows that the value of C_R inside the catalyst particle is always greater than the value of $C_{R,s}$, which is 0 for this example. Therefore, the reaction $R \rightarrow S$ proceeds at a finite rate. This causes $s(R/A)$ to be <1, compared with $s(R/A) = 1$ for $C_{R,s} = 0$ and no internal transport resistance.

The equations that describe the diffusion and reaction of A and R inside an isothermal catalyst particle have a simple solution when both reactions are first order and when the internal transport resistance is severe for both A and R. If $C_R(0) \cong 0$ and $C_A(0) \cong 0$.[12]

$$\frac{r_R}{-r_A} = s(R/A) = \frac{1}{1 + \sqrt{\dfrac{k_2 D_{A,eff}}{k_1 D_{B,eff}}}} \tag{9-36}$$

This result confirms the qualitative analysis performed above, since it shows that $s(R/A) < 1$ when the internal transport resistance is significant, even when $C_{R,s} = 0$. The above equation also shows that the loss of selectivity becomes progressively worse as k_2/k_1 increases.

---

[12] Froment, G. F. and Bischoff, K. B., *Chemical Reactor Analysis and Design,* 2nd edition, John Wiley & Sons (1990), p. 177.

9.4 EXTERNAL TRANSPORT

9.4.1 General Analysis—Single Reaction

Concentration gradients between the bulk fluid in the reactor and the external surface of a catalyst particle can also influence the rate of reaction, i.e., the apparent catalyst activity; and the product distribution, i.e., the apparent catalyst selectivity. First, let's deal with the rate of reaction.

Consider a spherical catalyst particle of radius R. There is a resistance to both mass and heat transfer that is concentrated at the interface between the fluid and the external surface of the catalyst particle. We represent this resistance as a film of fluid, i.e., a boundary layer, at the particle surface. The concentration of reactant A in the bulk fluid at any point in the reactor is $C_{A,B}$ and the concentration of A at the external surface of the particle is $C_{A,s}$. There is a concentration difference across the boundary layer ($C_{A,B} - C_{A,s}$). This concentration difference is the driving force for mass transfer of A from the bulk fluid to the external surface of the particle. The temperature of the bulk fluid is T_B and the temperature at the external surface is T_s. The temperature difference across the boundary layer is ($T_B - T_s$), and this temperature difference provides the driving force for heat transfer. If a single, exothermic reaction is taking place at steady state, and if the reaction is fast, the concentration profile of reactant A and the temperature profile through the boundary layer might look something like those shown in Figure 9-10.

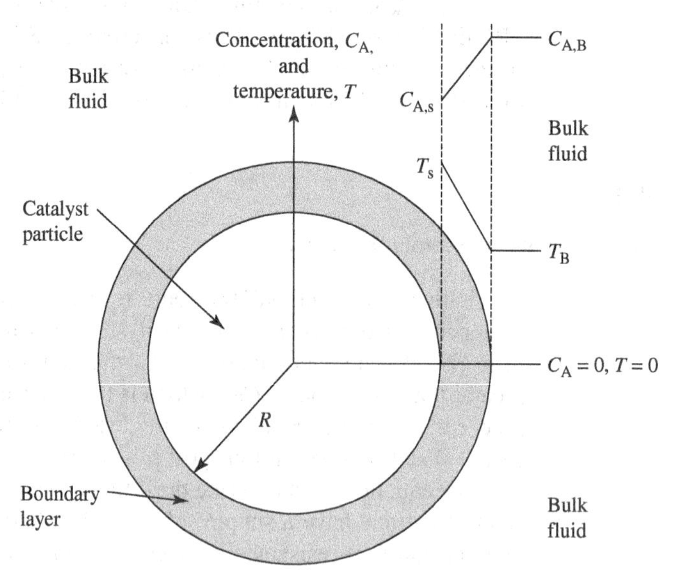

Figure 9-10 Profiles of temperature (T) and concentration of reactant A (C_A) through the boundary layer surrounding a catalyst particle in which an exothermic reaction is taking place at steady state. The width of the boundary layer is exaggerated relative to the size of the catalyst particle. The profiles of C_A and T are shown to be linear because the actual thickness of the boundary layer is small relative to the radius of the particle.

The magnitude of the resistances to heat and mass transfer through the boundary layer, i.e., the thickness of the boundary layer, depends on the velocity of the fluid *relative to the catalyst particle*. As this velocity increases, the heat-transfer coefficient between the bulk fluid and the surface of the catalyst particle increases and the mass-transfer coefficient between the bulk fluid and the catalyst surface increases. Therefore, the magnitude of the concentration and temperature differences between the bulk fluid and the particle surface will depend on the relative velocity, as well as on the properties of the fluid.

9.4.1.1 Quantitative Descriptions of Mass and Heat Transport

Mass Transfer Let $\vec{N}_{\mathrm{A}}$ be the flux (mol A/time area) of A from the bulk fluid through the boundary layer to the surface of the catalyst particle. Let k_c be the mass-transfer coefficient, based on concentration, between the bulk fluid and the external surface of the catalyst particle. The dimensions of k_c are length/time, and k_c will depend on the velocity of the fluid relative to the catalyst particle. The flux of A arriving at the external surface of the catalyst particle is given by

Flux of "A" to external surface of catalyst particle

$$\vec{N}_{\mathrm{A}} = k_c(C_{\mathrm{A,B}} - C_{\mathrm{A,s}}) \tag{9-37}$$

The rate at which A reaches the external surface of the whole catalyst particle is

$$\vec{N}_{\mathrm{A}}A_{\mathrm{G}} = A_{\mathrm{G}}k_c(C_{\mathrm{A,B}} - C_{\mathrm{A,s}})$$

In this expression, A_{G} is the *external* (geometric) surface area of the catalyst particle. If the catalyst particle were a sphere with radius R, $A_{\mathrm{G}} = 4\pi R^2$. If A is a reactant, $C_{\mathrm{A,B}} > C_{\mathrm{A,s}}$ and the flux of A is directed from the bulk fluid toward the particle, i.e., opposite to the radial direction.

At steady state, there is no accumulation or depletion of A inside the catalyst particle. Therefore, the rate at which A reaches the surface of the particle through the boundary layer must be equal to the rate at which A is consumed by reaction in the whole catalyst particle. Let $R_{\mathrm{A,P}}$ be the rate at which A reacts in the whole catalyst particle. Then,

$$R_{\mathrm{A,P}} = \vec{N}_{\mathrm{A}}A_{\mathrm{G}} = A_{\mathrm{G}}k_c(C_{\mathrm{A,B}} - C_{\mathrm{A,s}})$$

Rearranging,

$$C_{\mathrm{A,B}} - C_{\mathrm{A,s}} = R_{\mathrm{A,P}}/k_cA_{\mathrm{G}} \tag{9-38}$$

This equation shows that the concentration difference between the bulk fluid and the surface of the catalyst particle is directly proportional to the rate at which A is consumed in the catalyst particle. The faster the reaction, the larger the concentration difference.

For a given rate of reaction, i.e., a fixed value of $R_{\mathrm{A,P}}$, the concentration difference will depend on the velocity of the fluid relative to the catalyst particle, since the value of k_c depends on the relative velocity. The higher the relative velocity, the higher the value of k_c and the lower the value of $C_{\mathrm{A,B}} - C_{\mathrm{A,s}}$, for a given value of $R_{\mathrm{A,P}}$.

Heat Transfer Let q be the heat flux from the bulk fluid to the surface of the catalyst particle (energy/area-time) and let h be the heat-transfer coefficient (energy/area-time-temperature) between the bulk fluid and the external surface. The flux of energy arriving at the external surface of the catalyst particle is given by

Heat flux to external surface of catalyst particle

$$q = h(T_{\mathrm{B}} - T_{\mathrm{s}})$$

and the total rate of heat transfer to the particle is

$$qA_{\mathrm{G}} = hA_{\mathrm{G}}(T_{\mathrm{B}} - T_{\mathrm{s}})$$

If $T_B > T_s$, the heat flux is directed from the bulk fluid into the particle, i.e., opposite to the radial direction.

At steady state, there can be no accumulation or depletion of energy in the catalyst particle. The rate of energy "consumption" or "generation" resulting from the reaction in the catalyst particle is $R_{A,P}\Delta H_R$. This energy must be supplied or removed by heat transfer through the boundary layer. Therefore,

$$R_{A,P}\Delta H_R = hA_G(T_B - T_s) \tag{9-39}$$

If the reaction is endothermic, $\Delta H_R > 0$ and $T_B > T_s$. Heat is transferred from the bulk stream to the catalyst particle and is "consumed" by the endothermic reaction.

If the reaction is exothermic, the above equation can be written in a slightly more convenient form by multiplying both sides by -1:

$$R_{A,P}(-\Delta H_R) = hA_G(T_s - T_B)$$

For an exothermic reaction, $(-\Delta H_R) > 0$ and $T_s > T_B$. The heat that is "generated" by the reaction must be transferred from the particle into the bulk stream to keep the catalyst at steady state.

The above equation can be rearranged to give

$$T_s - T_B = R_{A,P}(-\Delta H_R)/hA_G \tag{9-40}$$

This equation shows that the temperature difference between the bulk fluid and the surface of the catalyst particle is directly proportional to $R_{A,P}$, the rate at which A is consumed in the catalyst particle. The faster the reaction, the larger the difference between T_B and T_s. For a given rate of reaction, the temperature difference will depend on the velocity of the fluid relative to the catalyst particle, since the value of h depends on this velocity. The higher the velocity, the lower the value of $T_B - T_s$ for a given value of $R_{A,P}$.

9.4.1.2 First-Order Reaction in an Isothermal Catalyst Particle—The Concept of a Controlling Step

To illustrate the use of Eqn. (9-37), and to expose some features of catalyst behavior in the presence of an external mass-transfer resistance, let's consider the reaction $A \rightarrow B$. The additional complication of heat transfer will be eliminated (at least temporarily) by assuming that $\Delta H_R = 0$. For $\Delta H_R = 0$, no energy is "consumed" or "released" by the reaction. Therefore, no heat is transferred between the particle and the bulk fluid, so that $T_s = T_B$. Moreover, when $\Delta H_R = 0$, the catalyst particle is isothermal.

Let's carry out a material balance on species A, using the whole catalyst particle as the control volume.

$$\left\{ \begin{array}{c} \text{rate of transfer of A from} \\ \text{bulk fluid to catalyst surface} \end{array} \right\} = \left\{ \begin{array}{c} \text{rate of disappearance of A due} \\ \text{to reaction in catalyst particle} \end{array} \right\}$$

$$A_G k_c(C_{A,B} - C_{A,s}) = R_{A,P} = V_G \eta(-r_{A,v})$$

In this equation, V_G is the geometric volume of the catalyst particle, η is the internal effectiveness factor, and $-r_{A,v}$ is the reaction rate with no internal gradients, on a volumetric basis.

Suppose that the reaction is first order in A and irreversible. Then,

$$-r_{A,v} = k_v C_{A,s}$$

This assumption will simplify the algebra required to analyze reaction behavior. However, the important concepts that will result from the following analysis do not depend on the form of the rate equation.

For a first-order, irreversible reaction, η does not depend on $C_{A,s}$ so that the above equations can be solved for $C_{A,s}$. The result is

$$C_{A,s} = \frac{C_{A,B}}{1 + (\eta k_v V_G / k_c A_G)} = \frac{C_{A,B}}{1 + (\eta k_v l_c / k_c)} \tag{9-41}$$

where l_c $(=V_G/A_G)$ is the characteristic dimension of the catalyst particle.

The rate of reaction in a single particle now can be expressed in terms of the bulk concentration of A ($C_{A,B}$). Since $R_{A,P} = V_G \eta(-r_{A,v}) = V_G \eta k_v C_{A,s}$,

$$R_{A,P} = V_G \eta k_v C_{A,B} / [1 + (\eta k_v l_c / k_c)] \tag{9-42}$$

As an aside, for use later in this discussion, let's convert $R_{A,P}$ into a rate per unit weight of catalyst ($-r_A$). This is done by dividing both sides of Eqn. (9-42) by the weight of the catalyst particle, $\rho_p V_G$, and recognizing that $k_v = \rho_p k$, where k is the rate constant based on mass of catalyst.

$$-r_A = \eta k C_{A,B} / [1 + (\eta \rho k l_c / k_c)] \tag{9-43}$$

Equation. (9-43) shows that the rate *per unit mass of catalyst* will depend on k_c, if the second term in the denominator of Eqn. (9-43) is significant compared to 1. In this case, $-r_A$ will depend on the velocity of the fluid relative to the catalyst particle, since k_c depends on this velocity.

Strictly speaking, Eqn. (9-43) applies only to an irreversible, first-order reaction in an isothermal catalyst particle. However, the conclusion that we have drawn from this equation can be generalized. For the present example, the term $\eta \rho k l_G / k_C$ is the ratio of the *maximum* rate of reaction inside the catalyst particle, allowing for an internal resistance that might cause η to be less than 1, to the *maximum* rate of mass transfer through the boundary layer to the external surface of the particle.

EXERCISE 9-5

Prove this assertion.

For any rate equation, if the ratio (maximum rate of reaction inside the catalyst particle/maximum rate of mass transfer to external surface) is significant compared to 1, the actual reaction rate per unit mass of catalyst will depend on the relative velocity.

EXERCISE 9-6

Write an expression for (maximum rate of reaction inside the catalyst particle/maximum rate of mass transfer to external surface) a second-order reaction.

Now let's examine the extremes of catalyst behavior using Eqns. (9-41) and (9-42).

$\boldsymbol{\eta k_v l_c / k_c \ll 1}$ If $\eta k_v l_c / k_c$ is very small, the *maximum* rate of reaction inside the catalyst particle is much smaller than the *maximum* rate of mass transfer through the boundary layer. Conceptually, the reaction in the catalyst particle is slow, and only a very small concentration difference across the boundary layer is required to supply the reactant at the rate it is consumed in the particle. In this case, external mass transfer has no significant effect on the reaction rate.

When $\eta k_{\mathrm{v}} l_{\mathrm{c}} / k_{\mathrm{c}} \ll 1$, this term can be neglected in the denominators of Eqns. (9-41) and (9-42). These equations then reduce to

$$C_{\mathrm{A,s}} = C_{\mathrm{A,B}}$$
$$R_{\mathrm{A,P}} = V_{\mathrm{G}} \eta k_{\mathrm{v}} C_{\mathrm{A,B}} \tag{9-44}$$

The concentration profile of reactant A is shown in Figure 9-11, for this case.
The decline of concentration inside the catalyst particle that is shown in Figure 9-11 occurs when the effectiveness factor is less than 1.

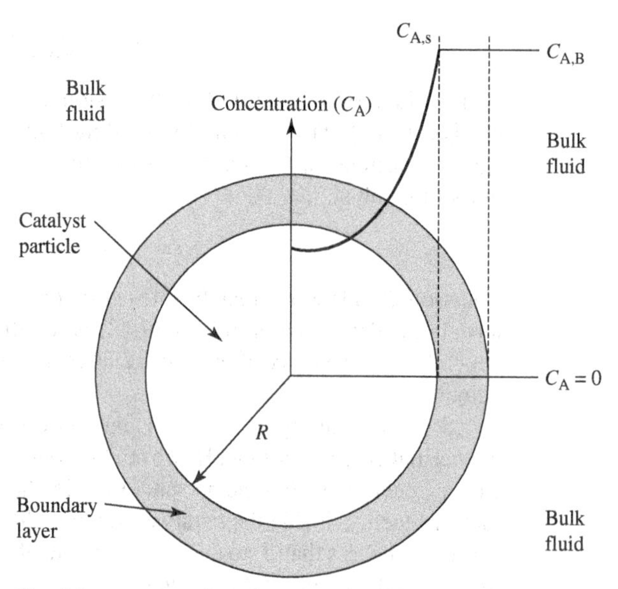

Figure 9-11 Profile of the concentration of reactant A (C_{A}) through the boundary layer and inside a catalyst particle when $\eta k_{\mathrm{v}} l_{\mathrm{c}} / k_{\mathrm{c}} \ll 1$. The resistance to external mass transport is very small compared with the resistance to reaction inside the particle. As a consequence, $C_{\mathrm{A,s}} \cong C_{\mathrm{A,B}}$. The concentration of A declines inside the catalyst particle if $\eta < 1$.

When $\eta k_{\mathrm{v}} l_{\mathrm{c}} / k_{\mathrm{c}} \ll 1$ *and* the effectiveness factor is close to unity ($\eta \cong 1$), the rate of reaction in the catalyst particle is $R_{\mathrm{A,P}} = k_{\mathrm{v}} V_{\mathrm{G}} C_{\mathrm{A,B}}$. The concentration profile of reactant A is shown in Figure 9-12. Neither the internal nor the external transport resistance has any influence on the rate of reaction, and no transport coefficients (k_{c} or $D_{\mathrm{A,eff}}$) appear in the expression for the reaction rate. In this case, the reaction is said to be *controlled by intrinsic kinetics*. Stated differently, intrinsic kinetics is the *controlling resistance*.

The word *control* is used *only* when the observed reaction rate is *completely* determined by a single step or resistance. *None* of the other steps or resistances have any effect on the reaction rate.

$\eta k_v l_c / k_c \gg 1$ If the quantity $\eta k_{\mathrm{v}} l_{\mathrm{c}} / k_{\mathrm{c}}$ is 1, the maximum rate of reaction inside the catalyst particle is much greater than the maximum rate of mass transfer through the boundary layer. Therefore, the reaction in the catalyst particle has the *potential* to be very fast relative to the maximum rate of external mass transfer. A large concentration difference

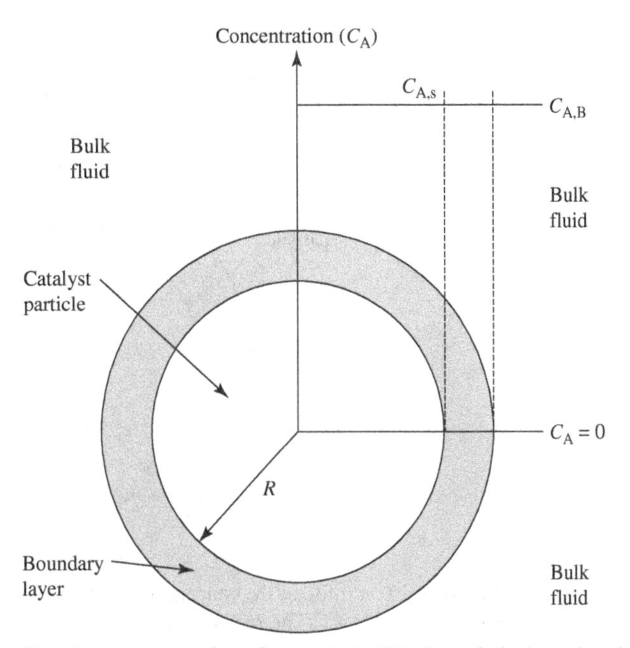

Figure 9-12 Profile of the concentration of reactant A (C_A) through the boundary layer and inside a catalyst particle when $\eta k_v l_c / k_c \ll 1$ *and* $\eta \cong 1$. Both the external and internal mass transport resistances are very small relative to the resistance to reaction inside the particle. As a consequence, $C_A \cong C_{A,B}$ throughout the particle. The reaction is *controlled* by intrinsic kinetics.

will be required across the boundary layer in order to supply reactant at the rate it is consumed in the particle. In this case, the external mass-transfer resistance will have a very significant effect on the reaction behavior.

When $\eta k_v l_c / k_c \gg 1$, Eqns. (9-41) and (9-42) reduce to

$$C_{A,s} \cong 0$$

$$R_{A,P} = k_c A_G C_{A,B} \tag{9-45}$$

The concentration of reactant A at the external surface of the catalyst particle is very close to zero. Moreover, Eqn. (9-45) shows that the rate of reaction does not depend on either the rate constant (k_v) or the effective diffusion coefficient ($D_{A,eff}$). The *only* rate or transport parameter that influences the reaction rate is the external mass-transfer coefficient, k_c. For this situation, the reaction is said to be *controlled by external mass transfer.* The concentration profile of reactant A is shown in Figure 9-13.

In this case, the *only* way to increase the reaction rate is to increase the value of k_c. This can be done by increasing the velocity of the fluid relative to the catalyst particle. Increasing the relative velocity reduces the thickness of the boundary layer, leading to an increase in k_c.

Equation (9-45) shows that the rate of reaction for the whole catalyst particle is proportional to the external surface area (A_G), not to the volume of the particle (V_G). The rate of reaction *per unit volume* of catalyst ($R_{A,P}/V_G = k_c C_{A,B}(A_G/V_G)$) is proportional to the surface-to-volume ratio of the catalyst particle. Since the characteristic dimension of the catalyst particle, l_c, is (V_G/A_G), the rate per unit volume or weight of catalyst is inversely proportional to the characteristic dimension, when the reaction is controlled by external mass transfer.

External mass-transfer control is an important limiting case of catalyst performance. For fixed hydrodynamic conditions, i.e., a fixed value of k_c, the reaction cannot go any faster

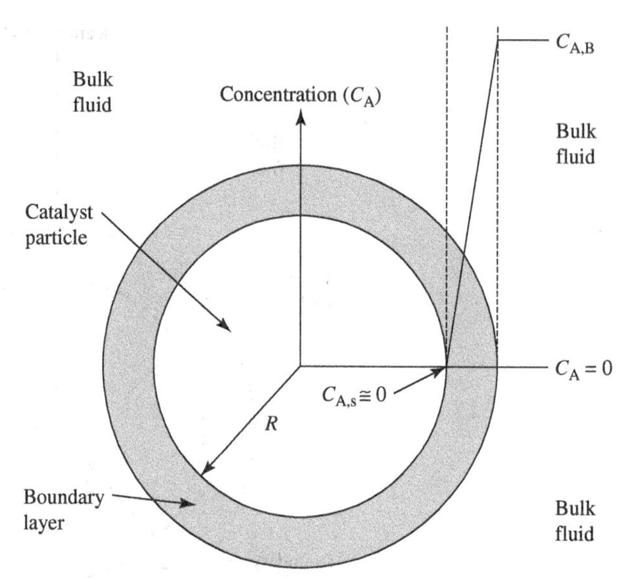

Figure 9-13 Profile of the concentration of reactant A (C_A) through the boundary layer for the case where $\eta k_v l_c / k_c \gg 1$. The reaction is *controlled* by transport of reactant A from the bulk fluid to the external surface of the catalyst particle, i.e., by external mass transfer.

than this limit. If the quantity of catalyst required to achieve a specified fractional conversion at a given feed rate and feed composition were calculated by assuming external mass-transfer control, this would be the *smallest* possible amount of catalyst that could do the specified job.

Question for discussion: Consider the situation shown in Figure 9-11, where $\eta k_v l_c / k_c \ll 1$ *and* $\eta < 1$. Is it legitimate to say that the reaction is *controlled* by internal transport, i.e., by pore diffusion, for this case?

Discussion: When the external transport resistance is insignificant, which is the case for Figure 9-11, the rate per pellet, $R_{A,P}$, is given by

$$R_{A,P} = V_G \eta k_v C_{A,B}$$

Consider a case where the resistance to internal diffusion is very large and consequently the concentration gradient inside the catalyst particle is very steep. For this situation,

$$\eta \cong 1/\phi = (D_{A,eff}/k_v)^{1/2}/l_c$$

Therefore,

$$R_{A,P} = A_G (D_{A,eff} k_v)^{1/2} C_{A,B}$$

A kinetic parameter (k_v) and a transport parameter ($D_{A,eff}$) *both* appear in the expression for the overall rate. Increasing *either* k_v or $D_{A,eff}$ increases the rate per particle, and the rate per unit weight or per unit volume of catalyst. Given this situation, we cannot say that the rate is *controlled* by internal transport. Both pore diffusion *and* the intrinsic reaction kinetics affect the reaction rate. For the situation shown in Figure 9-11, it is more appropriate to say that the reaction is *influenced* by both pore diffusion and the intrinsic kinetics.

EXERCISE 9-7

As demonstrated above, if $n = 1$, both pore diffusion and intrinsic kinetics affect the reaction rate, for the concentration profile shown in Figure 9-11. Show that this conclusion is valid for orders other than 1.

This analysis was carried out for an irreversible, first-order reaction with a very low effectiveness factor. However, for the situation shown in Figure 9-11, the rate per particle (or per unit weight or per unit volume) always depends on *both* the intrinsic rate constant and the effective diffusivity, no matter what rate equation is obeyed and no matter how steep the concentration gradients might be.

EXERCISE 9-8

Explain why k_v and $D_{A,eff}$ both influence the reaction rate for an nth-order reaction, as long as $\eta < 1$, even if $\eta \neq 1/\phi$.

9.4.1.3 Effect of Temperature

Earlier in Section 9.3.6.1, it was pointed out that k_v is much more sensitive to temperature than $D_{A,eff}$. The mass-transfer coefficient k_c is also much less sensitive to temperature than k_v. Over a limited range of temperature, the temperature sensitivity of k_c usually can be approximated by an Arrhenius relationship, with an activation energy in the region of 5–20 kJ/mol. The activation energy for k_v typically is between 50 and 300 kJ/mol. As a consequence, k_v increases much faster than k_c as the temperature is increased.

Let's illustrate the effect of temperature using the relationships that were derived in the previous section for an irreversible, first-order reaction in an isothermal catalyst particle. If the temperature is sufficiently low, k_v will be much less than k_c and Eqn. (9-42) will simplify to

$$R_{A,P} = V_G \eta k_v C_{A,B}$$

Moreover, at very low temperatures k_v will be so small that $\eta \cong 1$.

$$R_{A,P} = V_G k_v C_{A,B}$$

Therefore, when the temperature is sufficiently low, the reaction will be *controlled by intrinsic kinetics*. If the rate constant is measured as a function of temperature *in this region*, the true (kinetic) activation energy of the reaction will be observed.

As the temperature increases, k_v increases much faster than either k_c or $D_{A,eff}$. Under many circumstances, the effectiveness factor drops below 1 before the term $\eta k_v l_c / k_c$ becomes significant compared to 1. If the value of η is sufficiently low, the observed value of the activation energy will approach one half of the true value, as discussed in connection with Figure 9-8.

As the temperature is increased even further, the term $\eta k_v l_c / k_c$ eventually becomes much greater than 1, so that 1 can be neglected in the denominator of Eqn. (9-42). The reaction now is *controlled* by external mass transfer. The rate per pellet is given by

$$R_{A,P} = k_c A_G C_{A,B} \tag{9-45}$$

If the reaction rate is measured *in this regime*, the activation energy that is observed experimentally will reflect the temperature dependency of k_c, i.e., the measured activation energy will be of the order of 10 kJ/mol.

Figure 9-14 summarizes the preceding discussion and shows the regimes that can be observed as the temperature at which a heterogeneous catalytic reaction takes place is changed. This figure is an extension of Figure 9-8, to include the regime of external mass-transfer control.

Measured activation energies below about 30 kJ/mol should be regarded as a **RED FLAG**. Values this low usually arise because the catalyst particle is operating in a regime

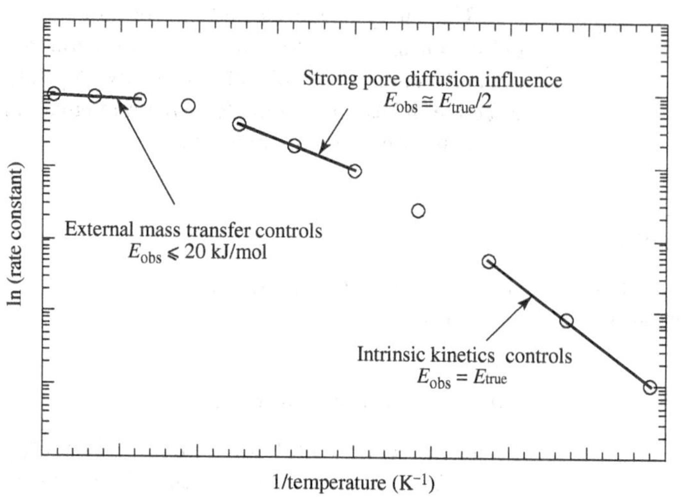

Figure 9-14 Arrhenius plot for a heterogeneous catalytic reaction, showing the behavior of the apparent activation energy in three regimes of operation: intrinsic kinetic control, strong internal (pore) diffusion influence, and control by external mass transfer.

where the reaction is influenced or controlled by either external or internal mass transfer (or possibly by both).

The analysis leading to Figure 9-14 was based on the behavior of an irreversible, first-order reaction in an isothermal catalyst particle. However, the results are quite general. Equation (9-45) will be obeyed by any *irreversible* reaction that is controlled by external mass transfer. This is true for any form of the rate equation, independent of whether the catalyst particle is isothermal, and independent of whether there is a temperature difference between the catalyst surface and the bulk fluid. The apparent activation energy will reflect the temperature sensitivity of k_c, and will be of the order of 10 kJ/mol.

When the temperature is sufficiently low, the reaction will be controlled by intrinsic kinetics. Therefore, there will always be a low-temperature regime where the true activation energy is reflected.

The shape of the Arrhenius plot at intermediate temperatures can vary from case to case. If the effectiveness factor becomes very low before the external transport resistances become appreciable, the Arrhenius plot will exhibit an intermediate linear region, where the apparent activation energy is approximately half of the true activation energy. This is the behavior shown in Figures 9-8 and 9-14. However, if the external transport resistances become appreciable before the effectiveness factor reaches its asymptotic behavior, then the distinct linear portion of the plot in the intermediate temperature range may be obscured.

9.4.1.4 Temperature Difference Between Bulk Fluid and Catalyst Surface

Equations (9-38) and (9-39) can be solved for $R_{A,P}$ and the results equated to give

$$k_c A_G (C_{A,B} - C_{A,s}) = R_{A,P} = (hA_G/\Delta H_R)(T_B - T_s) \tag{9-46}$$

The mass- and heat-transfer coefficients k_c and h can be written in terms of the Colburn j-factors for mass and heat transfer j_D and j_H, respectively:

$$j_D \equiv k_c^0 \rho \mathrm{Sc}^{2/3}/G$$
$$j_H \equiv h \mathrm{Pr}^{2/3}/c_{p,m} G$$

In these expressions, Sc is the Schmidt number, Pr is the Prandtl number, $c_{p,\mathrm{m}}$ is the mass heat capacity (energy/mass-temperature), G is the superficial mass velocity (mass/area-time), and k_c^0 is the mass-transfer coefficient when the net molar flux is zero. The difference between k_c and k_c^0 will be discussed in more detail later in this section. Temporarily, we will ignore the difference and assume that $k_\mathrm{c} = k_\mathrm{c}^0$.

Using these definitions in Eqn. (9-46) leads to

$$T_\mathrm{s} - T_\mathrm{B} = \left(\frac{j_\mathrm{D}}{j_\mathrm{H}}\right)\left(\frac{\mathrm{Pr}}{\mathrm{Sc}}\right)^{2/3}\left(\frac{(-\Delta H_\mathrm{R})C_{\mathrm{A,B}}}{\rho c_{\mathrm{p,m}}}\right)\left[\frac{C_{\mathrm{A,B}} - C_{\mathrm{A,s}}}{C_{\mathrm{A,B}}}\right]$$

The ratio Sc/Pr is the Lewis number (Le), introduced in Section 9.3.7. The quantity $(-\Delta H_\mathrm{R})C_{\mathrm{A,B}}/\rho c_{\mathrm{p,m}}$ is the adiabatic temperature change (ΔT_ad). For many types of reactor and regimes of flow, the Chilton–Colburn analogy between heat and mass transfer[13] leads to $j_\mathrm{H}/j_\mathrm{D} \cong 1$. Introducing these simplifications reduces the above equation to

$$T_\mathrm{s} - T_\mathrm{B} \cong \left(\frac{\Delta T_\mathrm{ad}}{\mathrm{Le}^{2/3}}\right)\left(\frac{C_{\mathrm{A,B}} - C_{\mathrm{A,s}}}{C_{\mathrm{A,B}}}\right) \tag{9-47}$$

Equation (9-47) shows that the temperature difference between the bulk fluid and the surface of the catalyst particle is directly proportional to the concentration difference between these points. As $C_{\mathrm{A,B}} - C_{\mathrm{A,s}}$ becomes larger, so does $T_\mathrm{s} - T_\mathrm{B}$.

Suppose that the reaction is *controlled* by external mass transfer, so that $C_{\mathrm{A,s}} \cong 0$. For this situation, Eqn. (9-47) becomes

$$T_\mathrm{s} \cong T_\mathrm{B} + \Delta T_\mathrm{ad}/\mathrm{Le}^{2/3} \tag{9-48}$$

For gases, where $\mathrm{Le} \cong 1$, Eqn. (9-48) shows that the surface of the catalyst can be very hot (or cold) when the reaction is controlled by external mass transfer. In fact, the temperature at the surface of the catalyst particle can approach the adiabatic reaction temperature, $T_\mathrm{B} + \Delta T_\mathrm{ad}$. This is especially troublesome for exothermic reactions where ΔT_ad is large.

When the catalyst particle is substantially hotter than the bulk fluid, several negative effects can occur. First, the reaction selectivity often will be much lower at high temperatures. One example is the gas-phase partial oxidation of ethylene (C_2H_4) to ethylene oxide (C_2H_4O), which is accompanied by the further oxidation of ethylene and ethylene oxide to CO, CO_2, and H_2O.

$$\begin{aligned} C_2H_4 + 1/2O_2 &\rightarrow C_2H_4O \\ C_2H_4O + 3/2O_2 &\rightarrow 2CO + 2H_2O \\ C_2H_4 + 2O_2 &\rightarrow 2CO + 2H_2O \\ CO + 1/2O_2 &\rightarrow CO_2 \end{aligned}$$

The activation energies of the second, third, and fourth reactions are higher than the activation energy of the partial oxidation of ethylene to ethylene oxide. Therefore, if a transport limitation causes the catalyst surface to be hotter than the bulk fluid, the selectivity to ethylene oxide will be lower than expected, based on the bulk fluid temperature.

In addition to selectivity loss, high temperatures may cause a catalyst to deactivate, due to phenomena such as collapse of the pore structure and growth of the particles of the active component. Particle growth can be particularly troublesome when the active component is

[13] See Bird, R. B., Stewart, W. E., and Lightfoot, E. N., *Transport Phenomena*, 2nd edition, John Wiley & Sons, Inc. (2002), p. 682.

metal nanoparticles that grow by thermal sintering, leading to a substantial loss of metal surface area. For example, "coke" can accumulate on certain metal-containing catalysts during use, causing the catalyst to lose activity over time. At some point, the reactor must be shut down so that the catalyst can be regenerated by burning off the coke.

$$\text{``coke''} + O_2 \rightarrow CO_2 + H_2O$$

The adiabatic temperature change for this reaction is very high. If air is used, ΔT_{ad} can be on the order of several thousand degree Celsius, depending on conditions and on the composition of the coke. If the coke combustion reaction is controlled or substantially influenced by external mass transfer, the surface of the catalyst will become very hot. The metal nanoparticles might agglomerate, leading to a loss of metal surface area, and perhaps to a loss of surface area of the support itself. One way to control the temperature of the catalyst surface is to reduce the O_2 concentration in the regenerating gas to a few percent, thus reducing ΔT_{ad}. This is a very common technique, especially at the start of regeneration. It is also important to carry out the regeneration at the highest feasible gas velocity, so that the values of h and k_c will be as high as possible.

Let's examine the significance of Eqn. (9-47) by means of a numerical illustration. Suppose that a gas-phase reaction with $\Delta T_{ad} = 100$ K is taking place on the external surface of a nonporous catalyst particle. We assume a nonporous particle in order to isolate the effect of external mass transfer, and to avoid having to analyze the effect of changes in T_s and $C_{A,s}$ on the effectiveness factor. Catalysts that are essentially nonporous are used in several commercial processes. For example, a Pt/Rh metal gauze is used for ammonia oxidation to nitric oxide and for the reaction of NH_3, O_2, and CH_4 to produce HCN, and a promoted Ag on α-Al_2O_3 is used in many ethylene oxide plants.

Assume that the difference between the concentration of reactant A in the bulk and at the catalyst surface is relatively small, for example, $C_{A,s} = 0.90\,C_{A,B}$. If Le $= 1$, Eqn. (9-47) predicts that $T_s - T_B = 100[(C_{A,B} - 0.90C_{A,B})/C_{A,B}] = 10$ K. The concentration decrease and the temperature increase will have opposite effects on the reaction rate. Depending on the reaction order, a concentration difference of 10% might cause the reaction rate at the surface to be 10–20% lower than if the concentration at the surface was $C_{A,B}$. However, if the catalyst surface is 10 K hotter than the bulk fluid, the actual rate at the surface can be roughly a factor of 2 greater than if the surface was at the temperature of the bulk fluid.

This simple example shows that the effect of temperature is usually more important than the effect of concentration when $(C_{A,B} - C_{A,s})/C_{A,B}$ is relatively small. A modest degree of external mass-transfer influence can cause the rate of an exothermic reaction to be higher than it would be if the external surface were at bulk conditions. However, as the external mass-transfer resistance becomes more severe, the effect of concentration becomes more and more important, until the reaction finally becomes mass-transfer controlled. This behavior will be illustrated in the next section.

9.4.2 Diagnostic Experiments

If a heterogeneous catalytic reaction is influenced or controlled by external transport, the actual rate of reaction will depend on the transport coefficients, h and k_c. These coefficients depend on the velocity of the fluid *relative to the catalyst particle*, which will be represented as $\vec{v}$.

Diagnostic experiments can be carried out to determine whether a heterogeneous catalytic reaction is being affected by external transport. These experiments are based on a very simple concept. If a heterogeneous catalytic reaction is controlled by intrinsic kinetics, or influenced *only* by intrinsic kinetics and internal transport, then the actual reaction rate will not depend on $\vec{v}$. In either circumstance, the outlet conversion from a continuous reactor, or the final conversion in a batch reactor, will not depend on $\vec{v}$, provided that all of the other operating conditions remain unchanged. On the other hand, if the reactor performance *does* depend on $\vec{v}$, at otherwise identical conditions, external transport probably influences or controls the reaction.

9.4.2.1 Fixed-Bed Reactor

To illustrate this idea, consider a reactor that is packed with catalyst particles. A fluid containing the reactant(s) flows through the fixed bed of particles. The reactor could be a differential reactor, as described in Chapter 6, or it could be an integral, ideal plug-flow reactor. Let's consider the PFR. The design equation is

$$\tau_0 = C_{A0} \int_0^{x_A(\text{out})} \frac{dx_A}{-r_A} \tag{3-35a}$$

We know that the rate $-r_A$ will depend on temperature and concentration. If the reaction is controlled or influenced by external transport, $-r_A$ also will depend on $\vec{v}$.

Suppose that we carry out two experiments in an ideal PFR at exactly the same value of τ_0, at exactly the same inlet conditions, and at exactly the same temperature conditions, but at different values of $\vec{v}$. If the outlet conversions are *different* for these two experiments, external transport must either control or influence the reaction rate.

Diagnostic experiments can be designed to determine whether a heterogeneous catalytic reaction is controlled or influenced by external transport, based on the above idea. Consider a tubular reactor with an inner diameter D_i loaded with a weight of catalyst W. The height of catalyst in the reactor is L. An experiment is carried out with a volumetric inlet flow rate of υ_0. The concentrations of the components of the feed stream are C_{i0}, the inlet temperature is T_0, and the reactor is operated either isothermally or adiabatically. The conversion of A leaving the reactor x_A is measured. This situation is shown schematically in Figure 9-15

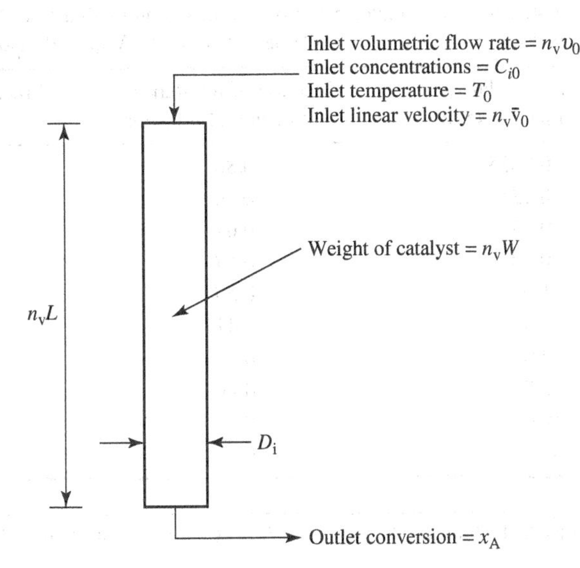

Figure 9-15 Schematic illustration of diagnostic experiments for external transport control/influence in a fixed-bed reactor. The linear velocity through the catalyst bed can be varied, without changing the space time, by changing the weight of catalyst in the reactor and the volumetric flow rate to the reactor in direct proportion, leaving the inner diameter of the reactor unchanged.

and corresponds to a base case, which is designated as $n_v = 1$. The space time at inlet conditions τ_0 is W/υ_0. The inlet superficial linear velocity is $\vec{v}_0 = 4\upsilon_0/\pi D_1^2$.

At this point, we might be tempted to just increase the volumetric feed rate to the reactor, leaving the amount of catalyst in the reactor constant. This certainly would increase the linear velocity though the catalyst bed. However, it would also reduce the space time, and the design equation tells us that changing the space time will change the outlet conversion.

Instead, a second experiment is carried out in the same tube with a different amount of catalyst, say $2W$. The height of the catalyst bed will be $2L$, since the inner diameter of the reactor has not changed. The inlet flow rate is increased to $2\upsilon_0$, so that the space time does not change ($\tau_0 = 2W/2\upsilon_0 = W/\upsilon_0$). However, the inlet superficial linear velocity $\vec{v}_0$ has doubled to $8\upsilon_0/\pi D_{\text{in}}^2$. The two experiments have the same space time, but the linear velocity in the second experiment is twice that in the first. Additional experiments are now carried out, each with a different value of n_v ($\vec{v}$), until a range of linear velocity has been mapped.

How wide a range of linear velocity must be studied? The following table shows the results of a calculation of the effect of linear velocity on the fractional conversion from an ideal PFR. The reaction considered is $A \rightarrow R$, which was assumed to be irreversible and first order in A, with $\Delta H_R = 0$. The reactor was assumed to be isothermal. The mass-transfer coefficient was assumed to be proportional to the square root of linear velocity, a relationship that is reasonably typical for flow-through packed beds.[14] Finally, to provide a starting point for the calculation, $\eta k_v l_c/k_c$ was taken to be 1.0 when the outlet fractional conversion of A was 0.50. When $\eta k_v l_c/k_c = 1$, the resistance to external transport is equal to the resistance to reaction inside the catalyst particle. At this condition, "n_v" arbitrarily was assigned a value of 1.

As the linear velocity increases (n_v increases), $\eta k_v l_c/k_c$ decreases and the resistance to external transport becomes a smaller fraction of the overall resistance, as shown in Table 9-1.

These calculations show that the outlet conversion is sensitive to linear velocity when the resistance to external mass transfer is a significant fraction of the total resistance. However, when the fraction of the overall resistance due to external transport is less than about 0.40, doubling the linear velocity produces such a small change in the fractional

Table 9-1 Calculated Fractional Conversion of Reactant A as a Function of Linear Velocity (Space-Time = Constant, $\Delta H_R = 0$, Isothermal Reactor)

n_v (=relative linear velocity)	Fraction of resistance due to external transport	Fractional conversion of A in outlet, x_A
0.0625	0.80	0.24
0.125	0.74	0.30
0.25	0.67	0.37
0.50	0.59	0.44
1.0	0.50	0.50
2.0	0.41	0.56
4.0	0.33	0.60
8.0	0.26	0.64
16.0	0.20	0.67
∞	0	0.75

[14] Geankopolis, C. J., *Transport Processes and Unit Operations*, 2nd edition, Allyn and Bacon, (1983), p. 436–437.

conversion that the difference between the two conversions may be within experimental error.

The results in Table 9-1 suggest that a reasonably wide range of linear velocities for example, a factor of at least 4, must be used in a well-designed set of diagnostic experiments. A wide range of velocities is necessary to ensure that changes in outlet conversion are large enough to be well outside the range of experimental error.

Some possible results from such diagnostic experiments are presented in the following two illustrations:

Illustration A $(\Delta H_R = 0)$

If the heat of reaction is essentially zero, heat transfer to and from the catalyst particle is not required in order for the reaction to take place at steady state. There will be no temperature gradients, either inside the catalyst particle or through the boundary layer. Therefore, the only external transport step that affects the reaction rate is mass transfer through the boundary layer.

Let's analyze this case and predict the shape of a plot of outlet conversion versus linear velocity. First, at a very high inlet linear velocity, k_c will be large and the *maximum* value of the rate of reaction inside the catalyst particle will be much less than the *maximum* value of the rate of mass transfer through the boundary layer, i.e., $\eta k_v l_c C_{A,B}^{n-1}/k_c \ll 1$. In this inequality, n is the true order of the reaction. There will be essentially no concentration difference across the boundary layer, and the concentration profile will resemble those shown in either Figure 9-11 or 9-12. The measured rate of reaction will not be sensitive to $\vec{v}_0$. For this situation,

$$\tau_0 = C_{A0} \int_0^{x_A(\text{out})} \frac{dx_A}{\eta k C_{A,B}^n}$$

Only the intrinsic kinetics $(kC_{A,B}^n)$ and the effectiveness factor (η) are required to calculate the outlet conversion, providing that $\vec{v}_0$ is sufficiently high that $C_{A,s} \cong C_{A,B}$ over the whole length of the reactor.

As the linear velocity is decreased, the value of k_c will decrease. As a result, the concentration difference $(C_{A,B} - C_{A,s})$ will become larger and $C_{A,s}$ will decrease. If n is greater than zero, the decrease in $C_{A,s}$ will cause the reaction rate to decrease. At some point, the measured outlet conversion will become sensitive to the value of k_c, and to the value of $\vec{v}_0$.

As $\vec{v}_0$ is decreased further, the value of k_c continues to decrease. The measured rate declines monotonically with $\vec{v}_0$. Eventually, k_c becomes so low that the reaction is *controlled* by external mass transport. At this point, the outlet conversion is most sensitive to changes in $\vec{v}_0$.

Figure 9-16 is a plot of the outlet fractional conversion of reactant A from a plug-flow reactor, at constant space time, τ_0, versus the linear or mass velocity in the catalyst bed. This figure is derived from the same calculations that were used to develop Table 9-1. The fraction of the total resistance due to external mass transport is also shown in Figure 9-16, as a function of velocity.

The behavior shown in Figure 9-16 is consistent with the above analysis. At high velocities, the conversion is not very sensitive to velocity. The sensitivity increases and the conversion declines as linear velocity decreases. At very low velocities, the conversion falls off very rapidly as the reaction approaches a condition of external mass-transfer control.

For this example, the conversion is 0.75 when the external mass-transfer resistance is negligible $(k_c \rightarrow \infty)$. However, the plot of conversion versus velocity becomes very flat before this asymptote is reached.

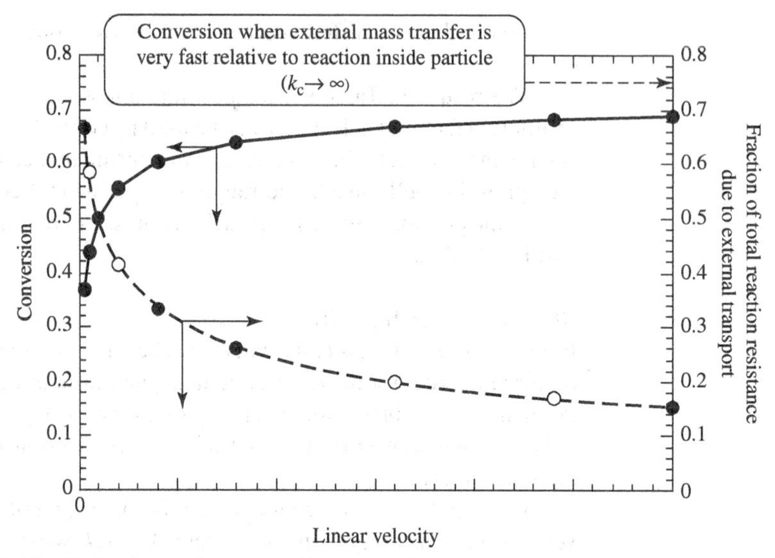

Figure 9-16 Outlet conversion versus linear velocity at constant space-time for the first-order reaction A → B in an isothermal, plug-flow reactor ($\Delta H_R = 0$).

In evaluating the results of diagnostic experiments, it is safe to conclude that the external mass-transfer resistance is significant when the conversion at constant space time varies with velocity. However, the converse is not true. Just because the conversion does not *appear* to depend on velocity does not mean that the resistance to external mass transport has been *totally* eliminated. The four points on the far right of the conversion curve in Figure 9-16 could easily be within experimental accuracy, and appear to be flat. Nevertheless, the external transport resistance is about 20% of the total resistance in this region, and the outlet conversions are 5–10% below the asymptotic value.

EXERCISE 9-9

Suppose that the reaction for which Figure 9-16 was developed had been endothermic, i.e., $\Delta H_R > 0$, instead of thermally neutral. What would a graph of outlet conversion versus linear velocity look like?

Illustration B ($\Delta H_R < 0$)

For an exothermic reaction, the catalyst surface will be *hotter* than the bulk fluid if the transport resistance through the boundary layer is significant. Moreover, as pointed out in connection with Eqn. (9-47), the temperature of the catalyst surface increases as the concentration of reactant at the surface decreases. The rate of reaction will be increased by the higher temperature, but decreased by the lower reactant concentration. Under some circumstances, especially when the external transport resistances just begin to become significant, the effect of temperature on the reaction rate will be more important than the effect of concentration.

Consider a situation where the reactor in Figure 9-15 is operating at such a high linear velocity that the external transport resistance is insignificant. In this region, the measured outlet conversion does not depend on $\vec{v}_0$, as shown by the flat portion of the line on the right-hand side of Figure 9-17. As the linear velocity is reduced, the surface concentration $C_{A,s}$ begins to drop and the surface temperature T_s begins to increase. If the activation energy is reasonably high, the increase in the rate due to the higher surface temperature will be larger than the decrease in rate due to the lower surface concentration. This will cause the actual rate to increase, and the outlet conversion to increase, as shown in Figure 9-17.

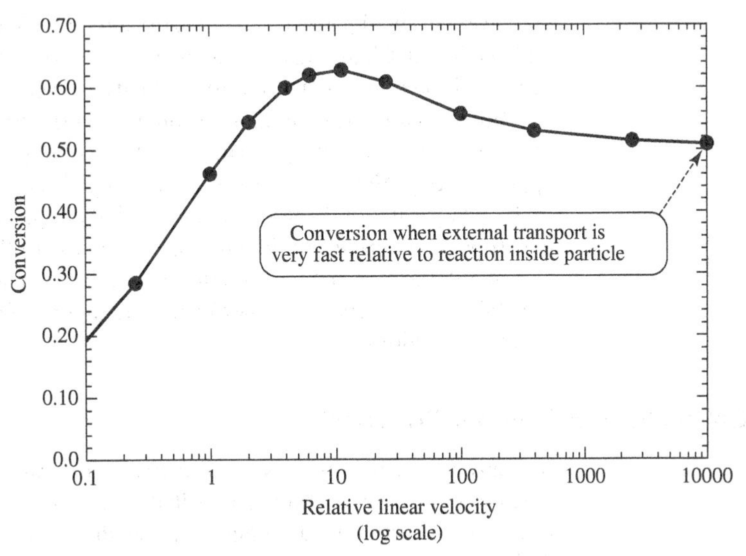

Figure 9-17 Outlet conversion versus linear velocity at constant space time for an exothermic, first-order reaction $A \rightarrow B$ in an isothermal, plug-flow reactor.

As the velocity is further reduced, the concentration at the surface of the catalyst particle eventually becomes so low that the actual reaction rate begins to fall, despite the high surface temperature. The outlet conversion falls with the decreasing rate. This decrease in conversion continues as $\vec{v}_0$ is reduced to very low values.

Figure 9-17 illustrates one possible outcome of a series of diagnostic experiments in an isothermal PFR in which an exothermic reaction takes place. The actual shape of a plot of outlet conversion versus linear or mass velocity through the PFR will depend on a number of factors, principally the values of the heat of reaction and the activation energy. If either of these values is very low, the maximum in the curve may not appear, and the plot may resemble Figure 9-16. However, if either ΔH_R or E is high, individual catalyst particles in the reactor can have multiple steady states. In this case, a plot of outlet conversion versus linear velocity may exhibit discontinuities and will depend on how the experiments are conducted.

In this discussion, the design of diagnostic experiments and the analysis of the resulting data have been illustrated for an ideal PFR operating at relatively high conversion. However, the concepts discussed apply equally well to a fixed-bed reactor operating in the differential mode.

9.4.2.2 Other Reactors

Diagnostic experiments such as those described above can be carried out in a straightforward manner when the velocity of the fluid relative to the catalyst can be controlled by the experimenter, and can be varied without changing any other parameters that affect the reaction rate. Unfortunately, this is not always the case.

Consider a batch reactor where very small particles of a solid catalyst are suspended in a liquid and kept in suspension by mechanical agitation. The reaction takes place between components of the liquid, and the products are soluble in the liquid. What is the velocity of the liquid relative to the catalyst particles, and how can it be varied in a systematic fashion?

We might be tempted to say that the velocity of the liquid relative to the particles was related to the design and rotational velocity of the agitator, and that the relative velocity could be changed by changing the rotational speed of the agitator. Unfortunately, this approach is not effective. For small particles (1–100 μm) and for typical liquid viscosities, the particles move with the liquid. Changing the rotational speed of the agitator changes the liquid velocity, but has very little effect on the *relative* velocity between the liquid and the catalyst particles.

For this case, gravitational settling is the primary cause of a relative velocity between the liquid and the catalyst particles. The settling velocity can be calculated, at least approximately. However, it cannot be changed without changing the physical properties of the catalyst and/or the liquid. It is essentially impossible to perform the kind of diagnostic experiments described in the previous section with this kind of reactor.

In order to evaluate the influence of external transport in a reactor that contains a slurry of catalyst, it is necessary to perform a calculation. This approach will be described in the following sections.

9.4.3 Calculations of External Transport

In many cases, it is necessary to *estimate* the rate at which a heterogeneous catalytic reaction will proceed, if it is controlled by external mass transfer. Alternatively, it may be necessary to estimate the concentration difference $(C_{A,B} - C_{A,s})$ and the temperature difference $(T_B - T_s)$ that are required to sustain a known or measured rate of reaction. Calculations of $C_{A,B} - C_{A,s}$ and $T_B - T_s$ are the only way to evaluate the influence of external transport when definitive diagnostic experiments are not feasible. Calculations such as these can be performed using Eqns. (9-38) and (9-40), provided that the transport coefficients k_c and h are known, or can be obtained from correlations.

The literature contains heat- and mass-transfer coefficient correlations for many important types of reactor, e.g., fixed beds, fluidized beds, slurry reactors, and straight-channel monoliths. This makes it possible to use Eqns. (9-38) and (9-40) in many different contexts.

9.4.3.1 Mass-Transfer Coefficients

Consider a catalyst in which the reaction

$$\sum_i \nu_i A_i = 0$$

is taking place *at steady state*. Visualize a small segment of the catalyst surface, as shown below. If there is a relative velocity between the fluid and the catalyst, a boundary layer will be established adjacent to the external surface of the catalyst. As an approximation, we will consider this boundary layer to be a stagnant film of fluid with a thickness δ.

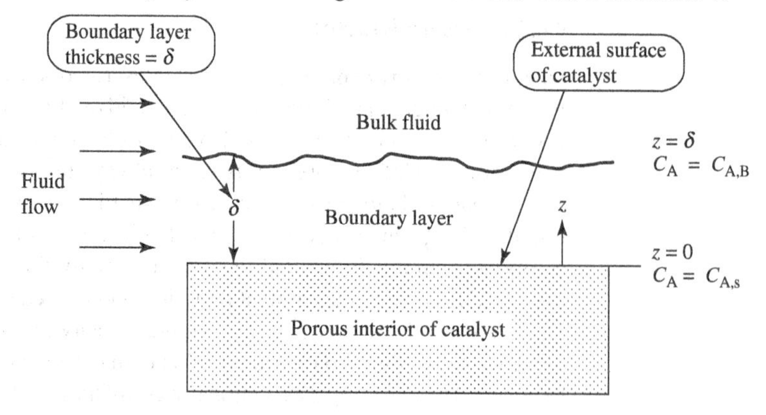

From the section "Bulk (Molecular) Diffusion" (Section 9.3.4.2); the flux of reactant A through the boundary layer, in the z-direction, is given by

$$\vec{N}_{A,z} = y_A \sum_{i=1}^{N} \vec{N}_{i,z} - D_{A,m}\frac{dC_A}{dz} \tag{9-19}$$

where $D_{A,m}$ is the diffusivity of A in mixture (area/time), y_A is the mole fraction of A, and $\vec{N}_{i,z}$ is the molar flux of species "i" in the z direction (moles/area-time).

The flux $\vec{N}_{i,z}$ is positive when it is directed in the $+z$ direction. When "i" is a reactant, $\vec{N}_{i,z}$ is negative, and when "i" is a product, $\vec{N}_{i,z}$ is positive. The following analysis is concerned only with transport normal to the catalyst surface, i.e., in the z-direction. For simplicity, let's drop the subscript "z".

Because of reaction stoichiometry, $\vec{N}_i/\nu_i = \vec{N}_A/\nu_A$ at steady state. Therefore,

$$\sum_{i=1}^{N} \vec{N}_i = \left(\sum_{i=1}^{N} \nu_i \vec{N}_A/\nu_A\right) = \left(\frac{\vec{N}_A}{\nu_A}\right)\sum_{i=1}^{N} \nu_i = \Delta\nu\left(\frac{\vec{N}_A}{\nu_A}\right)$$

where $\Delta\nu = \sum_{i=1}^{N} \nu_i$. At steady state, $\sum_{i=1}^{N} \vec{N}_i > 0$ when $\Delta\nu > 0$. In other words, there will be a net flux directed *away* from the catalyst surface, into the bulk fluid, when the number of moles of products is greater than the number of moles of reactants. Similarly, $\sum_{i=1}^{N} \vec{N}_i = 0$ when $\Delta\nu = 0$, and $\sum_{i=1}^{N} \vec{N}_i < 0$ when $\Delta\nu < 0$. The *net* molar flux is zero if there is no change in the number of moles on reaction, and the net flux is less than 0, i.e., it is directed *toward* the catalyst surface, if the number of moles decreases on reaction.

Substituting for $\sum_{i=1}^{N} \vec{N}_i$ in Eqn. (9-19) and rearranging

$$\frac{dC_A}{dz} - \left(\frac{\Delta\nu \vec{N}_A}{D_{A,m} C \nu_A}\right) C_A = -\frac{\vec{N}_A}{D_{A,m}} \tag{9-49}$$

In this equation, C is the total molar concentration of the mixture (moles/volume), such that $y_i = C_i/C$. This differential equation is subject to the boundary conditions:

$$C_A = C_{A,s}; \quad z = 0$$
$$C_A = C_{A,B}; \quad z = \delta$$

For a catalyst particle at steady state, $\vec{N}_A$ is constant, i.e., $\vec{N}_A \neq f(z)$. The total pressure in the boundary layer is essentially constant, so that $CD_{A,m}$ is approximately constant. For this situation, the solution to Eqn. (9-49) is

$$\ln\left(\frac{C + \gamma C_{A,B}}{C + \gamma C_{A,s}}\right) = -\left(\frac{\gamma \vec{N}_A}{CD_{A,m}}\right)\delta \tag{9-50}$$

where $\gamma = \Delta\nu/[-\nu_A]$.

EXERCISE 9-10

Prove Eqn. (9-50).

Now let $k_c^0 \equiv D_{A,m}/\delta$. Note that the value of k_c^0 is species dependent because $D_{A,m}$ is the diffusion coefficient *of species A* in the mixture. Substituting this relationship into Eqn. (9-50) and rearranging gives

$$-\vec{N}_A = \frac{k_c^0(C_{A,B} - C_{A,s})}{\dfrac{(1+\gamma y_{A,B}) - (1+\gamma y_{A,s})}{\ln\left(\dfrac{1+\gamma y_{A,B}}{1+\gamma y_{A,s}}\right)}} \tag{9-51}$$

Let the denominator of Eqn. (9-51) be defined as y_{fA}, which is referred to as the *film factor based on mole fraction*, or the mole fraction film factor.

$$y_{fA} = \frac{(1+\gamma y_{A,B}) - (1+\gamma y_{A,s})}{\ln\left(\dfrac{1+\gamma y_{A,B}}{1+\gamma y_{A,s}}\right)} \tag{9-52}$$

The parameter y_{fA} is just the log-mean value of the quantity $(1+\gamma y_A)$ and is dimensionless.

If there is no change in the number of moles on reaction, $\sum_{i=1}^{N} N_i = \Delta\nu = 0$ and $\gamma = 0$. For this situation, $y_{fA} = 1$. If $\Delta\nu > 0$, there is a net creation of moles as the reaction occurs $(\sum_{i=1}^{N} \vec{N}_i > 0)$. Then $\gamma > 0$ and $y_{fA} < 1$. Conversely, if $\gamma < 0$, there is a net reduction of moles as the reaction proceeds; $\sum_{i=1}^{N} \vec{N}_i < 0$ and $y_{fA} > 1$.

Equation (9-51) may be simplified to

$$-\vec{N}_A = \left(\frac{k_c^0}{y_{fA}}\right)(C_{A,B} - C_{A,s}) \tag{9-53}$$

Let k_c, the mass-transfer coefficient of A based on concentration, be defined as

Mass-transfer coefficient when net flux $\neq 0$

$$\boxed{k_c \equiv (k_c^0/y_{fA})} \tag{9-54}$$

Equations (9-53) and (9-54) can be combined to give

$$-\vec{N}_A = k_c(C_{A,B} - C_{A,s}) \tag{9-55}$$

This is just Eqn. (9-37), although the sign convention for $\vec{N}_A$ is not present in Eqn. (9-37).

From Eqns. (9-52) and (9-54), it is clear that *the value of the mass-transfer coefficient depends on the stoichiometry of the reaction*, i.e., on the value of y_{fA}, which in turn depends on γ. When $\gamma = 0$ and $y_{fA} = 1$, the mass-transfer coefficient k_c is equal to k_c^0. Recalling the definition $k_c^0 \equiv D_{Am}/\delta$, we can interpret k_c^0 as the mass-transfer coefficient when transport through the boundary layer is purely diffusive, i.e., when there is no *net* convective molar flux $(\sum_{i=1}^{N} \vec{N}_i = 0)$.

When there is a net molar flux either toward the catalyst surface or away from the catalyst surface, the value of the mass-transfer coefficient will depend on the magnitude and the sign of the flux. For example, when there is a net creation of moles as the reaction occurs, $\sum_{i=1}^{N} \vec{N}_i > 0$, $\Delta\nu > 0, \gamma > 0$ and $y_{fA} > 1$. Therefore, from Eqn. (9-54), $k_c < k_c^0$.

Physically, species A must diffuse toward the catalyst surface *against* a net convective flux in the opposite direction. If A is a reactant, some A is carried away from the catalyst surface as a result of the opposing convective flux. The diffusive flux of A toward the catalyst surface must be higher in order to compensate. The effect of the opposing convective flux is to reduce the apparent mass-transfer coefficient k_c.

When there is a net consumption of moles on reaction, $\sum_{i=1}^{N} \vec{N}_i < 0$, $\Delta\nu < 0$, $\gamma < 0$, and $y_{fA} < 1$. For this situation, $k_c > k_c^0$. The diffusive flux and the net convective flux are in the same direction; some reactant is transported to the catalyst surface by diffusion and some by convection. Consequently, the apparent mass-transfer coefficient k_c is greater than it would be in the absence of the flux.

9.4.3.2 Different Definitions of the Mass-Transfer Coefficient

All of the discussion so far has involved k_c, the mass-transfer coefficient based on a concentration driving force. However, the mass-transfer coefficient may be based on other driving forces, notably mole fraction and partial pressure (p). Thus, we can write

$$
\begin{aligned}
-\vec{N}_A &= k_c(C_{A,B} - C_{A,s}) \\
-\vec{N}_A &= k_y(y_{A,B} - y_{A,s}) \\
-\vec{N}_A &= k_p(p_{A,B} - p_{A,s})
\end{aligned}
$$

The units of the various k's are

$$
\begin{aligned}
k_c &= \text{volume/area-time} = \text{length/time} \\
k_y &= \text{moles/area-time} \\
k_p &= \text{moles/area-time-pressure}
\end{aligned}
$$

Relationships between the various mass-transfer coefficients can be derived by equating the fluxes. For example,

$$\vec{N}_A = k_y(y_{A,B} - y_{A,s}) = k_c(C_{A,B} - C_{A,s})$$

If differences in temperature and total pressure across the boundary layer are ignored, then the *total* mixture concentration C is the same at the catalyst surface as it is in the bulk, and C can be divided into both sides of the above equation.

$$\left(\frac{k_y}{C}\right)(y_{A,B} - y_{A,s}) = k_c\left(\frac{C_{A,B}}{C} - \frac{C_{A,s}}{C}\right) = k_c(y_{A,B} - y_{A,s})$$

$$k_y = k_c C$$

Using the relationship $k_c = k_c^0/y_{fA}$, we can write

$$k_y = C\left(\frac{k_c^0}{y_{fA}}\right) = \frac{k_y^0}{y_{fA}}$$

where $k_y^0 = Ck_c^0 = CD_{A,m}/\delta$

For an ideal gas, an exactly equivalent procedure can be used to relate the mass-transfer coefficient based on partial pressure to those based on mole fraction and concentration. If P is the total pressure

$$k_p = \left(\frac{k_y}{P}\right) = \left(\frac{k_y^0}{P y_{fA}}\right) = \left(\frac{k_p^0}{y_{fA}}\right)$$

$$k_p = \left(\frac{k_c}{RT}\right) = \left(\frac{k_c^0}{y_{fA} RT}\right) = \left(\frac{k_p^0}{y_{fA}}\right)$$

where $k_p^0 = k_c^0/RT = D_{A,m}/\delta RT$.

9.4.3.3 Use of Correlations

Most correlations of the mass-transfer coefficient involve one of two dimensionless groups, either the j-factor for mass-transfer j_D or the Sherwood number Sh. The j-factor for mass transfer (j_D) is defined as

$$j_D \equiv \frac{k_c^0 C M_m}{G} Sc^{2/3} = \frac{k_c^0 \rho}{G} Sc^{2/3}$$

An alternative expression for j_D is

$$j_D \equiv \frac{k_c^0 C}{G/M_m} Sc^{2/3} = \frac{k_c^0 C}{G_M} Sc^{2/3}$$

The Sherwood number is defined as

$$Sh \equiv \frac{k_c^0 d_p}{D_{A,m}}$$

The j-factor and the Sherwood number are related by

$$j_D = \frac{Sh}{Re Sc^{1/3}}$$

where Re is the Reynolds number (based on particle diameter), G is the superficial mass velocity (mass/area-time), G_M is the superficial molar velocity (moles/area-time), M_m is the molecular weight of the mixture, d_p is the particle diameter (length), Sc is the Schmidt number $= \mu/\rho D_{A,m}$, μ is the mixture viscosity (mass/length-time), and ρ is the mixture density (mass/volume).

Valid j-factor and Sh correlations are *always* based on k_c^0. This point is especially important when the analogy between heat and mass transfer, i.e., $j_D = j_H$, is used. To calculate the *actual* mass-transfer coefficient from a j-factor or Sh correlation, we must calculate k_c^0 (or k_y^0 or k_p^0) from the correlation, then use Eqn. (9-52) to calculate y_{fA}, and finally use Eqn. (9-54) to calculate k_c (or k_y or k_p).

Finally, when we are interested only in the absolute magnitude of N_A, the sign convention in Eqn. (9-55) can be dropped to give

$$\vec{N}_A = k_c(C_{A,B} - C_{A,s}) \tag{9-37}$$

However, the sign convention is critical to the correct calculation of $\sum_{i=1}^{N} \vec{N}_i$, γ, and y_{fA}, and cannot be dropped when calculating these parameters.

EXAMPLE 9-12
Estimating the Rate of Reaction for Mass-Transfer Control

The hydrogenation of nitrobenzene to aniline in a dilute solution of ethanol was studied at 30 °C and a H_2 pressure of 775 mmHg using a 3 wt.% Pd/carbon catalyst.[15] The average diameter of the catalyst was 16 μm, the particle density ρ_p was about 0.95 g/cm^3 and the porosity ε was about 0.50. In one experiment, a reaction-rate of 0.034 mol H_2/min-g catalyst was measured. At the conditions of the experiment, the solubility of H_2 in ethanol was about 3.5×10^{-7} mol/cm^3, the viscosity of ethanol was about 0.01 g/cm-s, the specific gravity of ethanol was about 0.78, and the diffusion coefficient of H_2 in ethanol was about 5×10^{-5} cm^2/s.

Estimate the rate of consumption of H_2 if the reaction was controlled by transport of H_2 from the liquid phase to the external surface of the catalyst particles. Is this step an important resistance under the actual conditions?

APPROACH

If the reaction is controlled by mass transfer of H_2 from the bulk liquid to the catalyst particles, then $-r_{H_2}$ (mol H_2/min-g-cat) $= k_c C_{H_2,B} A_G / V_G \rho$. All of these variables are specified in the problem statement, except for k_c. This parameter will be estimated from a correlation, and the value of $-r_{H_2}$ will be calculated. Finally, the experimental value of $-r_{H_2}$ will be compared with the predicted value.

SOLUTION

The mass-transfer coefficient can be estimated from a correlation that derives from the original work of Brian and Hales:[16]

$$Sh^2 = 16 + 4.84\text{Pe}^{2/3}$$

In this equation, Pe is the Peclet number, defined as $\text{Pe} = d_p \vec{v} / D_{A,m}$. As mentioned previously, $\vec{v}$ is the velocity of the fluid *relative to the catalyst particle*. This velocity is approximately equal to the terminal settling velocity of the particles in the fluid, so that,

$$\vec{v} = g d_p^2 (\rho_a - \rho_l) / 18\mu$$

Here, g is the acceleration due to gravity, ρ_a is the apparent density of the catalyst particle, ρ_l is the density of the liquid, d_p is the diameter of the catalyst particle, and μ is the viscosity of the liquid. Combining these relationships gives

$$Sh^2 = 16 + 4.84 \left(\frac{g d_p^3 (\rho_a - \rho_l)}{18 \mu D_{A,m}} \right)^{2/3} \tag{9-56}$$

The apparent density of the catalyst particle is the density of the particle, taking the density of the liquid in the pores of the catalyst into account. In this case, the apparent density is $0.95 + 0.5 \times 0.78 = 1.34$ g/cm^3. The mass-transfer coefficient of hydrogen, for a zero net flux, is obtained by substituting values into this equation.

$$\left(\frac{k_c^0 \times 16 \times 10^{-4}}{5 \times 10^{-5}} \right)^2 = 16 + 4.84 \left(\frac{981 \times 16^3 \times 10^{-12} (1.34 - 0.78)}{18 \times 0.01 \times 5 \times 10^{-5}} \right)^{2/3}$$

$$k_c^0 = 0.132 \text{ cm/s}$$

Let "A" denote H_2. For the hydrogenation of nitrobenzene to aniline, $\gamma = \Delta\nu / [-\nu_A] = (3-4)/[-3] = -0.33$. If the reaction is controlled by external mass transfer of hydrogen, $y_{A,s} = 0$. If the liquid is essentially pure ethanol and the concentration of H_2 in the liquid is 3.5×10^{-7} mol/cm^3, then $y_{A,B} \cong 2.0 \times 10^{-5}$. From Eqn. (9-52), the value of y_{fA} is very close to 1.0, so that $k_c = 0.13$ cm/s. According to Eqn. (9-37),

$$N_A = k_c (C_{A,B} - C_{A,s}) \tag{9-37}$$

---

[15] Roberts, G. W. The influence of mass and heat transfer on the performance of heterogeneous catalysts in gas/liquid/solid systems, *Catalysis in Organic Synthesis*, Academic Press, (1976), pp. 1–44.

[16] Brian, P. L. T. and Hales, H. B., *AIChE J.*, 15, 419 (1969).

However, if the reaction is controlled by transport of H_2 from the bulk liquid to the external surface of the catalyst particle,

$$N_A = k_c C_{A,B}$$

If $C_{A,B}$ is taken to be 3.5×10^{-7} mol/cm^3, the equilibrium solubility of H_2 in ethanol, then $N_A = 0.46 \times 10^{-7}$ mol/cm^2-s. In order to compare this value with the measured rate, the calculated rate must have units of moles H_2/min-g-cat. Thus,

$$(-r_{H_2})_{calc}\ (\text{mol } H_2/\text{min-g-cat}) = N_A[360/d_p\rho_p]$$

where d_p is the diameter of the catalyst particle, 16 μm. Substituting the values of d_p and ρ_p gives $(-r_{H_2})_{calc} = 0.011$ mol H_2/min-g-cat.

The value of the rate of H_2 consumption calculated by assuming that the reaction is controlled by external transport of H_2 is about a factor of 3 *lower* that the measured reaction rate. This is a physical impossibility, since the actual rate can never exceed the rate that would exist if the reaction were controlled by external transport. Nevertheless, the comparison suggests that external transport of H_2 is a very important resistance, and that this step may indeed control the overall reaction. Some of the approximations in the calculation may account for the fact that the measured rate is greater than the calculated rate. For example, correlations of transport coefficients are never perfect, and it may be that the value of k_c was underpredicted. Another possible source of error is the use of an "average" spherical particle diameter to describe what probably was a distribution of sizes of irregularly shaped particles. This could have led to an underprediction of the external surface area per gram of catalyst.

9.4.4 Reaction Selectivity

An *external* transport resistance will affect the apparent selectivity of a multiple-reaction system, in the same manner as an internal resistance. The effect of pore diffusion on selectivity is discussed in Section 9.3.8, and the effect of external transport can be understood qualitatively from that discussion.

In general, for parallel and independent reactions, the selectivity to the products that are formed in the slow reactions will be increased relative to the products formed in the fast reactions, if an external transport resistance is present. For series reactions, the selectivity to the intermediate product usually will be decreased if an external transport resistance is present.

The analysis of reaction selectivity in the presence of an external transport resistance is complicated by the fact that concentration differences through the boundary layer usually will be accompanied by a temperature difference. Therefore, changes in catalyst surface temperature must be considered, along with changes in the surface concentrations.

9.5 CATALYST DESIGN—SOME FINAL THOUGHTS

At the beginning of this chapter, it was noted that catalysts come in a wide range of shapes. It was also noted that the active catalytic component, e.g., a metal such as Pt, can be deposited into a particle so that the active component is concentrated near the external surface, in an "eggshell" configuration. Alternatively, the active component can be deposited into the interior of the particle, in an "egg yolk" configuration. During the discussion of the effect of pore diffusion on catalyst selectivity, in Section 9.3.8, potential applications of "eggshell" and "egg yolk" catalysts were suggested.

In the treatments of both internal and external transport, we encountered situations where the rate of a catalytic reaction was proportional to the external (geometric) area of the catalyst particle, rather than to its volume. This occurs, for example, when a reaction is controlled by external transport, or when the effectiveness factor is very low, i.e., in the asymptotic regime. In such cases, using an "eggshell" catalyst can avoid wasting an expensive active component that is located in the interior of the particle, where the concentration of reactant is very low, and the rate is low. However, the "eggshell" design is only a partial answer to the problem. The most desirable option is to increase the amount of external surface area per unit volume of reactor. Of course, using smaller catalyst particles could do this. The disadvantage of smaller particles is that the pressure drop per unit length of reactor increases as the particle size decreases. Depending on the particle Reynolds number, the pressure drop per unit length of reactor increases as the characteristic particle dimension is decreased, with a dependency somewhere between $1/l_c$ and $1/l_c^2$.

Many of the shapes shown in Figure 9-3, as well as the monolithic support shown in Figure 9-5, are designed to provide a higher external surface area per unit volume of reactor, without increasing the pressure drop. The stars and rings in Figure 9-3 are easy-to-visualize examples. The external area per unit volume of the monolithic structure is also very high, and the pressure drop characteristics are excellent because of the absence of form friction. However, monoliths are substantially more costly than typical particulate supports.

SUMMARY OF IMPORTANT CONCEPTS

- Most heterogeneous catalysts are very porous, and the reaction(s) being catalyzed take place predominantly on the walls of the pores, in the interior of the catalyst particle.
- In order for a catalytic reaction to occur, reactants must be transported from the bulk fluid to the surface of the particles, and then into the interior of the porous particle. Products must be transported in the opposite direction. These mass transport steps require a driving force, which is a concentration difference or gradient.
- Heat transfer between the bulk fluid and the catalyst particles will take place if the reaction(s) are exothermic or endothermic. In order for heat to be transported through the catalyst particle and the boundary layer, a temperature difference or gradient is required.
- The effectiveness factor η can be used to estimate the actual rate of reaction, including the effect of internal concentration gradients. For most nth-order reactions, η can be estimated by calculating the generalized Thiele modulus ϕ.
- Carefully planned "diagnostic" experiments can be used to evaluate the effect of both internal and external transport on catalyst performance.
- Kinetic studies on a heterogeneous catalyst should always be carried out under conditions where the reaction is *controlled* by the intrinsic kinetics. Otherwise, the results of the study will reflect transport parameters and will distort the picture of catalyst performance.

PROBLEMS

Internal Transport

Problem 9-1 (Level 1) The reaction $A \rightleftarrows B$ is taking place at steady state in a catalyst particle that can be represented as an infinite flat plate of thickness $2L$. Internal temperature gradients are negligible. The effective diffusivities of A and B are equal. The concentrations of A and B at the pellet surface are $C_{A,S}$ and $C_{B,S}$, respectively. The reaction rate is strongly influenced by pore diffusion, such that ϕ is large.

Sketch the concentration gradients (C_A and C_B) as a function of z (distance from the centerline of the slab) for the following cases:

1. irreversible reaction;
2. moderately reversible reaction ($K_{eq} = 1$);
3. highly reversible reaction ($K_{eq} = 0.10$).

Make a separate sketch for each case. Be sure that your curves contain all of the important qualitative and quantitative features of the concentration profiles.

Problem 9-2 (Level 2) Thiophene (C_4H_4S) has been used as a model compound to study the hydrosulfurization of petroleum naphtha. Estimate the effective diffusivity of thiophene at typical hydrosulfurization conditions of 660 K and 30 atm. The reaction taking place is $C_4H_4S + 3H_2 \rightarrow C_4H_8 + H_2S$. The gas mixture contains a very large excess of H_2.

The hydrodesulfurization catalyst has a BET surface area (S_p) of 180 m^2/g, a porosity (ε) of 0.40, and a pellet density (ρ_p) of 1.40 g/cm^3.

Problem 9-3 (Level 2) The irreversible gas-phase reaction $A \rightarrow R$ is being carried out in a fixed-bed catalytic reactor that operates as an ideal PFR. The reaction is second order in A, i.e., $-r_A = kC_A^2$. The reactor operates isothermally. The value of the rate constant is 2.5×10^{-3} m^6/mol-kg cat-s. The concentration of A in the feed to the reactor is 12 mol/m^3, and the volumetric flow rate is 0.50 m^3/s. The total pressure is 1 atm and pressure drop through the reactor can be neglected.

1. If the reaction is controlled by intrinsic kinetics, what weight of catalyst is required to achieve a fractional conversion of A of 0.90?
2. The catalyst particles are in the shape of rings. The outer diameter is 2 cm, the inner diameter is 1 cm, and the length is 2 cm. The particle density of the catalyst (ρ_p) is 3000 kg/m^3, and the effective diffusivity of A in the catalyst particle is 10^{-7} m^2/s. Estimate the value of the effectiveness factor at the very beginning of the bed. Estimate the value of the effectiveness factor at the very end of the bed. You may assume that external transport resistances are negligible, and that the catalyst particles are isothermal.
3. If the external transport resistances are negligible, what weight of catalyst is required to achieve a fractional conversion of A of 0.90 when the internal transport resistances are taken into account?

Problem 9-4 (Level 3)

1. The reversible synthesis of methanol (MeOH)

$$CO + 2H_2 \rightleftarrows CH_3OH$$

is occurring in a porous, heterogeneous catalyst at 523 K and 5.3 MPa. At these conditions, the equilibrium constant based on pressure is 0.156 MPa^{-2}. The catalyst has a BET surface area of 100 m^2/g, a pore volume of 0.34 cc/g, and a skeletal density of 5.73 g/cc. The composition of the gas surrounding the catalyst particle is H_2—55 mol%; CO—26 mol%; CH_4—2 mol%, N_2— 17 mol%. Calculate the effective diffusivity of carbon monoxide in the catalyst particle.

2. The net rate of methanol synthesis is given by

$$r_{MeOH} = k_f p_{CO} p_{H_2}^2 - k_r p_{MeOH}$$

where p_i is the partial pressure of species "i". The effective diffusion coefficient of H_2 inside the catalyst particles is much greater than that of carbon monoxide. Therefore, as an approximation, the partial pressure of H_2 inside the catalyst particle is essentially constant. With this approximation, show that

$$r_{MeOH} = k' p_{CO}\left(1 - \frac{p_{MeOH}/p_{CO}}{K'_{eq}}\right)$$

where $k' = k_f p_{H_2}^2$. Calculate the value of K'_{eq}.

3. If the measured reaction rate at the conditions given in Part 1 is 8.0×10^{-6} mol CO/cc (geometrical catalyst volume)-s, what is the value of the effectiveness factor? You may assume that $D_{MeOH,eff} \cong D_{CO,eff}$. The catalyst is a cylindrical tablet 1 cm in diameter and 1 cm long.
4. Suppose that the mole fraction of MeOH at the catalyst surface was 0.05 and that the mole fraction of N_2 was 0.12. All other conditions are the same. Estimate the value of the effectiveness factor. You may assume that the effective diffusivity of CO is not changed significantly by this composition change.

You may assume that the ideal gas laws are valid throughout the problem.

Problem 9-5 (Level 1) The reactions

$$A \xrightarrow{k_1} B$$
$$R \xrightarrow{k_2} S$$

are taking place in a porous catalyst particle that can be represented as an infinite flat plate of thickness $2L$. Both reactions are first order and irreversible. The rate constant k_1 is high and k_2 is low. The effective diffusivities of A and R in the catalyst particle are approximately the same. The concentrations of A and R at the surface of the particle are the same.

1. Sketch the concentration profiles of A and R in the particle for a situation where the effectiveness factor for the reaction $A \rightarrow B$ is approximately 0.5.
2. If S is an undesired by-product, is it desirable or undesirable to have a significant pore-diffusion resistance? Explain your answer.

Problem 9-6 (Level 2) Various catalysts are being tested in a differential reactor for the irreversible, first-order reaction:

$$A \rightarrow \text{products}$$

The catalyst particles are spherical and essentially isothermal. The effective diffusivity $D_{A,eff}$ is constant throughout each particle.

The effect of particle size was investigated for one catalyst, with the following results:

Particle radius (mm)	Measured reaction rate (mole A/L (cat)-s)
2	2.5
0.5	8.9
0.15	20.1

What is the value of the effectiveness factor for each particle size? You may assume the rate constant and the effective diffusivity do not depend on particle size. You may also assume that external transport resistances are insignificant.

Problem 9-7 (Level 1) The irreversible, first-order reaction $A \rightarrow$ products takes place over catalyst X, which is a porous solid. The intrinsic rate constant (based on geometric catalyst volume) is $1500\ s^{-1}$. The effective diffusivity of A is $2 \times 10^{-3}\ cm^2/s$.

The commercial version of catalyst X will be spherical in shape. What is the largest radius that can be used if the effectiveness factor must be 0.90 or greater?

Problem 9-8 (Level 1) The gas-phase hydrogenolysis of thiophene

$$C_4H_4S + 4H_2 \rightarrow C_4H_{10} + H_2S$$

was studied at 523 K and 1 atm total pressure using a "cobalt molybdate" catalyst. In all experiments, H_2 was in great excess over thiophene. The BET surface area (S_p) of the catalyst was 343 m^2/g, the pore volume per gram (V_p) was 0.470 cm^3/g, and the particle density (ρ_p) was 1.17 g/cm^3. The binary molecular diffusivity of C_4H_4S/H_2 at reaction conditions is 1.01 cm^2/s.

The reactor that was used was essentially a CSTR. In one steady-state experiment, a reaction rate of 6.13×10^{-6} mol C_4H_4S/g cat-min was measured. The concentration of thiophene in the bulk gas for this experiment was $1.73 \times 10^{-6}\ mol/cm^3$. The catalyst particles were cylinders with a length of 1.27 cm and a diameter of 0.277 cm.

1. Estimate the average radius of the pores in the catalyst.
2. What is the Knudsen diffusion coefficient of thiophene in a pore with the radius that you just calculated?
3. Assuming that diffusion is in the transition region, what is the *effective* diffusion coefficient of thiophene in the catalyst?
4. If the reaction is first order in thiophene, and if there is no significant external transport resistance, what is the value of the effectiveness factor?

External Transport

Problem 9-9 (Level 1) Air containing 1.0 mol % of an oxidizable organic compound (A) is being passed through a monolithic (honeycomb) catalyst to oxidize the organic compound before discharging the air stream to the atmosphere. Each duct in the monolith is square, and the length of a side is 0.12 cm. Each duct is 2.0 cm long. The inlet molar flow rate of A into each duct is 0.0020 mol A/h. The gas mixture enters the catalyst at 1.1 atm total pressure and a temperature of 350 K.

In order to determine a limit of catalyst performance, the conversion of A will be calculated for a situation where the reaction is *controlled* by external mass transfer of A from the bulk gas stream to the wall of the duct, over the whole length of the duct. Since the calculation is approximate, assume that

1. the gas flowing through the channel is in plug flow;
2. the system is isothermal;
3. the change in volume on reaction can be neglected;
4. the pressure drop through the channel can be neglected;
5. the ideal gas law is valid;
6. the rate of mass transfer of A from the bulk gas stream to the wall of the duct is given by

$$-r_A\left(\frac{\text{moles A}}{\text{area-time}}\right) = k_c\left(\frac{\text{length}}{\text{time}}\right)(C_{A,B} - C_{A,w}) \times \left(\frac{\text{moles A}}{\text{volume}}\right)$$

where k_c is the mass-transfer coefficient based on concentration, $C_{A,B}$ is the concentration of A in the bulk gas stream at any position along the length of the duct, and $C_{A,w}$ is the concentration of A at the wall at any position along the length of the duct.

1. If the reaction is *controlled* by mass transfer of A from the bulk gas stream to the duct wall over the whole length of the channel, what is the value of $C_{A,w}$ at every point on the wall of the duct?
2. For the situation described above, show that the design equation can be written as

$$\frac{A}{F_{A0}} = \int_0^{x_A} \frac{dx}{-r_A}$$

where A is the total area of the duct walls and x_A is the fractional conversion of A in the gas leaving the duct.

3. Show that

$$-\ln(1 - x_A) = \frac{k_c C_{A0} A}{F_{A0}}$$

provided that k_c does not depend on composition or temperature.

4. If $k_c = 0.25 \times 10^5$ cm/h, what is the value of x_A in the stream leaving the catalyst?
5. Is the value of x_A that you calculated a maximum or minimum value, i.e., will the actual conversion be higher or lower when the intrinsic reaction kinetics are taken into account? Explain your reasoning.

Problem 9-10 (Level 1) The oxidation of small concentrations of carbon monoxide (CO) in the presence of large concentrations of hydrogen (H_2) is an important reaction in the preparation of high-purity H_2 for use as a feed to a fuel cell. The following graph shows some results from testing a novel form of heterogeneous catalyst for this reaction. The catalyst was composed of a thin layer of Pt/Fe/Al_2O_3 deposited on the walls of a metal foam support. The gas flowed through the foam and the reaction took place on the catalyzed walls. Two different lengths of catalyst were tested. The catalyst diameter was 1 in. for both catalyst lengths. The reactor operated adiabatically, and the experimental conditions were the same for all experiments: inlet temperature = 80 °C; total pressure = 2 atm, feed composition: 1% CO, 0.50% O_2, 42% H_2, 9% CO_2, 12% H_2O, balance N_2.

The graph below shows the fractional conversion of CO at two different space velocities. At constant space velocity, the CO conversion was much higher for the 6-in. length of catalyst than for the 2-in. length of catalyst. Propose an explanation for this behavior.

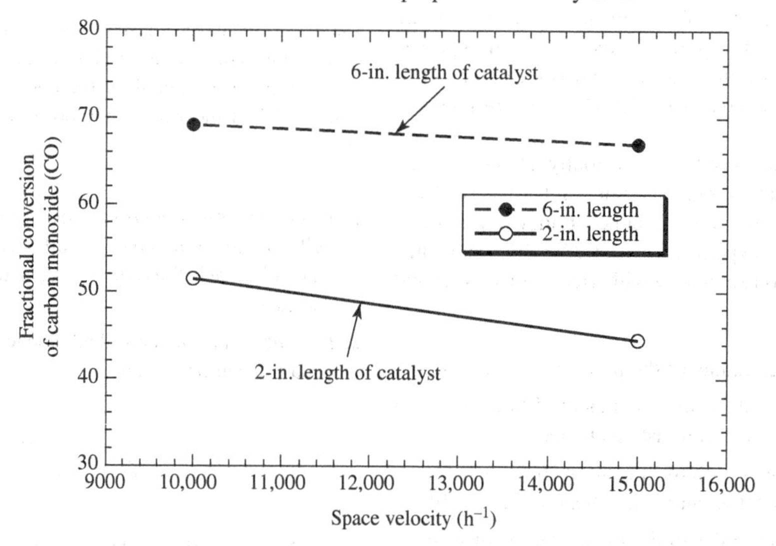

Problem 9-11 (Level 2) The *reversible* reaction

$$A \rightleftarrows B$$

takes place at steady state in a porous catalyst particle that is immersed in a flowing fluid. The equilibrium constant for the reaction is K_{eq}^{C}. The concentrations of "A" and "B" in the fluid are $C_{A,B}$ and $C_{B,B}$, respectively.

Derive an expression for the rate of reaction in a particle of radius R if the reaction is *controlled* by transport of "A" from the bulk fluid to the external surface of the catalyst particle. Assume that the mass-transfer coefficients for A and B are identical.

Problem 9-12 (Level 2) Formaldehyde is manufactured by the partial oxidation of methanol in a fixed-bed catalytic reactor at steady state. In one process, 50 mol% of methanol in air is passed over a nonporous silver catalyst at 600 °C and atmospheric pressure. What is the approximate value of T_S (the temperature at the surface of the catalyst) if the reaction is *controlled* by external mass transfer?

If the catalyst actually operates at the temperature you calculated above, what reaction(s) do you think will take place?

Hint: You will have to look up some data to answer this question. Since some approximations are required to calculate T_s, don't make any corrections to the data that are tedious.

Problem 9-13 (Level 1) The reaction $C_6H_6 + 3H_2 \rightarrow C_6H_{12}$ is occurring in a spherical catalyst particle. Initially, the reaction is *controlled* by external mass transfer. The mole fraction of benzene in the bulk gas stream that surrounds the particle is 0.10.

At some later time, the catalyst has deactivated to the point that the mole fraction of benzene at the *external surface* of the catalyst is 0.050. What is the ratio of the rate at this condition to the initial rate?

You may assume that the composition and temperature of the bulk gas stream do not change with time, and you may neglect any effect of composition and temperature on the physical properties of the system.

Problem 9-14 (Level 2) Satterfield and Cortez[17] studied the mass-transfer characteristics of woven-wire screen catalysts. This kind of catalyst is used for the oxidation of NH_3 to NO in the process for making nitric acid, and for the reaction of ammonia, oxygen, and methane to make hydrogen cyanide. A Pt/Rh wire usually is used for these reactions.

The configuration of the catalyst is shown in the following figure. The metal wires are *not* porous. The direction of gas flow is normal to the wire screen (i.e., into the page).

[17] Satterfield, C. N. and Cortez, D. H., Mass transfer characteristics of woven-wire screen catalysts, *Ind. Eng. Chem. Fundam.*, 9 (4), 613–620 (1970).

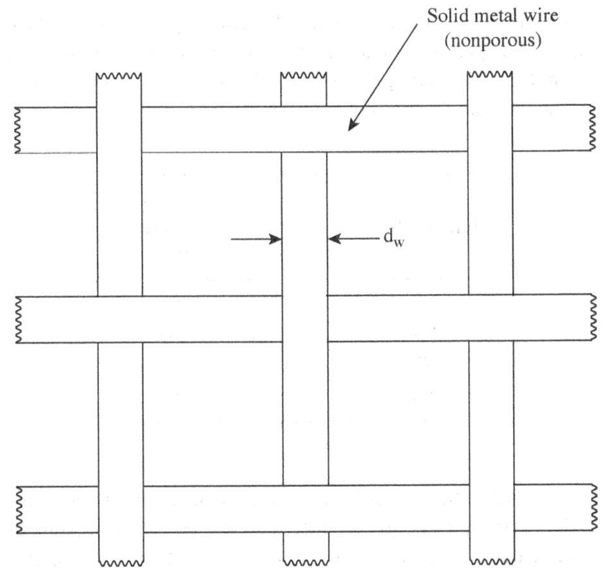

The following figure shows data on the catalytic oxidation of 1-hexene at three different gas flow rates using a single screen. At a fixed flow rate, the fractional conversion of hexene increases rapidly with the temperature of the feed gas when the temperature is low. However, at temperatures greater than about 420 °C, conversion is virtually independent of temperature.

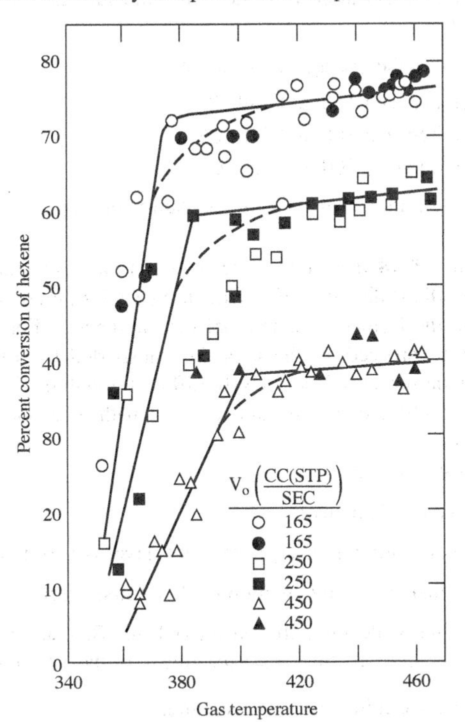

Reprinted with permission from *Ind. Eng. Chem. Fundamen.* 9, 613–620, (1970). Copyright 1970 American Chemical Society.

1. Explain the behavior shown in the figure, i.e., for a given flow rate, why is the fractional conversion of hexene very sensitive to feed gas temperature when the temperature is low, and almost independent of temperature when the temperature is high?
2. Data in the attached figure were taken with a feed gas consisting of 0.133 mol% 1-hexene in air. Estimate the *maximum possible* temperature of the wire when the temperature of the feed gas is 450 °C. (*Simplification*—if you need to estimate any physical properties, you may neglect the presence of 1-hexene and the reaction products (CO_2, H_2O, etc.). In other words, you may assume that the gas is pure air). The heat of combustion of 1 mol of 1-hexene is −900 kcal/mol, independent of temperature. You may assume that the heat capacity of air is 7.62 cal/mol-K, independent of temperature.

   Satterfield and Cortez found that the mass-transfer coefficient for wire screens could be correlated by

$$j_D = 0.94\,(N_{Re})^{0.717}, \quad j_D \equiv \left(\frac{k_c^0 \rho}{G_i}\right)(N_{sc})^{2/3}$$

where, N_{Re} is $d_W\, G_i/\mu$, d_W is the wire diameter (cm), G_i is the mass velocity based on open area of screen (g/cm^2-s), μ is the gas viscosity (g/cm-s), k_C^0 is the mass-transfer coefficient based on concentration (cm/s), ρ is the gas density; N_{Sc} is the Schmidt number ($\mu/\rho D$), and D is the diffusivity (cm^2/s).

The Schmidt number for 1-hexene in air is 0.58, independent of temperature, at 1 atm total pressure.

3. Assume that the rate of hexene oxidation is *controlled* by external mass transfer, i.e., by mass transfer of hexene from the bulk gas stream to the surface of the wire. Predict the value of the reaction rate (mol hexene/s-cm^2 of external wire area) for $G_i = 0.0710$ g/cm^2-s, an inlet gas temperature of 450 °C, an inlet 1-hexene concentration of 0.133 mol%, and a wire diameter of 0.010 cm. Assume that the hexene concentration in the bulk gas is constant, i.e., this concentration does not change as gas flows past the catalytic wire. The total pressure is 1 atm.
4. Based on your answer to Question 3, do you think that the assumption of a constant 1-hexene concentration is reasonable? Justify your answer. The total external surface area of the wires is 1.55 cm^2/cm^2 of screen open area.

Problem 9-15 (Level 1) Kabel and Johanson[18] studied the catalytic dehydration of ethanol to diethylether and water using a solid catalyst, a sulfonated copolymer of styrene and divinylbenzene in acid form. The reaction is

$$2C_2H_5OH \rightleftarrows (C_2H_5)_2O + H_2O$$

[18] Adapted from Kabel, R. L. and Johanson, L. N., Reaction kinetics and adsorption equilibrium in the vapor-phase dehydrogenation of ethanol, *AIChE J.*, 8 (5), 623 (2004).

The following experiments were carried out at 120 °C in an ideal plug flow reactor with a feed consisting of pure ethanol at 1.0 atm. The same reactor was used for all experiments.

Run	Weight of catalyst, W (g)	Ethanol conversion at $W/F_{A0} = 1000$ (g cat-min/ mole ethanol)	Ethanol conversion at $W/F_{A0} = 3000$ (g cat-min/ mole ethanol)
3–1	14.3	0.118	0.270
4–1	22.6	0.112	0.281

1. What was the probable purpose of these experiments? How do you interpret the results?
2. Comment on the design of the experiments. Do they provide a definitive answer to the question they were intended to explore?

Problem 9-16 (Level 1) The reaction $A \rightarrow \nu B$ takes place on the external surface of a spherical, nonporous catalyst particle. The reaction is *controlled* by mass transfer of A from the bulk fluid stream to the catalyst surface. The mole fraction of A in the bulk stream is 0.50.

Consider two cases, $\nu = 10$ and $\nu = 1/10$. What is the ratio of the reaction rates for these two cases? Everything except the stoichiometric coefficients is the same for the two cases, e.g., particle size, fluid velocity, and fluid properties.

Bridging Problems

Problem 9-17 (Level 1) Shingu (U.S. Patent #2,985,668) studied the catalytic partial oxidation of propylene to propylene oxide in a continuous gas-sparged reactor. The reaction is

$$C_3H_6 + 1/2O_2 \rightarrow C_3H_6O$$

A gaseous mixture of propylene and oxygen was fed to a reactor that contained a solid catalyst Ag/SiO_2 in the form of a very fine powder. The catalyst was suspended in liquid dibutyl phthalate. The unconverted oxygen and propylene, as well as the product propylene oxide and by-products such as carbon dioxide and water, left the reactor as a gas. The catalyst and the liquid remained in the reactor. The reactor was isothermal. The patent contains the following description of the reactor:

> Using a vertical reactor in which the liquid reaction phase containing the high-boiling solvent liquid and the finely divided catalyst powder was maintained in a suspensoidal state of efficient agitation by introducing the reactant gas mixture with a high speed through a jet placed at the bottom of the reactor, a series of continuous runs was made...

The reaction was studied at four different temperatures, using the same total pressure (1 atm), feed composition (21 mol% O_2, 79 mol% C_3H_6), and space velocity (35 cc-(gas)/g (cat)-h). Some of the data are shown in the following table.

Reactor temperature (°C)	Propylene conversion (%)
160	16
180	30
200	40
230	44

Assume that the reaction is first order in propylene, zero order in oxygen, and irreversible. You may also neglect volume change due to reaction, although this assumption is not strictly justified.

1. Which of the three ideal reactors best matches the actual reactor described above? Explain your answer.
2. Using your answer to Question 1, calculate the value of the rate constant at each of the four temperatures in the above table.
3. What is the average activation energy:
   (a) between 160 and 180 °C?
   (b) between 180 and 200 °C?
   (c) between 200 and 230 °C?
4. Explain the behavior of the activation energy.

Problem 9-18 (Level 3) The oxidation of elemental silicon (Si) to silicon dioxide (SiO_2) with dioxygen (O_2) is an important step in the formation of microelectronic devices. For the oxidation to proceed, O_2 must (a) be transported from the bulk gas to the surface of the SiO_2 (b) diffuse through the SiO_2 layer to the SiO_2/Si interface; and (c) react with Si at the SiO_2/Si interface.

Assume that

- the system is planar;
- the concentration of O_2 in the SiO_2 layer is very low;
- the kinetics of the oxidation of Si are first order;
- the time scale for diffusion through the SiO_2 layer is short compared to the time scale for growth of the SiO_2 layer

Use the following nomenclature:

T = thickness of SiO_2 layer
T_0 = initial ($t = 0$) thickness of SiO_2 layer
k = rate constant for O_2/Si reaction (length/time)

k_c = mass-transfer coefficient (based on concentration) between bulk gas and SiO_2 surface (length/time)
D_O = diffusion coefficient of O_2 in SiO_2
C_O = concentration of O_2 in bulk gas
C_s = concentration of O_2 in gas at gas/SiO_2 interface
C_e = concentration of O_2 in SiO_2 at gas/SiO_2 interface ($C_e = HC_s$)
C_i = concentration of O_2 in SiO_2 at SiO_2/Si interface
H = Henry's law constant for O_2 in SiO_2
N_1 = moles of O_2 in SiO_2/volume SiO_2

1. What simplifications result from assumptions 2 and 4?
2. Show that

$$T(t) = A\left\{\left[1 + \frac{B}{A^2}(t+\tau)\right]^{1/2} - 1\right\}$$

where

$$A = D_O\left(\frac{1}{k} + \frac{H}{k_c}\right)$$

$$B = 2HC_OD_O/N_1$$

$$\tau = (T_0^2 + 2AT_0)/B$$

3. Simplify the above expression for a gas phase that is pure O_2.

Problem 9-19 (Level 1) Two engineers are having an argument about how to test for the presence of an *external* mass-transfer resistance in an adiabatic, fixed-bed, catalytic reactor that behaves as an ideal PFR. The engineers agree that varying the *linear* velocity through the bed at constant space velocity will provide a definitive test. In other words, they agree that if the linear velocity is varied over a wide range and the outlet conversion changes, external mass transfer must control or influence the reaction rate.

Engineer A claims that there is a second way to test for the presence or absence of external mass transfer. Engineer A proposes to operate the reactor at constant space velocity *and* constant linear velocity, but to change the catalyst particle size. The mass-transfer coefficient and the external surface area per unit weight of catalyst both depend on particle size. Therefore, if the outlet conversion changes as the particle size is varied, it is safe to conclude that external transport controls or influences the reaction rate. Engineer B disagrees.

1. Which engineer is correct? Why?
2. From what school did engineer A graduate? Engineer B?

Problem 9-20 (Level 2) The following data were taken in an isothermal, ideal PFR reactor packed with spherical catalytic particles. The diameter of the particles was 0.39 cm. The reaction taking place was

$$A \rightarrow R$$

This reaction is irreversible at the conditions of the experiments and $\Delta H_R \cong 0$. The inlet concentration of A was 1.1×10^{-4} mol/cm^3 for all experiments. Experiments 1 through 4 were run at 400 °C:

Experiment no.	Space time (g-s/cm^3)	Fractional conversion of A	Superficial mass velocity (g-s/cm^2)*
1	0.18	0.50	0.19
2	0.36	0.75	0.19
3	0.72	0.94	0.19
4	1.08	0.98	0.19

1. Considering *only* the four experiments above, does a first-order rate equation fit the data? *Justify your answer.* If so, what is the value of the apparent rate constant?
2. Two additional experiments were carried out at 400 °C:

Experiment no.	Space time (g-s/cm^3)	Fractional conversion of A	Superficial mass velocity (g-s/cm^2)*
5	0.18	0.75	0.57
6	0.36	0.94	0.57

Based on these six experiments, does the process of *external* mass transfer have any influence on the performance of the catalysts? Explain your reasoning.

3. One additional experiment was carried out at 425 °C:

Experiment no.	Space time (g-s/cm^3)	Fractional conversion of A	Superficial mass velocity (g-s/cm^2)*
7	0.18	0.54	0.19

Is this result consistent with your answer to Part 2? Explain your reasoning. Be as quantitative as possible.

Problem 9-21 (Level 1) The irreversible reaction

$$A \rightarrow B$$

takes place at steady state in a porous catalyst particle immersed in a flowing fluid that contains "A". The intrinsic kinetics of the reaction are second order in A.

Sketch a graph of the *apparent* reaction order versus temperature.

APPENDIX 9-A SOLUTION OF EQUATION (9-4c)

$$\frac{d^2C_A}{dr^2}+\frac{2}{r}\frac{dC_A}{dr}=\frac{k_vC_A}{D_{A,eff}} \tag{9-4c}$$

Boundary conditions

$$\frac{dC_A}{dr}=0; \quad r=0 \tag{9-4a}$$

$$C_A=C_{A,s}; \quad r=R \tag{9-4b}$$

These equations are nondimensionalized by introducing the variables

$$\chi=C_A/C_{A,s}; \quad \rho=r/R$$

Using these new variables, Eqn. (9-4c) becomes

$$\frac{d^2\chi}{d\rho^2}+\frac{2}{\rho}\frac{d\chi}{d\rho}=\phi_{s,1}^2\chi \tag{9A-1}$$

In this equation, $\phi_{s,1}^2=R^2(k_v/D_{A,eff})$, and $\phi_{s,1}$ is the Thiele modulus for a first-order, irreversible reaction in a sphere. Introducing the new variables into the boundary conditions gives

$$\frac{d\chi}{d\rho}=0; \quad \rho=0 \tag{9A-2}$$

$$\chi=1; \quad \rho=1 \tag{9A-3}$$

Equation (9A-1) is a form of Bessel's equation and can be solved by comparison of its terms with those in the solution of the generalized Bessel's equation.

Another approach to solving Eqn. (9A-1) is to employ the variable transformation $\chi=\Gamma/\rho$. With this substitution, Eqn. (9A-1) becomes a second-order, ordinary differential equation *with constant coefficients.*

$$\frac{d^2\Gamma}{d\rho^2}-\phi_{s,1}^2\Gamma=0 \tag{9A-4}$$

$$\Gamma=0; \quad \rho=0 \tag{9A-5}$$

$$\Gamma=1; \quad \rho=1 \tag{9A-6}$$

The solution now proceeds as follows: Let Γ' denote $d\Gamma/d\rho$. Then Eqn. (9A-4) can be written as

$$(\Gamma'-\phi_{s,1})(\Gamma'+\phi_{s,1})=0 \tag{9A-7}$$

The solution of this equation is

$$\Gamma=C_1\exp(-\phi_{s,1}\rho)+C_2\exp(+\phi_{s,1}\rho) \tag{9A-8}$$

Here C_1 and C_2 are constants of integration that will be determined from the boundary conditions. Applying boundary condition (9A-5),

$$0=C_1+C_2; \quad C_2=-C_1$$

Applying boundary condition (9A-6),

$$1 = C_1\exp(-\phi_{s,1}) + C_2\exp(+\phi_{s,1}) = C_1\exp(-\phi_{s,1}) - C_1\exp(+\phi_{s,1})$$
$$C_1 = 1/[\exp(-\phi_{s,1})\} - \exp(+\phi_{s,1})] = -1/2\sinh(\phi_{s,1}) = -C_2$$

Substituting for C_1 and C_2 in Eqn. (9A-8),

$$\Gamma = [\exp(\phi_{s,1}\rho) - \exp(-\phi_{s,1}\rho)]/2\sinh(\phi_{s,1}) = \sinh(\phi_{s,1}\rho)/\sinh(\phi_{s,1}) = \chi\rho$$

Recalling the definitions of χ and ρ,

$$\frac{C_A(r)}{C_{A,s}} = \frac{R\sinh(\phi_{s,1}r/R)}{r\sinh(\phi_{s,1})} \qquad (9\text{-}6)$$

Chapter 10

Nonideal Reactors

LEARNING OBJECTIVES

After completing this chapter, you should be able to

1. explain how tracer injection techniques can be used to characterize mixing in a vessel;
2. check the quality of tracer data through a material balance;
3. calculate the distribution of times that fluid flowing through a vessel at steady state spends in the vessel, using the measured concentration of a tracer at the vessel exit;
4. derive mathematical expressions for the distribution of times that fluid spends in a CSTR and a PFR, both operating at steady state;
5. estimate the performance of a nonideal reactor from the measured concentration of tracer at the vessel exit as a function of time;
6. use the Dispersion model and the CSTRs-in-series models to estimate the performance of a nonideal reactor;
7. construct various compartment models and use them to estimate the performance of a nonideal reactor.

10.1 WHAT CAN MAKE A REACTOR "NONIDEAL"?

10.1.1 What Makes PFRs and CSTRs "Ideal"?

In Chapter 4, we defined the characteristics of two *ideal* continuous reactors, the ideal plug-flow reactor (PFR) and the ideal continuous stirred-tank reactor (CSTR). We use the term *ideal* to refer to these reactors because the conditions of mixing and fluid flow in them are defined very precisely. To recap:

- In the CSTR, mixing is so intense that the species concentrations and the temperature are the same at every point in the reactor. Moreover, mixing is complete down to a molecular scale. Every molecule that enters the reactor is immediately mixed with molecules that have been in the reactor for longer periods of time. There is no tendency for molecules that entered the reactor at the same time to remain associated.
- In the PFR, there is no mixing in the direction of flow. All of the molecules that enter the reactor at the same time stay together as they flow through the reactor, and they all leave the reactor at the same time. Moreover, there are no gradients of concentration or temperature normal to the direction of flow.

We might be tempted to explain the absence of gradients normal to flow in a PFR by invoking intense mixing normal to flow. However, it is difficult to visualize a mechanism that would create intense mixing *normal* to flow, but *no* mixing in the direction of flow.

Instead, the absence of gradients normal to flow is easier to rationalize when there are no driving forces that would create such gradients. For example, if there is no heat transfer normal to the direction of flow, then there should be no temperature gradients in that direction. The issue of temperature gradients normal to flow was discussed briefly in Chapter 8, Section 8.6.1. If there are temperature gradients normal to flow, concentration gradients probably will also exist, as a result of the effect of temperature on reaction kinetics. In the absence of temperature gradients normal to flow, there is no mechanism to create concentration gradients normal to flow, provided that the fluid velocity in the direction of flow is the same at every point in a plane normal to the flow.

The two continuous, ideal reactors represent *limiting cases* of fluid mixing. The behavior of a real reactor will often approximate that of one or the other of these ideal reactors. However, this is not always the case.

10.1.2 Nonideal Reactors: Some Examples

10.1.2.1 Tubular Reactor with Bypassing

Sometimes, conditions will exist that cause the performance of a continuous reactor to deviate from both of the ideal cases. For example, consider the catalytic reactor shown in Figure 10-1.

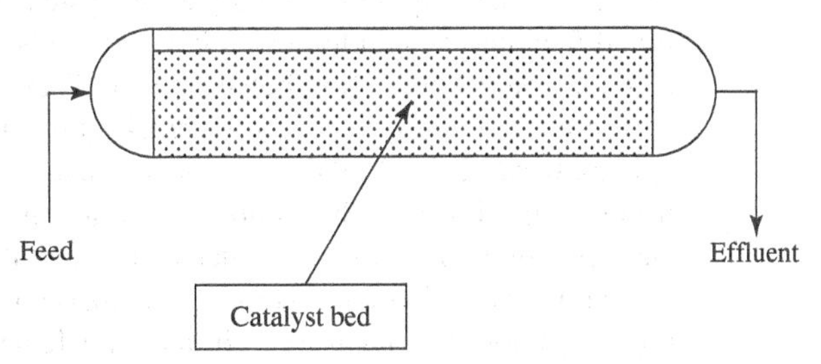

Figure 10-1 Packed tubular reactor with empty space above the catalyst-containing region.

EXERCISE 10-1

How you might go about creating *and maintaining* a full cross section along the whole length of a horizontally oriented, packed-bed reactor?

The catalyst bed occupies only a portion of the cross section normal to flow. Some of the fluid flows through the catalyst bed, but some "bypasses" the bed, flowing through the empty space at the top of the reactor. Even if the unpacked volume at the top of the reactor is fairly small, a significant portion of the flow can bypass the catalyst bed because the frictional resistance is much larger in the packed region than in the unpacked region.

As an aside, the situation described above provides a compelling reason for *never* mounting a packed-bed reactor in a horizontal position. Even if the reactor initially is packed perfectly, the catalyst bed can settle over time and create a short circuit such as the one shown above. Moreover, if the reactor is sizeable, i.e., a pilot plant or commercial reactor, obtaining a perfect packing initially is easier said than done.

Neither the PFR model nor the CSTR model will provide a reasonable description of the reactor in Figure 10-1. In essence, there are two "reactors" in parallel, one of which produces no reaction because of the absence of any catalyst.

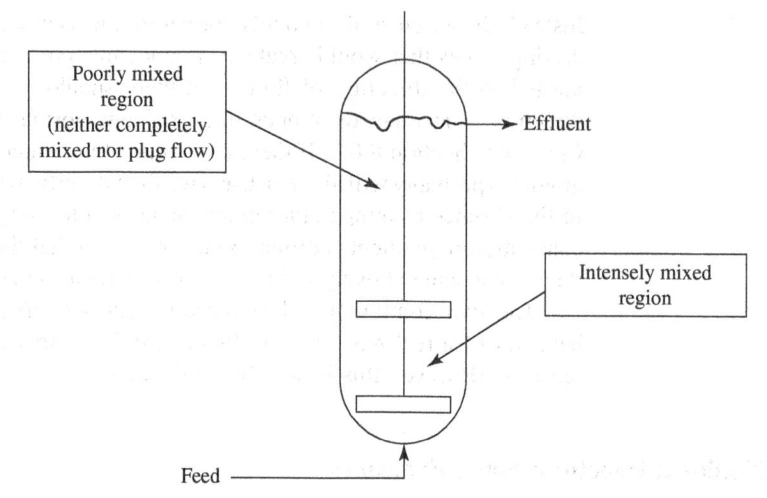

Figure 10-2 Stirred reactor with completely mixed and partially mixed regions.

10.1.2.2 Stirred Reactor with Incomplete Mixing

Let's consider the reactor shown in Figure 10-2. The length/diameter ratio of this stirred reactor is relatively high. The agitation system is such that the bottom region, where the liquid feed enters, is intensely mixed. There is no mechanical agitation in the top of the reactor. The lower section behaves as a CSTR. However, flow and mixing in the upper section are essentially uncharacterized. Liquid must flow through the region above the upper agitator because the fluid that enters the reactor at the bottom must leave through the outlet near the top. However, the nature of flow and mixing above the upper agitator is difficult to characterize. In the absence of mechanical agitation, there is no reason to presume that the upper region is well mixed. Moreover, there is no reason to presume that the region above the upper agitator is an ideal PFR. In fact, a calculation of the Reynolds number might even suggest that the upper region was in laminar flow.

At first glance, this example might seem a bit preposterous. Would a good engineer design a reactor with all of the agitation toward the bottom, and none at the top? Probably not. However, the internals of a reactor can change position over time, sometimes very quickly, as the reactor is started up and shut down, and operating conditions are adjusted. It may be that the upper agitator was in the correct position on day 1, but gradually (or suddenly) worked its way down the shaft, to the point where it was no longer effective in agitating the upper region of the reactor.

10.1.2.3 Laminar Flow Tubular Reactor (LFTR)

The two previous examples involved reactors where deviations from ideality were caused by flaws in design and/or construction. As a final example, let's consider a case where the deviation from ideal behavior is an inevitable consequence of the nature of flow through the reactor.

Picture a fluid flowing through and reacting in a cylindrical tube. If the Reynolds number is very high, flow will be highly turbulent and the behavior of the reactor should approximate that of an ideal PFR, unless there are significant radial temperature and concentration gradients resulting, for instance, from heat transfer through the wall of the reactor. However, the fluid flowing through the reactor may have a high viscosity, or the flow rate may be very low, or the diameter of the tube may be very small. The last condition inevitably occurs in so-called microfluidic reactors. When the Reynolds number is calculated for these cases, it may be that flow is in the laminar regime.

In laminar flow, the velocity profile across the tube diameter is not flat. If the fluid is Newtonian, and there are no radial variations in temperature or concentration, the velocity profile will be parabolic. In laminar flow, there will be radial concentration gradients at any point along the axis of the tube, since the fluid velocity at the wall approaches zero, whereas the velocity at the center of the tube is at a maximum. The fluid at the wall of the tube spends a long time in the reactor. Therefore, the concentration of reactant is relatively low in this region. The fluid at the centerline of the tube has the highest velocity, so that the reactant concentration is relatively high at this position.

In a similar manner, the temperature can vary with radial position, even if the reactor is adiabatic. If the reaction is exothermic, the temperature close to the wall will be relatively high, since the reactant conversion is high in this region. Conversely, the temperature will be relatively low at the center of the tube, where the residence time and the reactant conversion are lowest. The situation is shown schematically in Figure 10-3.

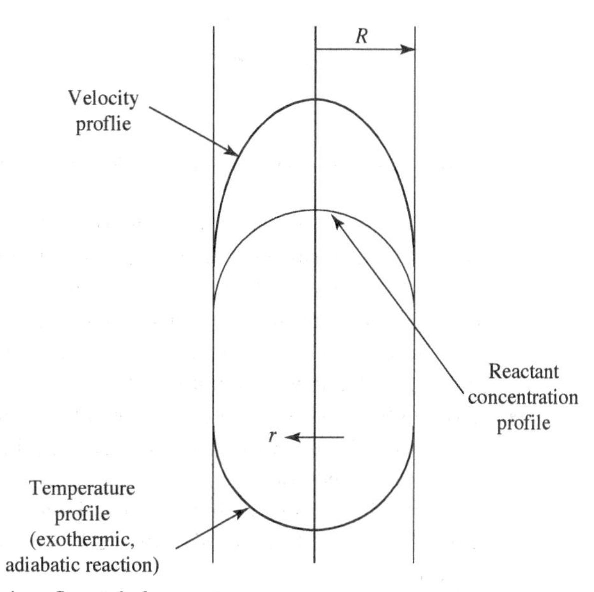

Figure 10-3 Laminar flow tubular reactor.

The exact shape of these profiles will depend on a number of factors, e.g., the reaction rate, the enthalpy change on reaction, and the sensitivity of the fluid viscosity to temperature. However, the behavior of an LFTR is very different from that of an ideal, plug-flow reactor.

Hopefully, the above discussion has established that every continuous reactor is not a CSTR or a PFR. The next thing we have to do is determine how to *tell* whether a given reactor is a PFR or a CSTR, or something in between, or something worse.

10.2 DIAGNOSING AND CHARACTERIZING NONIDEAL FLOW

10.2.1 Tracer Response Techniques

Suppose we were observing a vessel with fluid flowing through it at steady state. Suppose further that there was no change in density as the fluid flowed through the vessel. How would you go about trying to answer the question "How long does each molecule of fluid spend in the reactor?" What experiment(s) could be carried out to answer the question?

To begin, you might recognize that every molecule may not spend *exactly* the same time in the vessel. If there were mixing in the direction of flow, some molecules that entered the vessel at a time, $t = 0$, might "catch up" to molecules that entered at an earlier time, say $t = -\delta$. Similarly, some of the molecules that entered at $t = 0$ might be overtaken by ones that entered at a later time, say $t = \delta$. In general, individual molecules will spend different amounts of time in the vessel.

The *distribution* of times that individual elements of fluid spend in the vessel can be measured by means of *tracer injection techniques*. Consider the situation shown in Figure 10-4. A vessel with a volume of V is at steady state, with a volumetric flow rate, υ, going in and the same volumetric flow rate leaving.

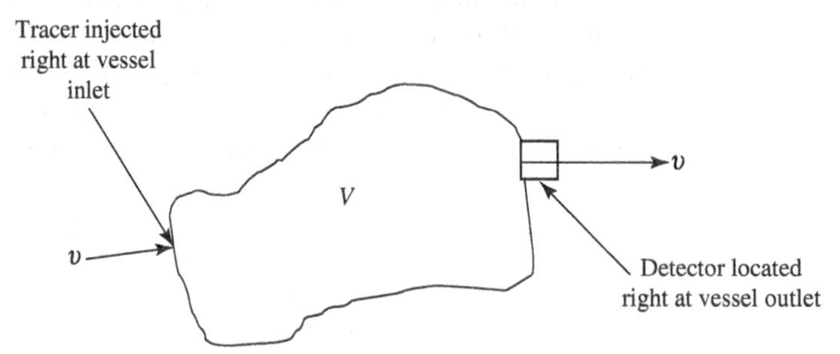

Figure 10-4 Schematic diagram of ideal tracer injection experiment.

Suppose we inject a small amount of a material into the inlet stream, right at the boundary of the vessel. The material behaves *exactly* like the fluid that is flowing through the vessel, and is called a *tracer*. The tracer must be chosen so that its concentration in the effluent from the vessel can be measured by a detector located right at the point that the fluid leaves the vessel.

The use of a tracer to study the flow of a fluid through a vessel is called a *tracer response* technique. A known amount of tracer is injected in a known pattern (such as an instantaneous pulse), and the *response* of the tracer to the flow conditions that exist in the vessel is measured. The use of tracer response techniques is common in medicine, as well as in chemical engineering.

Selection of a suitable tracer can be a challenging task. Since the tracer must move through the vessel *exactly* like the bulk fluid, the tracer cannot

- settle;
- phase separate;
- react;
- adsorb on the vessel walls, or on any internal components such as an agitator or baffle, or on a solid catalyst, if one is present in the vessel;
- diffuse relative to the bulk fluid;
- influence the flow of the bulk fluid in any way.

The tracer must also be easy to measure. Some commonly used measurement techniques are radioactivity, electrical conductivity, absorptivity (e.g., in the visible, ultraviolet, or infrared region), and refractive index. The tracer must also be injected so that it labels each element of the inlet fluid uniformly.[1]

[1] For a more complete discussion of tracer selection, injection and measurement, see Levenspiel, O., Lai, B. W., and Chatlynne, C. Y., Tracer curves and the residence time distribution, *Chem. Eng. Sci.*, 25, 1611–1613 (1970); Levenspiel, O. and Turner, J. C. R., The interpretation of residence-time experiments, *Chem. Eng. Sci.*, 25, 1605–1609 (1970).

The signal from the detector might look something like the one shown below.

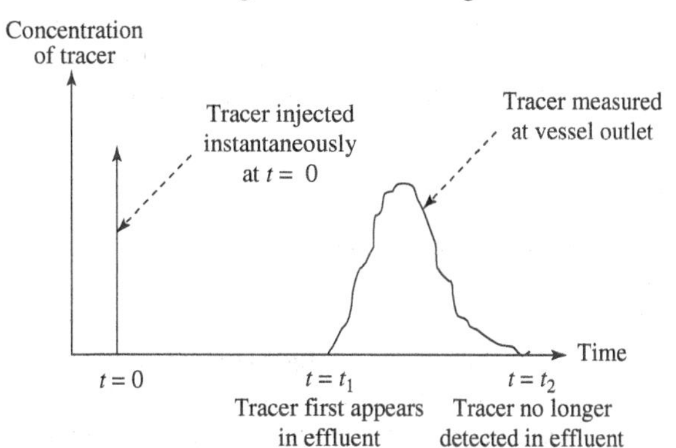

In this example, the tracer is injected as an ideal pulse, i.e., all of the tracer enters the vessel at the same time. However, the tracer leaves the vessel over a *range* of times. This indicates that fluid mixing takes place in the vessel. Tracer first appears in the effluent stream about t_1 after it is injected. Tracer cannot be detected after about t_2. Some of the tracer passes through the vessel relatively quickly, and emerges at a time close to t_1. However, some of the tracer mixes into elements of fluid that enter the vessel after the tracer is injected. These portions of the tracer emerge at later times.

Before discussing the mathematical analysis of tracer data, let's see if we can figure out what the tracer response would look like for some of the reactors that we discussed previously.

10.2.2 Tracer Response Curves for Ideal Reactors (Qualitative Discussion)

Consider a vessel at steady state with a constant-density fluid flowing through it. At $t = 0$, a sharp pulse of tracer is injected into the stream entering the vessel. Let's use what we know about the two ideal reactors to construct their tracer response curves, at least qualitatively. We will quantify these curves later in this chapter, after the necessary mathematical tools have been developed.

10.2.2.1 Ideal Plug-Flow Reactor

In an ideal PFR, elements of fluid pass through the reactor in single file. There is *no* fluid mixing in the direction of flow. Each element of fluid spends *exactly* the same time in the vessel. Therefore, every molecule of an ideal tracer will spend *exactly* that time in the vessel. The detector at the vessel exit will sense the entire quantity of injected tracer at the same time.

The tracer response curve will resemble the one shown below.

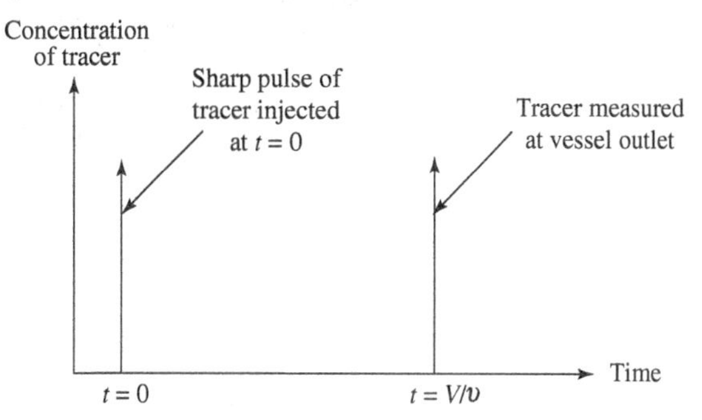

The time that the tracer spends in an ideal PFR is $t = \tau = V/\upsilon$. This is easy to see if the reactor has a constant cross section A in the direction of flow. In this case, the velocity of the fluid in the direction of flow is υ/A at every point in the reactor. If the length of the vessel in the direction of flow is L, the time required to traverse the vessel is $L/(\upsilon/A) = V/\upsilon$.

EXERCISE 10-2

Consider a radial-flow reactor (see Problem 3.1) that behaves as an ideal PFR. Show that the time required for a sharp pulse of tracer to emerge from the reactor is V/υ.

If there were a slight amount of mixing in the direction of flow, all of the tracer would not emerge at *exactly* the same time. A small amount might mix with fluid elements that were injected somewhat earlier, and a small amount might mix with elements that were injected somewhat later. This mixing would cause "spreading" of the tracer response curve, as shown below.

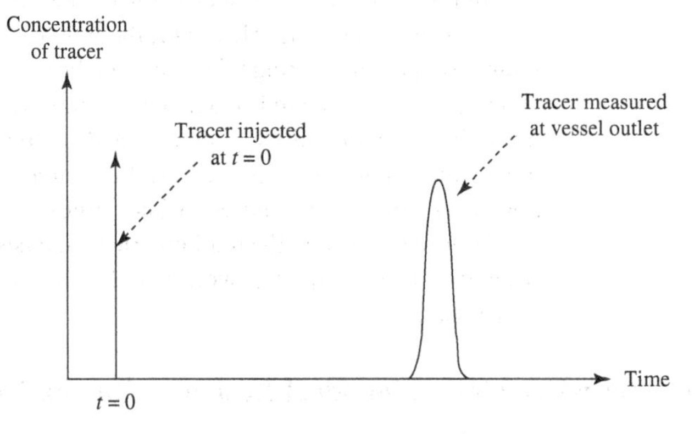

10.2.2.2 Ideal Continuous Stirred-Tank Reactor

In an ideal CSTR, the feed mixes *instantaneously* into the contents of the reactor, and the composition of the effluent stream is *exactly* the same as the composition of the fluid in the reactor. If a pulse of tracer is injected at $t = 0$, it will mix instantaneously with the fluid already in the reactor. The concentration of tracer in the reactor at $t = 0$ is as high as it ever will be. This is because the fluid that enters the reactor at later times does not contain any tracer, and because tracer begins to leave the reactor as soon as it is injected, since the composition of the effluent stream is the same as the composition of the fluid in the reactor.

The concentration of tracer in the stream leaving the CSTR has a maximum at $t = 0$, and it declines continuously thereafter. The tracer response curve for an ideal CSTR will resemble the one shown below.

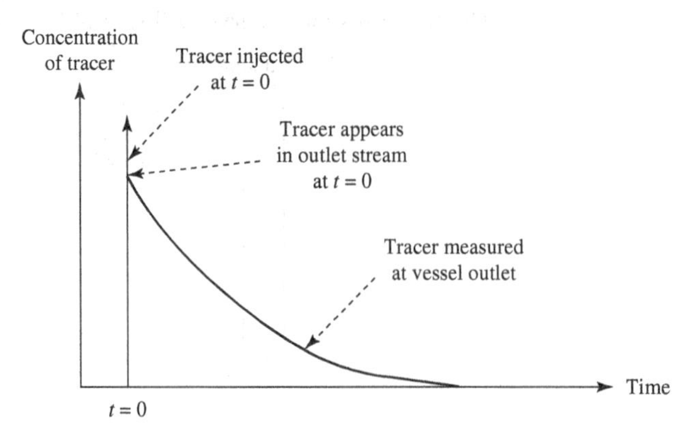

At this point, we cannot determine, based only on qualitative reasoning, the exact shape of the tracer response curve. That will be developed later in this chapter.

10.2.3 Tracer Response Curves for Nonideal Reactors

We discussed three nonideal reactors in Section 10.1.2. Let's try to figure out the qualitative behavior of the tracer response curves for these reactors.

10.2.3.1 Laminar Flow Tubular Reactor

The reactor shown in Figure 10-3 is perhaps the easiest of the three nonideal reactors to understand. This is because the flow is very well characterized, i.e., the velocity is known as a function of the radius. The velocity is maximum at $r = 0$, the center of the reactor. If the fluid is Newtonian, the velocity at $r = 0$ is twice the average velocity, the velocity at the wall is zero, and the velocity profile is parabolic.

If tracer is injected at $t = 0$, there will not be any tracer in the outlet until the tracer that is injected right at the centerline of the reactor ($r = 0$) emerges. This will require a finite time, say t_0. The concentration of tracer that emerges at t_0 will be high because the fluid velocity is highest at $r = 0$. The tracer that was injected at progressively larger radii will emerge at times longer than t_0. However, the concentration of tracer will decline with time because the fluid velocity is progressively smaller as the radius increases.

The tracer response curve for a laminar flow reactor will resemble the one shown below.

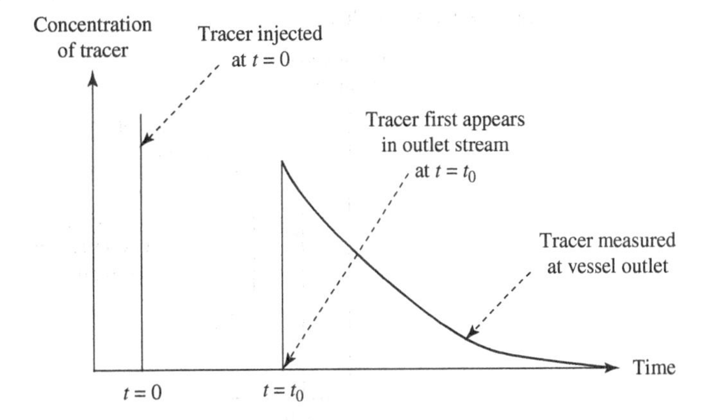

10.2.3.2 Tubular Reactor with Bypassing

The tracer curve for reactor shown in Figure 10-1 is more difficult to sketch than that of the LFTR because the flow is not as well characterized. A major portion of the flowing fluid probably will pass over the top of the catalyst bed, i.e., "bypass" the bed. This is because the resistance to flow is much greater in the packed region than in the unpacked region. As a result, a major portion of the *tracer* will also bypass the bed. In fact, since the tracer behaves exactly the same as the fluid, the fraction of tracer that bypasses the bed must be the same as the fraction of fluid that bypasses the bed. Of course, the *exact* flow distribution will depend on parameters such as the size and shape of the catalyst particles and the fraction of the cross-sectional area that is occupied by the catalyst.

Because of the difference in frictional resistance between the two regions of the bed, the fluid velocity in the upper (unpacked) region will be greater than in the packed region. Therefore, the time required for the tracer that bypasses the bed to emerge from the vessel will be substantially less than the time that it takes for the tracer that passes through the bed to emerge.

If the fluid that bypasses the catalyst bed as well as the fluid that passes through the bed were both in plug flow, and if there were no exchange of fluid between the two regions, the tracer response curve might resemble the one shown below.

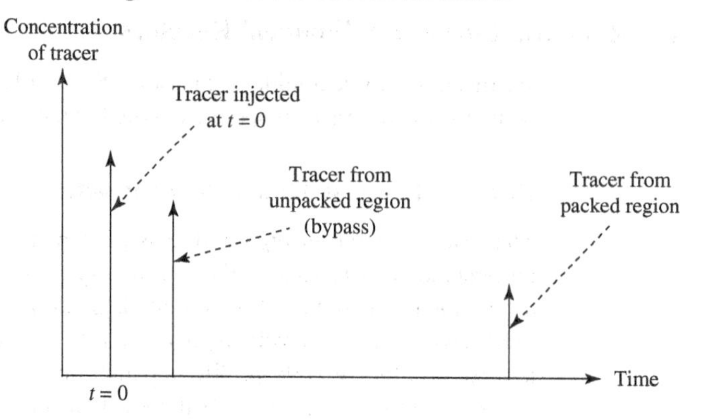

The response for the tracer that bypasses the bed is larger than the response for the tracer that passes through the packed section, since most of the tracer bypasses the bed. Moreover, the tracer from the unpacked region leaves much sooner because the fluid velocity is much higher in the unpacked region.

If flow through both regions is not quite ideal plug flow, or if there is a small amount of fluid exchange between the two regions, the tracer response curve might look something like the one shown below.

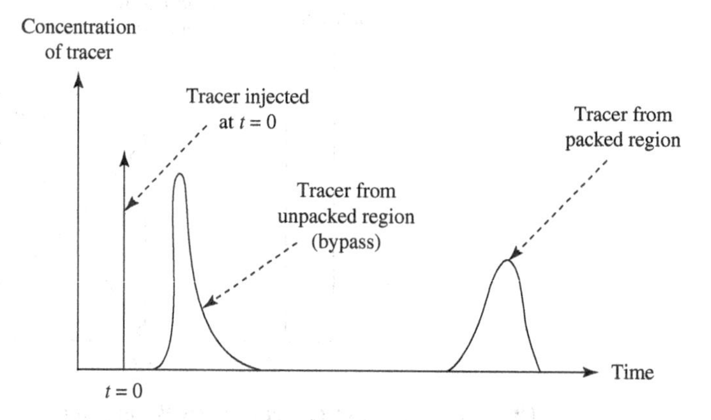

Mixing in the direction of flow causes some broadening of both peaks. Exchange of fluid between the regions causes a "tail" on the peak from the unpacked region and extends the leading edge of the peak from the packed region.

10.2.3.3 Stirred Reactor with Incomplete Mixing

The nature of flow through the reactor shown in Figure 10-2 is difficult to analyze. At the bottom of the reactor, there might be a zone (or possibly two zones in series) that is well mixed. However, the nature of flow through the upper, unagitated portion of the vessel is ill-defined. The extent of mixing in the upper section will depend on the Reynolds number in this region, i.e., on whether flow is laminar or turbulent. Given these uncertainties, it is not possible to make a reasonable sketch of the tracer response curve.

In this section, we have seen that tracer response techniques can be a very powerful diagnostic tool that can help to uncover the reason(s) for unanticipated reactor performance. The preceding discussion has focused on the use of tracer techniques in a qualitative and conceptual framework. However, tracer response curves can also be used to provide a quantitative description of flow through a reactor and can provide a basis for estimating reactor performance. We shall explore this quantitative side of tracer response techniques in the next section.

10.3 RESIDENCE TIME DISTRIBUTIONS

In the previous section, we learned that not all fluid elements spend *exactly* the same time in a reactor, except for the special case of an ideal, plug-flow reactor. *Residence time distribution* functions provide a quantitative way to describe how much time a flowing fluid spends in a reactor. Residence time distribution functions can be obtained from tracer response curves.

10.3.1 The Exit-Age Distribution Function, $E(t)$

Consider a vessel with a constant-density fluid flowing through it, *at steady state*. Fluid crosses the boundaries of the vessel *only* by convection; there is no diffusion across the system boundaries.

We will refer to such a vessel as a "closed" vessel, recognizing that this use of the word "closed" runs counter to its use in classical thermodynamics. In thermodynamics, the word "closed" means that there is no flow of mass or energy across the system boundaries. Here, the word "closed" is used to mean that mass cannot enter or leave the vessel by diffusion.

The *exit-age distribution* function, $E(t)$, is defined as

$E(t)\mathrm{d}t \equiv$ fraction of fluid leaving the vessel at time t that was in the vessel for a time between t and $t + \mathrm{d}t$

There are several other ways of saying the same thing:

$E(t)\mathrm{d}t \equiv$ fraction of fluid leaving the vessel at time t that had a *residence time* in the vessel between t and $t + \mathrm{d}t$

$E(t)\mathrm{d}t \equiv$ fraction of fluid leaving the vessel at time t that has an *exit* age between t and $t + \mathrm{d}t$

The exit-age distribution function is also known as the *external-age* distribution function. It is sometimes simply called the *residence time* distribution function. However, this can cause some confusion. As we shall see shortly, the exit-age distribution function is not the only function that is used to characterize the distribution of residence times.

The function $E(t)$ is represented graphically in the following figure.

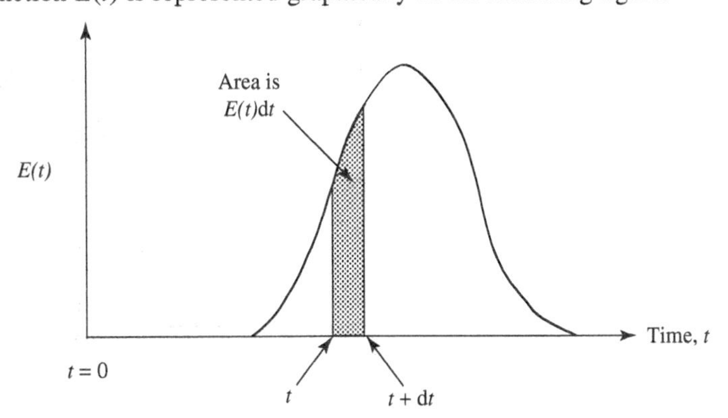

Since $E(t)\mathrm{d}t$ is a *fraction*, the units of $E(t)$ must be time inverse time, $(\text{time})^{-1}$. In addition, the fraction of fluid that leaves the vessel over all time, i.e., between $t = 0$ and $t = \infty$, must be 1. Therefore,

$$\int_0^\infty E(t)\mathrm{d}t = 1$$

The fraction of fluid *in the effluent stream* that was *in the vessel* for a time between $t = 0$ and $t = t$ is given by

$$\int_0^t E(t)\mathrm{d}t = \left\{\begin{matrix}\text{fraction of fluid in exit stream that} \\ \text{was in the vessel for a time less than } t\end{matrix}\right\} \tag{10-1}$$

Another way of saying the same thing is that $\int_0^t E(t)\mathrm{d}t$ is the fraction of fluid in the exit stream with an *exit age* less than t. This fraction can be represented graphically as shown below.

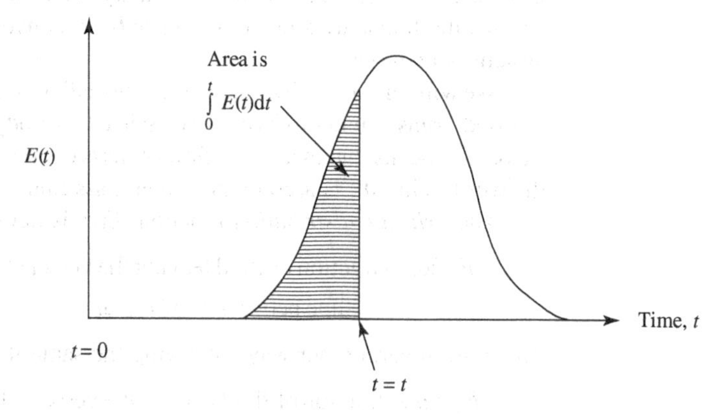

Similarly, the fraction of fluid in the effluent from the vessel that was in the vessel for a time t or longer is given by

$$\int_t^\infty E(t)\mathrm{d}t = \left\{\begin{matrix}\text{fraction of fluid in exit stream that} \\ \text{was in the vessel for a time greater than } t\end{matrix}\right\}$$

In other words, $\int_t^\infty E(t)\mathrm{d}t$ is the fraction of fluid in the stream leaving the vessel with an *exit age* of t or greater. This fraction can be represented graphically as shown below.

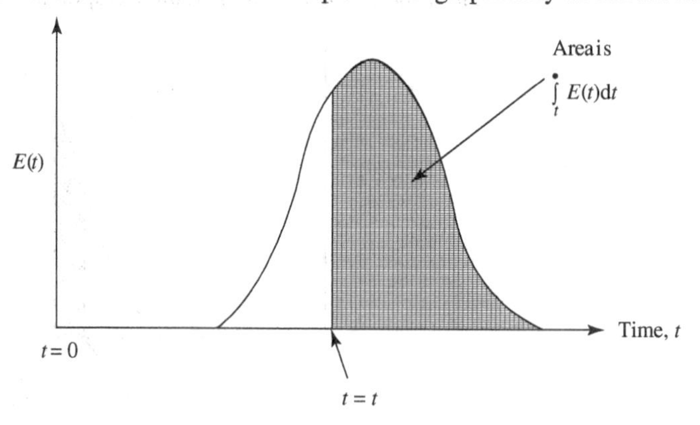

With the definition of $E(t)$ in hand, the next challenge is to connect this distribution function to tracer response curves.

10.3.2 Obtaining the Exit-Age Distribution from Tracer Response Curves

Consider an experiment where a sharp pulse of tracer is injected right at the entrance to a closed vessel, at a time designated as $t = 0$. The pulse contains M_0 units of tracer. As discussed previously, the tracer must behave *exactly* like the fluid, and it must be injected so that it labels each element of fluid proportionately.

The pulse of tracer that was injected can be described by means of the Dirac delta function, $\delta(t)$. The properties of the delta function are

$$\delta(t_0) = \infty,\ t = t_0$$
$$\delta(t_0) = 0,\ t \neq t_0$$

The Dirac delta function provides a mathematical description of the sharp pulses of tracer that were shown in the preceding figures.

The Dirac delta function is normalized, i.e.,

$$\int_0^\infty \delta(t)\mathrm{d}t = 1$$

Therefore, a pulse of tracer that contains M_0 units of tracer and is injected at $t = 0$ is described by

$$M(t) = M_0\delta(0)$$

Here, $M(t)$ is the amount of tracer injected at any time t.

Suppose that $C(t)$ is the concentration of tracer in the stream *leaving the vessel* at any time t. This is the concentration that would be measured in an experiment such as the one shown schematically in Figure 10-4. A material balance on the tracer over all time is

$$\text{tracer in} = \int_0^\infty M(t)\mathrm{d}t = \int_0^\infty M_0\delta(0)\mathrm{d}t = M_0 = \text{tracer out} = \upsilon\int_0^\infty C(t)\mathrm{d}t$$

As usual, υ is the volumetric flow rate through the vessel. The tracer material balance

Tracer material balance

$$M_0 = \upsilon\int_0^\infty C(t)\mathrm{d}t \qquad (10\text{-}2)$$

provides a very useful check on the quality of the data. If the *measured* amount of tracer leaving the vessel over all time, $\upsilon\int_0^\infty C(t)\mathrm{d}t$, is not equal to the amount injected, something is wrong. A careful investigation of the experimental technique and/or the analysis of the data is necessary.

EXAMPLE 10-1 A constant-density fluid is flowing through an experimental reactor at steady state. The flow rate is 165 cc/min. At $t = 0$, a pulse of tracer is injected into the fluid entering the reactor. The pulse contains 30 mmol of tracer. The following table shows the measured concentration of tracer in the effluent. Comment on the quality of the data. What are the possible sources of error?

Concentration of tracer in effluent at various times

Time (min)	Tracer concentration (μmol/cc)	Time (min)	Tracer concentration (μmol/cc)
0	0	9	13
1	0	10	9
2	0	11	5
3	1	12	3
4	10	13	1
5	19	14	0
6	26	15	0
7	24	16	0
8	19	17	0

APPROACH The "tracer balance" will be checked using Eqn. (10-2). This balance must be satisfied if the data are of high quality. The values of M_0 and υ are given. The value of $\int_0^\infty C(t)\mathrm{d}t$ will be evaluated by numerical integration of the data in the table.

SOLUTION The result of the numerical integration is $\int_0^\infty C(t)\mathrm{d}t = 131$ μmol-min/cc.[2] Multiplying by the flow rate, $\upsilon (= 165$ cc/min), gives $\upsilon \int_0^\infty C(t)\mathrm{d}t = 21.6$ mmol.

The measured amount of tracer in the vessel outlet is about 30% lower than the amount of tracer injected. Some possible reasons for the discrepancy include

1. A second peak might emerge at later times; the experiment was not allowed to run long enough.
2. The instrument used to measure the tracer concentration needs to be recalibrated; the readings are too low.
3. The volumetric flow rate is higher than stated.
4. The pulse that was injected contained less than 30 mmol of tracer.

The derivation of Eqn. (10-2) was designed to illustrate the use and properties of the Dirac delta function. Of course, Eqn. (10-2) is valid for *any* kind of tracer injection where the amount injected is M_0. The tracer does not have to be injected as a sharp pulse, i.e., a Dirac delta function. The amount of tracer that leaves the vessel over all time must be equal to the amount that was injected, independent of the shape of the input function.

Let's return to the question of how the exit-age distribution can be obtained from the tracer response curve. Since the tracer labels the fluid exactly

$$\left\{\begin{array}{c}\text{fraction of tracer in the effluent from the vessel}\\ \text{that was in the vessel for a time between } t \text{ and } t+\mathrm{d}t\end{array}\right\}$$

$$= \left\{\begin{array}{c}\text{fraction of fluid in the effluent from the vessel}\\ \text{that was in the vessel for a time between } t \text{ and } t+\mathrm{d}t\end{array}\right\} = E(t)\mathrm{d}t$$

The fraction of fluid that was in the vessel for a time between t and $(t + \mathrm{d}t)$ is just $E(t)\mathrm{d}t$. Since all of the tracer was injected exactly at $t = 0$, the fraction of tracer that was in the vessel for a time between t and $t + \mathrm{d}t$ is

$$\left\{\begin{array}{c}\text{fraction of tracer in the effluent from the vessel}\\ \text{that was in the vessel for a time between } t \text{ and } t+\mathrm{d}t\end{array}\right\} = \frac{\upsilon C(t)\mathrm{d}t}{\upsilon \int_0^\infty C(t)\mathrm{d}t}$$

[2] By Simpson's 1/3 Rule: value of integral $= (1\ \text{min}/3) \times [1 + 4 \times 10 + 2 \times 19 + 4 \times 26 + 2 \times 24 + 4 \times 19 + 2 \times 13 + 4 \times 9 + 2 \times 5 + 4 \times 3 + 1]$ (μmol/cc) $= 131$ mmol-min/cc.

This leads to

Calculation of exit-age distribution from measured response to a pulse input of tracer

$$E(t) = \frac{C(t)}{\int_0^\infty C(t)\mathrm{d}t} \tag{10-3}$$

Equation (10-3) permits the exit-age distribution function, $E(t)$, to be calculated from the tracer response curve that is measured after a *pulse* injection of tracer.

10.3.3 Other Residence Time Distribution Functions

10.3.3.1 *Cumulative* Exit-Age Distribution Function, $F(t)$

Sometimes it is not convenient, or even possible, to inject a sharp pulse of tracer right at the inlet to a vessel. An alternative approach is to use a *step* input of tracer. For example, consider a vessel with a constant-density fluid flowing through it at steady state. There is no tracer at all in the fluid entering the vessel. Then, at some time designated $t = 0$, the concentration of tracer in the feed is abruptly changed to a value of C_0 and is maintained at this concentration.

The concentration of tracer in the *effluent* stream is measured continuously. If we wait long enough, the effluent tracer concentration will be C_0. However, a good deal of information can be obtained from the measured tracer concentration during the period between $t = 0$ and the time required for the effluent tracer concentration to approach C_0. This type of *step input* experiment is illustrated in the following figure.

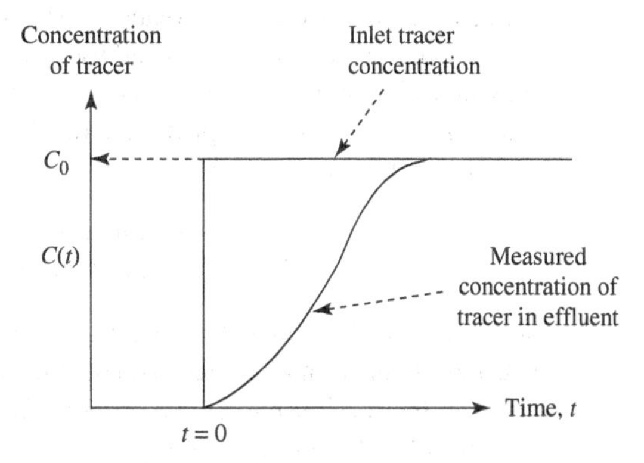

The *cumulative* exit-age distribution function, $F(t)$, is defined as the fraction of fluid in *the effluent from the vessel* that was in the vessel for a time *less than t*. Said differently, $F(t)$ is the fraction of fluid leaving the vessel that has an exit age less than t.

$$F(t) \equiv \left\{ \begin{array}{c} \text{fraction of fluid in the exit stream that was} \\ \text{in the vessel for a time less than } t, \text{ i.e., between 0 and } t \end{array} \right\} \tag{10-4}$$

The cumulative exit-age distribution function can be obtained from the curve of tracer concentration versus time that is shown above. Suppose that the concentration of tracer in the feed to the vessel was changed from 0 to C_0 *exactly* at $t = 0$. If the concentration of tracer in the effluent from the vessel is $C(t)$ at some time t, then the fraction of fluid that was in the vessel for a time less than t is simply $C(t)/C_0$. Therefore,

Calculation of cumulative exit-age distribution function from response to step input of tracer

$$F(t) = C(t)/C_0 \tag{10-5}$$

In Eqn. (10-5), $C(t)$ is the concentration of tracer in the effluent stream after a sharp step change in the inlet tracer concentration from 0 to C_0 at $t = 0$.

10.3.3.2 Relationship between *F*(*t*) and *E*(*t*)

From Eqn. (10-1),

$$\int_0^t E(t)\mathrm{d}t = \left\{ \begin{matrix} \text{fraction of fluid in exit stream that} \\ \text{was in the vessel for a time less than } t \end{matrix} \right\}$$

However, the right-hand side of this equation is the definition of $F(t)$. Therefore,

$$F(t) = \int_0^t E(t)\mathrm{d}t \tag{10-6}$$

Differentiating,

$$E(t) = \mathrm{d}F(t)/\mathrm{d}t \tag{10-7}$$

Equation (10-7) shows that $E(t)$ is the slope of the $F(t)$ curve at any point in time.

10.3.3.3 Internal-Age Distribution Function, *I*(*t*)

The internal-age distribution function $I(t)$ is not as important as $E(t)$ and $F(t)$ in the characterization of chemical reactors. However, it is quite important in medicine, where tracer response techniques are used for a variety of purposes such as measuring blood flow rates and characterizing the behavior of internal organs. The internal-age distribution function is discussed here primarily for the purpose of completeness.

The definition of $I(t)$ is

$$I(t)\mathrm{d}t \equiv \left\{ \begin{matrix} \text{fraction of fluid in vessel that has been in the} \\ \text{vessel for a time between } t \text{ and } t = \mathrm{d}t \end{matrix} \right\} \tag{10-8}$$

Notice the distinction between $I(t)$ on one hand and $F(t)$ and $E(t)$ on the other. The distribution function $I(t)$ is based on the fluid *in the vessel*. By contrast, both $F(t)$ and $E(t)$ are based on the fluid *in the stream leaving the vessel*.

From the definition of $I(t)$,

$$\int_0^t I(t)\mathrm{d}t = \left\{ \begin{matrix} \text{fraction of fluid in the vessel that} \\ \text{has been there between 0 and } t \end{matrix} \right\} \tag{10-9}$$

To relate $I(t)$ to $E(t)$ and $F(t)$, consider a vessel with a constant-density fluid flowing through it at steady state. The vessel volume is V and the volumetric flow rate through the vessel is υ. Let's do a balance on the molecules of fluid that have been in the vessel for a time between 0 and t:

$$\text{rate in} - \text{rate out} = \text{rate of accumulation}$$

$$\text{rate in} = \upsilon$$

$$\text{rate out} = \upsilon \int_0^t E(t)\mathrm{d}t$$

$$\text{rate of accumulation} = V\frac{\mathrm{d}}{\mathrm{d}t}\left[\int_0^t I(t)\mathrm{d}t\right]$$

$$VI(t) = \upsilon[1 - F(t)]$$

$$I(t) = \frac{1}{\tau}[1 - F(t)] \tag{10-10}$$

In this equation, τ is the space time, V/υ.

If $I(t)$ is evaluated at $t = 0$,

$$I(0) = 1/\tau \tag{10-11}$$

Equation (10-11) is a general result. It is valid for any closed vessel.

Finally, combining Eqns. (10-6) and (10-10),

$$I(t) = \frac{1}{\tau}\left[1 - \int_0^\infty E(t)\mathrm{d}t\right] = \frac{1}{\tau}\int_t^\infty E(t)\mathrm{d}t \tag{10-12}$$

The internal-age distribution function can be obtained from either $F(t)$ via Eqn. (10-10) or $E(t)$ via Eqn. (10-12).

10.3.4 Residence Time Distributions for Ideal Reactors

10.3.4.1 Ideal Plug-Flow Reactor

Consider an ideal PFR in the form of a tube with constant cross-sectional area A_c and a length L. A detector is located at that point. The material balance on a tracer passing through the reactor leads to the partial differential equation:

$$A_c\frac{\partial C}{\partial t} = -\upsilon\frac{\partial C}{\partial z} \tag{10-13}$$

Here, C is the concentration of tracer and z is the distance in the axial direction. The control volume for this balance is a differential slice of reactor normal to the direction of flow, as shown on page 50 and in Figure 7-4.

The concentration of tracer throughout the reactor at $t = 0$ is taken to be 0, so the initial condition for Eqn. (10-13) is

$$C = 0,\ t = 0,\ \text{all } z \tag{10-13a}$$

The tracer is injected as a sharp pulse containing M_0 units, exactly at the entrance to the reactor ($z = 0$). Therefore, Eqn. (10-13) is subject to the boundary condition

$$C = M_0\delta(0); \quad z = 0 \tag{10-13b}$$

Taking the Laplace transform of Eqn. (10-13) with respect to time gives

$$A_c s\overline{C} - A_c C(z, 0) = \upsilon\frac{\mathrm{d}\overline{C}}{\mathrm{d}z}$$

In this equation, $\overline{C}$ is the Laplace transform of C and s is the Laplace parameter. Since $C = 0$ at $t = 0$,

$$s\overline{C} = -\frac{\upsilon}{A_c}\frac{d\overline{C}}{dz}$$

Integrating from $z = 0$ to $z = L$

$$\ln \overline{C}|_0^L = -\frac{sA_cL}{\upsilon}$$

Taking the exponential of both sides, rearranging, and recognizing that $A_cL = V$, the reactor volume, and that $V/\upsilon = \tau$

$$\overline{C}(L) = \overline{C}(0)e^{-s\tau}$$

From Eqn. (10-13b) and the definition of the Laplace transform,

$$\overline{C}(0) = \int_0^{\infty} e^{-st} M_0\delta(0)dt = M_0$$

so that

$$\overline{C}(L) = M_0 e^{-s\tau} \quad (10\text{-}14)$$

Since the detector is located at L, the concentration of tracer at the reactor exit is given by the inverse transform of Eqn. (10-14).

$$C(t, L) = M_0\delta(\tau) \quad (10\text{-}15)$$

This is the concentration of tracer *at the reactor exit* that results from a pulse input of tracer right at the reactor inlet.

The exit-age distribution for an ideal PFR can be obtained by substituting Eqn. (10-15) into Eqn. (10-3),

$$E(t) = \frac{C(t)}{\int_0^{\infty} C(t)dt} = \frac{M_0\delta(\tau)}{\int_0^{\infty} M_0\delta(\tau)dt}$$

$E(t)$ for ideal PFR

$$\boxed{E(t) = \delta(\tau)} \quad (10\text{-}16)$$

This result agrees with the qualitative analysis that we performed in Section 10.2.2.1. It could have been deduced without going through the formality of solving Eqn. (10-13). In a PFR, each and every element of fluid spends exactly the same time in the reactor. For a constant-density fluid, that time is $V/\upsilon = \tau$. Therefore, if we inject a Dirac delta function of tracer at $t = 0$, a Dirac delta function will emerge at $t = \tau$. This is a necessary consequence of the fact that there is no mixing in the direction of flow in a PFR, and no gradients in the direction normal to flow.

From Eqn. (10-6),

$$F(t) = \int_0^t E(t)dt = \int_0^t \delta(\tau)dt$$

$$F(t) = 0,\ 0 < t < \tau$$

$$F(t) = 1,\ \tau \leq t$$

The cumulative age distribution function for a PFR can also be written as

$$\boxed{F(t) = U(\tau)} \tag{10-17}$$

The unit step function $U(t_1)$ is defined such that $U = 0,\ t < t_1,\ U = 1,\ t \geq t_1$.

Finally, the internal-age distribution function $I(t)$ for a PFR can be obtained from Eqn. (10-10).

$$\boxed{I(t) = \frac{1}{\tau}\left[1 - F(t)\right] = \frac{1}{\tau}\left[1 - U(\tau)\right]}$$

For an ideal PFR, $I(t) = 1/\tau$ for $t < \tau$, and $= 0$ for $t \geq \tau$.

10.3.4.2 Ideal Continuous Stirred-Tank Reactor

The various residence time distribution functions for the CSTR can also be derived from a material balance on the tracer that passes through the reactor. Initially, there is no tracer in the reactor, and the tracer is injected as a sharp pulse at $t = 0$.

For the whole CSTR,

$$\text{rate in} - \text{rate out} = \text{rate of accumulation}$$

For a pulse injection of M_0 units of tracer at $t = 0$,

$$M_0\delta(0) - \upsilon C = V\frac{\mathrm{d}C}{\mathrm{d}t} \tag{10-18}$$

where C is the concentration of the tracer in the vessel, and in the effluent, at any time t. This differential equation can be solved via the integrating factor approach. The result is

$$C(t) = \left(\frac{M_0}{V}\right)\mathrm{e}^{-t/\tau} \tag{10-19}$$

This expression agrees with the qualitative analysis that we performed in Section 10.2.2.2. The concentration of tracer in the effluent is highest at $t = 0$, and then it declines monotonically with time. The new feature that was obtained from the quantitative analysis is that the decline is exponential in time.

The various age distribution functions for an ideal CSTR can be derived from Eqn. (10-19). From Eqn. (10-3),

$$E(t) = \frac{C(t)}{\int_0^\infty C(t)\mathrm{d}t} = \frac{(M_0/V)\mathrm{e}^{-t/\tau}}{\int_0^\infty (M_0/V)\mathrm{e}^{-t/\tau}\mathrm{d}t}$$

$E(t)$ for ideal CSTR

$$\boxed{E(t) = \frac{1}{\tau}\mathrm{e}^{-t/\tau}} \tag{10-20}$$

From Eqn. (10-6),

$$F(t) = \int_0^t E(t)\mathrm{d}t = \int_0^t \frac{1}{\tau}\mathrm{e}^{-t/\tau}\mathrm{d}t = 1 - \mathrm{e}^{-t/\tau}$$

$$\boxed{F(t) = 1 - \mathrm{e}^{-t/\tau}} \tag{10-21}$$

Finally, from Eqn. (10-10),

$$I(t) = \frac{1}{\tau}\left[1 - F(t)\right] = \frac{1}{\tau}\left[1 - (1 - e^{-t/\tau})\right]$$

$$\boxed{I(t) = \frac{1}{\tau}e^{-t/\tau}} \tag{10-22}$$

Equations (10-20) and (10-22) show that $E(t) = I(t)$ for a CSTR. Because of the intense mixing in the vessel, the effluent from an ideal CSTR is a random sample of the fluid in the vessel. The probability of finding a molecule with an exit-age of t_1 in the effluent is the same as the probability of finding a molecule with an internal age of t_1 inside the reactor.

Figure 10-5 shows the external-age distribution $E(t)$ for the two ideal continuous reactors.

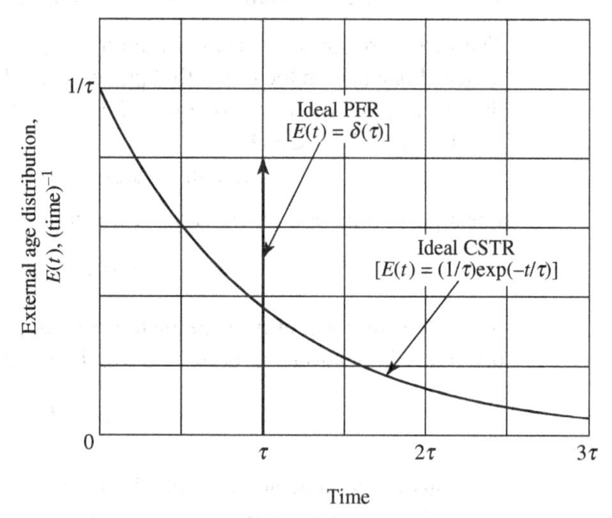

Figure 10-5 External-age distribution functions, $E(t)$, for the two ideal continuous reactors.

EXAMPLE 10-2
External-Age Distribution for a PFR and a CSTR in Series

A. What is the external-age distribution for a PFR with $V/v = \tau$, followed by a CSTR with the same residence time?

B. What is the external-age distribution for the same CSTR followed by the same PFR?

Part A: What is the external-age distribution for a PFR with $V/v = \tau$, followed by a CSTR with the same residence time?

APPROACH

This problem will be approached conceptually. The answer will be developed by using the known $E(t)$ functions for an ideal CSTR and an ideal PFR.

SOLUTION

Suppose that a pulse of tracer enters the first reactor (the PFR) as a Dirac delta function at $t = 0$. The tracer will emerge from the PFR and enter the CSTR as a delta function at $t = \tau$. Tracer will not appear in the effluent from the CSTR until $t = \tau$, since there is no tracer in the CSTR during the time period $0 \leq t \leq \tau$. During this period, all of the tracer is in the PFR.

For times longer than τ, the concentration of tracer in the effluent from the CSTR (and from the system as a whole) will be the same as that for an ideal CSTR, with a pulse of tracer that was injected at $t = \tau$, rather than at $t = 0$. The exit-age distribution for the system of a PFR with $V/\upsilon = \tau$, followed by a CSTR with $V/\upsilon = \tau$, follows from this analysis:

$$E(t) = 0; \quad 0 \le t < \tau$$

$$E(t) = \frac{1}{\tau} e^{-(t-\tau)/\tau}; \quad t \ge \tau$$

The overall exit-age distribution for the system of two reactors is shown in the following figure.

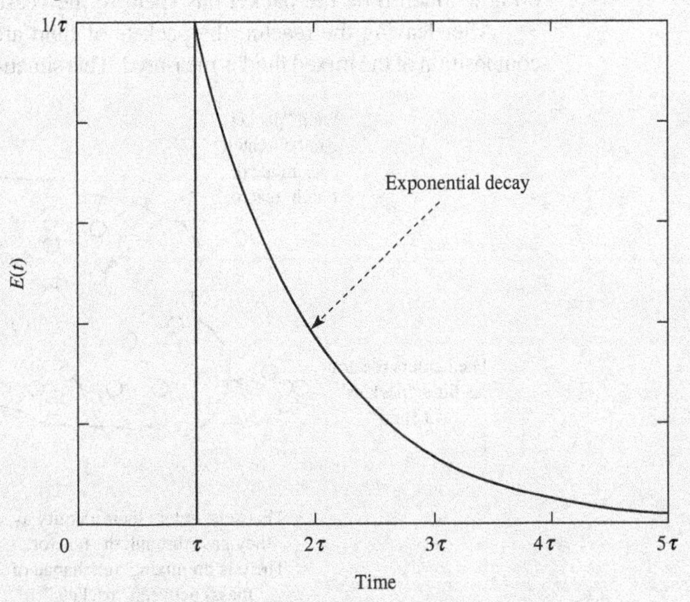

Part B: What is the external-age distribution for the same CSTR followed by the same PFR?

APPROACH The conceptual approach used in Part A will also be used here.

SOLUTION When the CSTR is located ahead of the PFR, and a pulse of tracer is injected at $t = 0$, the tracer will emerge from the CSTR with the exit-age distribution of an ideal CSTR. The tracer will enter the PFR as soon as it emerges from the CSTR. Each element of tracer will emerge from the PFR *exactly* τ time units after it enters. The effect of the PFR is to shift the exit-age distribution of the CSTR to later times by an amount τ. The exit-age distribution will be exactly the same for the two configurations.

The teachings of this simple example can be extended to any number of vessels in series. In general, *$E(t)$ for a series of independent vessels will not depend on the order of the vessels*. However, in Chapter 4, we learned that the final conversion from two different reactors in series may depend on the order of the two reactors. Although knowledge of the exit-age distribution is *necessary* to calculate the performance of a nonideal reactor, $E(t)$ alone is not always *sufficient* for this purpose.

10.4 ESTIMATING REACTOR PERFORMANCE FROM THE EXIT-AGE DISTRIBUTION—THE MACROFLUID MODEL

10.4.1 The Macrofluid Model

Suppose that a fluid flows through a vessel with no mixing between adjacent fluid elements. The feed enters the reactor as little "packets" of fluid. These packets retain their identity as they pass through the vessel; there is no exchange of mass between individual packets. However, the packets can mix in the reactor. The packets that enter at some time, t, will not all leave at the same time.

Each "packet" can be treated as a small ideal batch reactor. A reaction or reactions take place as the tiny packet moves through the vessel. The composition of each packet changes as it flows through the vessel, and the composition of a packet leaving the vessel will depend on how much time the packet has spent in the vessel, i.e., on its exit age.

After leaving the reactor, the packets of fluid are mixed on a molecular level, and the composition of the mixed fluid is measured. This situation is represented in the following figure.

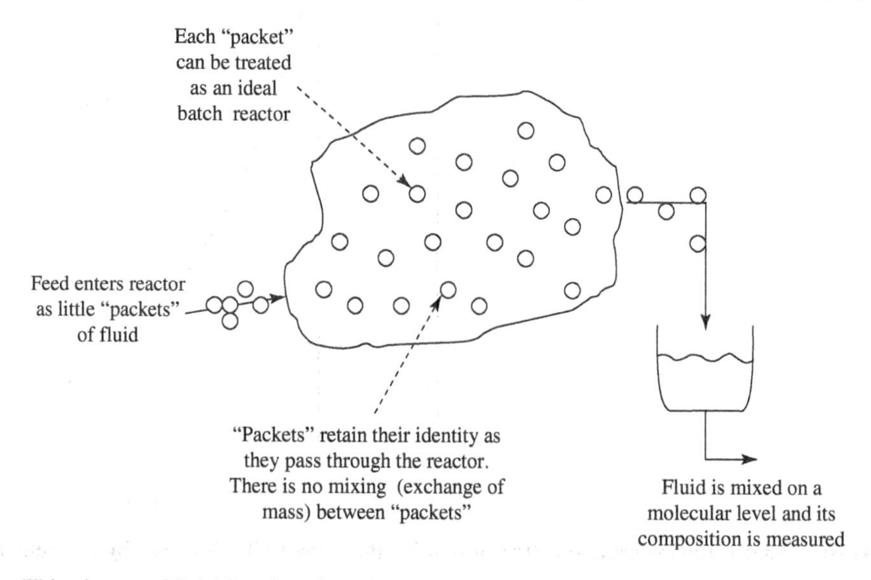

This picture of fluid flow is referred to as the "*macrofluid*" or "*segregated flow*" model. It is an idealization, i.e., a limiting case, when applied to a gas or a low viscosity liquid, because it is difficult to imagine that there will be *no* exchange of mass between fluid elements. However, it can be a very realistic model when applied to some situations involving two-phase flow. For example, if the "packets" were solid particles, and the reaction took place only in the solid phase, the macrofluid model should apply quite well.

The macrofluid model is important because it permits reactor behavior to be estimated directly from the exit-age distribution, E(t), and the reaction kinetics. No other information is necessary. If all of the reactions taking place are first order, the macrofluid model provides an exact result. If the reactions are not first order, the macrofluid model provides a *bound* of reactor behavior. We shall deal with these issues a bit later, after learning how to use the macrofluid model.

10.4.2 Predicting Reactor Behavior with the Macrofluid Model

Consider a continuous reactor at steady state. A single reaction A → products is taking place. The exit-age distribution $E(t)$ of the reactor is known, and this distribution function describes the behavior of the "packets" of fluid, as well as the behavior of the fluid as a whole.

The concentration of reactant A in a packet that was in the reactor for a time "t" is $C_A(t)$. This concentration can be calculated by solving the design equation for an ideal batch reactor, if a single reaction is taking place, or by solving the appropriate set of material balances (design equations), if multiple reactions are taking place.

The average concentration of A (or any other species) in the reactor effluent is obtained by taking a weighted average of $C_A(t)$ over the range of times that fluid elements spent in the reactor. The weighting function is $E(t)$.

$$\left\{\begin{matrix}\text{Average concentration of A} \\ \text{in fluid leaving vessel}\end{matrix}\right\} = \overline{C}_A =$$

$$\int_0^\infty \left\{\begin{matrix}\text{concentration of A in a packet} \\ \text{that was in the vessel for time } t\end{matrix}\right\} \times \left\{\begin{matrix}\text{fraction of packets that were in} \\ \text{the vessel for a time between } t \text{ and } t + \mathrm{d}t\end{matrix}\right\}$$

The "concentration of A in a packet that was in the vessel for time t" is just $C_A(t)$. The "fraction of packets that were in the reactor for a time between t and $t + \mathrm{d}t$" is just $E(t)\,\mathrm{d}t$. Therefore,

Macrofluid model

$$\overline{C}_A = \int_0^\infty C_A(t)E(t)\mathrm{d}t \qquad (10\text{-}23)$$

EXAMPLE 10-3
Use of Macrofluid Model—Series Reactions

A reactor has the external-age distribution shown in Figure 10-6. This kind of distribution might be found in a reactor that had a large number of tubes in parallel, if the pressures in the fluid distributor (inlet system) and fluid collector (outlet system) were not uniform spatially.

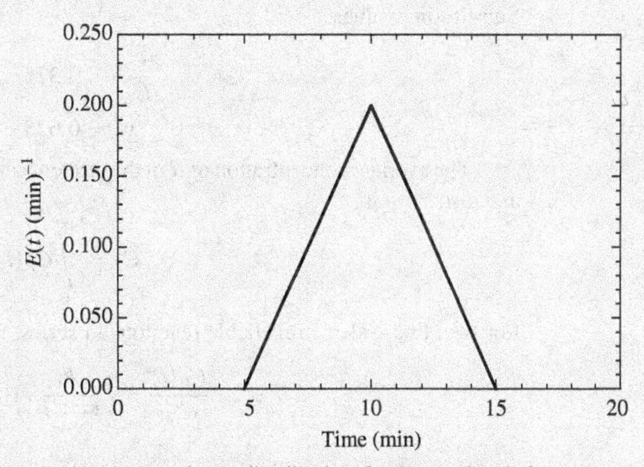

Figure 10-6 Hypothetical external age distribution for Example 10-3.

The liquid-phase reactions $A \rightarrow R \rightarrow S$ take place isothermally in this reactor. The reaction $A \rightarrow B$ is irreversible and first order in A, with a rate constant of $0.10\ \text{min}^{-1}$. The reaction $R \rightarrow S$ is irreversible and first order in R, with a rate constant of $0.30\ \text{min}^{-1}$. There is no R in the feed.

A. Predict the outlet conversion of A and the overall yield of R based on A ($Y(R/A)$).
B. Predict the conversion of A and the overall yield of R based on A for an ideal PFR with a space time of 10 min.

Part A: Predict the outlet conversion of A and the overall yield of R based on A.

APPROACH The conversion of A will be predicted with the macrofluid model, Eqn. (10–23). In order to predict $Y(R/A)$, the concentration of R in the reactor effluent must be known. This concentration will also be predicted with the macrofluid model. The concentrations $C_A(t)$ and $C_R(t)$ that are required to use the macrofluid model will be obtained by solving the design equations for an ideal batch reactor.

SOLUTION

From the figure showing $E(t)$ for this example (Figure 10-6)

$$E(t) = 0; \quad t < 5\,\text{min}$$

$$E(t) = 0.04 \times (t - 5) = 0.04t - 0.20\,(\text{min})^{-1}; \quad 5 \geq t \leq 10\,\text{min}$$

$$E(t) = 0.20 - 0.04 \times (t - 10) = 0.60 - 0.04t\,(\text{min})^{-1}; \quad 10 \leq t \leq 15\,\text{min}$$

$$E(t) = 0, \quad t > 15\,\text{min}$$

The concentration of reactant A in a packet of fluid can be calculated from the design equation for an ideal, constant-volume batch reactor

$$-\frac{dC_A}{dt} = kC_A$$

Integrating from $t = 0$, $C_A = C_{A0}$ to $t = t$, $C_A = C_A$.

$$C_A(t) = C_{A0}e^{-k_1 t}$$

From Eqn. (10-23),

$$\overline{C}_A = \int_0^\infty C_A(t)E(t)dt = \int_5^{10} C_{A0}e^{-k_1 t}[0.04t - 0.20]dt + \int_{10}^{15} C_{A0}e^{-k_1 t}[0.60 - 0.04t]dt$$

$$\frac{\overline{C}_A}{C_{A0}} = \left[0.04\frac{e^{-k_1 t}}{(-k_1)^2}(-k_1 t - 1) - \frac{0.20e^{-k_1 t}}{-k_1}\right]_5^{10} + \left[\frac{0.60e^{-k_1 t}}{-k_1} - \frac{0.04e^{-k_1 t}}{(-k_1)^2}(-k_1 t - 1)\right]_{10}^{15}$$

Substituting values

$$\frac{\overline{C}_A}{C_{A0}} = 0.375 = 1 - x_A$$

$$x_A = 0.625$$

The average concentration of R in the effluent can also be found by averaging over the external-age distribution

$$\overline{C}_R = \int_0^\infty C_R(t)E(t)dt$$

For two, first-order, irreversible reactions in series, with no intermediate (R) in the feed

$$\frac{C_R(t)}{C_{A0}} = \left(\frac{k_1}{k_2 - k_1}\right)(e^{-k_1 t} - e^{-k_2 t}) \tag{7-7}$$

$$Y(R/A) = \frac{\overline{C}_R}{C_{A0}} = \int_0^\infty \frac{C_R(t)}{C_{A0}}E(t)dt$$

$$= \left(\frac{k_1}{k_2 - k_1}\right)\left\{\int_5^{10}(e^{-k_1 t} - e^{-k_2 t})[0.04t - 0.20]dt + \int_{10}^{15}(e^{-k_1 t} - e^{-k_2 t})[0.60 - 0.04t]dt\right\}$$

Integrating,

$$\frac{\overline{C}_R}{C_{A0}} = \left\{\begin{matrix}\left[0.04\frac{e^{-k_1 t}}{(-k_1)^2}(-k_1 t - 1) - 0.20\frac{e^{-k_1 t}}{-k_1} - 0.04\frac{e^{-k_2 t}}{(-k_2)^2}(-k_2 t - 1) + 0.20\frac{e^{-k_2 t}}{-k_2}\right]_5^{10} \\ + \left[0.60\frac{e^{-k_1 t}}{-k_1} - 0.04\frac{e^{-k_1 t}}{(-k_1)^2}(-k_1 t - 1) - 0.60\frac{e^{-k_2 t}}{-k_2} + 0.04\frac{e^{-k_2 t}}{(-k_2)^2}(-k_2 t - 1)\right]_{10}^{15}\end{matrix}\right\}$$

Substituting values

$$Y(R/A) = \frac{\overline{C}_R}{C_{A0}} = 0.159$$

Part B: Predict the conversion of A and the overall yield of R based on A for an ideal PFR with a space time of 10 min.

APPROACH The design equations for an ideal, constant-density PFR will be solved.

SOLUTION For an ideal PFR,

$$\frac{C_A}{C_{A0}} = e^{-k_1\tau} = e^{-(0.10)(\text{min})^{-1}(10)(\text{min})} = 0.368 = 1 - x_A$$

$$x_A = 0.632$$

The conversion of A in the nonideal reactor is lower than in the ideal PFR, as expected. However, the difference is not large. The mixing associated with the broadened residence time distribution shown in Figure 10-6 is not sufficient to cause a significant conversion difference between the nonideal reactor and the ideal PFR.

Now consider the yield in the PFR. For two irreversible, first-order, liquid-phase reactions with no intermediate (R) in the feed, from Eqn. (7-7):

$$\frac{C_R}{C_{A0}} = \frac{k_1}{k_2 - k_1}(e^{-k_1\tau} - e^{-k_2\tau})$$

Substituting values

$$\frac{\overline{C}_R}{C_{A0}} = Y(R/A) = 0.159$$

The yield of R based on A is essentially the same for the two reactors. For the values used in this example, the yield is not a strong function of residence time, as shown in Figure 7-2. Therefore, the result is understandable.

EXAMPLE 10-4
Micromixing Versus Macromixing

The liquid-phase, second-order reaction $2A \rightarrow R$ will be run in a continuous, agitated reactor that has the same residence time distribution as an ideal CSTR. At the operating conditions of the reactor, $kC_{A0}\tau = 2.0$, where k is the second-order rate constant at reactor temperature, C_{A0} is the inlet concentration of A, and τ is the space time.

A. If the contents of the reactor are mixed on a molecular level (micromixed), what is the outlet conversion of A?

B. If the reactor obeys the macrofluid model, what is the outlet conversion of A?

Part A: If the contents of the reactor are mixed on a molecular level (micromixed), what is the outlet conversion of A?

APPROACH If the contents of the reactor are micromixed, i.e., mixed on a molecular level, the reactor behaves as an ideal CSTR. The design equation for an ideal CSTR can be solved for x_A.

SOLUTION The design equation is

$$\frac{V}{F_{A0}} = \frac{x_A}{-r_A} = \frac{x_A}{kC_{A0}(1-x_A)^2}$$

$$\frac{x_A}{(1-x_A)^2} = \frac{kC_{A0}^2 V}{F_{A0}} = kC_{A0}\tau = 2$$

Solving this quadratic equation for x gives

$$\boxed{x_A(\text{micromixed}) = 0.50}$$

Part B: If the reactor obeys the macrofluid model, what is the outlet conversion of A?

APPROACH The macrofluid model (Eqn. (10-23)) will be solved for x_A using $E(t)$ for an ideal CSTR.

SOLUTION Since the fluid passing through the reactor is a macrofluid,

$$\overline{C}_A = \int_0^\infty C_A(t)E(t)\mathrm{d}t \tag{10-23}$$

For a liquid-phase, second-order reaction in a batch reactor, $C_A(t) = C_{A0}/(1 + kC_{A0}t)$, and for an ideal CSTR, $E(t) = \mathrm{e}^{-t/\tau}/\tau$. Substituting these expressions into Eqn. (10-23),

$$\overline{C}_A = \frac{C_{A0}}{\tau}\int_0^\infty \frac{\mathrm{e}^{-t/\tau}\mathrm{d}t}{1+kC_{A0}t}$$

We might attempt to evaluate this integral via numerical integration. However, the upper limit of ∞ could cause a problem. At the very least, we would have to be very careful to carry the integration to a value of "t" that was high enough to ensure that the value of $\overline{C}_A$ did not depend on "t". A better alternative is to rearrange the above equation so that the integral can be evaluated analytically.

Let $y = 1/kC_{A0}\tau$. For this problem, y is a constant equal to 0.50. Let $z = y + (t/\tau)$. This converts the above equation to

$$\frac{\overline{C}_A}{C_{A0}} = y\mathrm{e}^y \int_y^\infty \frac{\mathrm{e}^{-z}\mathrm{d}z}{z}$$

The integral in the above equation is a form of the exponential integral, a tabulated function,[3] usually labeled Ei, E_1, or ei. Thus,

$$\frac{\overline{C}_A}{C_{A0}} = y\mathrm{e}^y\mathrm{Ei}(y)$$

Substituting values,

$$\frac{\overline{C}_A}{C_{A0}} = 0.50\mathrm{e}^{0.50}\mathrm{Ei}(0.50) = 0.50 \times 1.65 \times 0.560 = 0.461$$

Therefore,

$$\boxed{x_A(\text{macrofluid}) = 0.539}$$

For this example, where the apparent reaction order is greater than 1, the conversion is higher for a macrofluid than it is when the fluid is mixed on a molecular level.

[3] See, for example, Spiegel, M. R. and Liu, J., *Mathematical Handbook of Formulas and Tables,* 2nd edition, Schaum's Outline Series, McGraw-Hill (1999); Gautschi, W. and Cahill, W. F., "Exponential Integral and related functions", in: Abramowitz, M. and Stegun, I. E. (eds.), *Handbook of Mathematical Functions With Formulas, Graphs, and Mathematical Tables*, Applied Mathematics Series 55 U.S. Department of Commerce, National Bureau of Standards, (1964).

EXERCISE 10-3

Based on the discussion of mixing and reaction order in Chapter 4, is it reasonable the x_A(macrofluid) $>x_A$(microfluid) for Example 10-4?

10.4.3 Using the Macrofluid Model to Calculate Limits of Performance

In Chapter 4, we learned that the performance of a series of reactors could depend on the way in which the reactors were ordered. We rationalized this behavior through the concept of earliness or lateness of mixing. For a reaction with an effective order less than 1, the conversion is maximized when the reactors are arranged so that mixing takes place as early as possible during the course of the reaction. For a reaction with an effective order greater than 1, the reactors should be arranged so that mixing is delayed as long as possible in order to maximize the conversion. When the reaction is first order, the earliness or lateness of mixing does not affect the conversion.

All of the analyses in Chapter 4 were focused on the best arrangement of CSTRs and PFRs of different sizes in series. We did not recognize at the time that the external-age distribution for a given number, size, and type of reactors is the same, no matter how they are ordered. We saw a simple illustration of this in Example 10-2.

Although it was not stated explicitly, the question that we really addressed in Chapter 4 was: *If the external-age distribution is fixed, should mixing take place early in the reaction or late in the reaction, in order to maximize conversion?* The answer was

- as late as possible if the effective reaction order is greater than 1;
- as early as possible if the effective reaction order is less than 1;
- it doesn't matter if $n = 1$.

For a first-order system, earliness or lateness of mixing does not affect reactor performance, *for a given residence time distribution.* Therefore, when the residence time distribution is known, the *exact* performance of a system of first-order reactions can be calculated from the macrofluid model.

The macrofluid model represents the *latest* possible mixing *for a given residence time distribution.* There is no mixing between fluid elements until the reaction is over, i.e., until the fluid has left the reactor. For a reaction with an effective order greater than 1, the macrofluid model represents the *best* possible situation. It provides an *upper bound* on conversion. If some mixing takes place before the fluid has left the reactor, the actual conversion will be less than predicted by the macrofluid model.

Conversely, for a reaction with an effective order less than 1, the macrofluid model represents the *worst* possible situation. It provides a *lower bound* on conversion. For $n < 1$, if some mixing takes place before the fluid has left the reactor, the actual conversion will be greater than predicted by the macrofluid model.

This information is summarized in the following table.

Effective reaction order	Macrofluid (segregated flow) model provides
>1	Upper bound on conversion
$=1$	Exact result
<1	Lower bound on conversion

10.5 OTHER MODELS FOR NONIDEAL REACTORS

10.5.1 Moments of Residence Time Distributions

10.5.1.1 Definitions

The nth moment of a function, $f(x)$, about the origin, is designated μ_n and is defined as

$$\mu_n = \int_0^\infty x^n f(x)\mathrm{d}x \qquad (10\text{-}24)$$

The function $f(x)$ is a distribution function, just like $E(t)$ or $f(r)$.

Earlier in this chapter, we encountered the zeroth moment of the function $C(t)$, where $C(t)$ is the concentration of tracer in the effluent from a reactor at any time t. Equation (10-2) is a material balance on the tracer.

$$M_0 = \upsilon \int_0^\infty C(t)\mathrm{d}t \qquad (10\text{-}2)$$

Dividing both sides by the volumetric flow rate υ gives an expression for the zeroth moment of $C(t)$ about the origin.

$$M_0/\upsilon = \int_0^\infty C(t)\mathrm{d}t = \int_0^\infty t^0 C(t)\mathrm{d}t = \mu_0$$

As a distribution function $C(t)$ is a bit unusual, since it is not normalized. When a distribution function is normalized, the value of the zeroth moment must be 1. We saw this earlier with the exit-age distribution function, $E(t)$, where $\int_0^\infty E(t)\mathrm{d}t = 1$. Whenever a distribution function $f(x)$ is defined such that $f(x)\mathrm{d}x$ is the *fraction* of the values of x that lie between x and $x + \mathrm{d}x$, the value of the zeroth moment of that distribution function must be 1.

We also encountered moments in Chapter 9, Section 9.2.2.2, where the average pore radius in a catalyst particle was defined as

$$\bar{r} = \int_0^\infty r f(r)\mathrm{d}r \qquad (9\text{-}2)$$

The right-hand side of Eqn. (9-2) is simply the *first moment* about the origin of the distribution function for pore radii $f(r)$. Since $f(r)\mathrm{d}r$ is the *fraction* of the total pore volume that is in pores with radii between r and $r + \mathrm{d}r$, the distribution function $f(r)$ is normalized.

EXAMPLE 10-5 ***Calculation of Moments***

Calculate the zeroth, first, and second moments about the origin of the external-age distribution function shown in Figure 10-6.

APPROACH

From Example 10-3, $E(t)$ is given by

$$E(t) = 0; \quad t < 5\,\text{min}$$
$$E(t) = (0.04t - 0.20)\,\text{min}^{-1}; \quad 5 \le t \le 10\,\text{min}$$
$$E(t) = (0.60 - 0.04t)\,\text{min}^{-1}; \quad 10 \le t \le 15\,\text{min}$$
$$E(t) = 0; \quad t > 15\,\text{min}$$

This function will be used in Eqn. (10-24) to calculate μ_n for $n = 0$, 1, and 2.

SOLUTION Zeroth moment

$$\mu_0 = \int_0^\infty t^0 E(t)\mathrm{d}t = \int_0^\infty E(t)\mathrm{d}t = \int_5^{10} (0.04t - 0.20)\mathrm{d}t + \int_{10}^{15} (0.60 - 0.04t)\mathrm{d}t = 1$$

This calculation simply confirms that $E(t)$ is normalized.

First moment

$$\mu_1 = \int_0^\infty t^1 E(t)\mathrm{d}t = \int_0^\infty tE(t)\mathrm{d}t = \int_5^{10} t(0.04t - 0.20)\mathrm{d}t + \int_{10}^{15} t(0.60 - 0.04t)\mathrm{d}t = 10\,\mathrm{min}$$

The significance of μ_1 will be discussed in the next section of this chapter.

Second moment

$$\mu_2 = \int_0^\infty t^2 E(t)\mathrm{d}t = \int_5^{10} t^2(0.04t - 0.20)\mathrm{d}t + \int_{10}^{15} t^2(0.60 - 0.04t)\mathrm{d}t = 104\,\mathrm{min}^2$$

The physical significance of the second moment will also be discussed later, after we have dealt with the first moment.

In this example, the integrations required to calculate the moments were performed analytically, since a simple analytical expression was available for $E(t)$. However, these integrations could have been performed numerically if only discrete values of $E(t)$ (or $C(t)$) versus time had been available.

10.5.1.2 The First Moment of $E(t)$—The Average Residence Time

Average Residence Time Consider the first moment of $E(t)$ about the origin

$$\mu_1 = \int_0^\infty tE(t)\mathrm{d}t$$

and recall that $E(t)$ is just the fraction of fluid elements that stay in the reactor for a time between t and $t + \mathrm{d}t$. Therefore, the integrand of the above equation is just the time that an element of fluid spends in the reactor (t), weighted by the fraction of molecules with that residence time ($E(t)\mathrm{d}t$). Integrating over the whole range of possible residence times gives the *average* time that an element of fluid spends in the reactor. Let's designate this average residence time as $\bar{t}$, so that

$$\bar{t} = \int_0^\infty tE(t)\mathrm{d}t = \mu_1 \qquad (10\text{-}25)$$

Suppose that fluid enters and leaves the reactor *only* by convection. A vessel that meets this criterion is known as a "closed" vessel. For this case, it can be shown that[4]

$$\bar{t} = V/\upsilon = \tau \qquad (10\text{-}26)$$

[4] The proof of this relatively simple and intuitive relationship is fairly complicated. See, for example, Spalding, D. B., A note on mean residence times in steady flows of arbitrary complexity, *Chem. Eng. Sci.*, 9, 74 (1958); Danckwerts, P. V., *Chem. Eng. Sci.*, 2, 1 (1957).

In this equation, V and υ are, as usual, the reactor volume and the volumetric flow rate through the reactor, respectively.

Reactor Diagnosis Equation (10-26) can provide a useful check on operating data. In the laboratory, in a pilot plant, or in a full-scale plant, the volumetric flow rate usually is fairly easy to set and/or measure. However, the volume of reactor that is actually filled by fluid is not always so easy to determine. Mechanical drawings that can be used to determine a value of V when the reactor was installed may (or may not) be available. Moreover, things can change over time. Consider the following example.

EXAMPLE 10-6
Measurement of Reactor Volume

A small, continuous polymerization reactor has been in service for about 5 years, during which time it has been started up and shut down frequently, sometimes according to established procedure and sometimes not. The original drawings show the volume of the reactor to be 500 gallons.

The performance of the reactor appears to have deteriorated over time. There is some concern that solid polymer has built up in the reactor, reducing the volume in which the polymerization reaction takes place. Therefore, a tracer test was run, as follows.

Water was passed continuously through the reactor at a flow rate of 1000 gallons per hour. (Water is not a solvent for any polymer that might have accumulated.) When the flow of water was at steady state, a sharp pulse of tracer was injected right at the point where the water entered the reactor. The total amount of injected tracer was 100,000 units. The concentration of tracer was measured at the point where the water stream left the reactor. The results are shown in the following table.

Time after injection (min)	Tracer concentration in effluent (units/gal)	Time after injection (min)	Tracer concentration in effluent (units/gal)
0	0	25	92
1	205	30	76
2	225	40	53
3	222	50	35
4	215	60	24
5	205	70	16
10	165	80	10
15	138	90	4
20	111	100	0

Calculate the volume of fluid in the reactor and estimate the amount of polymer that has collected in the reactor.

APPROACH

A value of $\bar{t}$ will be calculated from the data in the table. The volume of fluid in the reactor will be calculated from Eqn. (10-26). The difference between 500 gallons and the calculated volume provides an estimate of the amount of polymer in the reactor.

SOLUTION

First, let's check the quality of the data using Eqn. (10-2).

$$M_0 = \upsilon \int_0^\infty C(t)dt \tag{10-2}$$

For this experiment, $M_0 = 100{,}000$ units and $\upsilon = 1000$ gal/h or 16.7 gal/min. The integral in Eqn. (10-2) must be evaluated numerically. Since the time interval between data points is not constant, the simplest way to perform the integration is to use the trapezoid rule.[5] The result is 6002 units-min/gal. Multiplying by υ (16.67 gal/min) gives $M_0 = 100{,}050$. This is an almost perfect check. The quality of the data appears to be acceptable.

The value of $\bar{t}$ is calculated from Eqn. (10-2). For a pulse injection of tracer,

$$E(t) = C(t) \Big/ \int_0^\infty C(t)\mathrm{d}t \tag{10-3}$$

so that, from Eqn. (10-25)

$$\bar{t} = \int_0^\infty tE(t)\mathrm{d}t = \int_0^\infty tC(t)\mathrm{d}t \Big/ \int_0^\infty C(t)\mathrm{d}t$$

The integral in the numerator of the right-hand side of this equation again is evaluated using the trapezoid rule[6]; the resulting value is 139,500 units-min^2/gal. Dividing this value by 6002, the value of $\int_0^\infty C(t)\mathrm{d}t$, gives $\bar{t} = 23.2$ min. Therefore, $V = \bar{t}\upsilon = 23.2(\text{min}) \times 16.7(\text{gal/min}) = 387$ gal. It appears that $500 - 387 = 113$ gallons of solid polymer may have collected in the reactor.

10.5.1.3 The Second Moment of $E(t)$—Mixing

Figure 10-5 shows the $E(t)$ curves for the two ideal, continuous reactors, the PFR and the CSTR. Clearly, the curve for the CSTR is very broad; $E(t)$ has nonzero values over the whole range of times from 0 to infinity. This breadth results from the intense mixing that takes place in the reactor. Conversely, the $E(t)$ curve for the PFR is very narrow. The external-age distribution has a nonzero value only when $t = \bar{t} = \tau = V/\upsilon$. This narrow distribution reflects the fact that there is no mixing in the direction of flow in a PFR.

Let's calculate the second moments for these two reactors. For the CSTR,

$$\mu_2 = \int_0^\infty t^2 E(t)\mathrm{d}t = \int_0^\infty t^2 \frac{1}{\tau} \mathrm{e}^{-t/\tau}\mathrm{d}t = 2\tau^2$$

For the PFR,

$$\mu_2 = \int_0^\infty t^2 E(t)\mathrm{d}t = \int_0^\infty t^2 \delta(\tau)\mathrm{d}t = \tau^2$$

For a given value of the mean residence time τ, the second moment about the origin for the CSTR is twice as large as that for the PFR. However, this difference does not reflect the difference in breadth that is visually evident in Figure 10-5, nor does it reflect the fact that there is *no* mixing in the direction of flow in a PFR and *complete* mixing in the CSTR.

The second moment *about the mean* is a much better indicator of mixing that the second moment *about the origin*. The second moment about the mean, also known as the *variance*, is defined as

$$\sigma^2 = \int_0^\infty (t - \bar{t})^2 E(t)\mathrm{d}t = \mu_2 - \tau^2 \tag{10-27}$$

---

[5] $\int_0^\infty C(t)\mathrm{d}t \cong \sum_{t_1=0}^{t_1=90\,\text{min}} (C(t_2) + C(t_1)) \times (t_2 - t_1))/2.$

[6] $\int_0^\infty tC(t)\mathrm{d}t \cong \sum_{t_1=0}^{t_1=90\,\text{min}} (t_2 C(t_2) + t_1 C(t_1)) \times (t_2 - t_1)/2.$

For a CSTR, $\bar{t} = \tau$ and $\sigma^2 = \tau^2$. For a PFR, $\bar{t} = \tau$ and $\sigma^2 = 0$. These values of σ^2 clearly reflect the major difference in mixing between the two ideal, continuous reactors.

EXERCISE 10-4

Prove that $\int_0^\infty (t-\tau)^2 E(t)\mathrm{d}t = \mu_2 - \tau^2$.

EXAMPLE 10-7 ***Calculation of the Variance of E(t)***

Calculate the value of σ^2 for the external-age distribution in Figure 10-6.

APPROACH

From Example 10-3,

$$E(t) = 0; \quad t < 5\,\text{min}$$
$$E(t) = (0.04t - 0.20)\,\text{min}^{-1}; \quad 5 \le t \le 10\,\text{min}$$
$$E(t) = (0.60 - 0.04t)\,\text{min}^{-1}; \quad 10 \le t \le 15\,\text{min}$$
$$E(t) = 0; \quad t > 15\,\text{min}$$

The values of μ_1 and μ_2 for this distribution function were found to be 10 min and 104 min^2, respectively, in Example 10-5. Moreover, $\mu_1 = \bar{t} = \tau$. The variance will be calculated from Eqn. (10-27).

SOLUTION

From Eqn. (10-27),

$$\sigma^2 = \mu_2 - \tau^2 = [104 - (10)^2] = 4\,\text{min}^2$$
$$\sigma = 2\,\text{min}$$

The value of σ is proportional to the width of the $E(t)$ curve. For an ideal CSTR, $\sigma/\tau = 1$, whereas for an ideal PFR, $\sigma/\tau = 0$. For this example, $\sigma/\tau = 0.20$. The distribution of residence times is not very broad compared to the mean residence time.

For the distribution shown in Figure 10-6, the fraction of fluid with an exit-age between $\tau - \sigma$ and $\tau + \sigma$ (i.e., between 8 and 12 min) is

$$\text{fraction}\,(\tau - \sigma \le t \le \tau + \sigma) = \int_{\tau-\sigma}^{\tau+\sigma} E(t)\mathrm{d}t$$

$$\text{fraction}\,(8\,\text{min} \le t \le 12\,\text{min}) = \int_{8\,\text{min}}^{10\,\text{min}} (0.04t - 0.20)\mathrm{d}t + \int_{10\,\text{min}}^{12\,\text{min}} (0.60 - 0.04t)\mathrm{d}t = 0.64$$

This calculation helps to explain some of the results obtained in Example 10-3. In particular, the conversion x_A and the yield $Y(R/A)$ were essentially the same for the actual reactor and for an ideal PFR. This is not surprising in view of the fact that the exit ages of 64% of the fluid fall within ±20% of the average residence time of the fluid.

10.5.1.4 Moments for Vessels in Series

Consider two independent vessels in series, as shown in the following figure.

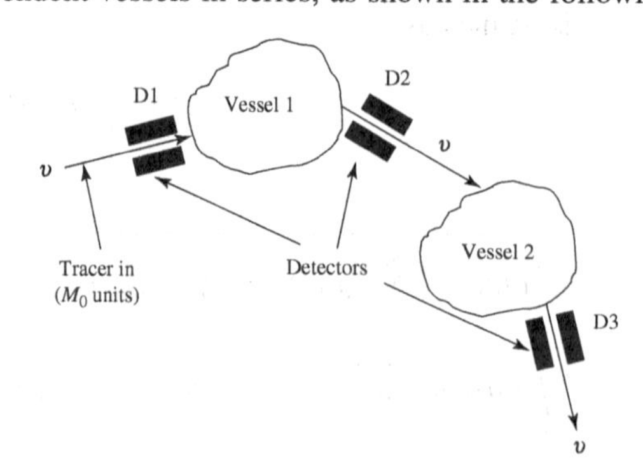

"Independent" means that the residence time of a fluid element in a downstream vessel, e.g., vessel 2 in the above sketch, does not depend on its residence time in the upstream vessel. Flow through the system is at steady state, and the volumetric flow rate is υ.

A quantity of tracer M_0 is injected upstream of vessel 1 and the shape of the pulse is measured by the first detector (D1) located right at the entrance to this vessel. The second detector (D2) measures the shape of the pulse as it leaves the first vessel, and the third detector measures the shape of the pulse as it leaves the second vessel. We will assume that the pulse entering the second vessel is the same as the one leaving the first vessel, i.e., there is no delay or distortion of the tracer in the line connecting the two vessels. If tracer is injected upstream of detector 1, the signals from the three detectors might look something like those shown in the following figure.

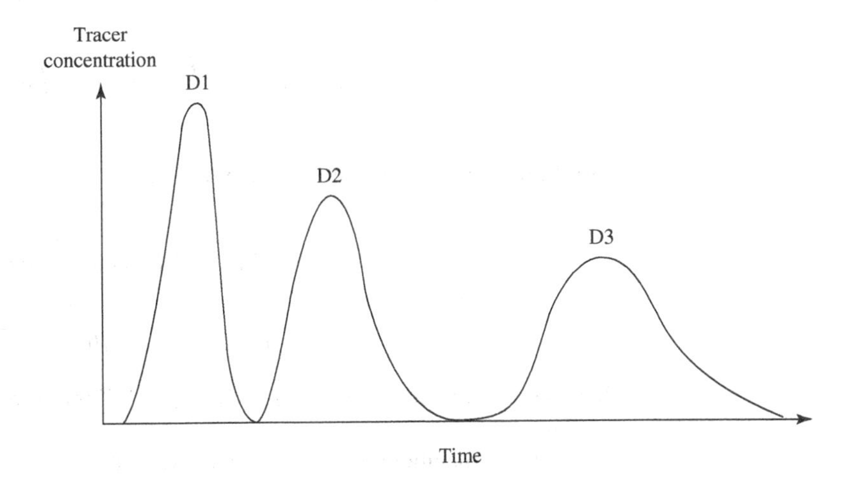

First, consider the *combination* of vessel 1 and vessel 2. The external-age distribution of this *combination* can be calculated if the external-age distributions of the two vessels are known. Suppose that an element of fluid in the stream leaving vessel 2 spent a *total* time of t in flowing through *both* vessels, and that this element spent a time t' in the first vessel, where t' is less than t. Consequently, that element of fluid spent a time $t - t'$ in the second vessel. The fraction of fluid *in the effluent from the second vessel* that has spent t' in the first vessel and $t - t'$ in the second is

$$E(t - t')\mathrm{d}t = E_1(t')\mathrm{d}t' x E_2(t - t')\mathrm{d}t \tag{10-28}$$

In this expression, E_1 is the external-age distribution function for vessel 1, E_2 is the external-age distribution function for vessel 2, and E is the external-age distribution function for the *combination* of the two vessels. This equation simply states that the probability of one event being followed by a second event is just the probability of the first event multiplied by the probability of the second. This is true as long as the events are *independent*, i.e., the probability of the second does not depend on the probability of the first.

In order to calculate $E(t)$ for the *combination* of vessel 1 and vessel 2, we must take into account the fact that t' can have any value between 0 and t. To do this, Eqn. (10-28) must be integrated over the allowable range of t'.

$$E(t) = \int_0^t E_1(t - t')E_2(t')\mathrm{d}t' \tag{10-29}$$

The integral in Eqn. (10-29) is known as the "convolution" integral. It is relatively easy to use this equation to calculate E for the combination of two vessels, provided that E_1 and E_2 are known. The converse, calculating an unknown function, say E_1, from known functions E and E_2, is known as "deconvolution" and is more difficult mathematically. Fortunately, in some cases, "deconvolution" can be avoided.

Suppose that we knew E_1 and E, and that we were willing to settle for the *moments* of E_2.[7] Let's take the Laplace transform of Eqn. (10-29), which can be found in most tables of Laplace transforms.

$$\overline{E} = \overline{E}_1 \times \overline{E}_2 \tag{10-30}$$

Here, the overbar again denotes the Laplace transform.

Recalling the definition of the Laplace transform, this equation can be written as

$$\int_0^\infty e^{-st}E(t) = \int_0^\infty e^{-st}E_1(t) \times \int_0^\infty e^{-st}E_2(t)$$

Taking the limit of this expression as $s \to 0$,

$$\int_0^\infty E(t) = \int_0^\infty E_1(t) \times \int_0^\infty E_2(t)$$

$$\mu_{0,E} = \mu_{0,E_1} \times \mu_{0,E_2}$$

If E_1 and E_2 are normalized, $\mu_{0,E_1} = \mu_{0,E_2} = 1$. Therefore, $\mu_{0,E}$ must also be normalized, i.e., $\mu_{0,E} = 1$.

In these manipulations, we have shown that

$$\lim \overline{E}_i \xrightarrow[s \to 0]{} \mu_{0,E_i}$$

Here, the subscript "i" indicates any single vessel or combination of vessels.

Now, let's take the derivative *with respect to s* of Eqn. (10-30)

$$\frac{d\overline{E}}{ds} = \overline{E}_1 \times \frac{d\overline{E}_2}{ds} + \frac{d\overline{E}_1}{ds} \times \overline{E}_2 \tag{10-31}$$

The next step is to take the limit of the above expression as $s \to 0$. Let's first evaluate the derivative of $\overline{E}$ with respect to s and then take its limit

$$\frac{d\overline{E}_i}{ds} = \frac{d}{ds}\int_0^\infty e^{-st}E_i dt = -\int_0^\infty t\, e^{-st}E_i dt$$

$$\lim \frac{d\overline{E}_i}{ds} \xrightarrow[s \to 0]{} -\int_0^\infty tE_i dt = -\mu_{1,E_i} = -\bar{t}_i$$

[7] The moment-generating property of the Laplace transform will be used in the following development. An excellent application of this technique to the analysis of mixing can be found in van der Laan, E. T., Notes on the diffusion-type model for the longitudinal mixing in flow, *Chem. Eng. Sci.*, 7, 187 (1958).

Since $\lim_{s\to 0} \overline{E}_i \longrightarrow \mu_{0,E_i}$, taking the limit of Eqn. (10-31) as $s \to 0$ gives

$$\mu_{1,E} = \mu_{0,E_1}\mu_{1,E_2} + \mu_{0,E_2}\mu_{1,E_1}$$

Since $\mu_0 = 1$,

$$\mu_{1,E} = \bar{t} = \mu_{1,E_2} + \mu_{1,E_1} = \bar{t}_1 + \bar{t}_2$$

This equation shows that the mean residence time for the combination of vessels 1 and 2 is just the sum of the mean residence time for each vessel. If two of the three mean residence times in this equation are known, the remaining one can be calculated.

Now, Eqn. (10-31) is differentiated with respect to s, and the limit as $s \to 0$ is taken. The result is

$$\sigma_E^2 = \sigma_{E_1}^2 + \sigma_{E_2}^2 \tag{10-32}$$

This equation shows that the variance for two vessels in series is the sum of the variances for the individual vessels.

EXERCISE 10-5

First, show that $\lim_{s\to 0} \mathrm{d}^2\overline{E}_i/\mathrm{d}s^2 \longrightarrow \mu_{2,E_i}$. Then prove Eqn. (10-32).

Now let's turn our attention to the signals at detectors 1 and 2, i.e., to the behavior of the combination of the tracer and the first vessel. Let $C_0(t)$ be the tracer concentration that is measured by detector 1, just as the tracer is entering vessel 1. The normalized age distribution of tracer entering vessel 1 is

$$E_0^*(t) = C_0(t) \bigg/ \int_0^\infty C_0(t)\mathrm{d}t = \upsilon C_0(t)/M_0$$

The age distribution of the tracer entering the vessel is labeled E^* to differentiate it from a "true" exit-age distribution for a vessel. The function E^* will depend on the extent of mixing in the lines between the injection point and detector 1. It will also depend on the way in which tracer is injected, e.g., how long the injection takes, whether the injection rate varies with time, etc. This latter characteristic is unique to E^*; the exit-age distribution functions for vessels *per se* do not depend on the way in which tracer is injected.

The mean residence time and the variance of the tracer entering vessel 1 can be calculated from the curve of $C_0(t)$ versus time that is measured at detector 1. Thus,

$$\bar{t}_0 = \mu_{1,0} = \int_0^\infty tE_0^*(t)\mathrm{d}t$$

$$\sigma_0^2 = \mu_{2,0} - (\bar{t}_0)^2 = \int_0^\infty t^2E_0^*(t)\mathrm{d}t - (\bar{t}_0)^2$$

The results obtained previously can now be generalized. Suppose there are N independent vessels in series, with an arbitrary tracer input entering the first vessel. Then,

$$\bar{t}_{\text{overall}} = \bar{t}_0 + \bar{t}_1 + \bar{t}_2 + \cdots + \bar{t}_N = \bar{t}_0 + \sum_{i=1}^{N} \bar{t}_i \tag{10-33}$$

$$\sigma_{\text{overall}}^2 = \sigma_0^2 + \sigma_1^2 + \sigma_2^2 + \cdots \sigma_N^2 = \sigma_0^2 + \sum_{i=1}^{N} \sigma_i^2 \tag{10-34}$$

In these equations, the subscript "overall" denotes the combination of all of the vessels *plus* the injection system. In other words, $\bar{t}_{\text{overall}}$ and $\sigma^2_{\text{overall}}$ are the mean residence time and the variance that would be computed from the signal of the detector on the stream leaving the last (N^{th}) vessel.

In the next section, some models of nonideal flow through reactors will be introduced. Each model will contain a parameter (or parameters) that may have to be determined from tracer response experiments. For a given model, the unknown parameters usually can be determined from the moments of the tracer curve.

10.5.2 The Dispersion Model

10.5.2.1 Overview

The Dispersion model, or dispersed plug-flow model, is an extension of the ideal plug-flow model that allows for some mixing in the direction of flow. The Dispersion model is one dimensional. Like the ideal plug-flow model, all of the concentration and temperature gradients are in the direction of flow. There are no concentration or temperature gradients normal to the direction of flow.

Consider a tubular reactor with a cross-sectional area A_c and a length L such as the one shown in Figure 7-4. The reactor contains a packing (spheres, cylinders, saddles, etc.). Fluid flows through the interstices between the particles, and the fractional interstitial volume ε_i is defined as

$$\varepsilon_i \equiv \frac{\text{volume of interstices between particles of packing}}{\text{total reactor volume}}$$

If the packing is porous, e.g., porous catalyst particles, the volume of the pores is *not* included in ε_i.

The fraction of the reactor that is occupied by the packing is $1 - \varepsilon_i$. In other words, $1 - \varepsilon_i$, is the *geometrical* volume of the packing per unit geometrical volume of reactor.

We will assume that the bed is isotropic. Therefore, the fraction of the *cross-sectional area* that is occupied by the interstices also is ε_i, and the cross-sectional area through which fluid *actually flows* is $\varepsilon_i A_c$. Similarly, the volume of reactor through which fluid *actually flows* is $\varepsilon_i A_c L\,(=\varepsilon_i V)$.

If the reactor were an ideal PFR, the material balance on reactant A over a differential slice of reactor, such as the one shown in Figure 7-4, would be

$$\upsilon \mathrm{d}C_A + (-r'_A)A_c \mathrm{d}z = 0$$

In this equation, z is the dimension in the direction of flow, i.e., the axial direction in a tubular reactor. As usual, υ is the volumetric flow rate. We will assume that the density of the fluid flowing through the reactor is constant, so that υ is the same at every cross section along the length of the reactor. Finally, $-r'_A$ is the rate of disappearance of A per unit of *geometric* reactor volume.

Rearranging the above equation,

$$\left(\frac{\upsilon}{A_c}\right)\frac{\mathrm{d}C_A}{\mathrm{d}z} - r'_A = 0 \qquad (10\text{-}35)$$

For plug flow in a reactor that contains no packing, $u = \upsilon/A_c$ is the actual velocity at every point in the reactor. If the reactor contains packing, then υ/A_c is the *superficial* velocity, and the actual velocity is $\upsilon/\varepsilon_i A_c$.

To form the dispersed plug-flow model, the term $-D(\mathrm{d}^2C_A/\mathrm{d}z^2)$ is added to Eqn. (10-35) to describe mixing in the direction of flow.

$$-D\frac{\mathrm{d}^2C_A}{\mathrm{d}z^2}+u\frac{\mathrm{d}C_A}{\mathrm{d}z}-r'_A=0 \tag{10-36}$$

The model of Eqn. (10-36) is based on the assumption that mixing of species A in the direction of flow (z-direction) is proportional to the concentration gradient ($\mathrm{d}C_A/\mathrm{d}z$), i.e., mixing is Fickian in nature. The parameter D is known as the axial *dispersion* coefficient. In general, D is not the same as the molecular diffusion coefficient $D_{A,m}$. Except in rare cases, D is much greater than the molecular diffusion coefficient because D includes the effects of radial and temporal velocity fluctuations, in addition to molecular diffusion.

Equation (10-36) can be made partially dimensionless by introducing the variables $Z = z/L$ and $\tau^* = V/\upsilon = L/u$, where $u = \upsilon/A_c$.

$$\left(\frac{D}{uL}\right)\frac{\mathrm{d}^2C_A}{\mathrm{d}Z^2}+\frac{\mathrm{d}C_A}{\mathrm{d}Z}+(\tau^*)(-r'_A)=0 \tag{10-37}$$

The parameter τ^* is the mean residence time of fluid *in an empty reactor, i.e., one that does not contain any packing*. If the reactor contains packing, then the mean residence time, as measured by tracer techniques, will be less than τ^* because some of the total volume of the reactor will be occupied by the solid material of the packing, causing the actual velocity of the fluid to be greater than u.

The dimensionless group (D/uL) is known as the *Dispersion number*. It can have values between 0 and ∞. If D/uL is 0, there is no backmixing and the reactor is an ideal PFR. As D/uL approaches ∞, the reactor approaches the behavior of an ideal CSTR. The inverse of the Dispersion number is referred to as the axial Peclet number, $\mathrm{Pe_a} = uL/D$.

Equation (10-37) raises two questions:

1. How does one get values of D/uL?
2. How should the term describing the rate of disappearance of A, $-r'_A$, be formulated?

Let's deal with the second question first.

10.5.2.2 The Reaction Rate Term

Equation (10-37) has an analytical solution only for a first-order reaction, although numerical and/or approximate solutions are available for other rate equations.[8] The present discussion will be focused on the first-order case, with careful attention to the questions of reversibility, and whether the reaction is homogeneous or heterogeneous.

Homogeneous Reaction Consider a reversible reaction $A \rightleftarrows B$, that takes place only in the fluid phase. In this discussion of homogeneous reactions, we will assume that any packing in the reactor is not porous. The reversible reaction obeys the rate equation

$$-r_A = k_fC_A - k_rC_B$$

[8] Levenspiel, O., *The Chemical Reactor Omnibook*, Oregon State University Bookstores, Corvallis, O.R. (1989), 64–21, 22; Westerterp, K. R., van Swaaij, W. P. M., and Beenackers, A. A. C. M., *Chemical Reactor Design and Operation*, John Wiley & Sons, (1984), pp. 199–202.

The units of $-r_A$ are moles A/volume of *fluid*-time, and the units of k_f and k_r are time^{-1}. The rate per unit volume of *reactor*, $-r'_A$, is given by $-r'_A = \varepsilon_i(-r_A)$. Since $k_r = k_f/K_{eq}^C$ and $C_B = (C_{A0} + C_{B0}) - C_A$,

$$-r_A = \frac{k_f(1 + K_{eq}^C)}{K_{eq}^C} C_A - \frac{k_f(C_{A0} + C_{B0})}{K_{eq}^C}$$

Let $\alpha = k_f(1 + K_{eq}^C)/K_{eq}^C$ and $\beta = k_f(C_{A0} + C_{B0})/K_{eq}^C$. Then,

$$-r_A = \alpha C_A - \beta$$

and

$$-r'_A = \varepsilon_i(\alpha C_A - \beta) \tag{10-38}$$

If the reaction is essentially irreversible, $K_{eq}^C \to \infty$, $\alpha \to k_f$, and $\beta \to 0$, so that $-r'_A = \varepsilon_i k_f C_A$. The term α/β is the concentration of A when the reaction is in equilibrium and $-r'_A = 0$, i.e., $C_{A,eq} = \alpha/\beta$.

Substituting Eqn. (10-38) into Eqn. (10-37),

$$-\left(\frac{D}{uL}\right)\frac{d^2C_A}{dZ^2} + \frac{dC_A}{dZ} + (\varepsilon_i\tau^*)(\alpha C_A - \beta) = 0$$

Let $\Psi_A = (C_A/C_{A0}) - (\beta/\alpha C_{A0})$. This transformation makes the above equation completely dimensionless

$$-\left(\frac{D}{uL}\right)\frac{d^2\Psi_A}{dZ^2} + \frac{d\Psi_A}{dZ} + (\varepsilon_i\tau^*)\alpha\Psi_A = 0 \tag{10-39}$$

Since β/α is the concentration of A at equilibrium, the parameter Ψ_A can be interpreted as the difference between the dimensionless concentration of A at any point in the reactor (C_A/C_{A0}) and the dimensionless *equilibrium* concentration of A $(\beta/\alpha C_{A0})$. The parameter $1 - \beta/\alpha C_{A0}$ is the *maximum* possible change in the dimensionless concentration of A, i.e., the difference between the dimensionless feed and equilibrium concentrations.

Note that $\varepsilon_i\tau^*$ $(= \varepsilon_i V/\upsilon)$ is the *actual* space time for the reactor, i.e., the reactor volume that is *not* occupied by packing divided by the volumetric flow rate.

The boundary conditions for Eqn. (10-39) have been the subject of considerable discussion.[9] The so-called "Danckwerts boundary conditions" are perhaps the most commonly used. These conditions apply to a closed vessel, i.e., one with no diffusion across the system boundaries.

$$C_{A0} = C_A(0^+) - \frac{D}{u}\frac{dC_A}{dz}(0^+)$$

$$1 - \frac{\beta}{\alpha C_{A0}} = \Psi_A(0^+) - \frac{D}{uL}\frac{d\Psi_A(0^+)}{dZ} \tag{10-39a}$$

$$\frac{dC_A}{dz}(L^-) = 0$$

$$\frac{d\Psi_A}{dZ}(1^-) = 0 \tag{10-39b}$$

[9] See, for example, Froment, G. F. and Bischoff, K. B., *Chemical Reactor Analysis and Design*, 2nd edition, John Wiley & Sons, New York, (1990), p. 535.

The solution to Eqns. (10-39), (10-39a), and (10-39b) will be developed after their equivalents have been formulated for a heterogeneous catalytic reaction.

Heterogeneous Catalytic Reaction Consider the same reversible reaction $A \rightleftarrows B$ that now takes place on a heterogeneous catalyst. The reaction obeys the rate equation

$$-r_A = k_f C_A - k_r C_B$$

The units of $-r_A$ now are moles/*weight catalyst*-time, and the units of k_f and k_r are volume fluid/*weight catalyst*-time. The rates $-r'_A$ and $-r_A$ are related by

$$-r'_A = \rho_B(-r_A)$$

Here ρ_B is the bulk density of the catalyst (weight catalyst/geometrical reactor volume).

The equivalent of Eqn. (10-39) for a heterogeneous catalytic reaction is

$$-\left(\frac{D}{uL}\right)\frac{d^2\Psi_A}{dZ^2} + \frac{d\Psi_A}{dZ} + (\rho_B\tau^*)\alpha\Psi_A = 0 \tag{10-40}$$

The term $\rho_B\tau^*$ is just the space time τ for a heterogeneous catalytic reaction, i.e., $\tau = \rho_B\tau^* = W/\upsilon$.

10.5.2.3 Solutions to the Dispersion Model

Rigorous Equations (10-39) and (10-40) have exactly the same form and are subject to the same boundary conditions, Eqns. (10-39a) and (10-39b). Let the Dispersion number be designated Δ.

$$\Delta = D/uL \tag{10-41a}$$

Further, let a dimensionless parameter "a" be defined as

$$\begin{aligned} a &\equiv \sqrt{1 + (4\Delta\varepsilon_i\tau^*\alpha)} \quad \text{(homogeneous)} \\ a &\equiv \sqrt{1 + (4\Delta\rho_B\tau^*\alpha)} \quad \text{(heterogeneous)} \end{aligned} \tag{10-41b}$$

Equations (10-39) and (10-40) can now be solved for the value of Ψ_A in the effluent from an isothermal reactor:

$$\frac{\Psi_A}{1 - (\beta/\alpha C_{A0})} = \frac{4a\exp(1/2\Delta)}{(1+a)^2\exp(a/2\Delta) - (1-a)^2\exp(-a/2\Delta)} \tag{10-42}$$

If the reaction is irreversible, $\Psi_A = C_A/C_{A0}$ and $\beta = 0$, so that

$$\frac{C_A}{C_{A0}} = \frac{4a\exp(1/2\Delta)}{(1+a)^2\exp(a/2\Delta) - (1-a)^2\exp(-a/2\Delta)} \tag{10-43}$$

EXAMPLE 10-8
Use of the Dispersion Model

The reversible catalytic reaction $A \rightleftarrows B$ takes place isothermally in a packed tubular reactor. The equilibrium constant for the reaction, based on concentration, is 1.0, and the forward rate constant k_f is 10 l/g cat-min. The concentration of A in the feed to the reactor is 1.0 mol/l and there is no B in the feed. The Dispersion number for the reactor is 0.20, and the space time, $\tau\,(=W/\upsilon)$, is 0.10 (g-min/l).

A. What is the value of C_A in the stream leaving the reactor?
B. What value of C_A would be expected for an ideal PFR operating at the same value of τ?
C. If the Dispersion number remained constant,[10] what value of τ would be required in order for the dispersed plug-flow reactor to produce the same outlet concentration of A as the PFR?

Part A: What is the value of C_A in the stream leaving the reactor?

APPROACH The units of k_f show that the reaction is heterogeneous. The values of α, β, and a will be calculated from the information provided in the problem statement. The value of Δ is given. The value of C_A in the stream leaving the reactor will be calculated using Eqn. (10-43).

SOLUTION First, calculate values of α, β, $\beta/\alpha C_{A0}$, and a.

$$\alpha = k_f(1 + K_{eq}^C)/K_{eq}^C = 10(\text{l/g-min}) \times (1+1)/1 = 20\,\text{l/g-min}$$

$$\beta = k_f(C_{A0} + C_{B0})/K_{eq}^C = 10(\text{l/g-min}) \times (1+0)(\text{mol/l})/1 = 10\,\text{mol/g-min}$$

$$\beta/\alpha C_{A0} = 10(\text{mol/g-min})/[20(\text{l/g-min}) \times 1(\text{mol/l})] = 0.50$$

$$a = \sqrt{1 + (4 \times [D/uL] \times \tau \times \alpha}$$

$$a = \sqrt{1 + (4 \times [0.20] \times 0.10(\text{g-min/l}) \times 20(\text{l/g-min})} = 1.61$$

Substituting these values into Eqn. (10-42) gives $\Psi_A = 0.102$. Since $\Psi_A = (C_A/C_{A0}) - (\beta/\alpha C_{A0})$,

$$C_A(\text{out}) = 0.602\,\text{mol/l}.$$

For perspective, the equilibrium concentration of A (β/α) is $C_{A,eq} = 0.50\,\text{mol/l}$.

Part B: What value of C_A would be expected for an ideal PFR operating at the same value of τ?

APPROACH This question could be answered by starting with the design equation for a heterogeneous catalytic reaction taking place in an ideal PFR with a constant-density fluid passing through it, i.e., Eqn. (3-31). An alternative approach is to recognize that the behavior of an ideal PFR with no axial dispersion is described by Eqn. (10-40), with the first term on the left-hand side (the axial dispersion term) removed. This question will be answered by adopting the latter approach.

SOLUTION Removing the axial dispersion term from Eqn. (10-40) gives

$$\frac{d\Psi_A}{dZ} + (\rho_B \tau^*)\alpha\Psi_A = 0; \quad \frac{d\Psi_A}{\Psi_A} = -\rho_B\tau^*\alpha dZ = -\alpha\tau dZ$$

Integrating this equation from $Z = 0$, $\Psi_A = [1 - (\beta/\alpha C_{A0})]$ to $Z = 1$, $\Psi_A = \Psi_A(\text{out})$ gives

$$\Psi_A(\text{out}) = [1 - (\beta/\alpha C_{A0})]\exp(-\alpha\tau)$$

Substituting values,

$$\Psi_A(\text{out}) = [0.50]\exp(-2.0) = 0.0677; \; C_A(\text{out}) = 0.568\,(\text{mol/l})$$

As expected, the outlet concentration from the PFR is lower, and closer to equilibrium, than the outlet concentration from the dispersed plug-flow reactor.

Part C: If the Dispersion number remained constant,[10] what value of τ would be required in order for the dispersed plug-flow reactor to produce the same outlet concentration of A as the PFR?

[10] Generally, the dispersion number will *not* remain constant when additional catalyst is added to an existing reactor. This behavior will be discussed in more detail later in this chapter.

APPROACH

The value of τ that produces a value of Ψ_A of 0.0677 must be found for the dispersed plug-flow reactor. To do this, the value of Ψ_A in Eqn. (10-42) will be set to 0.0677 and this equation will be solved for "a". The value of τ then will be calculated from "a".

SOLUTION

An EXCEL spreadsheet was set up to do this calculation. GOALSEEK was used to find the value of "a" that produced the desired value of Ψ_A. The result is

$$a = 1.77; \quad \tau = (a^2 - 1)/4\alpha(D/uL) = 0.133$$

About 33% more catalyst is required in the dispersed plug-flow reactor to reach the same outlet reactant concentration as the ideal PFR, for this particular example. This extra catalyst requirement results from the axial mixing (backmixing) that occurs in the reactor.

Approximate (Small Values of D/uL) For small deviations from plug flow, i.e., for low values of D/uL, the exponentials in Eqn. (10-42) can be expanded and higher-order terms can be dropped. This procedure gives some approximate results that can be useful in analyzing reactor behavior. For example,

$$\frac{\Psi_A}{\Psi_A(\text{PFR})} = 1 + \left(\frac{D}{uL}\right)(\alpha\tau)^2 \tag{10-44}$$

This equation shows that the conversion in an ideal PFR will always be greater than in a dispersed plug-flow reactor operating at the same τ.

Another useful presentation of the results for small values of D/uL is to compare the reactor volumes (or weights of catalyst) required to achieve a specified outlet concentration. For a homogeneous reaction,

$$\frac{V}{V_{\text{PFR}}} = 1 + \left(\frac{D}{uL}\right)\ln\left(\frac{1 - (\beta/\alpha C_{A0})}{\Psi_A}\right) \tag{10-45a}$$

For a heterogeneous catalytic reaction,

$$\frac{W}{W_{\text{PFR}}} = 1 + \left(\frac{D}{uL}\right)\ln\left(\frac{1 - (\beta/\alpha C_{A0})}{\Psi_A}\right) \tag{10-45b}$$

Not surprisingly, these equations show that the required reactor volume or catalyst weight is greater for a dispersed plug-flow reactor than for an ideal PFR.

10.5.2.4 The Dispersion Number

Values of D/uL for a given reactor configuration and operating conditions can be measured using tracer response techniques. This approach is discussed in some detail later in this section. However, one of the most powerful features of the Dispersion model is the availability of correlations that can be used to estimate values of D/uL. Over the years, the Dispersion model has been used to analyze the behavior of tracers in a number of reactor geometries, e.g., packed and empty tubes, and over a wide range of operating conditions. These studies have led to the development of correlations for common situations. We will begin by discussing these correlations.

Estimating D/uL from Correlations

Packed beds

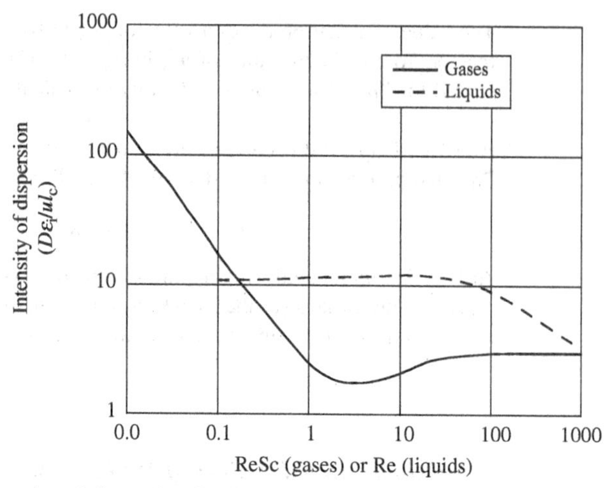

Figure 10-7 Intensity of dispersion for liquids and gases flowing through packed beds.

Figure 10-7 is a correlation for dispersion in packed tubular reactors.[11] The parameter $D\varepsilon_i/ul_c$ that is plotted on the y-axis is known as the "intensity of dispersion" or the "local" Dispersion number. For the situation shown in Figure 10-7, the intensity of dispersion depends on the particle Reynolds number, Re $(= l_c u\rho/\mu)$. For gases, $D\varepsilon_i/ul_c$ also depends on the Schmidt number, $\mu/\rho D_{A,m}$.

In this plot, l_c is the characteristic dimension of the packing or catalyst particle, as defined previously by Eqn. (9-10).

$$l_c = \text{characteristic dimension} = \frac{\text{geometric volume of particle}}{\text{geometric surface area of particle}} \tag{9-10}$$

Although the correlations in Figure 10-7 are shown as lines, there is considerable scatter in the raw data on which these correlations are based. Visual pictures of the uncertainty in the correlations are available in the references from which this figure was developed.

The Dispersion number for the reactor is the product of the "intensity of dispersion" and a geometric factor, $l_c/\varepsilon_i L$, i.e.,

$$D/uL = (D\varepsilon_i/ul_c) \times (l_c/\varepsilon_i L)$$

The "intensity of dispersion" depends on local conditions, e.g., particle size and superficial velocity. However, the geometric factor is inversely proportional to the reactor length L. Therefore, for fixed *local* conditions, the Dispersion number decreases as the reactor becomes longer.

In many previous correlations, the "intensity of dispersion" was expressed in terms of the equivalent diameter of a spherical particle. In fact, spherical particles were used in many of the studies on which Figure 10-7 is based. In this chapter, for consistency, the correlations have been converted to the same "characteristic dimension" that was used in Chapter 9.

Empty pipes—turbulent flow

A correlation for turbulent flow in empty pipes is shown in Figure 10-8. The structure of this correlation is the same as that shown previously, i.e., an "intensity of dispersion" is

[11] Adapted from Wen, C. Y. and Fan, L. T., *Models for Flow Systems and Chemical Reactors,* Marcel Dekker, Inc. (1975), p. 171 (for gases) and from Levenspiel, O., *Chemical Reaction Engineering*, 3rd edition, John Wiley & Sons (1999), p. 311 (for liquids).

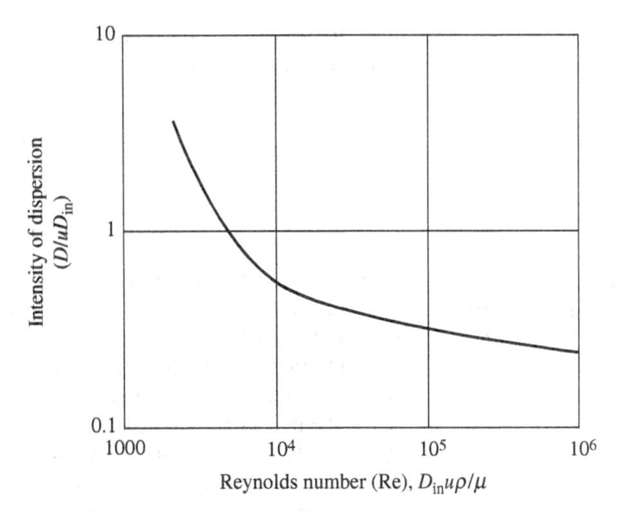

Figure 10-8 Intensity of dispersion for fluids flowing through empty pipes in turbulent flow.

correlated against a local Reynolds number. In this case, the characteristic dimension that is used in both parameters is the inner diameter of the empty tube, D_{in}.

Wen and Fan[12] present a graph showing the data on which Figure 10-8 is based. Again, there is considerable scatter in the raw data. Over the range of Re covered by the above plot, the data are correlated well by

$$D/uD_{in} = \frac{3.0 \times 10^7}{(\mathrm{Re})^{2.1}} + \frac{1.35}{(\mathrm{Re})^{1/8}}$$

Laminar flow tubular reactors
Under some circumstances, the Dispersion model can be applied to laminar flow tubular reactors. If either $4Dt/(D_{in})^2 \geq 0.80$ or $D/uD_{in} \leq 1$,[13] the Dispersion number for laminar flow in long, empty tubes is given by

$$D/uL = (1/\mathrm{Re} \times \mathrm{Sc}) + (\mathrm{Re} \times \mathrm{Sc}/192)$$

and values of D/uL calculated from this equation can be used in the Dispersion model to calculate reactor performance.

Finally, the Dispersion model has been applied to catalytic reactors where the particles are in motion, e.g., fluidized bed reactors and slurry bubble column reactors. These situations can be complicated because the Dispersion model may have to be applied to both the fluid and solid phases separately. Analysis of such reactors is beyond the scope of this text, but further details can be found in the literature.[14]

Criterion for Negligible Dispersion In analyzing the behavior of a reactor, we may wish to know whether the effect of axial dispersion needs to be taken into account at all. From the preceding discussion, it is clear that the value of D/uL will be negligibly small when the reactor is sufficiently long. In this case, the ideal PFR model will be sufficient.

---

[12] Wen, C. Y. and Fan, L. T., *Models for Flow Systems and Chemical Reactors*, Marcel Dekker, Inc. (1975), p. 146.

[13] This combination of criteria is somewhat conservative. See Wen, C. Y. and Fan, L. T., *Models for Flow Systems and Chemical Reactors*, Marcel Dekker, Inc. (1975), p. 106.

[14] See, for example, Wen, C. Y. and Fan, L. T., *Models for Flow Systems and Chemical Reactors,* Marcel Dekker, Inc. (1975), pp. 150–167, 175–181.

Suppose that δ_{err} is the allowable fractional error in V (or W). Then, for a first-order reaction, the effect of axial dispersion can be neglected if[15]

$$\frac{L}{l_c} > \frac{1}{\delta_{err}}\left(\frac{D}{ul_c}\right)\ln\left(\frac{1-(\beta/\alpha C_{A0})}{\Psi_A}\right) \tag{10-46}$$

EXAMPLE 10-9
Test for Negligible Dispersion

A circular tube is packed with spherical catalyst particles that have a diameter of 1.0 mm. The fractional interstitial volume ε_i is 0.35. The particle Reynolds number in the tube is 200. The fluid flowing in the tube is a liquid.

The reversible, first-order reaction $A \rightleftarrows B$ will be carried out in the reactor, which must be sized so that the reaction is at least 99% of the way to equilibrium in the outlet. The concentrations of A and B in the feed to the reactor are 2.0 and 0.10 mol/l, respectively, and the value of the equilibrium constant K_{eq}^{C} is 2.0. The reactor will operate isothermally.

How long does the reactor have to be in order to neglect axial dispersion, if a 1% error in the calculated catalyst requirement is acceptable?

APPROACH

The value of L will be calculated from Eqn. (10-46). The value of Ψ_A will be calculated from the statement that the reaction must be at least 99% of the way to equilibrium, and the value of D/uL will be estimated via the correlation in Figure 10-7. All of the other required values are either given or can be calculated directly from the given values.

SOLUTION

The value of δ_{err} is 0.01. From Figure 10-7, the value of $\varepsilon_i D/ul_c$ is about 6 for a flowing liquid with a particle Reynolds number of 200. Therefore, $D/ul_c = 6/0.35 = 17$. The equilibrium concentration of A, β/α, is $C_{A,eq} = (C_{A0} + C_{B0})/(1 + K_{eq}^{C}) = 0.70$ mol/l. The value of the term $1 - (\beta/\alpha C_{A0})$ is $1 - (0.70/2) = 0.65$. The concentration of A leaving the reactor must be such that $C_{A0} - C_A(\text{out}) = 0.99 \times [C_{A0} - (\beta/\alpha)]$. Using the definition of Ψ_A, this can be rearranged to

$$\Psi_A(\text{out}) = 0.01[1 - (\beta/\alpha C_{A0})] = 0.0065$$

Substituting values into Eqn. (10-46),

$$L/l_c > (1/0.01)(17)\ln(0.65/0.0065) = 7800$$

The value of l_c is $0.10/6 = 0.0167$ mm. Therefore, the effect of axial dispersion can be neglected if the packed bed is at least 130 mm (0.13 m) long.

Measurement of D/uL The correlations discussed previously cover a wide range of important conditions and configurations. Nevertheless, a reactor configuration may be encountered that has not been studied previously, and for which correlations are not available. In this event, D/uL must be measured using tracer response techniques.

The *unsteady-state* material balance on a nonreactive, nonadsorbing tracer that obeys the Dispersion model is

$$-D\frac{\partial^2 C}{\partial z^2} + u\frac{\partial C}{\partial z} = \varepsilon_i\frac{\partial C}{\partial t} \tag{10-47}$$

providing that the packing is not porous. The solution of this partial differential equation depends on the boundary conditions. Fortunately, for small values of D/uL, the solution is not very sensitive to the choice of boundary conditions.

[15] Adapted from Mears, D. E., *Ind. Eng. Chem. Fundam.*, 15, 20 (1976).

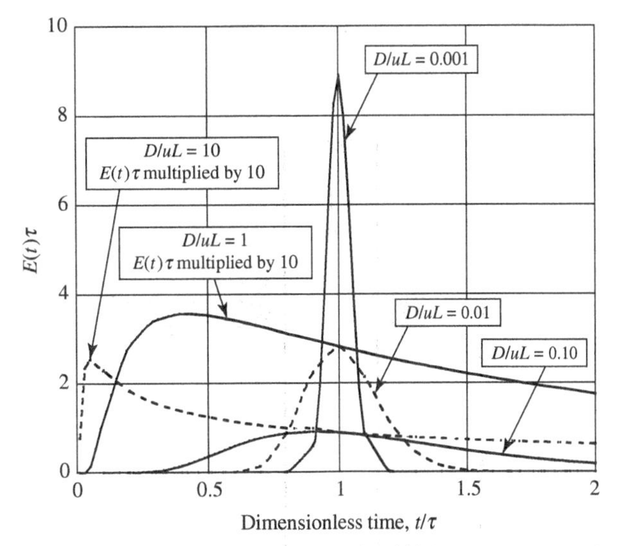

Figure 10-9 The dimensionless exit-age distribution ($\tau E(t)$) for the dispersed plug-flow model as a function of the dimensionless time, t/τ, for various values of the dispersion number, D/uL. "Open" vessel boundary conditions.

Figure 10-9 shows the dimensionless exit-age distribution that is obtained by solving Eqn. (10-47) for an "open" vessel.

This figure shows the behavior of the *dimensionless* exit-age distribution function as a function of the dimensionless time, $\Theta = t/\tau$.

The dimensionless exit-age distribution depends only on the Dispersion number. At very low values of D/uL, the shape of the distribution approximates that of an ideal PFR. However, the distribution becomes broader as the value of D/uL increases. At high values of D/uL, the exit-age distribution approaches that of an ideal CSTR.

An "open" vessel configuration is convenient to use for measuring the dispersion of a tracer in an empty or packed tube. A schematic of an idealized experimental setup is shown in Figure 10-10.

If the two detectors are located a distance L apart, and if the tracer is injected upstream of the first detector, the average residence time of the tracer between the two detectors is given by

$$\bar{t} = \tau + 2(D/uL)$$

Note that $\bar{t} \neq \tau$ *because the system being studied, i.e., the section of the bed between the two detectors, is not "closed."* Tracer can diffuse across both system boundaries, i.e., the planes at which the detectors are located.

The difference in the variance between the two detectors is given by

$$\Delta\sigma^2 = 2(D/uL) + 8(D/uL)^2$$

For small values of D/uL, the difference between $\bar{t}$ and τ may not be large enough to permit an accurate estimate of D/uL. The measured variance difference allows a more accurate calculation. Note that the second term in the expression for the variance difference is small compared with the first, if D/uL is small.

If the use of an "open" configuration is not feasible or practical, other options can be considered. van der Laan[16] provides relationships between D/uL, $\bar{t}$ and σ^2 for different configurations that involve different boundary conditions. In all cases, when D/uL is small,

[16] van der Laan, E. Th., *Chem. Eng. Sci.*, 7, 187 (1957).

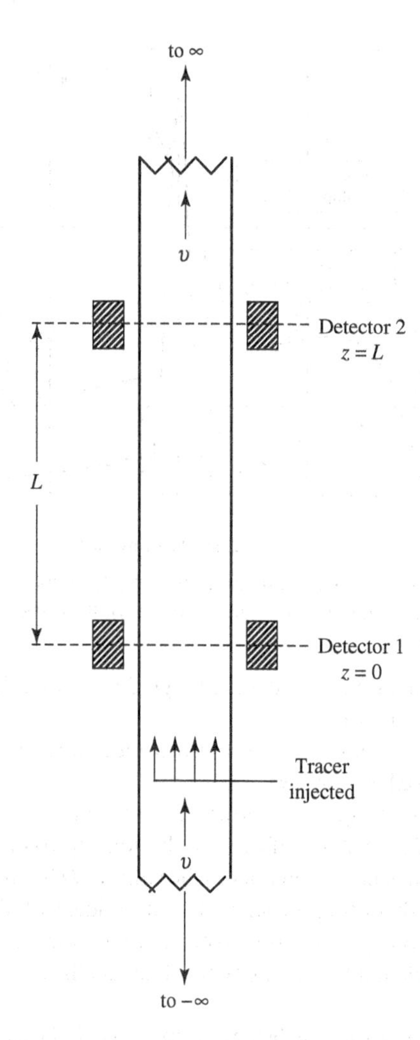

Figure 10-10 Diagram of "infinite" packed-bed system for measuring the dispersion coefficient.

the variance difference is independent of the specific boundary conditions and is well approximated by

$$\Delta\sigma^2 \cong 2(D/uL)$$

10.5.2.5 The Dispersion Model—Some Final Comments

The Dispersion model has been widely used, especially for describing relatively small deviations from plug flow in packed beds and empty tubes. The availability of correlations that can be used to estimate D/uL for common reactor configurations makes this model especially convenient. Nevertheless, there are many situations, primarily high values of D/uL, for which the Dispersion model is not appropriate. Two alternative approaches to describing nonideal reactors are considered in the final sections of this chapter.

10.5.3 CSTRs-In-Series (CIS) Model

10.5.3.1 Overview

In Chapter 4, we learned that the reaction rate disadvantage associated with a single CSTR can be reduced by putting several CSTRs in series. In fact, if the number of CSTRs is sufficiently large, the behavior of the series of reactors closely approximates plug flow. The CIS model, also referred to as the "tanks-in-series" model, is built on this observation.

The CIS model is quite simple conceptually. A number "N" of CSTRs are arranged in series. Each reactor has the same volume V or contains the same weight of catalyst W. This configuration is shown schematically in the following figure.

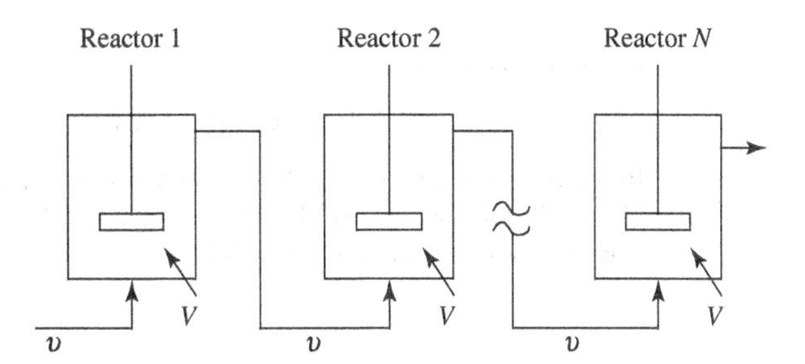

The CIS model is very flexible. It can describe backmixing behavior that ranges from complete mixing, i.e., a single CSTR ($N = 1$), to a situation that approaches ideal plug flow ($N \rightarrow \infty$). The number of CSTRs in series (N) is the only variable in the CIS model; "N" must be adjusted to match the mixing behavior of the actual reactor. If the actual vessel closely approaches plug flow, N will be relatively large. If there is considerable mixing in the direction of flow, N will be small.

The CIS model and the Dispersion model are referred to as "one-parameter" models, since only a single parameter, D/uL or N, is used to characterize mixing. When there is very little mixing in the direction of flow, i.e., when N is large or D/uL is small, the physical basis of the Dispersion model is stronger than that of the CIS model. Moreover, when the number of CSTRs in series is large, it can be tedious to calculate the performance of the series of reactors. On the other hand, it is relatively straightforward to use even the most complex rate equations in the CIS model. It is not necessary to restrict the analyses to first-order rate equations.

One disadvantage of the CIS model is that correlations do not exist that permit N to be *predicted* for given reactor configurations and flow conditions. For practical purposes, "N" must be determined experimentally, via tracer response techniques, for each situation.

10.5.3.2 Determining the Value of "N"

The *total* volume of the "N" CSTRs ($N \times V$) should be the same as the total volume of the nonideal reactor that is being characterized. The total volume can be checked via tracer response techniques. If the mean residence time $\bar{t}$ in the vessel is measured, and if the vessel is "closed,"

$$NV = \upsilon\bar{t}$$

The number of CSTRs in series, N, must be determined via tracer response techniques. The number of CSTRs is related to the difference in the *variance* of the exit-age distribution between the reactor inlet and the reactor outlet.

$$\Delta\sigma^2 = \bar{t}^2/N \tag{10-48}$$

Clearly, the larger the variance of the exit-age distribution, the smaller the value of N.

EXAMPLE 10-10
Calculation of the Number of CSTRs in Series from E(t)

Calculate the value of "N" in the CIS model for the external-age distribution shown in Figure 10-6.

APPROACH

The value of μ_1 $(= \bar{t})$ was calculated to be 10 min in Example 10-5. The value of σ^2 was calculated to be 4 min^2 in Example 10-7. Therefore, N will be calculated directly from Eqn. (10-48).

SOLUTION

From Eqn. (10-48),

$$N = \bar{t}^2/\sigma^2 = (10)^2\,\text{min}^2/4\,\text{min}^2 = 25$$

Based on this high value of N, the behavior of a reactor with the exit-age distribution shown in Figure 10-6 should be very close to the behavior of an ideal PFR. This is consistent with the results that were obtained in Example 10-3.

10.5.3.3 Calculating Reactor Performance

Procedures for calculating the performance of a series of CSTRs were developed in Chapter 4. However, there are three questions that can arise when using the CIS model:

- What should be done when N, as obtained from tracer experiments, is not an integer?
- Should the fluid flowing through the series of CSTRs be treated as a macrofluid or a microfluid?
- Does mixing occur in the lines connecting the CSTRs, in the case of a macrofluid?

These issues can be explored by performing calculations for the various limiting cases. However, before considering an example, let's expand on the last question. Suppose that the fluid flowing through each CSTR is a macrofluid. If the fluid that leaves the first CSTR becomes mixed *on a molecular level* before it enters the second CSTR, then the composition of the feed to the second reactor will be uniform, and will be equal to the mixing-cup average composition of the stream leaving the first reactor. In this case, the performance of the second CSTR can be calculated by applying the macrofluid model with a feed stream that has the average composition of the stream leaving the first reactor.

However, the fluid leaving the first CSTR may not become mixed on a molecular level before it enters the second reactor. In the limit, the fluid elements in this stream may remain completely segregated between the two reactors. For this situation, the procedure described in the preceding paragraph is not appropriate because the stream entering the second CSTR is not uniform on a molecular level. Rather, the feed to the second CSTR consists of packets of fluid with different compositions. In this case, the macrofluid model must be used, with the measured RTD for the *whole* reactor, i.e., the reactor that was being modeled as a series of equal-volume CSTRs. In the case of a fluid that remains as a macrofluid, there is no need to fit the CIS model to the measured RTD. The macrofluid model must be applied directly.

EXAMPLE 10-11
Limiting Cases of the CIS Model

The second-order, liquid-phase reaction $2A \rightarrow B$ takes place in an isothermal, nonideal reactor with a total volume of 1000 l. The volumetric flow rate through the reactor is 100 l/min and the concentration of A in the feed, C_{A0}, is 4.0 mol/l. The rate constant at the temperature of operation is 0.25 l/mol-min.

Tracer tests have been performed to determine the nature of mixing in the reactor. These tests showed that $\bar{t} = 10$ min and that $\Delta\sigma^2 = 40$ min^2.

Use the CIS model to estimate the conversion of A in the stream leaving the reactor. Assume that the fluid becomes micromixed between reactors.

APPROACH

The number of CSTRs in series will be calculated from the values of $\bar{t}$ and $\Delta\sigma^2$. The microfluid and macrofluid models will then be used to establish bounds of reactor behavior. If the number of CSTRs in series, as calculated from $\bar{t}$ and $\Delta\sigma^2$, is not an integer, calculations will be performed for integral numbers of reactors that bracket the calculated value of N.

SOLUTION

The number of CSTRs in series is

$$N = \bar{t}^2/\Delta\sigma^2 = (10)^2\,\text{min}^2/40\,\text{min}^2 = 2.5$$

The four calculations shown in the following table will be performed to explore the importance of N and the scale of mixing.

	$N = 2$	$N = 3$
Microfluid		
Macrofluid		

Microfluid model

$N = 2$:

Since the total reactor volume is 1000 l and the volumetric flow rate is 100 l/min, each reactor will have a volume of 500 l and a space time τ of 5 min. Let $x_{A,1}$ be the fractional conversion of A in the stream leaving the first reactor and let $x_{A,2}$ be the conversion of A in the stream leaving the second reactor, and the final conversion from the whole nonideal reactor.

Design equation—first reactor:

$$\frac{V}{F_{A0}} = \frac{x_{A,1}}{-r_A(x_{A,1})} \Rightarrow kC_{A0}\tau = \frac{x_{A,1}}{(1 - x_{A,1})^2} = 0.25\left(\frac{1}{\text{mol-min}}\right)4.0\left(\frac{\text{mol}}{1}\right)5(\text{min}) = 5$$

$$x_{A,1} = 0.642$$

Design equation—second reactor:

$$\frac{V}{F_{A0}} = \frac{x_{A,2} - x_{A,1}}{-r_A(x_{A,2})} \Rightarrow kC_{A0}\tau = \frac{x_{A,2} - 0.642}{(1 - x_A)^2} = 5$$

$$x_{A,2} = 0.814$$

In these equations, $-r_A(x_{A,i})$ indicates that the reaction rate is evaluated at the concentrations in the stream leaving the "i"th reactor, i.e., at $x_{A,i}$.

$N = 3$:

Each reactor now will have a volume of 333 l and a space time τ of 3.33 min. Applying the procedure shown above yields $x_{A,1} = 0.582$, $x_{A,2} = 0.765$, and $x_{A,3} = 0.845$.

Macrofluid model

$N = 2$:

Each reactor will have a volume of 500 l and a space time τ of 5 min.

First reactor

For a macrofluid,

$$\overline{C}_A = \int_0^\infty C_A(t)E(t)dt \tag{10-23}$$

Let $\overline{C}_{A,1}$ be the concentration of A leaving the first reactor and entering the second reactor. Let $\overline{C}_{A,2}$ be the concentration of A leaving the second reactor. For an ideal CSTR, $E(t) = (1/\tau)\exp(-t/\tau)$ and for a second-order reaction, $C(t) = C_{A0}/(1 + kC_{A0}t)$. Then, as shown in Example 10-4,

$$\overline{C}_{A,1} = \frac{C_{A0}}{\tau}\int_0^{\infty}\frac{e^{-t/\tau}dt}{1 + kC_{A0}t}$$

$$\frac{\overline{C}_{A,1}}{C_{A0}} = y_1 e^{y_1}\mathrm{Ei}(y_1)$$

Here, $y_1 = 1/kC_{A0}\tau = 1/[0.25(\text{l/mol-min}) \times 4.0(\text{mol/l}) \times 5(\text{min})] = 0.20$.
Therefore,

$$\frac{\overline{C}_{A,1}}{C_{A0}} = (0.20)e^{0.20}\mathrm{Ei}(0.20) = 0.20 \times 1.22 \times 1.22 = 0.298$$

$$\overline{C}_{A,1} = 0.298 \times 4.0 = 1.19$$

The fractional conversion of A in the stream leaving the first reactor is $x_{A,1} = (C_{A0} - \overline{C}_{A,1})/C_{A0} = (4.0 - 1.19)/4.0 = 0.702$.

The concentration of A in the effluent from the second CSTR is given by

$$\overline{C}_{A,2} = \frac{\overline{C}_{A,1}}{\tau}\int_0^{\infty}\frac{e^{-t/\tau}dt}{1 + k\overline{C}_{A,1}t}$$

Letting $y_2 = 1/k\overline{C}_{A,1}\tau = 0.671$,

$$\frac{\overline{C}_{A,2}}{\overline{C}_{A,1}} = (y_2)e^{y_2}\mathrm{Ei}(y_2) = 0.671 \times 1.956 \times 0.395 = 0.518$$

$$\overline{C}_{A,2} = 0.518 \times 1.19 = 0.616$$

The final conversion is $x_{A,2} = (C_{A0} - \overline{C}_{A,2})/C_{A0} = (4 - 0.616)/4.0 = 0.846$.

$N = 3$:

Each reactor will have a volume of 333 l and a space time τ of 3.33 min. Using the procedure shown above yields $x_{A,1} = 0.633$, $x_{A,2} = 0.795$, and $x_{A,3} = 0.864$.

The results of these four calculations are summarized below.

	$N = 2$	$N = 3$
Microfluid	$x_A = 0.81$	$x_A = 0.85$
Macrofluid	$x_A = 0.85$	$x_A = 0.86$

Since the effective reaction order for this example is greater than 1, we should have known *a priori* that the macrofluid model would predict a higher conversion than the microfluid model. Moreover, based on our discussion of CSTRs in series in Chapter 4, we should have known that $N = 3$ would give a higher conversion that $N = 2$. Therefore, we could have bracketed the above results with two calculations: $N = 2$/microfluid (lowest conversion) and $N = 3$/macrofluid (highest conversion).

10.5.4 Compartment Models

10.5.4.1 Overview

The last approach to characterizing nonideal reactors is the use of *compartment* models. This methodology is based on the idea that a real reactor can be described as an assembly of vessels in which the flow is well characterized. The types of vessels that are used to construct compartment models are CSTRs, PFRs, and well-mixed stagnant zones (WMSZs).

Once again, the use of tracer response experiments is a critical element in the development of a compartment model. As with the Dispersion and CIS models, the parameters of a compartment model may have to be calculated from the moments of the external-age distribution function. Moreover, the shapes of the $E(t)$ curve must be used to help choose the types and arrangement of compartments that will comprise the model. In order to conceptualize a compartment model from the measured $E(t)$ curve at the reactor outlet, it is important to inject the tracer as a sharp pulse, as close to the reactor inlet as possible. If the tracer that enters the reactor is too dispersed, the shape of the tracer curve that leaves the reactor will not reflect its behavior in sufficient detail to formulate an accurate model.

Compartment models are "multiparameter" models, in that more than one parameter is required to characterize flow and mixing. The exact number of parameters depends on the number of compartments in the model.

At this point, our understanding of CSTRs and PFRs is well developed, so that the discussion of compartment models can begin with combinations of these two elements.

10.5.4.2 Compartment Models Based on CSTRs and PFRs

Reactors in Parallel For two vessels *of any kind* in parallel,

$$E(t) = f_1 E_1(t) + f_2 E_2(t) \tag{10-49}$$

In this equation, $E_1(t)$ and $E_2(t)$ are the exit-age distributions of the two vessels, and f_1 and f_2 are the fractions of the total flow υ that pass through each vessel. If υ_1 is the flow rate through vessel 1, then $f_1 = \upsilon_1/\upsilon$. Only two vessels will be considered here, so that $f_2 = 1 - f_1$.

If the two vessels in parallel are "closed," then $\bar{t}_1 = \tau_1 = V_1/\upsilon_1$ and $\bar{t}_2 = \tau_2 = V_2/\upsilon_2$. Further, since $\upsilon = \upsilon_1 + \upsilon_2$, and $V = V_1 + V_2$,

$$f_1 = (\tau_2 \upsilon - V)/(\tau_2 - \tau_1) \tag{10-50}$$

and

$$V_1 = \tau_1(\tau_2 \upsilon - V)/(\tau_2 - \tau_1) \tag{10-51}$$

Suppose that the mean residence time in each vessel ($\bar{t}_1 = \tau_1$ and $\bar{t}_2 = \tau_2$) can be determined from a tracer response experiment. Then, if the value of the total volume V is known, and if the total volumetric flow rate υ is known, the fraction of flow passing through each vessel, and the volume of each vessel, can be calculated from Eqns. (10-50) and (10-51). Equations (10-51) and (10-52) are valid for *any* kind of vessels in parallel, as long as they are "closed."

EXERCISE 10-6

Prove Eqns. (10-50) and (10-51).

Figure 10-11 shows the shape of the $E(t)$ curves that result from various combinations of CSTRs and PFRs in parallel. The comments beside each figure indicate how the values of τ_1 and τ_2 that are required to quantify the model and to calculate reactor performance can be extracted from the $E(t)$ curves.

(a) Two PFRs in parallel

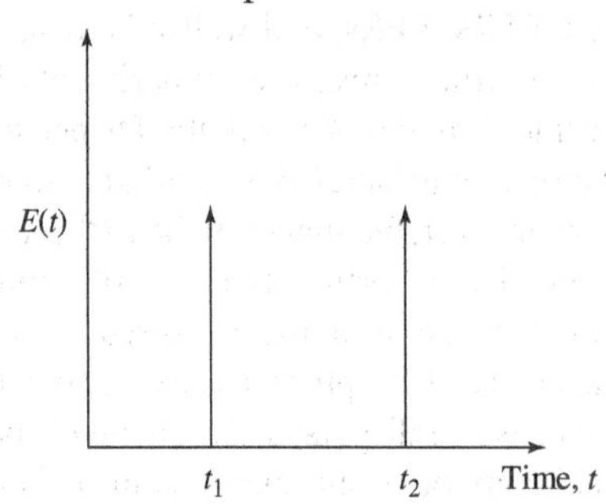

For two ideal PFRs in parallel, Eqn. (10-49) becomes

$$E(t) = f_1\delta(t_1) + f_2\delta(t_2)$$

The values of τ_1 (=t_1) and τ_2 (=t_2) are obtained from the positions of the two "spikes" (delta functions).

(b) Two CSTRs in parallel (with different values of τ)

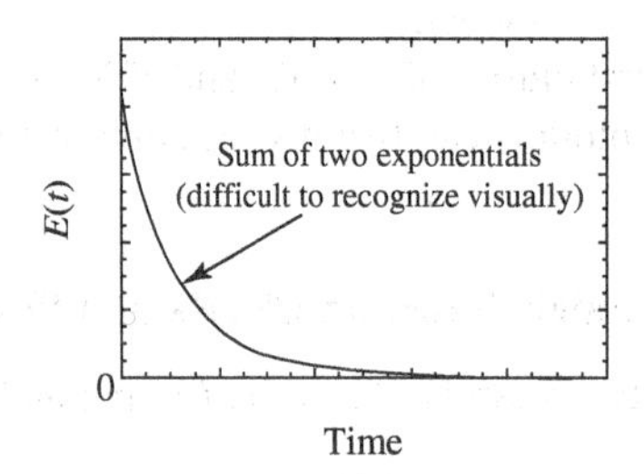

From Eqn. (10-49),

$$E(t) = (f_1/\tau_1) \times \exp(-t/\tau_1) + (f_2/\tau_2) \times \exp(-t/\tau_2).$$

The curve of $E(t)$ versus t is the sum of these two exponentials.

The existence of two CSTRs in parallel is easier to detect, and analysis of the $E(t)$ data is easier, if $\ln[E(t)]$ is plotted against time. If the values of τ for the two reactors are sufficiently different, the value of $(f/\tau) \times E(t)$ for the reactor with the smaller value of τ will be much greater than $(f/\tau) \times E(t)$ for the reactor with the larger value, at short times. The converse will be true at long times.

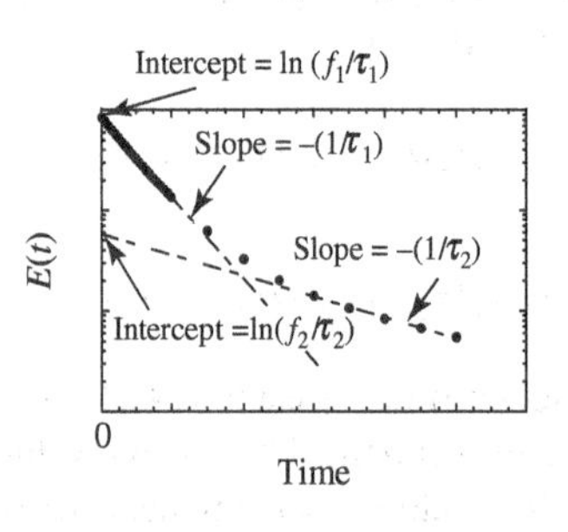

The values of τ_1 and τ_2 can be obtained from the slopes of the short-time and long-time fits to the semi-log plot of $E(t)$ versus t. The values of the intercepts can be used to check the values of f_1 and f_2 calculated from Eqn. (10-50).

If τ_1 and τ_2 are not sufficiently different to permit an accurate analysis from a semi-log plot, nonlinear regression can be used to determine τ_1 and τ_2.

(c) PFR and CSTR in parallel

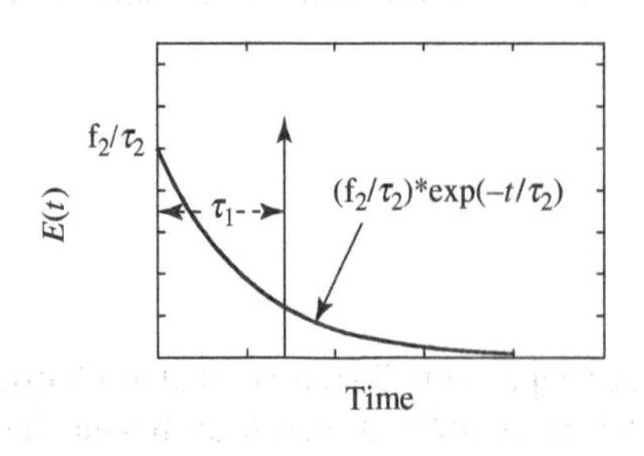

From Eqn. (10-49), for closed vessels,

$$E(t) = f_1\delta(\tau_1) + (f_2/\tau_2)\exp(-t/\tau_2)$$

The value of τ_1 is obtained from the position of the sharp peak associated with the PFR. The value of τ_2 is obtained from the slope of a semi-log plot of $f_2E(t)$ versus t. The slope of the semi-log plot is $-1/\tau_2$. The intercept of the semi-log plot is f_2/τ_2 and can be used to check the value of f_1 calculated from Eqn. (10-50).

Figure 10-11 The exit-age distribution for various combinations of CSTRs and PFRs in parallel.

Reactors in Series The CIS model is an example of this class of compartment models. In the CIS model, all of the CSTRs are the same size, and any number of reactors can be arranged in series. The following discussion is confined to only two reactors in series. However, the volume of the vessels is not necessarily the same.

When two independent vessels are arranged in series, the external-age distribution for the combination of vessels is given by Eqn. (10-29).

$$E(t) = \int_0^t E_1(t-t')E_2(t')\mathrm{d}t' \tag{10-29}$$

As with the reactors-in-parallel models, we will presume that the volumetric flow rate υ and the total reactor volume V are known. The two space times τ_1 and τ_2 are the parameters that must be determined from the measured external-age distribution. One relationship is

$$\bar{t} = \tau = \tau_1 + \tau_2$$

Figure 10-12 shows the shape of the $E(t)$ curves for various combinations of CSTRs and PFRs in series. The comments beside each figure show how the information required to quantify the model and to calculate reactor performance can be extracted.

(a) Two unequal-volume CSTRs in series

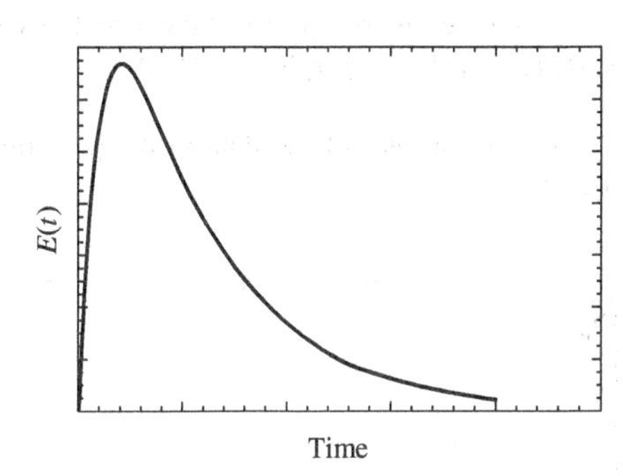

Since the total flow rate υ is known and since the total flow passes through each reactor, the unknowns are τ_1 and τ_2 (or V_1 and V_2). The external-age distribution function for the reactor *system* is

$$E(t) = [\exp(-t/\tau_1) - \exp(-t/\tau_2)]/(\tau_1 - \tau_2)$$

Values of τ_1 and τ_1 can be obtained from the slope of $E(t)$ at $t = 0$ and from the position of the maximum value of $E(t)$.

$$\mathrm{d}E(t)/\mathrm{d}t\ (t = 0) = 1/\tau_1\tau_2$$
$$t_{\max} = [\tau_1\tau_2/(\tau_1 - \tau_2)] \times \ln(\tau_2/\tau_1)$$

(b) PFR and CSTR in series

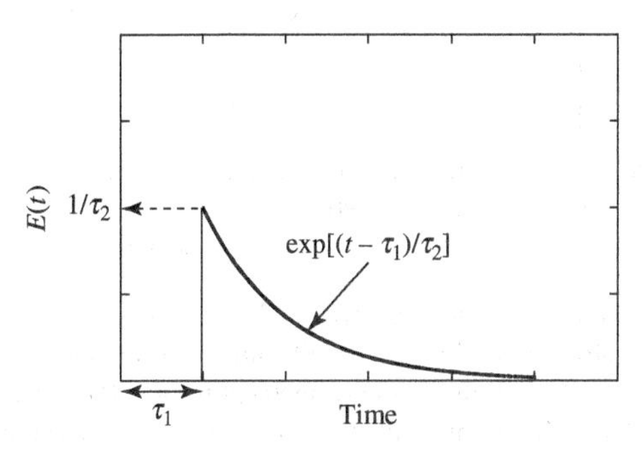

The space time for the PFR is τ_1. The space time for the CSTR τ_2 can be determined from either the value of $E(t)$ when $t = \tau_1$ or from the slope of a semi-log plot of the CSTR portion of the tracer curve.

Figure 10-12 The exit-age distribution for various combinations of PFRs and CSTRs in series.

EXAMPLE 10-12
CSTRs in Parallel

The information in the table below was obtained for a vessel with a constant-density fluid flowing through it at steady state. It is has been suggested that the vessel can be modeled as two CSTRs in parallel, each with a different space time. Determine the "best" values of the unknown parameters in this model and compare the model with the values in the table.

Time (min)	$E(t)$ (min^{-1})	Time (min)	$E(t)$ (min^{-1})
1	0.076	16	0.019
2	0.069	20	0.0134
3	0.063	30	0.0062
4	0.057	40	0.0033
5	0.052	60	0.0014
6	0.047	80	0.00083
8	0.039	100	0.00054
10	0.033	120	0.00036
12	0.027		

APPROACH

A semi-log plot of $E(t)$ versus t will be constructed. Straight lines will be fitted to the "short-time" and "long-time" portions of the data. Values of τ_1 and τ_2 will be estimated from the slopes of the "short-time" and "long-time" lines. The value of f_1 will be estimated from the intercepts of the "short-time" and "long-time" lines. Finally, nonlinear regression will be used to refine the estimates of τ_1, τ_2, and f_1 by minimizing the sum of squares of the deviations between the experimental data and $E(t) = (f_1/\tau_1)\exp(t/\tau_1) + ((1 - f_1)/\tau_2)\exp(-t/\tau_2)$.

SOLUTION

The following figure is a semi-log plot of the data in the preceding table, as suggested by the discussion in Figure 10-11b.

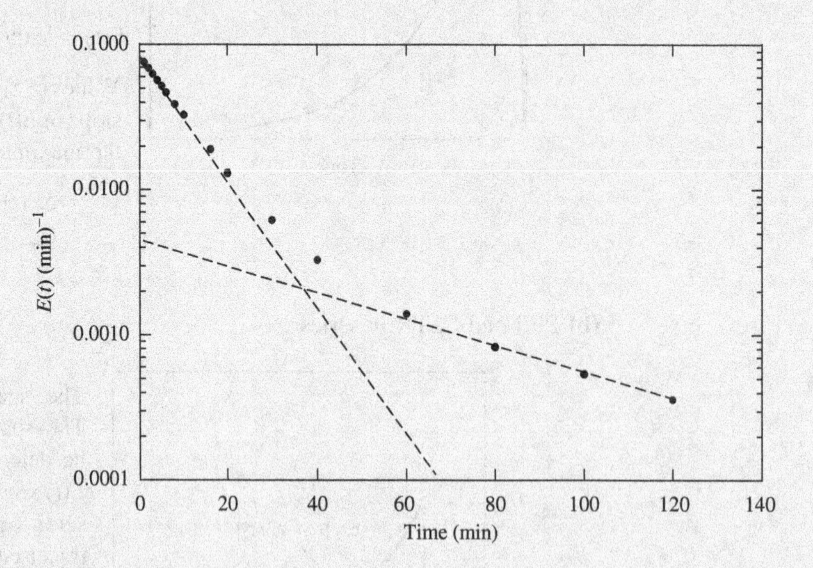

Straight lines have been fitted "by eye" to the "short-time" and "long-time" portions of the data. The slope of the "short-time" line is -0.098 min^{-1} and the intercept is ln(0.083). The slope of the "long-time" line is -0.021 min^{-1} and the intercept is ln(0.0045). Figure 10-11b shows that the space time of each CSTR is the negative of the inverse of its slope. The estimated space time values are $\tau_1 = 10$ min (τ_1 is the "short-time" space time) and $\tau_2 = 48$ min (τ_2 is the "long-time" space time).

Figure 10-11b shows that f_1 is the "short-time" intercept multiplied by τ_1, and that f_2 (= $1 - f_1$) is the "long-time" intercept multiplied by τ_2. The estimated values of f_1 from the two intercepts are 0.83 and 0.78, respectively.

A nonlinear regression was then performed in an EXCEL spreadsheet using SOLVER, beginning with these estimates.[17] The resulting values of τ_1, τ_2, and f_1 were 10 min, 49 min, and 0.80. These values were then used to calculate $E(t)$. A comparison of the $E(t)$ data from the table above with the calculated values of $E(t)$ from the CSTRs in parallel compartment model are shown in the following table. The fit of the model to the data is excellent.

Time (min)	$E(t)$ (min^{-1}) (measured)	$E(t)$ (min^{-1}) (model)
1	0.076	0.076
2	0.069	0.069
3	0.063	0.063
4	0.057	0.057
5	0.052	0.052
6	0.047	0.047
8	0.039	0.039
10	0.033	0.033
12	0.027	0.027
16	0.019	0.019
20	0.013	0.013
30	0.0062	0.0062
40	0.0033	0.0033
60	0.0014	0.0014
80	0.00083	0.00083
100	0.00054	0.00054
120	0.00036	0.00036

10.5.4.3 Well-Mixed Stagnant Zones

Consider a situation where a reactor is mechanically agitated, but top-to-bottom mixing is not sufficient to ensure that the composition is identical at every point in the reactor. This situation might occur with either a batch reactor[18] or a continuous agitated reactor. The situation is represented in Figure 10-13. Each of the zones in the vessel shown is well mixed, i.e., there are no concentration or temperature gradients *within each zone*.

Although both zones are well mixed, the compositions of the two zones are not necessarily the same. Suppose that a batch reactor is charged initially with a mixture of A and B, which react with each other only in the presence of a soluble catalyst C. The initial concentrations of A and B will be the same in both zones. Now, catalyst is added into the top zone. The reaction starts to take place in the upper zone, and the concentrations of A and B begin to decline.

[17] Since the $E(t)$ values cover a range of more than two orders of magnitude, the sum of the squares of the *percentage* deviations was minimized.

[18] Up to this point, the discussion of nonideal reactors has dealt exclusively with continuous reactors. However, both batch and semibatch reactors can also be nonideal if the concentrations and the temperature are not spatially uniform at every time.

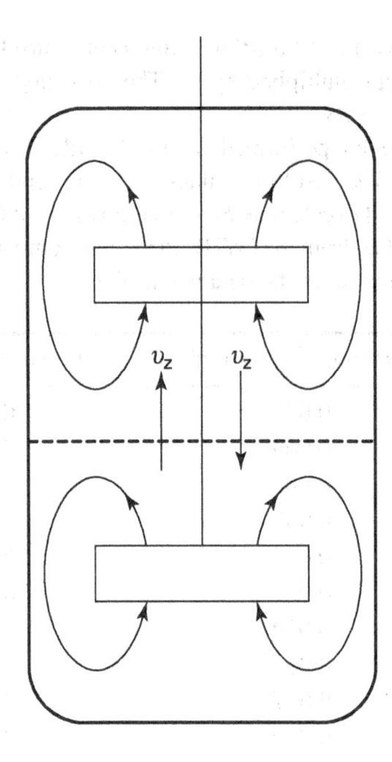

Figure 10-13 Schematic diagram of batch reactor with two well-mixed zones.

If there is no exchange of fluid between the two zones, and if there is no diffusion across the imaginary boundary between the two zones, then catalyst will never enter the bottom zone, and no reaction will take place in that portion of the reactor. This leads us to refer to the bottom zone as *stagnant*.

In reality, there will be *some* exchange of fluid between the two zones. Figure 10-13 shows a flow rate v_z leaving the top zone and entering the bottom zone, and an identical flow rate leaving the bottom zone and entering the top zone. This exchange of fluid will reduce the concentration differences between the two zones. However, the rate of exchange may be insufficient to *eliminate* concentration differences between the two reactor sections. If the reaction is fast, and the rate of fluid exchange is slow, the concentration differences will be quite significant.

EXAMPLE 10-13
Use of Tracer Techniques to Characterize WMSZs

Consider a batch reactor, such as the one shown in Figure 10-13. It is proposed to model this reactor as two WMSZs with a volumetric flow rate v_z between them. How can the model be tested, and how can the fraction of the total volume that is in each zone, and the value of the flow between zones, v_z, be determined using tracer techniques?

APPROACH

A known quantity of tracer will be injected into the top zone of the reactor and the tracer concentration in this zone will be measured as a function of time. Material balances will be written on the tracer in *each* zone. These balances will be solved for the tracer concentration in the top zone as a function of time. The resulting equation will be tested against the data. If the model fits the data, the volumes of the two zones, and the flow rate between zones, will be extracted from the data.

SOLUTION

Let C_t and C_b be the concentrations of tracer in the top and bottom zones, respectively. Let v_z be the flow rate between zones, and let V_t and V_b be the volumes of the top and bottom zones, respectively. The material balance on tracer in the top zone is

$$v_z C_b - v_z C_t = V_t \frac{dC_t}{dt} \tag{10-52}$$

The material balance on tracer in the bottom zone is

$$v_z C_b - v_z C_t = -V_b \frac{dC_b}{dt}$$

Dividing the first equation by the second (the time-independent method!) gives

$$\frac{dC_b}{dC_t} = -\frac{V_t}{V_b}$$

Suppose that M_0 moles of tracer are injected instantaneously into the top zone at $t = 0$. Then C_t at $t = 0$ will be M_0/V_t, while C_b at $t = 0$ will be 0. Integrating the above equation from $t = 0$ to $t = t$ gives

$$C_b = \frac{M_0}{V_b} - \left(\frac{V_t}{V_b}\right) C_t$$

Substituting this expression for C_b into Eqn. (10-52)

$$v_z \left[\frac{M_0}{V_b} - \left(\frac{V_t}{V_b}\right) C_t\right] - v_z C_t = V_t \frac{dC_t}{dt}$$

Let $\alpha = v_z(V_{tot}/V_b V_t)$ and let $\beta = v_z(M_0/V_b V_t)$, where V_{tot} is the total filled volume of the reactor, i.e., $V_{tot} = V_t + V_b$. With these definitions, the above equation becomes

$$\frac{dC_t}{dt} + \alpha C_t = \beta$$

This differential equation can be solved via the "integrating factor" method. The solution is

$$C_t = \frac{M_0}{V_{tot}} + M_0 \left(\frac{V_b}{V_t V_{tot}}\right) e^{-\alpha t}$$

Rearranging and taking the log of both sides

$$\ln\left(\frac{C_t V_{tot}}{M_0} - 1\right) = \ln\left(\frac{V_b}{V_t}\right) - \alpha t$$

If this model fits the data, a plot of

$$\ln\left(\frac{C_t V_{tot}}{M_0} - 1\right)$$

versus time will be a straight line. The intercept will be $\ln(V_b/V_t)$, and the slope will be $-\alpha$. Values of V_b and V_t can be calculated from the intercept plus the known value of V_{tot} $(= V_b + V_t)$. Since $\alpha = v_z(V_{tot}/V_b V_t)$, the value of v_z can also be calculated.

As an aside, it would have been better to measure the concentration of tracer in *both* zones as a function of time. This would have permitted a material balance to be used to check the quality of the data.

EXAMPLE 10-14 ***Performance of a Continuous Reactor Consisting of Two WMSZs***

A *continuous* reactor is divided into two zones, as shown in Figure 10-13. The feed enters the top zone at a rate v of 500 l/min, and the effluent leaves from the top zone. The total volume of the reactor is 1000 l, and the volume of the bottom zone is 300 l. The rate of fluid exchange between zones v_z is 50 l/min. The irreversible, first-order, liquid-phase reaction $A \rightarrow B$ takes place isothermally in the reactor. The rate constant is 0.80 min^{-1}. At steady state, what is the fractional conversion of A in the stream leaving the reactor? What would the fractional conversion of A be if the whole reactor behaved as an ideal CSTR?

APPROACH

Material balances will be written on reactant A for both zones. These balances will be solved simultaneously to obtain the concentration of A in the top zone, i.e., the concentration of A leaving the reactor. The design equation for an ideal CSTR will be solved to compare the performance of the two reactors.

SOLUTION

Let $C_{A,b}$ be the concentration of A in the bottom zone, $C_{A,t}$ be the concentration of A in the top zone, V_b be the volume of the bottom zone, and V_t be the volume of the top zone. The material balance on reactant A in the bottom zone is

$$\upsilon_z C_{A,t} - \upsilon_z C_{A,b} = V_b k C_{A,b}$$

$$C_{A,b} = [\upsilon_z/(V_b k + \upsilon_z)]C_{A,t}$$

$$C_{A,b} = [50\,(1/\text{min})/\{300(1) \times 0.80\,(\text{min}^{-1}) + 50\,(1/\text{min})\}]C_{A,t} = 0.17C_{A,t}$$

The concentration of A in the bottom zone is substantially less than in the top zone. Therefore, the rate of reaction in the bottom zone will be substantially lower than in the top zone.

The material balance on A in the top zone is

$$\upsilon C_{A0} + \upsilon_z C_{A,b} - \upsilon C_{A,t} - \upsilon_z C_{A,t} = V_t k C_{A,t}$$

Here, C_{A0} is the concentration of A in the feed to the reactor. Substituting for $C_{A,b}$ and rearranging

$$C_{A,t}\{kV_t + (\upsilon + \upsilon_z) - \upsilon_z^2/(kV_b + \upsilon_z)\} = \upsilon C_{A0}$$

Since $x_A = [1 - (C_{A,t}/C_{A0})]$,

$$x_A = 1 - \frac{\upsilon}{kV_t + (\upsilon + \upsilon_z) - [\upsilon_z^2/(kV_b + \upsilon_z)]}$$

Substituting values

$$x_A = 1 - \frac{500(\text{l/min})}{700\,(1) \times 0.80\,(\text{min}^{-1}) + (500 + 50)(\text{l/min}) - \dfrac{(50)^2\,(\text{l/min})^2}{(300\,(1) \times 0.80\,(\text{min}^{-1}) + 50\,(\text{l/min})}}$$

$$x_A = 0.55$$

For an ideal CSTR with a volume of 1000 l,

$$x_A = \frac{k\tau}{1 + k\tau} = \frac{0.80\,(\text{min}^{-1}) \times [(1000(1)/500\,(1/\text{min})]}{1 + 0.80\,(\text{min}^{-1}) \times [(1000(1)/500\,(1/\text{min})]} = 0.62$$

The reactant conversion in the actual reactor is lower than in the ideal CSTR.

10.6 CONCLUDING REMARKS

Not all reactors are "ideal" in the sense that mixing conforms to one of the limiting cases represented by the ideal CSTR, the ideal PFR, and the ideal batch reactor. To completely characterize mixing, three pieces of information are necessary:

- the exit-age or residence time distribution (RTD);
- the scale of mixing (macrofluid? microfluid? something in between?);
- the "earliness" or "lateness" of mixing.

The performance of a reactor generally depends on each of these variables.[19] Unfortunately, for an existing reactor, or one in the design stage, sufficient information may not be available in each of these three areas.

[19] For first-order reactions, performance depends only on the residence time distribution. The scale of mixing and the "earliness" or "lateness" of mixing do not affect either the conversion or the product distribution, if *all* reactions are first order.

The objective of this chapter has been to develop some relatively simple approaches to estimating the performance of nonideal reactors. In many cases, one or more of these approaches can provide a reasonable approximation to the behavior of the real reactor.

The Dispersion model can be used to predict the performance of a nonideal reactor in the absence of a measured RTD. However, the geometric parameters and the flow conditions of the nonideal reactor must fall within the range of existing correlations for the "intensity of dispersion."

If the Dispersion model cannot be used, or is not physically appropriate, the external-age distribution function $E(t)$ must be measured using tracer response techniques. Once $E(t)$ is available, the macrofluid model can be used to establish bounds on reactor performance. Moreover, the shape of the external-age distribution function can suggest various "compartment" models," including the CSTRs-in-series model, which can be used to explore reactor performance.

SUMMARY OF IMPORTANT CONCEPTS

- Not all reactors are "ideal".
- Tracer-response experiments can be used to diagnose problems, such as "bypassing" and accumulated solids, associated with flow and mixing in a reactor.
- The external-age distribution $E(t)$ of a reactor can be measured using tracer response techniques.
- The performance of a nonideal reactor can be estimated from the external-age distribution, either directly via the macrofluid model, or indirectly via the Dispersion model, CSTRs-in-series model, or a compartment model.
- The Dispersion model can be used to estimate the performance of a nonideal reactor, without first measuring $E(t)$. However, the geometry and operating conditions of the reactor must be within the range of the "intensity of dispersion" correlations.

PROBLEMS

Conceptual Questions

1. Can the Dispersion model be used for the analysis or design of a radial-flow, fixed-bed reactor, based on the existing "intensity of dispersion" correlations?
2. Explain qualitatively why $E(t)$ should be the same as $I(t)$ for an ideal CSTR.
3. A reaction with an effective order of 0.5 is taking place in a series of three equal-volume CSTRs. The fluid in each reactor is a macrofluid. Is the final conversion higher if the fluid remains segregated between the reactors or if the fluid is mixed on a molecular level between reactors?

Problem 10-1 (Level 1) The irreversible reaction $A \rightarrow B$ obeys the rate equation: $-r_A = kC_A/(1 + KC_B)$. At 50 °C, $k = 4.08\ \text{min}^{-1}$ and $K = 10$ l/mol.

A viscous liquid containing A at a concentration C_{A0} of 1.0 mol l and no B is fed to a tubular reactor with a volume of 200 l at a volumetric flow rate υ of 20 l/min.

1. If the reactor behaves as an ideal plug-flow reactor and operates isothermally at 50 °C, what is the fractional conversion of A?
2. Suppose that flow in the reactor is laminar, and the fluid is a macrofluid. If the reactor operates isothermally at 50 °C, what is the fractional conversion of A in the effluent?
3. Is the conversion that you calculated in Part 2 an upper or a lower limit?

Note: The external-age distribution function for a laminar flow tubular reactor with no radial or axial diffusion is

$$E(t) = 0; \quad t < \tau/2;$$
$$E(t) = \tau^2/2t^3; \quad t \geq \tau/2.$$

Problem 10-2 (Level 2) A nonadsorbing tracer was injected as a sharp pulse into a tubular vessel filled with an experimental, nonporous packing designed to promote radial mixing. The volumetric flow rate through the vessel during the test was 150 l/min. The concentration of tracer in the vessel effluent is given in the following table.

1. How much tracer (in moles) was injected into the reactor?
2. What was the volume of fluid in the reactor during the tracer test?

The same vessel with the same packing will be used to carry out a first-order, noncatalytic, irreversible, liquid-phase reaction with a rate constant $k = 0.040\ \text{min}^{-1}$. The volumetric flow rate through the reactor will be 50 l/min.

3. Assume that the reactor obeys the Dispersion model. Estimate the conversion that will be observed.

Time after injection (min)	Tracer concentration (mmol/l	Time after injection (min)	Tracer concentration (mmol/l
0	0	16	22
2	0	18	16
4	0	20	11
6	2.0	22	7.0
8	12	24	5.0
10	37	26	3.8
12	35	28	2.7
14	28	30	2.0

4. If the reactor were to behave as an ideal PFR and it were to operate at the conversion calculated in Part 3, how much smaller would it be compared to the actual reactor?

Problem 10-3 (Level 2) The following abstract concerning the behavior of laminar flow tubular reactors was taken directly from a well-respected technical journal.

"By expressing the residence time of annular elements of the fluid in a laminar flow reactor as a function of reactor length and radial position it is possible to relate the fractional conversion of reactant to dimensionless groups containing the rate constant, inlet concentrations, reactor volume, and flow rate. The functional dependence will vary with the order of the kinetic expression relating the rate of disappearance of reactant to concentration. For kinetic orders other than zero, the fractional conversion obtained in the laminar flow reactor will be less than that calculated by the plug-flow assumption. This analysis extends the previous treatments to include the general *n*th-order rate expression and first-order consecutive reaction rate expressions."

Does this abstract suggest a problem in the analysis? What problem?

Problem 10-4 (Level 1) Two identical, isothermal tubular reactors are arranged in parallel, as shown in the following figure.

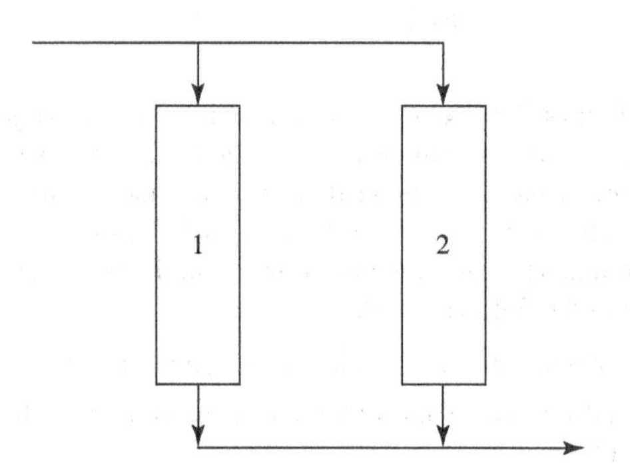

The gas-phase reactions $A \xrightarrow{k_1} B \xrightarrow{k_2} C$ take place in the reactors. Both reactions are first order. The values of the rate constants are $k_1 = 0.10\ \text{min}^{-1}$ and $k_2 = 0.050\ \text{min}^{-1}$. The concentration of A in the feed is 0.030 mol/l. There is no B in the feed.

A tracer test has been run on the reactor system, with the results shown below.

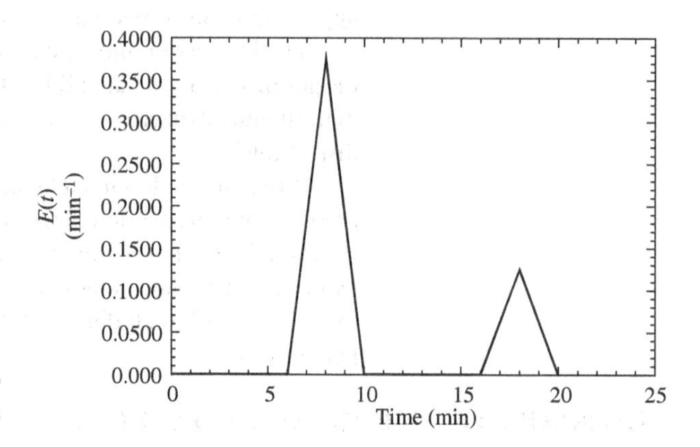

1. What is the average residence time in the reactor *system*?
2. What concentration of A would you expect for an ideal plug-flow reactor operating at this residence time?
3. What concentration of B would you expect in the outlet from an ideal plug-flow reactor operating at this residence time?
4. What is the concentration of A in the combined effluent from the two reactors?
5. What is the concentration of B in the combined effluent from the two reactors?
6. Which reactor, 1 or 2, gives rise to the first peak (the one centered on 8 min) in the $E(t)$ plot? Explain your reasoning.

Problem 10-5 (Level 2) Styrene is being polymerized at steady state in a 1000 l continuous reactor that can be modeled as two perfectly mixed regions, as shown in the following figure. A solution of styrene monomer and the initiator 2, 2′-azobisisobutyronitrite (AIBN) in toluene is fed to the larger region.

The flow rate of the feed is 1500 l/h, the concentration of styrene in the feed is 4.0 mol/l, and the concentration of the initiator is 0.010 mol/l. The temperature of both zones of the reactor is 200 °C.

The initiator decomposes by a first-order process, with a rate constant at 200 °C of $9.25\ \text{s}^{-1}$. The rate equation for styrene disappearance is

$$-r_{\text{p}}(\text{mol/l-s}) = k[\text{S}][\text{I}]^{1/2}$$

where [S] is the concentration of styrene, [I] is the concentration of undecomposed initiator, and $k = 0.925\ \text{l}^{1/2}/\text{mol}^{1/2}\text{-s}$ at 200 °C.

1. What are the concentrations of undecomposed initiator in V_1 and V_2?

2. What is the fractional conversion of styrene in the effluent from the reactor?
3. Each initiator molecule decomposes into two free radicals. Each free radical starts a polymer chain, and the chain termination process is such that there is one fragment of initiator in each molecule of polymer that leaves the reactor. What is the average number of monomer units in each molecule of polymer?

Reactor schematic

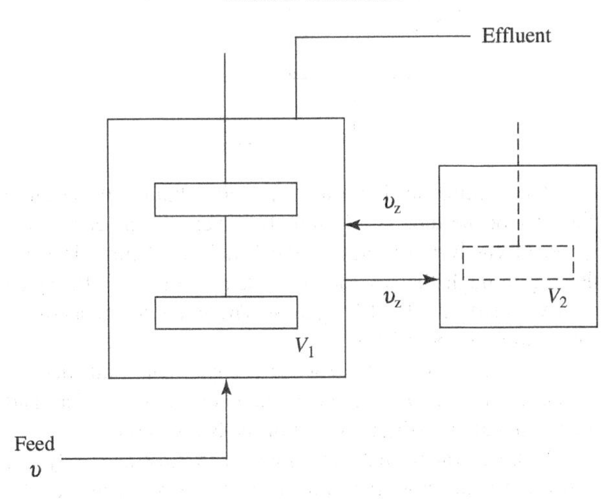

Values	Feed concentrations
$V_1 = 700\,\mathrm{l}$	Styrene: 4.0 mol/l
$V_2 = 300\,\mathrm{l}$	AIBN: 0.010 mol/l
$\upsilon = 1500\,\mathrm{l/h}$	
$\upsilon_z = 100\,\mathrm{l/h}$	

Problem 10-6 (Level 2) A homogeneous, liquid-phase reaction A $\rightarrow$ products takes place in a tubular, isothermal reactor packed with spheres of uniform diameter d_p. The reactor obeys the Dispersion model. The fractional conversion of A in the reactor effluent is 99%. At these conditions, the Dispersion number is 0.0625.

Suppose that the diameter of the particles in the tube is increased by a factor of 4, and that the void fraction ε_i of the particle bed does not change as a result. The tube diameter stays constant, as does the volumetric feed rate and the feed composition.

If the fractional conversion of A is to be maintained at 99%, must the packed tube length be increased or decreased? Why?

If the reaction is first order and irreversible, what fractional change in reactor length is required to keep the conversion of A at 99%?

Problem 10-6a (Level 2) The following e-mail is in your inbox at 8 AM on Monday.

To: U. R. Loehmann
From: I. M. DeBosse
Subject: Cost Saving/Debottlenecking Possibility

One of Cauldron Chemical Company's most profitable products is made via a homogeneous, irreversible, liquid-phase reaction A $\rightarrow$ products using a tubular, isothermal reactor packed with noncatalytic, nonporous spheres of uniform diameter d_p. The spheres are believed to promote radial mixing and heat transfer. The fractional conversion of A in the effluent from the current reactor is 99%.

The inert spheres in the reactor always are replaced during the annual turnaround, since a small amount of attrition of these spheres is known to take place. The company that supplies the spheres has suggested that we use a larger diameter particle the next time we change. The supplier has spheres with a diameter that is a factor of 4 larger than our present spheres. The larger spheres are significantly cheaper, and the supplier claims that they are more attrition resistant, so that we may not have to replace the larger spheres every year. The supplier has also suggested that the larger spheres will permit the production rate to be increased, with no decrease in conversion. This last issue is particularly important since the product that we make in this reactor always is in short supply, and we occasionally have to allocate the product to some customers.

Please analyze the effect of substituting the larger spheres. Assume that the void fraction ε_i of the particle bed does not change as a result of the substitution, and that the Dispersion model is obeyed. The tube diameter and tube length will remain constant, as will the feed composition and the temperature. At the present operating conditions, with the present spheres, the Dispersion number is 0.0625 and the particle Reynolds number is about 10.

At the same volumetric flow rate that we have been using, how much will the fractional conversion of A increase above the current 99%, if the existing spheres are replaced with the larger ones?

If the conversion is kept at 99% by increasing the feed rate, by what percentage will the production rate increase?

Please report your results to me in a one-page memo. Attach your calculations in case someone wants to review them.

Problem 10-7 (Level 1) A *step* input of tracer is started into a vessel at t (time) $= 0$. The concentration of tracer in the feed was zero initially; at $t = 0$ the inlet concentration of tracer was changed to C_0. The concentration of tracer in the outlet from the vessel is shown in the following figure.

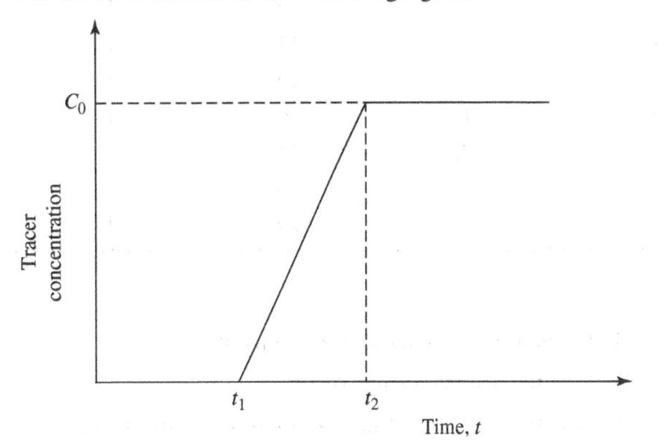

1. Derive an expression for the external-age distribution $E(t)$ of the vessel. Plot $E(t)$ versus time.
2. What kind of a vessel/flow condition might give rise to an external-age distribution that resembles the one you have drawn.

Problem 10-8 (Level 2) A Newtonian fluid is flowing in fully developed, isothermal, laminar flow through a tube of radius R and length L. The volumetric flow rate is υ. The velocity distribution in the pipe is

$$u(r) = 2\left(\frac{\upsilon}{\pi R^2}\right)\left(1 - \left(\frac{r}{R}\right)^2\right)$$

1. Show that the external-age distribution $E(t)$ is given by

$$E(t) = 0; \quad t < \pi R^2 L/2\upsilon$$
$$E(t) = (\pi R^2 L/\upsilon)^2/2t^3; \quad t \geq \pi R^2 L/2\upsilon$$

2. Prove that the average residence time of fluid in the pipe is $\bar{t} = \pi R^2 L/\upsilon$. Do *not* start with the relationship $\bar{t} = \tau$.
3. Derive an expression for σ^2, the second moment of the external-age distribution about the mean residence time $\bar{t}$.
4. A second-order reaction is occurring in the tube. Derive an expression for the conversion of Reactant A as a function of the space time, $\tau\ (\tau = \pi R^2 L/\upsilon)$, assuming that the fluid flowing through the tube is a macrofluid.

Problem 10-9 (Level 1) A fluid is flowing through a vessel at steady state. The flow rate is 0.10 l/min. The response to a pulse input of tracer at $t = 0$ is given in the following table.

Time (min)	Outlet concentration of tracer (mmol/l)	Time (min)	Outlet concentration of tracer (mmol/l)
0	0	10	16
1	0	11	13
2	0	12	10
3	1	13	7
4	2	14	5
5	8	15	3
6	18	16	2
7	25	17	1
8	22	18	0
9	19	19	0

1. How much tracer (mmoles) was injected?
2. What is the volume of the vessel?
3. What is the variance of the external-age distribution?

Problem 10-10 (Level 1) An experimental catalyst is being tested in a microreactor, designed as shown below.

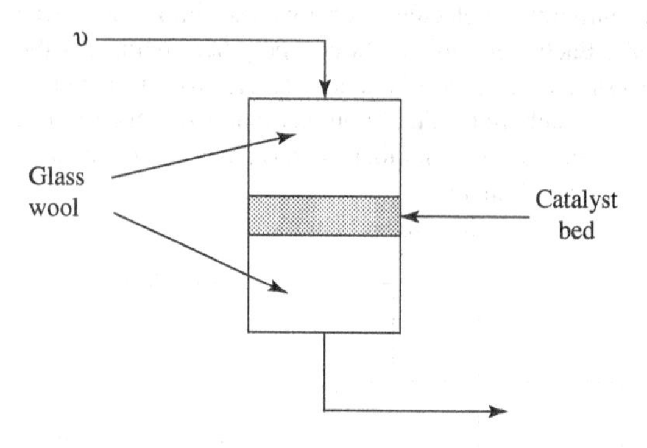

The volumetric flow rate υ is 350 cc/min and the inner diameter of the tube is 7.7 mm. The average diameter of the spherical catalyst particles in the bed is 0.20 mm. The gas flowing through the reactor has a density of 2.0×10^{-4} g/cc and a viscosity of 3.0×10^{-4} g/cm-s. The Schmidt number of the gas mixture is about 0.70.

The bed contains 0.20 g. of catalyst. The bulk density (g-cat/cc. of *reactor volume*) of the catalyst is 1.5 g/cc, and the fractional interstitial volume of the bed is 0.40.

In order to analyze the data for the experimental catalyst, it is necessary to know how much axial mixing occurs in the catalyst bed. Can the catalyst bed be treated as an ideal plug-flow reactor? Justify your answer quantitatively.

If this reactor operates in a differential mode (outlet fractional conversion of reactant ≤ 0.05), will axial mixing have a significant effect on the value of the rate constant that is calculated from the measured outlet conversion? Justify your answer by calculating the ratio $k(\text{CSTR})/k(\text{PFR})$ for an irreversible, first-order reaction and an outlet conversion of 0.05. In this ratio, k(PFR) is the rate constant calculated using the PFR model and k(CSTR) is the rate constant using the CSTR model.

Suppose that this reactor operates in an integral mode (e.g., at an outlet fractional conversion of $\cong 0.90$), and that an irreversible, first-order reaction is taking place. A rate constant can be calculated from the outlet conversion by assuming either plug-flow, complete backmixing (i.e., that the reactor is an ideal CSTR), or the dispersed plug-flow model. Calculate the values of the ratios $k(\text{CSTR})/k(\text{PFR})$ and $k(\text{DPF})/k(\text{PFR})$ for a measured outlet conversion of 0.90.

Problem 10-11 (Level 2) The exit-age distribution of a nonideal reactor is shown in the following table.

1. Can the reactor be modeled as two CSTRs with different volumes in series? If so, what are values of the space times for the two CSTRs?
2. The irreversible, liquid-phase reaction $A + B \rightarrow C + D$ will be carried out in this reactor. The reaction obeys the rate equation: $-r_A = kC_A^2 C_B$; $k = 10{,}000$ gal^2/lb-mol^2-h. The concentrations of both A and B in the feed to the reactor

Time (min)	$E(t)$ (min)$^{-1}$	Time (min)	$E(t)$ (min)$^{-1}$
0	0	40	0.0108
2	0.0036	50	0.0090
4	0.0063	60	0.0075
6	0.0085	70	0.0061
8	0.0101	80	0.0050
10	0.0113	90	0.0041
12	0.0121	100	0.0034
14	0.0127	110	0.0028
16	0.0131	120	0.0023
18	0.0133	130	0.0019
20	0.0134	140	0.0015
22	0.0133	150	0.0012
24	0.0132	160	0.0010
26	0.0130	170	0.00083
28	0.0128	180	0.00068
30	0.0125	190	0.00056
		200	0.00046

are 0.025 lb mol/gal. Estimate the fractional conversion of A if the flow rate into the reactor is the same as it was when the exit-age distribution was measured. Assume that both reactors are mixed at the molecular level.

Problem 10-12 (Level 1) An ideal pulse input of radioactive tracer is injected into the stream entering a closed reactor at $t = 0$. The *total* amount of radioactivity *in the reactor* then is monitored as a function of time. An example of some raw data is given in the following table:

Experiment #4

$Q = 15$ gal/min

Total radioactivity in pulse (arbitrary units) $= 300$

Time (min)	Radioactivity in vessel (arbitrary units)
5	250
10	208
15	174
20	147
30	104
40	75
50	55
60	40
80	23
100	13

1. *Explain* how the external-age distribution $E(t)$ can be obtained from data such as that above. The method you propose should be general and should not rely on any specific features of the above data.
2. What are the advantages and disadvantages of this technique for measuring $E(t)$, compared with the "standard" technique involving injection of a pulse tracer into the inlet stream and measurement of the *exit* tracer concentration?

Nomenclature

English Letters

a_h	heat-exchange area per unit length of heat exchanger (length)
a_i	activity of species "i"
A	preexponential (frequency) factor in rate constant (same units as rate constant)
A_c	cross-sectional area (length^2)
A_h	total area through which heat is transferred (length^2)
A_i	chemical species or compound
A_G	geometric external area of catalyst particle (length^2)
$c_{p,i}$	constant-pressure heat capacity of species "i" (energy/mole-degree)
$c_{p,m}$	mass heat capacity (energy/mass-degree)
$\bar{c}_{p,m}$	average mass heat capacity (energy/mass-degree)
C	total molar concentration (moles/volume)
C_i	concentration of species "i" (moles/volume or (rarely) mass/volume)
C_{cat}	catalyst concentration, on mass basis (mass catalyst/volume)
d	equivalent molecular diameter (length)
d_p	particle diameter (length)
D	dispersion coefficient (Chapter 10 only) (length^2/time)
D	dilution rate or dilution (Chapter 4 only) (time^{-1})
D_i	inner diameter (length)
D_o	outer diameter (length)
D_{ij}	binary molecular diffusion coefficient of species "i" in species "j" (length^2/time)
$D_{i,k}$	Knudsen diffusion coefficient of species "i" (length^2/time)
$D_{i,t}$	diffusion coefficient of species "i" in the transition regime (length^2/time)
$D_{A,m}$	diffusivity of A in mixture (length^2/time)
$D_{i,eff}$	effective diffusivity of species "i" inside catalyst particle (length^2/time)
D_{in}	inner diameter (length)
$D_{ip}(r)$	diffusion coefficient of species "i" in an assembly of straight, round pores with varying radii r (length^2/time)
e	energy/molecule (energy/molecule)
e^*	minimum energy required for a molecule to cross over the reaction energy barrier (energy/molecule)
E	activation energy (energy/mole)
E_{app}	apparent activation energy (energy/mole)
E_{diff}	activation energy for $D_{A,eff}$ (energy/mole)
E_{el}	number of elements in a system
E_{kin}	true or intrinsic activation energy (energy/mole)
E^*	$e^* N_{av}$ (energy/mole)
$E(t)$	dimensional exit-age distribution function (time^{-1})
f_i	fugacity of species "i" (pressure)

f_i	fraction of total flow passing through vessel "i" (Chapter 10 only)
f_i^0	fugacity of species "i" in the standard state (pressure)
$f(e)$	distribution function for molecular energies (molecules/energy)
$f(r)$	distribution function for pore radii ($length^{-1}$)
F_i	molar flow rate of species "i" (moles/time)
$F(t)$	cumulative exit-age distribution function
$F(\text{all}C_i)$	concentration-dependent term in rate equation (units depend on reaction locus, e.g., moles/volume) or (moles/weight of catalyst)
$F(e > e^*)$	fraction of molecules with energy e, greater than e^*
$F(E > E^*)$	fraction of molecules with energy E, greater than E^*
g	acceleration due to gravity (length/$time^2$)
G	superficial mass velocity (mass/$length^2$-time)
G_i	rate of generation of species "i" (moles/time)
G_M	superficial molar velocity (moles/$length^2$-time)
h	heat-transfer coefficient (energy/$length^2$-time-degree)
h_{in}	inside heat-transfer coefficient (energy/$length^2$-time-degree)
H	enthalpy (energy)
$\overline{H}$	partial molar enthalpy (energy/mole)
$\dot{H}$	rate of enthalpy transport (energy/time)
I	intercept
$I(t)$	dimensional internal-age distribution function ($time^{-1}$)
j_D	Colburn *j*-factor for mass transfer
j_H	Colburn *j*-factor for heat transfer
k	rate constant (units depend on the concentration dependence of the reaction rate and on the reaction locus)
k_B	Boltzmann constant (energy/molecule-absolute temperature)
k_c	mass-transfer coefficient based on concentration (length/time)
k_c^0	mass-transfer coefficient based on concentration when net molar flux = 0 (length/time)
k_{eff}	effective thermal conductivity of catalyst particle (energy/time-length-degree)
k_f	rate constant for forward reaction
k_p	mass-transfer coefficient based on partial pressure (moles/area-time-pressure)
k_r	rate constant for reverse reaction
k_v	rate constant on a volume of catalyst basis
k_y	mass-transfer coefficient based on mole fraction (moles/$length^2$-time)
k_{th}	thermal conductivity (energy/time-length-degree)
K_A	parameter in Eqn. (2-25) (volume/mole), and in the denominator of other rate equations, e.g., Eqn. (4-13)
K_s	constant in the Monod eqation (mass/volume)
K_{eq}	equilibrium constant based on activity
K_{eq}^C	equilibrium constant based on concentration $(\text{moles}/\text{volume})^{\delta v}$
K_{eq}^P	equilibrium constant based on pressure ($\text{pressure}^{\delta v}$)
K_m	Michaelis constant (parameter in Eqn. (2-25a) (volume/mole)
l_c	characteristic length of catalyst particle (length)
L	length (length)
Le	Lewis number

L_p	length of pore (length)
m_i	mass of molecule "i" (mass)
$\overline{m}$	numerical average mass of colliding molecules (mass)
$M(t)$	amount of tracer injected at time t (mass or moles)
M_i	molecular weight of species "i" (mass/mole)
M_m	molecular weight of mixture (mass/mole)
M_{SE}	mean squared error
$\dot{M}$	mass flow rate (mass/time)
n	reaction order
n_p	number of pores per gram of catalyst ($mass^{-1}$)
n_v	relative linear velocity
N	number of species
N	number of data points (Chapter 6)
N_{av}	Avogadro's number (molecules/mole)
N_i	number of moles of species "i"
Nu_h	Nusselt number for heat transfer
$\vec{N}_i$	molar flux of species "i" (moles/$length^2$/time)
p_i	partial pressure of species "i" (force/$length^2$)
P	total pressure (force/$length^2$)
Pe	Peclet number
Pe_a	axial Peclet number
Pr	Prandtl number
q	heat flux (energy/$length^2$-time)
Q	rate of heat transfer (energy/time)
r	pore radius (length)
r	radial coordinate in cylindrical or spherical coordinates (length)
r	species independent reaction rate (Chapter 1 only) (e.g., moles/time-volume[1])
$\bar{r}$	numerical average radius of colliding molecules (Chapter 2) (length)
$\bar{r}$	average radius of catalyst pores (Chapter 9) (length)
$r_{A,r}$	rate of formation of A in reverse reaction
r_i	rate of formation of species "i" (single reaction) (e.g., moles/time-volume[1])
r_i	radius of molecule "i" (length)
r_{ki}	rate of formation of species "i" in Reaction "k" (e.g., moles/time-volume[1])
$-r_{A,f}$	rate of disappearance of A in forward reaction (e.g., moles/time-volume[1])
$r_{A,r}$	rate of formation of A in reverse reaction (e.g., moles/time-volume[1])
$-r_A(\text{net})$	net rate of disappearance of A in a reversible reaction (e.g., moles/time-volume[1])
$-r_i^{bulk}$	reaction rate with no gradients inside catalyst particle or through boundary layer, i.e., reaction rate evaluated at temperature and concentrations in the bulk fluid (moles "i"/mass-time)
$-r_i^{surf}$	reaction rate with no gradients inside catalyst particle, i.e., reaction rate evaluated at temperature and concentrations at the external surface of the catalyst particle (moles "i"/mass-time)
R	gas constant (energy/mole-absolute temperature)
R	radius (length)
R	recycle ratio (Chapter 4)

[1]see pages 8 and 9 for a more comprehensive difinition of the units of r_1.

R	number of independent reactions
R_0	inside radius (length)
Re	Reynolds number
$-R_{A,v}$	rate of disappearance of species A per unit geometric volume of catalyst (moles A/time-length^3)
$-R_{A,P}$	rate of disappearance of species A in a whole catalyst particle (mole A/time)
s	Laplace parameter (Chapter 10 only) (time^{-1})
s_I	standard estimated error (intercept)
s_s	standard estimated error (slope)
s(I/J)	instantaneous or point selectivity to species "I" based on species "J")
S	slope
S_{cc}	number of chemical compounds in a system
S_p	BET surface area of catalyst (length^2/mass)
Sc	Schmidt number
Sh	Sherwood number
S(I/J)	overall selectivity to species "I" based on species "J"
SS_E	error sum of squares
t	time (time)
$\bar{t}$	average residence time (time)
T	temperature (degree)
U	overall heat-transfer coefficient (energy/length^2-time-degree)
$U(t)$	unit step function at t
v	linear velocity (length/time)
$\vec{v}$	velocity of fluid relative to catalyst particle (length/time)
V	volume (length^3)
V_p	specific pore volume of catalyst (length^3/mass)
V_G	geometric volume of catalyst particle (length^3)
V(I)	variance of intercept
V(S)	variance of slope
V_m	maximum rate of reaction (same dimensions as reaction rate)
V_{max}	maximum rate of reaction (same dimensions as reaction rate)
W	weight of catalyst (mass)
$\dot{W}_s$	shaft work (energy/time)
$\dot{W}$	mass flow rate (mass/time)
x_i	fractional conversion of Reactant "i"
y_i	mole fraction of species "i"
$y_{f,A}$	film factor for species "A" based on mole fraction
Y(I/J)	yield of species "I" based on consumption of species "J"
z	length (length)
Z_{AB}	frequency of collisions between molecules A and B (time^{-1}, volume^{-1})

Greek Letters

α_i	order of reaction with respect to species "i"
α_s	ratio of concentration of species "B" at surface of catalyst particle to concentration of species "A" at surface of catalyst particle, i.e., $C_{B,s}/C_{A,s}$ (Chapter 9)
$\alpha_{f,i}$	order of forward reaction with respect to species "i"

$\alpha_{r,i}$	order of reverse reaction with respect to species "i"
β	ratio of effective diffusivity of species "B" to effective diffusivity of species "A", i.e., $D_{B,eff}/D_{A,eff}$ (Chapter 9)
β_i	order of reverse reaction with respect to species "i" (Example 2-4)
γ	$\Delta\nu/[-\nu_A]$
δ	boundary layer thickness (length)
$\delta(t)$	Dirac delta function (time^{-1})
Δ	Dispersion number
ΔE_k	height of energy barrier (energy/mole)
ΔE_p	energy difference between reactants and products (energy/mole)
$\Delta G^0_{f,i}$	standard Gibbs free energy of formation of species "i" (energy/mole)
ΔG^0_R	standard Gibbs free energy change for a reaction (energy/mole)
ΔH_R	enthalpy change of a reaction (energy/mole)
$\Delta H^0_{f,i}$	standard enthalpy of formation of species "i" (energy/mole)
ΔH^0_R	standard enthalpy change of a reaction (energy/mole)
ΔT_{ad}	adiabatic temperature change (degrees)
ΔT_{ex}	temperature difference between hot and cold streams in a heat exchanger (degrees)
$\Delta\nu$	sum of the stoichiometric coefficients of all species in a reaction ($=\sum_{i=1}^{N}\nu_i$)
ε	porosity of a catalyst particle
ε_i	interstitial volume of catalyst bed
η	effectiveness factor
Θ	angular coordinate in cylindrical coordinates
Θ	dimensionless time t/τ (Chapter 10 only)
Θ_{ij}	C_{i0}/C_{j0}
λ	inverse of adiabatic temperature difference (Chapter 8 only) ($=1/\Delta T_{ad}$) (degrees^{-1})
λ_m	mean free path of a gas molecule (length)
Λ	ratio of the rates of disappearance of reactants in two independent reactions
Λ^0	ratio of the rates of disappearance of reactants in two independent reactions, evaluated at the conditions at the surface of the catalyst particle
μ	specific growth rate of cells (Chapter 4 only)
μ	viscosity (mass/length-time)
μ_i	ith moment of a distribution function (units depend on units of distribution function)
ν	kinematic viscosity (length^2/time)
ν_i	stoichiometric coefficient of species "i" (single reaction)
ν_{ki}	stoichiometric coefficient of species "i" in Reaction "k"
ρ	density (mass/volume)
ρ_a	apparent density (mass/length^3)
ρ_B	bulk density of catalyst (mass/volume of reactor) (mass/length^3)
ρ_l	density of liquid (mass/length^3)
ρ_p	density of catalyst particle (mass/length^3)
ρ_s	skeletal density of catalyst particle (mass/length^3)
σ^2	variance of E(t) (time^2)
τ	space time (time (for homogeneous reaction); time-weight catalyst/volume of fluid (for heterogeneous reaction))
τ_p	tortuosity of catalyst particle
ϕ	generalized Thiele modulus

$\phi_{s,1}$	Thiele modulus for an irreversible, first-order reaction in a sphere
Φ	generalized Weisz modulus
ξ	extent of reaction (single reaction) (moles or moles/time)
ξ_k	extent of Reaction "k" (multiple reactions) (moles or moles/time)
ξ_{max}	maximum value of ξ for a single reaction (moles or moles/time)
ξ_e	equilibrium extent of reaction (moles or moles/time)
υ	volumetric flow rate (volume/time)
Ψ	parameter in Thiele modulus for a reversible reaction; defined by Eqn. (9-13a)
Ψ_i	difference between dimensionless concentration of species "i" and the dimensionless equilibrium concentration of species "i"
Ω(T, A,B,C)	function in Eqn. (2–6), defined approximately by Eqn. (2-7) (length^2-time^{-1})

Subscripts

ad	adiabatic
B	bulk fluid
c	refers to cooling (or heating) fluid
D	desired
e	effluent or exit
eq	equilibrium
f	final or outlet
i	index denoting species "i"
lm	log mean
k	index denoting Reaction "k"
out	outlet
ref	reference
s	external surface of catalyst particle
U	undesired
0	initial or feed

Superscripts

0	reference conditions
$\overline{C}$	(overbar) average value of variable C (Chapters 1 through 9)
$\overline{C}$	(overbar) Laplace transform of the variable C (Chapter 10 only)

Index

ATOMIC WEIGHTS AND NUMBERS

Atomic weights apply to naturally occurring isotopic compositions and are based on an atomic mass of $^{12}C = 12$

Element	Symbol	Atomic number	Atomic weight	Element	Symbol	Atomic number	Atomic weight
Actinium	Ac	89	—	Iridium	Ir	77	192.2
Aluminum	Al	13	26.9815	Iron	Fe	26	55.847
Americium	Am	95	—	Krypton	Kr	36	83.80
Antimony	Sb	51	121.75	Lanthanum	La	57	138.91
Argon	Ar	18	39.948	Lawrencium	Lr	103	—
Arsenic	As	33	74.9216	Lead	Pb	82	207.19
Astatine	At	85	—	Lithium	Li	3	6.939
Barium	Ba	56	137.34	Lutetium	Lu	71	174.97
Berkelium	Bk	97	—	Magnesium	Mg	12	24.312
Beryllium	Be	4	9.0122	Manganese	Mn	25	54.9380
Bismuth	Bi	83	208.980	Mendelevium	Md	101	—
Boron	B	5	10.811	Mercury	Hg	80	200.59
Bromine	Br	35	79.904	Molybdenum	Mo	42	95.94
Cadmium	Cd	48	112.40	Neodymium	Nd	60	144.24
Calcium	Ca	20	40.08	Neon	Ne	10	20.183
Californium	Cf	98	—	Neptunium	Np	93	—
Carbon	C	6	12.01115	Nickel	Ni	28	58.71
Cerium	Ce	58	140.12	Niobium	Nb	41	92.906
Cesium	Cs	55	132.905	Nitrogen	N	7	14.0067
Chlorine	Cl	17	35.453	Nobelium	No	102	—
Chromium	Cr	24	51.996	Osmium	Os	75	190.2
Cobalt	Co	27	58.9332	Oxygen	O	8	15.9994
Copper	Cu	29	63.546	Palladium	Pd	46	106.4
Curium	Cm	96	—	Phosphorus	P	15	30.9738
Dysprosium	Dy	66	162.50	Platinum	Pt	78	195.09
Einsteinium	Es	99	—	Plutonium	Pu	94	—
Erbium	Er	68	167.26	Polonium	Po	84	—
Europium	Eu	63	151.96	Potassium	K	19	39.102
Fermium	Fm	100	—	Praseodymium	Pr	59	140.907
Fluorine	F	9	18.9984	Promethium	Pm	61	—
Francium	Fr	87	—	Protactinium	Pa	91	—
Gadolinium	Gd	64	157.25	Radium	Ra	88	—
Gallium	Ga	31	69.72	Radon	Rn	86	—
Germanium	Ge	32	72.59	Rhenium	Re	75	186.2
Gold	Au	79	196.967	Rhodium	Rh	45	102.905
Hafnium	Hf	72	178.49	Rubidium	Rb	37	84.57
Helium	He	2	4.0026	Ruthenium	Ru	44	101.07
Holmium	Ho	67	164.930	Samarium	Sm	62	150.35
Hydrogen	H	1	1.00797	Scandium	Sc	21	44.956
Indium	In	49	114.82	Selenium	Se	34	78.96
Iodine	I	53	126.9044	Silicon	Si	14	28.086